Karen Eich Drummond
Ed.D., R.D., F.A.D.A., F.M.P.

Lisa M. Brefere
C.E.C., A.A.C.

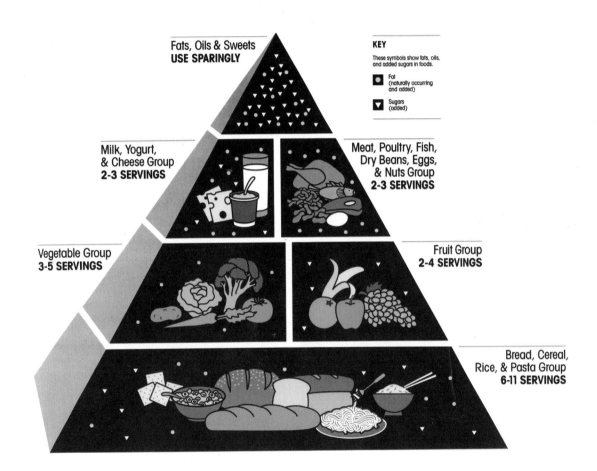

Fats, Oils & Sweets
USE SPARINGLY

KEY
These symbols show fats, oils, and added sugars in foods.

● Fat (naturally occurring and added)

▼ Sugars (added)

Milk, Yogurt, & Cheese Group
2-3 SERVINGS

Meat, Poultry, Fish, Dry Beans, Eggs, & Nuts Group
2-3 SERVINGS

Vegetable Group
3-5 SERVINGS

Fruit Group
2-4 SERVINGS

Bread, Cereal, Rice, & Pasta Group
6-11 SERVINGS

Nutrition for Foodservice and Culinary Professionals

Nutrition for Foodservice and Culinary Professionals

Fourth Edition

John Wiley & Sons, Inc.

New York   Chichester   Weinheim   Brisbane   Singapore   Toronto

Body Mass Index table on inside back cover from *Shape Up America,* National Institutes of
Health.

Food Guide Pyramid table on inside front cover from United States Department of
Agriculture.

Library of Congress Cataloging-in-Publication Data
Drummond, Karen Eich.
   Nutrition for foodservice and culinary professionals / Karen Eich Drummond, Lisa M.
Brefere.—4th ed.
     p. cm.
   Rev. ed. of: Nutrition for the foodservice professional.
   Includes index.
   ISBN 0-471-34777-9 (cloth : acid-free paper)
    1. Nutrition. 2. Food service. I. Brefere, Lisa M. II. Drummond, Karen Eich. Nutrition
for the foodservice professional. III. Title.

TX353.D78 2000
613.2—dc21

                                                                         00-026053

Printed in the United States of America.

10 9 8 7 6 5 4 3 2

In memory of my father,
Frank J. Eich

Contents

Part Two

Developing and Marketing Healthy Recipes and Menus

Chapter 8

Developing Healthy Menus and Recipes

Chapter 12
Weight Management and Exercise — 385

Chapter 13
Nutrition Over the Life Cycle — 409

Preface

This book is written for students in hotel, restaurant, and institution management programs and culinary programs. Practicing management and culinary professionals will find it useful as well. As with previous editions, this is meant to be a practical how-to book tailored to the needs of students and professionals. It is written for those who need to use nutritional principles to evaluate and modify menus and recipes, as well as to respond knowledgeably to customers' questions and needs.

This fourth edition has a new title, *Nutrition for Foodservice and Culinary Professionals,* because the addition of Lisa Brefere, C.E.C., A.A.C., as co-author has resulted in a greater emphasis on applying nutrition to selecting, cooking, and menuing healthy foods in restaurants and foodservices. After all, we eat foods, not nutrients!

What's New

In addition to a chef co-author, following are the elements that make this book new and different.

- New emphasis on **healthy ingredients.** Chapters 3 to 7 now include an in-depth look at the nutrition, purchasing, receiving, storage, cooking, and menuing of appropriate ingredients.
  - Chapter 3—Carbohydrates   Grains, Legumes, Pasta
  - Chapter 4—Lipids   Meats, Poultry, Fish
  - Chapter 5—Protein   Milk, Dairy Products, Eggs
  - Chapter 6—Vitamins   Fruits and Vegetables
  - Chapter 7—Minerals   Nuts and Seeds
- A new heading, **"Chef's Tips,"** is used in many chapters. Chef's Tips include experienced advice on which foods go together, how to flavor foods, how to use foods' natural colors to create an attractive dish, which foods work well in which dishes, availability of new and different

products, and other culinary techniques to make healthy food taste wonderful.

■ The **Dietary Guidelines for Americans (2000)** is included, and, along with a more in-depth look at the **Food Guide Pyramid,** lays a foundation for menu planning in Chapter 2.

■ Chapter 8, Developing Healthy Menus and Recipes, includes more than 25 **recipes** developed and tested by Lisa Brefere, C.E.C., A.A.C.

■ The popular **Food Facts** and **Hot Topics** have been revised and updated to include current topics such as biotechnology and functional foods.

■ The end of each chapter now includes a comprehensize **quiz** for students to see what they know, as well as a new activity entitled **Nutrition Web Explorer.** Each chapter gives two or more specific assignments to complete by going to designated reliable websites.

New illustrations have also been added.

Organization

The book is organized into three major parts, as before, but the section on healthy recipes and menus now appears after the first part, Fundamentals of Nutrition and Foods, instead of at the end of the book.

I. Fundamentals of Nutrition and Foods (Chapters 1–7)
II. Developing and Marketing Healthy Recipes and Menus (Chapters 8–10)
III. Nutrition's Relationship to Health and Lifespan (Chapters 11–13)

Part I has two introductory chapters followed by chapters on specific nutrients: carbohydrate, lipids, proteins, vitamins, and water and minerals, and the foods in which they are found. The first two chapters introduce basic nutrition concepts and tell how to use dietary recommendations, the Food Guide Pyramid, and food labels to plan menus.

Part II discusses how to use ingredients, flavoring principles, healthy cooking methods and techniques, and presentation to produce tasty, healthy food with eye appeal. Two chapters go further to explain how to market light beverages and foods in restaurants, foodservices, and beverage operations.

Part III looks at nutrition and health issues, including how nutrition affects heart disease, cancer, diabetes, and obesity. Vegetarian diets and diets for weight loss are discussed in detail. The final chapter looks at nutrition over the lifespan, from pregnancy to the infant, child, adolescent, and older adult.

Pedagogical Aids

In addition to tables, charts, illustrations, and a glossary, the book includes many other pedagogical aids.

1. Each chapter begins with an **outline** to help students get the big picture first.
2. The outline is followed by an introduction and a bulleted list of **learning objectives.**
3. **Sidebars** appear in all chapters to give definitions of key terms.
4. Instead of a summary at the end of each chapter, a **Mini-Summary** is given after each chapter heading. This is intended to help students focus on what is important in each section of the chapter.
5. **Food Facts** and **Hot Topics** continue to appear in most chapters. Food Facts discusses relevant food-related topics, such as an in-depth discussion of oils and margarines. Hot Topics discuss often controversial subjects related to nutrition, such as biotechnology and functional foods.
6. At the end of each chapter, the **Check-Out Quiz** allows students to check their comprehension of the important concepts presented. Answers to this quiz are found in Appendix E.
7. After the quiz, you will find **Activities and Applications.** This section encourages students to analyze, evaluate, create, problem-solve, and apply chapter concepts. One set of activities, **Nutrition Web Explorer,** involves getting specific food and nutrition-related information from carefully selected websites.

Lastly an instructor's manual (ISBN 0-471-35735-9) is available that includes class outlines, additional exercises/activities, visual aids, and test questions and answers. Please contact your John Wiley & Sons representative for a copy.

Acknowledgments

We are grateful for the help of all the educators who have contributed to this edition through their constructive comments. They include:

Kevin Monti, Western Culinary Institute
Richard Roberts, Wake Technical Community College
Joan Vogt, Kendall College
Jane Ziegler, Cedar Crest College

Part One
Fundamentals of Nutrition and Foods

Chapter 1

Introduction to Nutrition

Americans are fascinated with food: choosing foods, reading newspaper articles on food, perusing cookbooks, preparing and cooking foods, checking out food in a new restaurant, and, of course, eating foods. Why are we so interested in food? There are a number of possibilities. Most Americans know that their choice of diet strongly influences certain diseases (such as heart disease and cancer), and they are genuinely concerned about the role food plays in their long-term health. Indeed, high costs are associated with poor eating patterns: $71 billion per year in medical bills, lost productivity, and other costs associated with chronic disease conditions in which diet plays an important role. Food is also important for many other reasons, such as the pleasure of eating with others.

This introductory chapter explores why you choose the foods you eat, and then goes into important nutrition concepts that will build a foundation for the remaining chapters. It will help you to:

■ consider factors that influence food selection.
■ define nutrition, calorie, nutrient, and nutrient density.
■ identify the classes of nutrients and their characteristics.
■ describe four characteristics of a nutritious diet.
■ define Dietary Reference Intakes.
■ compare and contrast the EAR, RDA, AI, and UL.
■ explain the importance and function of the Dietary Reference Intakes.
■ describe the processes of digestion, absorption, and metabolism, and how the digestive system works.

Factors Influencing Food Selection

Why do people choose the foods they do? This is a very complex question, and there are many possibilities involving you and the food itself, as you can see from this list:

■ your cultural background and religious beliefs
■ your values or beliefs
■ your age, gender, occupation, education, and income
■ family influences
■ social influences
■ emotional influences
■ peer pressure
■ how familiar you are with the food
■ what you associate the food with (such as cake on birthdays)
■ nutritional value of the food

- how the food smells, looks, and tastes
- the cost, convenience, and availability of the food
- your nutritional knowledge and concerns
- desire to improve your health
- the media, including advertising
- environmental concerns

Now we will look at many of these factors in depth.

Cultural and Family Influences

Culture—the behaviors and beliefs of a certain social, ethnic, or age group.

The food a child grows up with depends largely on his/her cultural background. **Culture** can be defined as the behaviors and beliefs of a certain social, ethnic, or age group. Culture strongly influences the eating behavior of its members. People never eat all that is edible in their environment. Culture will determine what are acceptable and unacceptable foods. Many common American practices seem strange or illogical to persons from other cultures. For example, what could be more unusual than the person who boils water to make tea, adds ice to make it cold again, adds sugar to sweeten it, and then adds lemon to sour it?

Religious Influences

Religion affects the food decisions of many people. For some people, religion affects their day-to-day foods. For others, religion influences what they eat during religious holidays and celebrations. For example, many Jewish people abide by the Jewish dietary laws, called the Kashrut. They do not eat pork, nor do they eat meat and dairy products together. Muslims, who believe in the Islamic religion, also have their own dietary laws. Like the Jews, they will not eat pork. Their religion also prohibits drinking alcoholic beverages.

Social and Emotional Influences

People have historically eaten meals together, making meals important social occasions. Our food choices are influenced by the social situations we find ourselves in, whether in the comfort of our home or eating out in a restaurant. For example, social influences are involved when several members of a group of college friends are vegetarian. Peer pressure no doubt influences many food choices for children and young adults. Even as adults, we tend to eat the same foods that our friends and neighbors eat. This is due to cultural influences as well.

Food is often used to convey social status. For example, in a trendy, upscale New York City restaurant, you will find prime cuts of beef and high-priced wine. Likewise, foods such as beans and potatoes are associated with less well-to-do people.

Emotions are closely tied to some of our food selections. You may have been given something sweet to eat, such as cake or candy, whenever you were unhappy or upset. As an adult, you may gravitate to those kinds of foods when under stress.

Desire to Improve Health and Appearance

Have you tried to diet to lose weight? At some point in our lives almost all of us have wanted to change our weight. Surveys report that up to a quarter of American men and as many as 40 percent of American women are on a diet at any given time. We know that what we eat and what we weigh matters. Obesity and overweight can increase our risk of cancer, coronary heart disease, diabetes, and many other health problems. Likewise, eating too much fat or too little calcium may also cause health problems. Many consumers choose certain foods based on their desire to improve their health and/or appearance.

Food Cost, Convenience, and Availability

Food cost is a major consideration. For example, breakfast cereals were inexpensive for many years. Then prices jumped, and it seemed that most boxes of cereal cost over $3.00. Some consumers switched from cereal to bacon and eggs because the bacon and eggs became less expensive. Cost is a factor in many of the purchasing decisions at the supermarket, whether one is buying dry beans at $0.39 per pound or fresh salmon at $7.99 per pound.

Convenience is more of a concern now than at any time in the past. Just think about the variety of foods you can purchase today that are already cooked and can simply be microwaved. Even if you desire ready-to-eat fruits and vegetables, most supermarkets offer cut-up fruits, vegetables, and salads that need no further preparation. Of course, convenience foods are more expensive than their raw counterparts, and not every budget can afford them.

Everyone's food choices are affected by availability. Whether it is a wide choice of foods at an upscale supermarket or a choice of only two restaurants within walking distance of where you work, you can eat only what is available. The availability of foods is very much influenced by how food is produced and distributed. Fresh fruits and vegetables are perfect examples of foods that are most available (and at their lowest prices) when in season.

Flavor

Flavor—An attribute of a food that includes its appearance, smell, taste, feel in the mouth, texture, temperature, and even the sounds made when it is chewed.

Taste—Sensations perceived by the taste buds on the tongue.

Taste buds—Clusters of cells found on the tongue, cheeks, throat, and roof of the mouth. Each taste bud houses 60 to 100 receptor cells. The body regenerates taste buds about every three days. These cells bind food molecules dissolved in saliva and alert the brain to interpret them.

Flavor is a combination of all five senses: taste, smell, touch, sight, and sound. It also includes food characteristics such as temperature and texture (e.g., soft or crispy). From birth, we have the ability to smell and taste. Most of what we call taste is really smell, a fact we realize when a cold hits our nasal passages. Even though the taste buds are working fine, the smell cells are not, and this dulls much of food's flavor.

First, let's look at taste. Taste comes from 10,000 **taste buds**—clusters of cells resembling the sections of an orange. Taste buds, found on the tongue, cheeks, throat, and the roof of the mouth, house 60 to 100 receptor cells each. The body regenerates taste buds about every three days. They are most numerous in children under six, which may explain why youngsters are such picky eaters. These cells bind food molecules dissolved in saliva, and alert the brain to interpret them.

Although the tongue is often depicted as having regions that specialize in particular taste sensations, such as the tip that detects sweetness, researchers know that taste buds for each sensation (sweet, salty, sour, and bitter) are actually scattered around the tongue. In fact, a single taste bud can have receptors for all four types of taste.

If you could taste only sweet, salt, sour, and bitter, how could you taste the flavor of cinnamon or chicken, or any other food? This is where smell comes in. Your ability to identify the flavors of specific foods requires smell.

The ability to detect the strong scent of a fish market, the antiseptic odor of a hospital, the aroma of a ripe melon—and thousands of other smells—is possible thanks to a yellowish patch of tissue the size of a quarter high up in your nose. This patch is actually a layer of 12 million specialized cells, each sporting 10 to 20 hairlike growths called cilia that bind with the smell and send a message to the brain. Our sense of smell may not be as refined as that of dogs, who have billions of olfactory cells, but we can distinguish among about 10,000 scents.

You can smell foods in two ways. If you smell coffee brewing while you are getting dressed, you smell it directly through your nose. But if you are drinking coffee, the smell of the coffee goes to the back of your mouth and then up into your nose.

The senses of smell and taste begin with detection by receptors in the nose and tongue. Nerve impulses generated by the taste buds and olfactory receptors travel to the areas of the brain that interpret the messages as smell and taste. To some extent, what you smell or taste is genetically determined.

Environmental Concerns

Some people have environmental concerns, such as the use of synthetic fertilizers, so they often, or always, choose organically grown foods (which are grown without synthetic fertilizers—see Food Facts on page 23 for more information). For ecological reasons, vegetarians won't eat meat or chicken because livestock and poultry require so much land, energy, water, and plant food, which they consider wasteful.

Now that you have a better understanding of why we eat the foods we do, we can look at some basic nutrition concepts and terms.

■ **MINI-SUMMARY**

Factors such as cultural background, family, religion, social and emotional influences, desire to improve health and appearance, environmental concerns, and a food's cost, convenience, availability, and flavor affect our food choices.

Basic Nutrition Concepts

Nutrition

Nutrition—A science that studies nutrients and other substances in foods and in the body and how these nutrients relate to health and disease. Nutrition also explores why you choose particular foods and the type of diet you eat.

Nutrients—The nourishing substances in food that provide energy and promote the growth and maintenance of your body.

Diet—The food and beverages you normally eat and drink.

Nutrition is a science, which means that it is a branch of knowledge dealing with a body of facts. Compared with some other sciences, such as chemistry, nutrition is a young science. Many nutritional facts revolve around nutrients, such as sugar. **Nutrients** are the nourishing substances in food that provide energy and promote the growth and maintenance of your body. In addition, nutrients aid in regulating body processes, such as heart rate and digestion, and supporting the body's optimum health and growth.

Nutrition researchers look at how nutrients relate to health and disease. Almost daily we are bombarded with news reports that something in the food we eat, such as fat, is not good for us—that it may indeed cause or complicate conditions such as heart disease or cancer. Researchers look closely at the relationships between nutrients and disease, as well as the processes by which you choose what to eat and the balance of foods and nutrients in your diet.

In summary, nutrition is a science that studies nutrients and other substances in foods and in the body, and how these nutrients relate to health and disease. Nutrition also explores why you choose the foods you eat and the type of diet you eat. *Diet* is a word that has several meanings. Anyone who has tried to lose weight has no doubt been on a diet. In this sense, diet means weight-reducing diet. But a more general definition of **diet** is: the foods and beverages you normally eat and drink.

Calories

Your diet probably averages a certain number of calories every day. Food energy, as well as the energy needs of the body, is measured in units of energy called **calories.** The number of calories in a particular food can be determined by burning a weighed portion of the food and measuring the amount of heat (or calories) it produces. One calorie heats one gram of water 1 degree Celsius. Although we use the term calorie in our speech, and will use it in this book, the correct term is *kilocalorie,* a unit of 1,000 calories.

The number of calories you need is based on three factors: your energy needs when your body is at rest and awake (referred to as **basal metabolism**), your level of physical activity, and the energy you need to digest and absorb food (referred to as the **thermic effect of food**). Basal metabolic needs include energy needed for vital bodily functions when the body is at rest but awake. For example, your heart is pumping blood to all parts of your body, your cells are making proteins, and so on. Your basal metabolic rate (BMR) depends on the following factors:

Calorie—A measure of the energy in food, specifically the energy-yielding nutrient. The correct name for calorie is *kilocalorie.*
Basal metabolism—The minimum energy needed by the body for vital functions when at rest and awake.
Thermic effect of food—The energy needed to digest and absorb food.

1. **Gender.** Men have a higher BMR than women because men have a higher proportion of muscle tissue (muscle requires more energy for metabolism than fat does). The BMR for women is about 10 percent lower than for men.
2. **Age.** As people age, they generally gain fat tissue and lose muscle tissue, so BMR declines about 2 to 3 percent per decade after age 30.
3. **Growth.** Children, pregnant women, and lactating women have higher BMRs.
4. **Height.** Tall, thin people have more body surface and lose body heat faster. Their BMR is therefore higher.
5. **Temperature.** BMR increases in both hot and cold environments, in order to keep the temperature inside the body constant.
6. **Fever and stress.** Both of these increase BMR. Fever raises BMR by 7 percent for each 1 degree Fahrenheit above normal.

Basic metabolic rate also decreases when you diet or eat fewer calories than normal. Basal metabolic rate accounts for the largest percentage of energy expended—about two-thirds for individuals who are not very active.

Your level of physical activity strongly influences how many calories you need. Table 1-1 shows the calories burned per hour for a variety of

TABLE 1-1 Calories Spent per Hour in Physical Activity

Activity	Calories Burned*
Bicycling, 6 mph	240 calories
Bicycling, 12 mph	410 calories
Cross-country skiing	700 calories
Jogging, 5-1/2 mph	740 calories
Jogging, 7 mph	920 calories
Jumping rope	750 calories
Running in place	650 calories
Running, 10 mph	1280 calories
Swimming, 25 yards/minute	275 calories
Swimming, 50 yards/minute	500 calories
Tennis, singles	400 calories
Walking, 2 mph	240 calories
Walking, 3 mph	320 calories
Walking, 4-1/2 mph	440 calories

* The calories burned in a particular activity vary in proportion to one's body weight. For example, a 100-pound person burns 1/3 fewer calories, so you would multiply the number of calories by 0.7. For a 200-pound person, multiply by 1.3.

Source: National Heart, Lung, and Blood Institute and the American Heart Association. 1993. *Exercise and Your Heart: A Guide to Physical Activity.*

activities. The number of calories burned depends on the type of activity, how long and how hard it is performed, and the individual's size. The larger your body, the more energy you use in physical activity. Aerobic activities such as walking, jogging, cycling, and swimming are excellent ways to burn calories if they are brisk enough to raise heart and breathing rates. Physical activity accounts for 25 to 40 percent of total energy needs.

The thermic effect of food is the smallest contributor to your energy needs: from 5 to 10 percent of the total. In other words, for every 100 calories you eat, 5 to 10 are used for digestion, absorption, and metabolism of nutrients, our next topic.

Nutrients

As stated, nutrients provide energy or calories, and promote the growth and maintenance of the body. There are about 50 nutrients that can be arranged into six classes, as follows:

1. Carbohydrates
2. Fats (the proper name is lipids)
3. Proteins
4. Vitamins
5. Minerals
6. Water

Carbohydrates, lipids, and proteins are called **energy-yielding nutri-ents** because they can be burned as fuel to provide energy for the body. They provide calories as follows:

Carbohydrates: 4 calories per gram
Lipids: 9 calories per gram
Proteins: 4 calories per gram

(A gram is a unit of weight in the metric system; there are about 28 grams in 1 ounce.) Vitamins, minerals, and water do not provide energy or calories.

Carbohydrates include the sugars, starches, and fibers found in foods. Sugar is most familiar in its refined forms, such as table sugar or high-fructose corn syrup, and is used in soft drinks, cookies, cakes, pies, candies, jams, jellies, and other sweetened foods. Sugar is also present naturally in fruits and milk (even though milk does not taste sweet!). Starch is found in breads, breakfast cereals, pastas, potatoes, and beans. Both sugar and starch are important sources of energy for the body. Fiber can't be broken down or digested in the body, so it is excreted. It therefore does not provide energy for the body. Good sources of fiber include legumes (dried beans and peas), whole grains such as whole wheat bread or cereal, and fruits, vegetables, nuts, and seeds.

The most familiar **lipids** are fats and oils, which we find in butter, margarine, vegetable oils, mayonnaise, and salad dressings. Lipids are also found in the fatty streaks in meat, the fat under the skin of poultry, the fat in milk and cheese (unless it is made without fat), baked goods such as cakes, fried foods, nuts, and many processed foods, such as canned soups and frozen dinners. Most breads, cereals, pasta, fruits, and vegetables have little or no fat. Triglycerides are the major form of lipids. They provide energy for the body as well as a way to store energy as fat.

Most of the calories we eat are from carbohydrate or fat. Only about 15 percent of total calories are from **protein.** This doesn't mean that protein is less important. On the contrary, protein is the main structural component of all the body's cells. Protein is present in significant amounts in foods from animal sources, such as beef, pork, chicken, fish, eggs, milk, and cheese. Protein appears in plant foods, such as grains and beans, in smaller quantities.

Energy-yielding nutrients—Nutrients that can be burned as fuel to provide energy for the body, including carbohydrates, fats, and protein.

Carbohydrates—A nutrient group containing carbon, hydrogen, and oxygen that includes sugars, starches, and fibers.

Lipids—A group of fatty substances, including triglycerides and cholesterol, present in blood and body tissues.

Proteins—Major structural parts of animal tissue made up of nitrogen-containing amino acids.

Vitamins—Noncaloric, organic nutrients found in foods that are essential in small quantities for growth and good health.

Minerals—Naturally occurring, inorganic chemical elements that form a class of nutrients.

There are 13 different **vitamins** in food. Although they appear in foods in small amounts, they are vital to life. Instead of being burned to provide energy for the body, vitamins work as helpers. They assist in all the processes of the body that keep you healthy and alive. For example, vitamin A is needed by the eyes for vision in dim light. Unlike other nutrients, many vitamins are susceptible to being destroyed by heat, light, or other agents.

The final class of nutrients, **minerals,** are also required by the body in small amounts and do not provide energy. Like vitamins, they work as helpers in the body. Some minerals, such as calcium and phosphorus, become part of the body's structure by building bones and teeth. Unlike vitamins, minerals are indestructible.

Although deficiencies of energy or nutrients can be sustained for months or even years, a person can survive only a few days without water. Experts rank water second only to oxygen as essential to life. Water plays a vital role in all bodily processes and makes up just over half of your body's weight. It supplies the medium in which various chemical changes of the body occur and aids digestion and absorption, circulation, and lubrication of body joints. For example, as a major component of blood, water helps deliver nutrients to body cells and removes waste to the kidneys for excretion.

It's been said many times, "You are what you eat." This is certainly true; the nutrients you eat can be found in your body. As mentioned, water is the most plentiful nutrient in the body, accounting for about 60 percent of your weight. Protein accounts for about 15 percent of your weight, fat for 20 to 25 percent, and carbohydrates only 0.5 percent. The remainder of your weight includes minerals, such as calcium in bones, and traces of vitamins.

Essential nutrients— Nutrients that either cannot be made in the body or cannot be made in the quantities needed by the body; therefore, we must obtain them through food.

Most, but not all, nutrients are considered **essential nutrients.** Essential nutrients either cannot be made in the body or cannot be made in the quantities needed by the body; therefore, we must obtain them through food. Carbohydrate (in the form of glucose), vitamins, minerals, water, and some lipids and some parts of protein are considered essential.

Nutrient Density

All foods were not created equal in terms of the calories and nutrients they provide. Some foods, such as milk, contribute much calcium to your diet, especially when you compare them to other beverages, such as soft drinks. Your typical can of cola (12 fluid ounces) contributes large amounts of sugar

(40 grams, or about 10 teaspoons), no vitamins, and virtually no minerals. When you compare calories, you will find that skim milk (at 86 calories per cup) packs fewer calories than cola (at 97 calories per cup). Therefore, we can say that milk is more "nutrient dense" than cola, meaning that milk contains more nutrients per calorie than colas.

The **nutrient density** of a food depends upon the amount of nutrients it contains and the comparison of that to its caloric content. In other words, nutrient density is a measure of the nutrients provided per calorie of the food. As Figure 1-1 shows, broccoli offers many nutrients for its few calories. Broccoli is considered to have a high nutrient density because it is high in nutrients relative to its caloric value. Vegetables and fruits are examples of high-density foods. In comparison, the cupcake contains many more calories and few nutrients. By now, you no doubt recognize that "junk foods," such as the cupcake or candy bars, have a low nutrient density, meaning that they are low in nutrients and high in calories. The next section will tell you more about what a nutritious diet is.

Nutrient density—A measure of the nutrients provided in a food per calorie of the food.

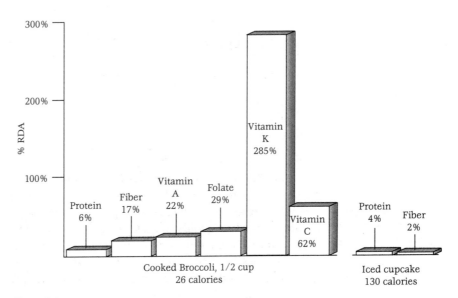

Figure 1-1

Nutrient density comparison*

*The nutrients examined were protein, fiber, and vitamins that contributed at least 10% of the RDA.

■ **MINI-SUMMARY**

Nutrition is a science that studies nutrients and other substances in foods and in the body, and how these nutrients relate to health and disease. The number of calories (a measure of the energy in food) you need is based on three factors—your energy needs when your body is at rest and awake (referred to as basal metabolism), your level of physical activity, and the energy you need to digest and absorb food (referred to as the thermic effect of foods). Nutrition also explores the reasons you choose the foods you eat and the type of diet you eat. Nutrients are the nourishing substances in food, providing energy and promoting the growth and maintenance of the body. In addition, nutrients regulate the many body processes. The six classes of nutrients are carbohydrates, fats (properly called lipids), protein, vitamins, minerals, and water. The first three are energy-yielding nutrients. Vitamins and minerals act as helpers in the body. Protein is a major structural component of the body's cells. Carbohydrate (in the form of glucose), vitamins, minerals, water, and some lipids and some parts of protein are considered essential. Nutrient density is a measure of the nutrients provided per calorie of a food.

Characteristics of a Nutritious Diet

A nutritious diet has four characteristics. It is:

1. adequate
2. balanced
3. moderate
4. varied

Adequacy—Describes a diet that provides enough calories, essential nutrients, and fiber to keep a person healthy.

Moderation—Describes a diet that avoids excessive amounts of calories or any particular food or nutrient.

Your diet must provide enough nutrients, but not too many. This is where adequate and moderate diets fit in. An **adequate diet** provides enough calories, essential nutrients, and fiber to keep you healthy, whereas a **moderate diet** avoids excessive amounts of calories or any particular food or nutrient. In the case of calories, for example, consuming too many leads to obesity. The concept of moderation allows you to occasionally indulge in foods that some consider harmful, such as French fries, premium ice cream, or chocolate.

Although it may sound simple to eat enough, but not too much, of the necessary nutrients, surveys show that most adult Americans find this hard to do. One national study showed that only 2 percent of participants ate a nutritionally adequate and calorically moderate diet.

One of the best ways to overcome this problem is to select nutrient-dense foods. As stated earlier, **nutrient-dense foods** contain many nutrients for the calories they provide.

Next, you need a **balanced diet.** Eating a balanced diet means choosing foods from all five food groups (see the Food Guide Pyramid on page 35). For example, if you drink a lot of soft drinks, you will be getting too much sugar and possibly not enough calcium, a mineral found in milk.

Last, you need a **varied diet**—in other words, you need to eat a wide selection of foods to get the protein, vitamins, minerals, and fiber needed in a healthful diet. If you imagine everything you eat for one week piled in a grocery cart, how much variety is in your cart from week to week? Do you eat the same bread, the same brand of cereal, the same types of fresh fruit, and so on, every week? Do you constantly eat favorite foods? Do you try new foods? A varied diet is important because it is more likely that you will get the essential nutrients in the right amounts. Our next topic, the Dietary Reference Intakes, gets specific about the right amounts we need of most nutrients.

Balanced—Describes a diet that includes foods from all five food groups every day.

Varied—Describes a diet in which a wide selection of foods is eaten to get the protein, vitamins, minerals, and fiber needed for health.

■ MINI-SUMMARY

A nutritious diet is adequate, moderate, balanced, and varied.

Nutrient Recommendations: Dietary Reference Intakes

The **Dietary Reference Intakes (DRI)** expand and replace what you may have known as the Recommended Dietary Allowances (RDA) in the United States and the Recommended Nutrient Intakes in Canada. The RDA is the amount of a nutrient that meets the known nutrient needs of practically all healthy persons. The RDA have been published since 1941 by a government-sponsored committee of nutrition experts, the Food and Nutrition Board. The Food and Nutrition Board works within the National Academy of Sciences.

The DRI are greatly expanded from the original RDA, and include the original RDA as well as three new values.

1. **Estimated Average Requirement (EAR):** The dietary intake value that is estimated to meet the requirement of half the healthy individu-

Dietary Reference Intake (DRI)—Nutrient standards that include four lists of values for dietary nutrient intakes of healthy Americans and Canadians.
Estimated Average Requirement (EAR)—The dietary intake value that is estimated to meet the requirement of half the healthy individuals in a group.
Recommended Dietary Allowance (RDA)—The dietary intake value that is sufficient to meet the nutrient requirements of 97 to 98 percent of all healthy individuals in a group.
Adequate Intake (AI)—The dietary intake value that is used when a RDA cannot be based on an estimated average requirement.
Tolerable Upper Intake Level (UL)—The maximum intake level above which risk of toxicity would increase.

als in a group. At this level of intake, the remaining 50 percent would not have its needs met. The EAR is used to assess nutritional adequacy of intakes of groups or populations and in nutrition research. An EAR is only set when there is conclusive scientific research.

2. **Recommended Dietary Allowance (RDA):** The dietary intake value that is sufficient to meet the nutrient requirements of 97 to 98 percent of all healthy individuals in a group. The RDA is based on the EAR. If there is not enough scientific evidence to justify setting an EAR, no RDA value is given. The RDA is a goal for individuals.

3. **Adequate Intake (AI):** The dietary intake that is used when a RDA cannot be based on an estimated average requirement (EAR). AI is based on an approximation of nutrient intake for a group (or groups) of healthy people. For example, there is no EAR or RDA for calcium, only an AI. An AI is given when there is insufficient scientific research to support a RDA. Both the RDA and the AI may be used as goals for individual intake or to assess individual intake.

4. **Tolerable Upper Intake Level (UL):** The maximum intake level above which risk of toxicity would increase. Intakes below the UL are unlikely to pose a risk of adverse health effects in healthy people. For most nutrients, this figure refers to total intakes from food, fortified food, and nutrient supplements. UL cannot be established for some nutrients, due to inadequate research.

The DRI vary depending upon age and gender, and there are DRI for pregnant and lactating women. The DRI are meant to help healthy people maintain health and prevent disease. They are not designed for seriously ill people, whose nutrient needs may be much different.

The Dietary Reference Intakes project has been divided into seven nutrient groups.

1. Calcium, vitamin D, phosphorus, magnesium, fluoride
2. Folate and other B vitamins
3. Antioxidants (e.g., vitamins C and E, selenium)
4. Macronutrients (e.g., protein, fat, carbohydrates)
5. Trace elements (e.g., iron, zinc)
6. Electrolytes and water
7. Other food components (e.g., fiber, phytoestrogens)

The first three reports (on calcium; vitamins C, D, and E; phosphorus, magnesium, selenium, and fluoride; and folate and other B vitamins) are used in this book and appear in Appendix A. It is hoped that the entire set of nutrients will be reviewed and DRIs released by the year 2003. The RDAs from 1989 are being used for calories and the following nutrients until new recommendations come out: protein; vitamins A and K, iron, zinc, and iodine. A discussion on how nutrients in food are digested and absorbed is next.

■ **MINI-SUMMARY**

The DRI include four dietary intake values: EAR (value estimated to meet requirements of half the healthy individuals in a group), RDA (value estimated to meet requirements of 97 to 98 percent of healthy individuals in a group), AI (the dietary intake used when there is not enough scientific basis for an EAR or RDA), and UL (maximum upper limit). Both the RDA and AI can be used as goals for individual intake. The EAR is used to assess the nutritional adequacy of the intake of groups. DRIs have been released for calcium, vitamins C, D, and E, phosphorus, magnesium, selenium, fluoride, folate, and other B vitamins. The RDAs from 1989 are used for calories and other nutrients until new DRIs are released.

What Happens When You Eat

Digestion, Absorption, and Metabolism

Digestion—The process by which food is broken down into its components in the mouth, stomach, and small intestine with the help of digestive enzymes.

Absorption—The passage of digested nutrients through the walls of the intestines or stomach into the body's tissues. Nutrients are then transported through the body via the blood or lymph systems.

Enzymes—Catalysts in the body.

To become part of the body, food must first be digested and absorbed. **Digestion** is the process by which food is broken down into its components in the mouth, stomach, and small intestine with the help of digestive **enzymes.** Complex proteins are digested, or broken down, into their building blocks called amino acids; complicated sugars are reduced to simple sugars such as glucose; and fat molecules are broken down into fatty acids.

Before the body can use any nutrients present in food, they must pass through the walls of the stomach or intestines into the body's tissues, a process called **absorption.** Nutrients are absorbed into either the blood or the lymph, two fluids that circulate throughout the body delivering needed products to the cells and picking up wastes. Blood is composed mostly of water, red blood cells (which carry and deliver oxygen to the cells), white blood cells (which are important in resistance to disease, called immunity), nutrients, and other components. Lymph is similar to blood but has no red blood cells. It goes into areas where there are no blood vessels to feed the cells.

Within each cell, **metabolism** takes place. Metabolism refers to all the chemical processes by which nutrients are used to support life. Metabolism has two parts: the building up of substances (called **anabolism**) and the breaking down of substances (called **catabolism**). Within each cell, nutri-

Metabolism—All the chemical processes by which nutrients are used to support life.

Anabolism—The metabolic process by which body tissues and substances are built.

Catabolism—The metabolic processes by which large, complex molecules are converted to simpler ones.

Gastrointestinal tract—A hollow tube running down the middle of the body in which digestion of food and absorption of nutrients takes place.

Oral cavity—The mouth.

Saliva—A fluid secreted into the mouth by the salivary glands, which contains important digestive enzymes and lubricates the food so that it may readily pass down the esophagus.

Bolus—A ball of chewed food that travels from the mouth through the esophagus to the stomach.

Epiglottis—The flap that covers the larynx (voicebox) and the opening to the trachea (windpipe) so that food does not enter during swallowing.

ents such as glucose are split into smaller units in a catabolic reaction that releases energy. The energy is either converted to heat to maintain body temperature or used to perform work within the cell. During anabolism, substances such as proteins are built from their amino-acid building blocks.

Gastrointestinal Tract

Once we have smelled and tasted our food, our meal goes on a journey through the **gastrointestinal tract** (also called the digestive tract), a hollow tube running down the middle of your body (Figure 1-2). The top of the tube is your mouth, which is connected in turn to your esophagus, stomach, small intestine, large intestine, rectum, and anus, where solid wastes leave the body.

The digestive system starts with the mouth, also called the **oral cavity.** Your tongue and teeth help with chewing. The tongue, which extends across the floor of the mouth, moves food around the mouth during chewing. Your

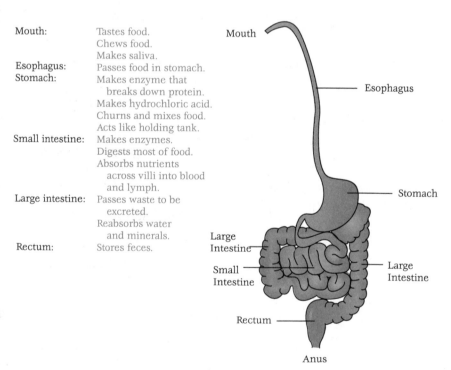

Mouth: Tastes food.
                Chews food.
                Makes saliva.
Esophagus: Passes food in stomach.
Stomach: Makes enzyme that
                   breaks down protein.
                Makes hydrochloric acid.
                Churns and mixes food.
                Acts like holding tank.
Small intestine: Makes enzymes.
                Digests most of food.
                Absorbs nutrients
                   across villi into blood
                   and lymph.
Large intestine: Passes waste to be
                   excreted.
                Reabsorbs water
                   and minerals.
Rectum: Stores feces.

Figure 1-2
Human digestive tract

Esophagus—The approximately 10-inch muscular tube that connects the pharynx to the stomach.

Peristalsis—Involuntary muscular contraction that forces food through the entire digestive system.

Cardiac sphincter—A muscle that relaxes and contracts to move food from the esophagus into the stomach.

Stomach—A digestive-tract organ that prepares food chemically and mechanically so that it can be further digested and absorbed.

Hydrochloric acid—A strong acid made by the stomach that aids in protein digestion, destroys harmful bacteria, and increases the ability of calcium and iron to be absorbed.

Chyme—A liquid mixture in the stomach that contains partially digested food and stomach secretions.

Small intestine—The digestive-tract organ that extends from the stomach to the opening of the large intestine.

Duodenum—The first segment of the small intestine, about one foot long.

32 permanent teeth grind and break down food. Chewing is important because it breaks up the food into smaller pieces so it can be swallowed. **Saliva,** a fluid secreted into the mouth from the salivary glands, contains important digestive enzymes and lubricates the food so that it may readily pass down the esophagus. Enzymes are substances that speed up chemical reactions. Digestive enzymes help break down food into forms of nutrients that can be used by the body. The tongue rolls the chewed food into a **bolus** (or ball) to be swallowed.

When swallowing occurs, a flap of tissue, the **epiglottis,** covers the trachea, or windpipe, so that food does not get into the lungs. Food now enters the **esophagus,** a muscular tube that leads to the stomach. Food is propelled down the esophagus by **peristalsis,** rhythmic contractions of muscles in the wall of the esophagus. You might think of this involuntary contraction that forces food through the entire digestive system as squeezing a marble (the bolus) through a rubber tube. Peristalsis also helps break up food into smaller and smaller particles.

Food passes from the esophagus through the **cardiac sphincter,** a muscle that relaxes and contracts (in other words, opens and closes) to move food from the esophagus into the stomach. The **stomach,** a muscular sac that holds about four cups, or one liter, of food, is lined with a mucous membrane. Within the folds of the mucous membrane are digestive glands that make **hydrochloric acid** and an enzyme to break down proteins. Hydrochloric acid aids in protein digestion, destroys harmful bacteria, and increases the ability of calcium and iron to be absorbed. Because hydrochloric acid can damage the stomach, the stomach protects itself with a thick mucous lining. Also, acid is produced only when we are eating or thinking about eating.

From the top part of the stomach, food is slowly moved to the lower part, where the stomach churns it with the hydrochloric acid and digestive enzymes. The food is ready to move on when it reaches a liquid consistency known as **chyme.** Food is then passed into the first part of the small intestine in small amounts (the small intestine can't process too much food at one time). Liquids leave the stomach faster than solids, and carbohydrate or protein foods leave faster than fatty foods. The stomach absorbs few nutrients, but it does absorb alcohol. It takes one to two hours for the stomach to empty.

The **small intestine,** about 20 feet long, has three parts: the **duodenum,** the **jejunum,** and the **ileum.** The small intestine was so named because its diameter is smaller than the large intestine, not because it is shorter. Actually, the small intestine is longer.

The duodenum, about one foot long, receives the digested food from the stomach as well as enzymes from other organs in the body, such as the liver and pancreas. The liver also provides **bile,** a substance that helps

Jejunum—The second portion of the small intestine, between the duodenum and the ileum.
Ileum—The final segment of the small intestine.
Villi—Tiny, fingerlike projections in the wall of the small intestines that are involved in absorption.
Large intestine—The part of the gastrointestinal tract between the small intestine and the rectum.
Rectum—The last section of the large intestine, in which feces, the waste products of digestion, are stored until elimination.
Anus—The opening of the digestive tract through which feces travel out of the body.

in fat digestion, and the pancreas provides bicarbonate, a substance that neutralizes stomach acid. The small intestine itself produces digestive enzymes.

In the duodenal wall (and throughout the entire small intestine) are tiny, fingerlike projections called **villi.** The muscular walls mix the chyme with the digestive juices and bring the nutrients into contact with the villi. Most nutrients pass through the villi of the duodenum and jejunum into either the blood or lymph vessels, where they are transported to the liver and to the body cells. The duodenum connects with the second section, the jejunum, which connects to the ileum. Most digestion is completed in the first half of the small intestine; whatever is left goes into the large intestine. Food is in the small intestine for 7 to 8 hours, and about 12 hours in the large intestine.

The **large intestine** (also called the **colon**) is four to five feet long and extends from the end of the ileum to a cavity called the rectum. One of the functions of the large intestine is to receive and store the waste products of digestion; in other words, the material that was not absorbed into the body. The large intestine stores this material in the **rectum** until it is released as solid feces through the **anus,** which opens to allow elimination. The large intestine also absorbs water and some minerals.

■ **MINI-SUMMARY**

Before the body can use the nutrients in food, the food must be digested and the nutrients absorbed through the walls of the small or large intestine into either the blood or lymph system. Within each cell, metabolism (all the chemical processes by which nutrients are used to support life) takes place. Metabolism has two parts: anabolism (building up) and catabolism (breaking down).

Check–Out Quiz

1. Match the nutrients with their functions/qualities. The functions/qualities may be used more than once.

Nutrients	Functions
Carbohydrate	Provide energy
Lipid	Work as helpers
Protein	Supplies the medium in which chemical changes of the body occur
Vitamins	Work as main structure of cells
Minerals	
Water	

2. Match the Dietary Reference Intake values with their definition.

DRI Value	Definition
RDA	Maximum safe intake level
AI	Value that meets requirements of 50 percent of individuals in a group
UL	Value that meets requirements of 97-98 percent of individuals
EAR	Value used when there is not enough scientific data to support an RDA

3. Match the terms on the left with their definitions on the right.

Term	Definition
Absorption	Process of building substances
Enzyme	Involuntary muscular contraction
Anabolism	Substance that speeds up chemical reactions
Peristalsis	Process of breaking down substances
Catabolism	Process of nutrients entering the tissues from the gastrointestinal tract

4. Which digestive-system organ passes waste to be excreted and reabsorbs water and minerals?

   a. stomach   b. small intestine

   c. large intestine   d. liver

5. Which nutrient supplies the highest number of calories per gram?

   a. carbohydrate   b. fat   c. protein   d. vitamin pills

6. Flavor is a combination of all five senses.

   a. True   b. False

7. Women have a higher basal metabolic rate than men.

   a. True   b. False

8. Hydrochloric acid aids in protein digestion, destroys harmful bacteria, and increases the ability of calcium and iron to be absorbed.

   a. True   b. False

9. The nutrient density of a food depends upon the amount of nutrients it contains and the comparison of that to its caloric content.

   a. True   b. False

10. The DRI are designed for both healthy and sick people.

   a. True   b. False

Activities and Applications

1. How Many Calories Do You Need Each Day?

Use the following two steps to calculate the number of calories you need.

A. To determine your basal metabolic needs, multiply your weight in pounds by 10.9 if you are male, 9.8 if you are female. (These numbers are based on a BMR factor of 1.0 calorie per kilogram of body weight per hour for men and 0.9 calorie for women.) Example: 150-pound woman × 9.8 = 1470 calories

B. To determine how much you use each day for physical activity, first determine your level of activity as one of these four.

Very light activity—You spend most of your day seated or standing.

Light activity—You spend part of your day up and about, such as in teaching or cleaning house.

Moderate activity—You engage in an exercise activity for an hour or so at least every other day, or your job requires some physical work.

Heavy activity—You engage in manual labor, such as construction.

Once you have picked your activity level, you need to multiply your answer in A by one of the following numbers.

Very light (men and women): Multiply by 1.3.

Light (men): Multiply by 1.6.

Light (women): Multiply by 1.5.

Moderate (men): Multiply by 1.7.

Moderate (women): Multiply by 1.6.

Heavy (men): Multiply by 2.1.

Heavy (women): Multiply by 1.9.

EXAMPLE: A woman with light activity.

1470 calories × 1.5 = 2205 calories needed daily

2. Factors That Influence What Foods You Eat

Determine which of the following factors influence what you eat, and why or why not:

■ your age, gender, occupation, education, and income
■ your cultural background and religious beliefs
■ your values or beliefs
■ family and social influences
■ peer pressure
■ nutritional value of food
■ your nutritional knowledge and concerns
■ desire to improve your health
■ the food itself: how familiar you are with it; what you associate with it; how it smells, looks, and tastes

- the cost, convenience, and availability of foods
- the media, including advertising
- environmental concerns

3. Taste and Smell

Pick one of your favorite foods, eat it normally, and then take a bite of it while holding your nose closed. How does it taste when you can't smell very well? What influence does smell have on taste?

4. Nutrient-Dense Foods

Pick one food that you ate yesterday that could be considered nutrient dense. Also pick one food that would *not* be considered nutrient dense. Compare the nutrition labels. Explain why you chose these foods.

Nutrition Web Explorer

U.S. Government Healthfinder: www.healthfinder.gov

This government site can help you find information on virtually any health topic. Click on "Cancer" on this web page, then click on "Questions and Answers." Find one nutrient they recommend eating more of, and one nutrient they recommend eating less of, to lower your cancer risk.

McDonald's: www.mcdonalds.com

Visit the McDonald's website to add up the calories in this meal combination: Big Mac, Super Size Fries, and Super Size Coke (double the small drink). To get nutrition information, click on "Restaurant Locator," then "Food," then "Nutrition Facts." Compare the number of calories in this meal to the number of calories you need daily from the first activity. Or you may visit any fast-food website to add up the calories in a meal you might order.

National Academy of Sciences: www.nas.edu

As noted in this chapter, the Food and Nutrition Board, which develops the DRI, is located with the National Academy of Sciences. Visit its web page and click on "Health and Medicine." Next, click on "Food and Nutrition," which lists its most recent nutrition publications. Determine if any new DRIs have been published and, if so, click on the new publication and read the Summary.

Food Facts *Food Basics*

Whole foods are foods as we get them from nature. Examples include milk, eggs, meats, poultry, fish, fruits, vegetables, dried beans and peas, and grains. Many whole foods are also considered **fresh foods.** Fresh foods are raw foods that have not been processed (such as canned or frozen), or heated. Fresh foods also do not contain any preservatives. Examples of fresh foods include fresh fruits and vegetables, fresh meats, poultry, and fish.

Organic foods are generally foods that have been grown without synthetic pesticides, fertilizers, herbicides, antibiotics, or hormones, and without genetic engineering or irradiation (see the Hot Topic at the end of this chapter). Organic farmers use, as examples, animal and plant manures to increase soil fertility and crop rotation to decrease pest problems. The goal of organic farming is to preserve the natural fertility and productivity of the land. Common organic foods include fruits, vegetables, and cereals. Meat and poultry can also be organic, if the animals are raised without hormones or antibiotics, live in an environment that closely approximates their normal habitat, and are fed organic feed. Table 1-2 details the Proposed National Organic Standards released in early 2000.

Processed foods have been prepared using a certain procedure: cooking (such as frozen pancakes), freezing (frozen dinners), canning (canned vegetables), dehydrating (dried fruits), milling (white flour), culturing with bacteria (yogurt), or adding vitamins and minerals (enriched foods). In some cases, processing removes and/or adds nutrients.

When processing adds nutrients, the resulting food is either an **enriched** or a **fortified food.** For example, when whole wheat is milled to produce white flour, nutrients are lost. By law, white flour must be enriched with four vitamins and iron to make up for some of these lost nutrients. Milk is often fortified with vitamin D because there are few good food sources of this vitamin. A food is considered enriched when nutrients are added to it to replace the same nutrients that are lost in processing. A food is considered fortified when nutrients are added that were not present originally, or nutrients are added that increase the amount already present. For example, orange juice does not contain calcium, so when calcium is added to orange juice, the product is called calcium-fortified orange juice. Probably the most notable fortified food is iodized salt. Iodized salt was introduced in 1924 to combat iodine deficiencies.

Whereas the food supply once contained mostly whole farm-grown foods, today's supermarket shelves are stocked primarily with processed foods. Some of these are minimally processed, such as canned peaches. But many processed foods contain parts of whole foods, and often have added ingredients such as sugars, or sugar or fat substitutes. For instance, cookies are made with eggs and white flour from grains. Then sugar, shortening, and nutrients are added. Highly processed foods, such as most breakfast cereals, cookies, crackers, sauces, soups, baking mixes, frozen entrees, pasta, snack foods, and condiments, are staples nowadays.

As you can probably guess, your best bet nutritionally is to consume more whole foods and fewer highly processed foods.

TABLE 1-2 Proposed National Organic Standards (2000)

The Proposed National Organic Standards (2000) address the methods, practices, and substances used in producing and handling crops, livestock, and processed agricultural products. The requirements apply to the way the product is created, not to measurable properties of the product itself. Although specific practices and materials used by organic operations may vary, the proposed standards require every aspect of organic production and handling to comply with the provisions of the Organic Foods Production Act (OFPA).

TABLE 1-2 *(continued)*

Crop Standards

The proposed organic crop production standards say that:

■ Land would have no prohibited substances applied to it for at least three years before the harvest of an organic crop.
■ Crop rotation would be implemented.
■ The use of genetic engineering (included in excluded methods), irradiation, and sewage sludge is prohibited.
■ Soil fertility and crop nutrients would be managed through tillage and cultivation practices, supplemented with animal and crop waste materials and allowed synthetic materials.
■ Preference would be given to the use of organic seeds and other planting stock, but a farmer could use nonorganic seeds and planting stock under certain specified conditions.
■ Crop pests, weeds, and diseases would be controlled primarily through management practices, including physical, mechanical, and biological controls. When these practices are not sufficient, a biological, botanical, or allowed synthetic substance may be used.

Livestock Standards

These standards apply to animals used for meat, milk, eggs, and other animal products represented as organically produced.
The proposed livestock standards say that:

■ Animals for slaughter must be raised on an organic operation from birth, or no later than the second day of life for poultry.
■ Producers would be required to feed 100 percent organically produced feeds to livestock but could also provide allowed vitamin and mineral supplements.
■ Organically raised animals could not be given hormones or antibiotics.
■ Preventive management practices, including the use of vaccines, would be used to keep animals healthy. Producers would be prohibited from withholding treatment from a sick or injured animal; however, animals treated with a prohibited medication would be removed from the organic operation.
■ All organically raised animals would have to have access to the outdoors, including access to pasture for ruminants. They could be temporarily confined only for reasons of health, safety, or to protect soil or water quality.

Handling Standards

The proposed handling standards say that:

■ All nonagricultural ingredients, whether synthetic or nonsynthetic, must be included on the National List of Allowed Synthetic and Prohibited Non-Synthetic Substances.
■ Handlers must prevent the commingling of organic with nonorganic products and protect organic products from contact with prohibited substances.

Hot Topic Biotechnology

Since 1994, a growing number of foods developed using the tools of the science of biotechnology have come onto the domestic and international markets. **Biotechnology** is a collection of scientific techniques, including genetic engineering, that are used to create, improve, or modify plants, animals, and microorganisms. **Genetic engineering** is a process that allows plant breeders to modify the genetic makeup of a plant species precisely and predictably, creating improved varieties faster and more easily than can be done using more traditional plant-breeding techniques. Modern techniques now enable scientists to move genes (and therefore desirable traits) in ways they could not before—and with greater ease and precision.

For hundreds of years, scientists have been using conventional techniques, such as selective breeding, to improve plants and animals for human benefit. Ever since the latter part of the nineteenth century, when Gregor Mendel discovered that characteristics in pea plants could be inherited, scientists have been improving plants by changing their genetic makeup. Typically, this was done through hybridization, in which two related plants were cross-fertilized and the resulting offspring had characteristics of both parent plants. Breeders then selected and reproduced the offspring that had the desired traits. These conventional methods are not as precise or as fast as newer methods.

Today, to change a plant's traits, scientists are able to use the tools of modern biotechnology to insert a single gene—or, often, two or three genes—into the crop to give it new, advantageous characteristics. The genes direct the production of a specific protein, such as one that makes it resistant to drought or pests. Most genetic modifications make it easier to grow the crop.

For example, about half of the American soybean crop planted in 1999 carries a gene that makes it resistant to a herbicide used to control weeds. The herbicide normally poisons an enzyme essential for plant survival. Other forms of this normal plant enzyme have been identified that are unaffected by the herbicide. Putting the gene for this resistant form of the enzyme into the plant protects it from the herbicide. That allows a farmer to treat a field with the herbicide to kill the weeds without harming the crop. In addition, about a quarter of U.S. corn planted in 1999 contains a gene that produces a protein toxic to certain caterpillars, eliminating the need for certain conventional pesticides. Modifications have also been made to canola and soybean plants to produce oils with a different fatty acid composition, so they can be used in new food-processing systems.

Bioengineered foods are regulated by three federal agencies: the Food and Drug Administration (FDA), the U.S. Department of Agriculture (USDA), and the Environmental Protection Agency (EPA). The USDA's Animal and Plant Health Inspection Service (APHIS) oversees the agricultural environmental safety of planting and field testing genetically engineered plants. Besides regulating the field testing of these plants, APHIS is responsible for protecting American agriculture against pests and diseases. A company or other organization that wishes to field test a biotechnology-derived plant must generally get APHIS approval before proceeding. Also, before a genetically engineered crop can be produced on a wider scale and sold commercially, its creators must petition APHIS for a "determination of non-regulated status," which requires the submission of field test reports, information about indirect effects on other plants, and other reports.

Over 5,000 field trials have been safely conducted with APHIS approval since 1987. About 40 new agricultural products have completed all the federal regulatory requirements and may be sold commercially. They range from longer-lasting tomatoes to pest-resistant corn. Acreage for biotech soybean, cotton, and corn has increased dramatically since the introduction of these crops in the mid-1990s. They accounted for 20 to 44 percent of acreage planted in 1998.

The FDA is responsible for the safety and labeling of all foods and animal feeds derived from crops, including biotech plants. The Food and Drug Administration ensures that any foods sold, whether conventional or bioengineered, meet the safety standards set by the federal Food, Drug, and Cosmetic Act. In the case of bioengineered foods, the FDA does not review biotech products under the same regulations that it reviews food additives. Instead, it uses a voluntary consultation process to help companies meet the requirement.

During the consultation process, companies provide documents summarizing the information and data they have generated to demonstrate that a bioengineered food is as safe as the conventional food. The documents describe the genes they use: whether they are from a commonly allergenic plant, the characteristics of the proteins made by the genes, their biological function, and how much of them will be found in the food. They also provide information on whether the new food contains the expected levels of nutrients or toxins, and any other information about the safety and use of the product. FDA scientists review the information and generally raise questions.

The FDA's biotechnology policy does treat genetically engineered substances as food additives if they are significantly different in structure, function, or amount than substances currently found in food. According to the FDA, many of the food crops currently being developed using biotechnology do not contain substances that are significantly different from those already in the diet and thus had no pre-market approval.

Legal authority for food labeling also rests with the FDA. Foods derived from biotechnology currently must be labeled only if they differ significantly from their conventional counterparts—for example, if their nutritional content or potential to cause allergic reactions is altered. Companies must tell consumers on the food label when a product includes a gene from one of the common allergy-causing foods, unless it can show that the protein produced by the added gene does not cause allergies in the particular food. For example, tomatoes bred to contain a peanut protein would need to be labeled to disclose the presence of the peanut protein, unless it were conclusively demonstrated that the new tomato was not allergenic to those allergic to peanuts.

If a food had a new allergy-causing protein introduced into it, the label would have to state that it contained the allergen. Foods using oils such as soybean and canola oil that have been bioengineered to have a different fatty acid composition must also be labeled. The label must include a new standard name that indicates the bioengineered oil's difference from conventional soy and canola oils.

On the positive side, biotechnology has helped farmers and consumers in many ways, such as the following:

■ Combating human diseases. The first biotechnology products were medicines designed to address human diseases. Insulin, used to treat diabetics, and blood-clot-busting enzymes for heart-attack victims are now produced easily and cheaply as a result of biotechnology.

■ Fighting hunger by resisting plant diseases and increasing crop yields. Biotechnology can help farmers increase crop yields and feed more people. For example, a scientist used biotechnology to pinpoint a gene that could help wheat, a major food staple, grow on millions of acres worldwide that are now hostile to the crop. Scientists have also developed an experimental potato hybrid that contains genes to resist a new, more virulent strain of the so-called "late blight," the disease that caused the Irish potato famine in the 1840s.

■ Producing more nutritious foods, such as rice with vitamin A for underdeveloped countries.

■ Helping the environment by reducing pesticide use. Biotechnology can help farmers reduce their reliance on insecticides and herbicides.

For example, Bt cotton, a widely grown biotech crop, kills several important cotton pests.

On the negative side, there is concern about the safety of the proteins produced by the genes inserted into the plant. The FDA asserts that all of the proteins placed into foods on the market are nontoxic and rapidly digestible, and do not have the characteristics of proteins known to cause allergies. As for the genes, the chemical that encodes genetic information—DNA—is present in all foods, and its ingestion is not associated with human illness. Breeders do extensive field-testing of plants to make sure the plant looks and grows normally, and that the food tastes right and has the expected levels of nutrients and toxins. But some scientists are concerned that putting genes from bacteria or viruses into fruit, vegetables, or any food may not be safe; and at the very least, these foods need to be tested extensively before being put on the market.

Up to now, many of the proteins introduced into plants have been enzymes. The enzymes have been considered to be "generally recognized as safe" by the FDA, so they have not required testing such as that done for food additives. Not all consumers agree with the FDA's assessment. In particular, there is concern about genetically engineered proteins causing allergy problems.

Other concerns about bioengineered foods revolve around the following.

■ Some people are worried about potential environmental problems when the genes pass on to other plants. This could cause, for example, the unintentional creation of weeds resistant to some herbicides and pests resistant to certain pesticides. There is evidence of genes from biotech plants that moved quickly and easily from crops to wild relatives. Luckily, corn, soybeans, and cotton have no wild relatives with which to interbreed, but other crops, such as canola, do have wild relatives in the United States. Organic farmers have complained that blown pollen from fields of genetically engineered plants can transfer traits to their fields.

■ There is also the possibility of problems with treating human infections because of antibiotic-resistant genes in food.

■ Corn, cotton, and potato crops that produce the insecticidal toxin known as Bt may have adverse effects on beneficial insects, such as monarch butterfly caterpillars. Beneficial, predatory insects often get sick when they eat Bt or pests that have eaten Bt corn.

■ Some individuals are opposed to genetically engineered foods based on their religious, ethical, or philosophical beliefs.

Chapter 2

Using Dietary Recommendations, Food Guides, and Food Labels to Plan Menus

Dietary recommendations have been published for the healthy American public for almost 100 years. Early recommendations centered on encouraging intake of certain foods to prevent deficiencies, fight disease, and enhance growth. Although deficiency diseases have been virtually eliminated, they have been replaced by diseases of dietary excess and imbalance—problems that now rank among the leading causes of illness and death in the United States. Diseases such as heart disease and cancer touch the lives of most Americans and generate substantial health-care costs. More recent dietary guidelines have therefore centered on modifying the diet, in most cases cutting back on certain foods, to prevent lasting degenerative conditions such as heart disease.

This chapter looks at dietary recommendations and food guides and labels. It will help you to:

- list and discuss the Dietary Guidelines for Americans.
- list the goals of the Food Guide Pyramid.
- describe how the Food Guide Pyramid encourages variety, proportionality, and moderation.
- use the Food Guide Pyramid and food labels to plan menus.
- discuss how the Exchange System is used, and how its advantages and disadvantages compare with the Food Guide Pyramid.
- read with understanding the food label, including the Nutrition Facts panel, nutrient claims, and health claims.

Dietary Recommendations and Food Guides

Dietary recommendations— Guidelines that discuss specific foods and food groups to eat for optimal health.

Food guides—Guidelines that tell us the kinds and amounts of foods that constitute a nutritionally adequate diet.

Food guides tell us the amounts of foods we need to eat for a nutritionally adequate diet. Their primary role, whether in the United States or around the world, is to communicate an optimal diet for overall health of the population. The use of the Food Guide Pyramid has been very successful in the United States. The pyramid shape easily conveys the concept of variety and the relative amounts to eat of the various food groups. However because of cultural differences in communication and other cultural norms, the pyramid is not necessarily the graphic of choice for food guides worldwide. Other countries, such as Canada, have a four-banded rainbow, with each color representing one of its four food groups.

The Food Guide Pyramid is based on the Dietary Guidelines for Americans and nutrient recommendations currently found in the Dietary Reference Intakes. To better understand the Food Guide Pyramid, let's first look at the Dietary Guidelines for Americans.

Dietary Guidelines for Americans

Dietary recommendations are quite different from the Dietary Reference Intakes (DRI). Whereas the DRI deal with specific nutrients, dietary recommendations discuss specific foods and food groups that will help individuals meet the DRI. The DRI also tend to be written in a technical style. Dietary recommendations are generally written in easy-to-understand terms.

The most recent set of U.S. dietary recommendations, the **Dietary Guidelines for Americans** (fifth edition), was published in 2000. Updated every five years by a joint advisory committee of the U.S. Department of Agriculture and the Department of Health and Human Services, the Dietary Guidelines for Americans are used by the federal government to plan food programs. These guidelines are for healthy Americans aged 2 years and over—not for younger children and infants, whose dietary needs differ. Dietary Guidelines for Americans reflect the recommendations of nutrition authorities who agree that enough is known about diet's effect on health to encourage certain dietary practices. The guidelines should be applied to diets consumed over several days rather than to single meals or foods. Following are the Dietary Guidelines for Americans.

Dietary Guidelines for Americans—A set of dietary recommendations for Americans that is periodically revised.

Aim for Fitness

- Aim for a healthy weight.
- Be physically active each day.

Build a Healthy Base

- Let the Pyramid guide your food choices.
- Eat a variety of grains daily, especially whole grains.
- Eat a variety of fruits and vegetables daily.
- Keep food safe to eat.

Choose Sensibly

- Choose a diet that is low in saturated fat and cholesterol, and moderate in total fat.
- Choose beverages and foods to moderate your intake of sugars.
- Choose and prepare foods with less salt.
- If you drink alcoholic beverages, do so in moderation.

The new "ABC" approach for the Dietary Guidelines focuses on "Aim, Build, Choose—for Good Health." By following these guidelines, you can promote your health and reduce your risk of chronic diseases such as heart disease, certain cancers, diabetes, stroke, and osteoporosis. These diseases are leading causes of death and disability among Americans. Your food choices, your lifestyle, your environment, and your genes all affect your well-being.

Aiming for fitness involves two guidelines: Aim for a healthy weight, and become physically active each day. To evaluate your weight, the Dietary Guidelines recommend that you evaluate your weight-for-height or Body Mass Index (BMI). This concept is explained and a chart given in Chapter 12. If your BMI is above the healthy range, you may benefit from weight loss, especially if you have other health risk factors. To make it easier to manage your weight, choose a healthful assortment of foods that includes vegetables, fruits, grain (especially whole grains), skim milk, and fish, lean meat, poultry, or beans. Most of the time, choose foods that are low in fat and added sugars. Eating mainly vegetables, fruits, and grains helps you feel full, achieve good health, and manage your weight. Whatever the food, eat a sensible portion size. Also aim for at least 30 minutes of physical activity daily. To maintain a healthy weight after weight loss, it helps for adults to do at least 45 minutes of moderate physical activity daily.

The first guideline for building a healthy base is to let the Food Guide Pyramid, the next topic, guide your food choices. Also, choose a variety of grains daily, especially whole grains. Foods made from grains (like wheat, rice, and oats) are the foundation of a nutritious diet. They provide vitamins, minerals, carbohydrates, and other substances that are important for good health. Grain products are low in fat, unless fat is added in processing, in preparation, or at the table. Whole grains, such as whole-wheat bread and brown rice, contain more fiber and nutrients than do refined products, such as white bread and white rice.

Choose a variety of fruits and vegetables daily. Eating plenty of fruits and vegetables of different kinds may help protect you against heart disease, stroke, and some types of cancer. It also promotes healthy bowel function. Fruits and vegetables provide essential vitamins and minerals, fiber, and other substances important for good health. Most people eat fewer servings of fruits and vegetables than are recommended—at least two servings of fruits and three servings of vegetables each day.

The final guideline under building a healthy base is to keep food safe to eat. Foods that are safe from harmful bacteria, viruses, parasites, and chemical contaminants are vital for healthful eating. Safe means that the food poses little risk of foodborne illness.

The last four guidelines involve choosing sensibly. First, choose a diet that is low in saturated fat and cholesterol and moderate in total fat. Fats supply energy and essential fatty acids, and they help absorb the fat-soluble vitamins A, D, E, and K. You need some fat in the food you eat, but choose sensibly. Some kinds of fat, especially saturated fats, increase the risk of coronary heart disease by raising blood cholesterol. In contrast, unsaturated fats (found mainly in vegetable oils) do not increase blood cholesterol.

Choose beverages and foods to moderate your intake of sugars and salt. Foods containing sugars and starches can promote tooth decay. High sugar levels can also contribute to overweight. You can reduce your chances of developing high blood pressure by consuming less salt.

The final guideline recommends that if you drink alcoholic beverages, do so in moderation. Alcoholic beverages are harmful when consumed in excess. Taking more than one drink per day for women or two drinks per day for men can raise the risk for auto accidents, other accidents, high blood pressure, stroke, violence, suicide, birth defects, and certain cancers. Too much alcohol may cause social and psychological problems, cirrhosis of the liver, inflammation of the pancreas, and damage to the brain and heart. Heavy drinkers also are at risk of malnutrition, because alcohol contains calories that may substitute for more nutritious foods.

Food Guide Pyramid

Food Guide Pyramid—A food guide developed by the U.S. Department of Agriculture to help healthy Americans follow the Dietary Guidelines for Americans.

The **Food Guide Pyramid** (Figure 2-1) was developed to help healthy Americans follow the Dietary Guidelines for Americans. This guide was developed by the U.S. Department of Agriculture to:

1. promote overall health for Americans 7 years of age and older (there is a separate Food Guide Pyramid for Young Children for ages 2 to 6).
2. be based on up-to-date research,
3. focus on the total diet, and
4. be useful, flexible, and practical.

Many years of research and testing went into the development of the Food Guide Pyramid.

The Food Guide Pyramid was a major change from the previous guide, known as the "Basic Four." The Basic Four was a foundation diet. This means that it was only intended to meet part of the caloric needs and part of the Dietary Reference Intakes for nutrients. The assumption was that people would eat more food than was recommended by the Basic Four. In this way, they would get enough calories and nutrients. At that time, not much was known about the importance of fiber or the relationship between high intakes of certain food components and disease. Little was said about the selection of fat and added sugars or about appropriate caloric intake in the Basic Four food guide.

Instead of being based on just a foundation diet, the Pyramid is based on the total diet. This means that the Pyramid shows how to *get enough*

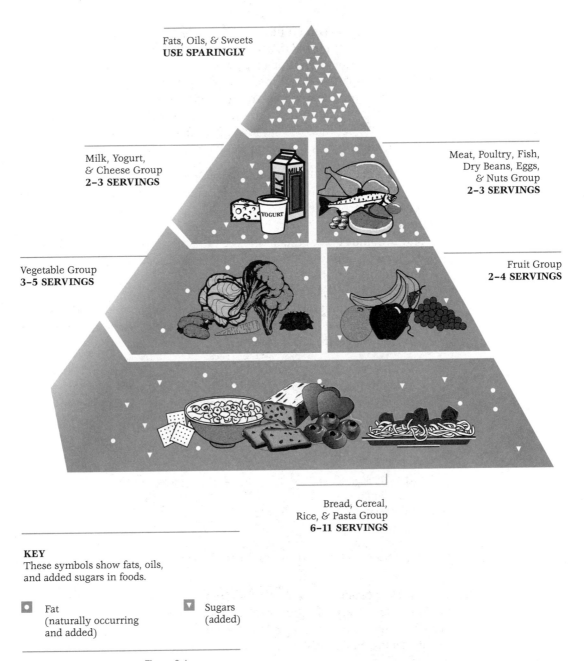

Fats, Oils, & Sweets
USE SPARINGLY

Milk, Yogurt,
& Cheese Group
2–3 SERVINGS

Meat, Poultry, Fish,
Dry Beans, Eggs,
& Nuts Group
2–3 SERVINGS

Vegetable Group
3–5 SERVINGS

Fruit Group
2–4 SERVINGS

Bread, Cereal,
Rice, & Pasta Group
6–11 SERVINGS

KEY
These symbols show fats, oils,
and added sugars in foods.

Fat
(naturally occurring
and added)

Sugars
(added)

Figure 2-1

The food guide pyramid—a guide to daily food choices

Source: U.S. Department of Agriculture and U.S. Department of Health and Human Services.

nutrients and, at the same time, how to *avoid excesses of certain food components.* Those components high in fat and/or added sugars and low in nutrient density — e.g., butter, margarine, oils, sugars, jam, soft drinks — are separated from the Pyramid's five major food groups. They are placed in the Pyramid's small tip. In addition, symbols are used within the major food groups to show that some of these foods are also high in fat and/or added sugars. The Pyramid is now the official food guide for the United States.

The nutritional goals of the Pyramid are as follows:

1. **Energy:** provide 1,300 to 3,000 calories.
2. **Protein, vitamins, and minerals:** provide 100% of the DRI for people over two years of age.
3. **Fiber:** increase intake.
4. **Total fat:** limit to 30 percent or less of total calories.
5. **Saturated fat:** limit to less than 10 percent of total calories.
6. **Cholesterol:** limit to 300 mg or less.
7. **Sodium:** limit to 2,400 mg or less.
8. **Added sugars:** not to exceed caloric needs. When enough food has been eaten from the major food groups, the rest of the calories can come from fat and added sugars.

As you can see, the first three goals are related to nutritional adequacy, whereas the remaining goals are related to moderation.

The Food Guide Pyramid emphasizes moderation and adequacy to reduce your risk of conditions or diseases such as the following:

1. high blood pressure
2. stroke
3. heart disease
4. certain cancers
5. osteoporosis
6. diabetes

Chapter 11 discusses nutrition and health issues.

Variety, Proportionality, and Moderation. The key concepts of the Pyramid include variety, proportionality, and moderation. The Pyramid was designed around the variety of food eaten by many Americans. To meet the Pyramid's nutritional goals, it is important to eat from all five food groups and to choose a variety of foods within each group. Certain ethnic groups and people such as vegetarians, who do not eat foods from all of the food groups, may need special help on how to meet their nutrient needs.

Proportionality—A concept of eating relatively more foods from the larger food groups at the base of the Food Guide Pyramid and fewer foods from the smaller food groups nearer the top of the Pyramid.

Proportionality means eating relatively more foods from the larger food groups at the base of the Pyramid and fewer foods from the smaller food groups nearer the top of the Pyramid. The Pyramid illustrates proportionality by the shape of the Pyramid itself, the relative size of the food group sections, and the recommended number of servings for each group. The Pyramid emphasizes grains, fruit, and vegetables, and de-emphasizes animal products such as meat and milk. Proportionality does not mean that some food groups are more important than others. Each of the food groups provides some, but not all, of the necessary nutrients. To meet nutrient requirements while eating a typical American diet, no one group can replace another group.

Moderation means not eating any foods to excess. The Pyramid illustrates moderation by recommending a certain number of servings for each food group, encouraging variety (which helps one avoid eating any foods to excess), and advising to "use sparingly" the foods at the tip of the Pyramid (fats, oils, and sweets).

Food Groups. The five nutrient-dense food groups in the Pyramid are the following:

1. Bread, Cereal, Rice, and Pasta Group
2. Vegetable Group
3. Fruit Group
4. Milk, Yogurt, and Cheese Group
5. Meat, Poultry, Fish, Dry Beans, Eggs, and Nuts Group

When placing foods in these groups, the USDA grouped foods primarily by the nutrients they provide. Table 2-1 lists the major nutrients in each food group. Typical use of a food in meals and how it was grouped in past guides were also considered.

Subgroups within the major food groups (Table 2-2) emphasize foods that are particularly good sources of dietary fiber or of certain vitamins and minerals that are low in the diets of many Americans. Eating more foods from certain subgroups is recommended. For example, the bread, cereal, rice, and pasta group has two subgroups: enriched and whole grains. USDA recommends several servings a day of whole grains from this group. Whole grains provide more fiber, vitamins, and minerals than enriched grains.

The milk, yogurt, and cheese group is divided into two subgroups: low-fat milk products and other milk products with more fat or sugar. It is recommended to choose primarily low-fat or nonfat items from this group. It is also recommended to choose meat alternates, such as beans, several times a week, because they are rich in dietary fiber and minerals.

TABLE 2-1 Major Nutrients in Food Groups

Bread, Cereal, Rice, and Pasta

Complex carbohydrates
B vitamins—thiamin, riboflavin, niacin, and folate
Minerals—iron
Fiber

Vegetables

Carbohydrate
Vitamins—A, C, and folate
Minerals—iron and magnesium
Fiber

Fruit

Carbohydrate
Vitamins—A and C
Minerals—potassium
Fiber

Milk, Yogurt, and Cheese

Protein
Fat
Vitamins—riboflavin, A, and D (if fortified)
Minerals—the best source of calcium

Meat, Poultry, Fish, Dry Beans, Eggs, and Nuts

Protein
Fat
B vitamins—niacin, thiamin, and B_{12}
Minerals—iron and zinc

The vegetable group has five subgroups: dark green, deep yellow, starchy, dry beans and peas, and other. Dark-green leafy vegetables and dry beans and peas are particularly important. The fruit group has two subgroups: citrus, melons, and berries, and other. Citrus, melons, and berries are emphasized for their high vitamin C content. The USDA recommends a dark-green vegetable for vitamin A and a vitamin-C-rich fruit every day. Table 2-2 lists some food examples in each food group and subgroup.

TABLE 2-2 Variety from the Food Groups

BREAD, CEREAL, RICE, PASTA

Whole-Grain		Enriched		Grain Products with More Fat and Sugar	
Brown rice	Pumpernickel bread	Bagels	Italian bread	Biscuits	Danish
Buckwheat groats	Whole-grain cereals	Cornmeal	Macaroni	Cake (unfrosted)	Doughnuts
Bulgur	Rye bread and crackers	Crackers	Noodles	Cookies	Muffins
Corn tortillas	Whole-wheat bread rolls, crackers	English muffins	Pancakes and waffles	Cornbread	Pie crust
Graham crackers		Farina	Pretzels	Croissants	Tortilla chips
Granola	Whole-wheat pasta	Flour tortillas	Ready-to-eat cereals		
Oatmeal	Whole-wheat cereals	French bread	White rice		
Popcorn		Grits	Spaghetti		
		Hamburger and hot dog rolls	White bread and rolls		

FRUITS

Citrus, Melons, Berries			Other Fruits		
Blueberries	Honeydew melon	Strawberries	Apples	Guavas	Pineapples
Cantaloupe	Kiwifruit	Tangerines	Apricots	Grapes	Plantains
Citrus juices	Lemons	Watermelons	Asian pears	Mangoes	Plums
Cranberries	Oranges	Ugli fruit	Bananas	Nectarines	Prickly pears
Grapefruit	Raspberries		Cherries	Papayas	Prunes
			Dates	Passion fruit	Raisins
			Figs		Rhubarb
			Fruit juices	Peaches	Star fruit
				Pears	

VEGETABLES

Dark-Green Leafy			Deep Yellow	Starchy	
Beet greens	Dandelion greens	Romaine lettuce	Carrots	Breadfruit	Lima beans
Broccoli	Endive	Spinach	Pumpkin	Corn	Potatoes
Chard	Escarole	Turnip greens	Sweet potatoes	Green peas	Rutabagas
Chicory	Kale	Watercress	Winter squash	Hominy	Taro
Collard greens	Mustard greens				

TABLE 2-2 *(continued)*

Dry Beans and Peas (Legumes)		Other Vegetables			
Black beans	Lima beans (mature)	Artichokes	Cauliflower	Green peppers	Snow peas
Black-eyed peas	Mung beans	Asparagus	Celery		Summer squash
Chickpeas (garbanzos)	Navy beans	Bean and alfalfa sprouts	Chinese cabbage	Lettuce	
	Pinto beans			Mushrooms	Tomatoes
Kidney beans	Split peas	Beets	Cucumbers	Okra	Turnips
Lentils		Brussels sprouts	Eggplants	Onions (mature and green)	Vegetable juices
			Green beans		Zucchini
		Cabbage		Radishes	

MEAT, POULTRY, FISH, AND ALTERNATES

Meat, Poultry, and Fish				Alternates	
Beef	Ham	Pork	Veal	Eggs	Peanut butter
Chicken	Lamb	Shellfish	Luncheon meats, sausage	Dry beans and peas (legumes)	Tofu
Fish	Organ meats	Turkey			
				Nuts and seeds	

MILK, YOGURT, AND CHEESE

Lowfat Milk Products		Other Milk Products with More Fat or Sugar			
Buttermilk	Lowfat or nonfat plain yogurt	Cheddar cheese	Frozen yogurt	Ice milk	Swiss cheese
Lowfat cottage cheese	Skim milk	Chocolate milk	Fruit yogurt	Process cheeses and spreads	Whole milk
1% and 2% milk		Flavored yogurt	Ice cream		
				Puddings made with whole milk	

FATS, SWEETS, AND ALCOHOLIC BEVERAGES

Fats		Sweets			Alcoholic Beverages
Bacon, salt pork	Mayonnaise	Candy	Jam	Popsicles and ices	Beer
Butter	Mayonnaise-type salad dressing	Corn syrup	Jelly		Liquor
Cream (dairy, nondairy)	Salad dressing	Frosting (icing)	Maple syrup	Sherbets	Wine
	Shortening	Fruit drinks	Marmalade	Soft drinks and colas	
Cream cheese			Molasses		

40

TABLE 2-2 *(continued)*

FATS, SWEETS, AND ALCOHOLIC BEVERAGES				
Fats		Sweets		Alcoholic Beverages
Lard	Sour cream	Gelatin desserts	Table syrup	Sugar (white and brown)
Margarine	Vegetable oil	Honey		

Source: "Using the Food Guide Pyramid: A Resource for Nutrition Educators" by A. Shaw, L. Fulton, C. Davis, and M. Hogbin. U.S. Department of Agriculture: Food, Nutrition, and Consumer Services; Center For Nutrition Policy and Promotion.

The tip of the Pyramid includes fats, oils, and sweets. This group does not count as a major food group because these foods provide energy but little else nutritionally. Servings of these foods are optional. Examples of these foods include butter, margarine, salad dressing, sugar, jelly, honey, regular soft drinks, and candy bars. Foods such as doughnuts, cakes, and cookies are counted in the bread group. Potato chips are counted in the vegetable group. The extra fat and sugar found in these foods is counted as additional fat and/or added sugar in the diet.

Some food items can be difficult to classify. For example, grouping of corn products depends on the form in which corn is used. Sweet corn is counted as a starchy vegetable; popcorn and cornmeal products such as corn tortillas are counted as grain products; hominy is grouped with starchy vegetables, and hominy grits with grain products. Snack and dessert items such as cakes, cookies, ice cream, French-fried potatoes, potato chips, and so forth count with the food group of their major ingredient, e.g., the bread, dairy, or vegetable group. However, use of these higher-fat items must be limited to keep total fat intake to the recommended level.

Dry beans and peas (called legumes) can count either as a meat alternate or as a starchy vegetable (they should not be double counted in the same menu). These foods are good sources of protein and other nutrients provided by the meat group, such as iron and zinc, and have long been recommended as inexpensive alternatives to meat. Dry beans and peas are also high in carbohydrate and are good sources of vitamins, minerals, and dietary fiber. To increase use of these nutrient-dense foods, the Food Guide Pyramid suggests including dry beans and peas as a vegetable selection several times a week, instead of considering them only as meat alternates.

Symbols for fat and/or added sugar are included in all food groups on the Pyramid to show that fat and/or added sugar is found in all the food groups. Small circles represent added or naturally occurring fat. Upside-down triangles represent added sugar, but no naturally occurring sugar as found in fruit. The relative concentrations of fat and added sugars in the food groups are shown by the number of symbols in each group. The greatest number of symbols are in the tip.

Number of Servings. The Food Guide Pyramid suggests foods and number of servings for the total diet. If more calories are needed than are provided by the lower numbers of servings in the ranges, additional servings from the major food groups are suggested, along with modest increases in amounts of total fat and added sugars. Increasing amounts of grain products, vegetables, and fruit helps keep higher-calorie diets moderate in fat and also provides additional vitamins, minerals, and dietary fiber—nutrients that are low in many American diets.

Table 2-3 shows sample food patterns for a day at three calorie levels, covering the ranges of servings suggested by the Pyramid. It also indicates some age/sex groups for whom those calorie levels may be appropriate. The sample food patterns are not prescriptions, but illustrations of healthy proportions in the diet. Specific numbers of servings may vary somewhat from day to day.

TABLE 2-3 Sample Food Patterns for a Day at Three Calorie Levels

1,600 calories is about right for many sedentary women and some older adults.	**2,200 calories** is about right for most children, teenage girls, active women, and many sedentary men. Women who are pregnant or breast-feeding may need somewhat more.	**2,800 calories** is about right for teenage boys, many active men, and some very active women.	
	About 1,600	About 2,200	About 2,800
Bread Group Servings	6	9	11
Fruit Group Servings	2	3	4
Vegetable Group Servings	3	4	5
Meat Group	5 ounces	6 ounces	7 ounces
Milk Group Servings	2–3*	2–3*	2–3*
Total fat (grams)[a]	53	73	93
Total added sugars (teaspoons)[a]	6	12	18

* Women who are pregnant or breast-feeding, teenagers, and young adults to age 24 need 3 servings.
[a] Values for total fat and added sugars include fat and added sugars that are in food choices from the five major food groups as well as fat and added sugars from foods in the Fats, Oils, and Sweets group.

Source: "Using the Food Guide Pyramid: A Resource for Nutrition Educators" by A. Shaw, L. Fulton, C. Davis, and M. Hogbin. U.S. Department of Agriculture: Food, Nutrition, and Consumer Services; Center For Nutrition Policy and Promotion.

Table 2-4 shows one day's menu adapted for three calorie levels — 1,600, 2,200, and 2,800 calories. Those with higher calorie needs take larger portions of some meal items and can supplement their meals with more snacks.

Portion or Serving Sizes. Serving sizes specified by the Pyramid are listed in Table 2-5, and an expanded list of serving sizes appears in Appendix C. For ease of use, the number of different serving sizes for foods in each food group was kept to a minimum. Food guide servings are based on food "as eaten." That is, meats are cooked, and trimmed of fat and bone. Vegetables are rinsed, trimmed, and cooked, or eaten raw as appropriate. Rice, pasta, and cereal grains such as oatmeal are cooked.

The serving size for all fruit juices is 3/4 cup, rather than varying from 1/3 to 3/4 cup based on the carbohydrate content of the specific juice. For most food groups, the amount to count as a serving is comparable to the amount typically reported in food-consumption surveys — for example, 1/2 cup of cooked vegetable, or 1 cup of leafy raw salad greens. For foods in the bread group, portions typically reported (e.g., 1 cup of rice or pasta, 1 whole hamburger bun) more nearly equate to two servings from the Pyramid. For this group, the familiar serving size used in previous guides (e.g., 1 slice of bread or 1/2 cup of rice or pasta) was retained for the Pyramid.

For meat, poultry, and fish, the portion sizes reported in surveys vary widely depending on the type of meat and the eating occasion. For example, dinner portions are typically 3 ounces or more, while amounts used in a sandwich are 1 to 2 ounces. Common portions of meat alternates, such as one egg, or 2 tablespoons of peanut butter, or 1/2 cup of cooked dry beans or peas, are equivalent in protein and most vitamins and minerals to 1 ounce of lean meat. Thus, the Pyramid suggests that the two to three servings from the meat group should total 5 to 7 ounces per day.

For foods in the Fats, Oils, and Sweets category, no serving size or numbers of servings are listed. The amounts of these foods that can be included depend on the fat and added sugars provided as part of the specific food items selected from the major food groups. For example, a medium croissant counts as two servings from the bread group but provides 12 grams of fat, compared with 2 grams of fat provided by two slices of plain bread. Thus, if a croissant is selected, the amount of spreads and dressings used should be reduced to compensate for the extra fat provided by the croissant (equivalent to about 2 teaspoons of butter or margarine) to keep total fat in the menu below 30 percent of total calories.

Table 2-6 shows how to count food group servings in one day's menu. Note the following points:

■ A large portion of a food item counts as more than one serving. For example, the whole toasted raisin English muffin at breakfast counts as

TABLE 2-4 One Day's Menu and Food Group Servings at Three Calorie Levels

Item	Calorie Level		
	1,600	2,200	2,800
BREAKFAST			
Cantaloupe	1/4 medium	1/4 medium	1/4 medium
Whole-wheat pancakes	2	2	3
Blueberry sauce	1/4 cup	1/4 cup	6 tablespoons
Margarine		1 teaspoon	2 teaspoons
Turkey patty		1-1/2 ounces	1-1/2 ounces
Milk	skim, 1 cup	skim, 1 cup	2%, 1 cup
LUNCH			
Chili-stuffed baked potato	3/4 cup chili, 1 potato	3/4 cup chili, 1 potato	3/4 cup chili, 1 potato
Low-fat, low-sodium cheddar cheese		3 tablespoons	3 tablespoons
Spinach-orange salad	1 cup	1 cup	1 cup
Wheat crackers	6	6	6
Grapes			12
Fig bars			2
Milk		skim, 1 cup	2%, 1 cup
DINNER			
Apricot-glazed chicken	1 breast half	1 breast half	1 breast half
Rice-pasta pilaf	3/4 cup	3/4 cup	3/4 cup
Steamed zucchini			1/2 cup
Tossed salad	1 cup	1 cup	1 cup
Reduced-calorie Italian dressing	1 tablespoon	1 tablespoon	
Regular Italian dressing			1 tablespoon
Hard roll(s)	1 small	2 small	2 small
Margarine		2 teaspoons	2 teaspoons
Vanilla ice milk	1/2 cup	1/2 cup	1/2 cup
SNACKS			
Fig bar	1		
Skim milk	3/4 cup		
Apple		1/2 medium	1/2 medium
Soft pretzel		1 large	1 large
Lemonade			1 cup
2% fat milk			1 cup

Source: ''Using the Food Guide Pyramid: A Resource for Nutrition Educators'' by A. Shaw, L. Fulton, C. Davis, and M. Hogbin. U.S. Department of Agriculture: Food, Nutrition, and Consumer Services; Center For Nutrition Policy and Promotion.

TABLE 2-5 The Pyramid Guide to Daily Food Choices

Food Group	Suggested Daily Servings	What Counts as a Serving
Bread, Cereal, Rice, Pasta Whole-grain Enriched	6 to 11 servings from entire group (Include several servings of whole-grain products daily.)	1 slice of bread 1/2 hamburger bun or English muffin a small roll, biscuit, or muffin 5 to 6 small or 3 to 4 large crackers 1/2 cup cooked cereal, rice, or pasta 1 ounce ready-to-eat cereal
Fruits Citrus, melon, berries Other fruits	2 to 4 servings from entire group	a whole fruit such as a medium apple, banana, or orange a grapefruit half a melon wedge 3/4 cup juice 1/2 cup berries 1/2 cup chopped, cooked, or canned fruit 1/4 cup dried fruit
Vegetables Dark-green leafy Deep-yellow Dry beans and peas (legumes) Starchy Other vegetables	3 to 5 servings (Include all types regularly; use dark-green leafy vegetables and dry beans and peas several times a week.)	1/2 cup cooked vegetables 1/2 cup chopped raw vegetables 1 cup leafy raw vegetables, such as lettuce or spinach 3/4 cup vegetable juice
Meats, Poultry, Fish, Dry Beans and Peas, Eggs, and Nuts	2 to 3 servings from entire group	Amounts should total 5 to 7 ounces of cooked lean meat, poultry without skin, or fish per day. Count 1 egg, 1/2 cup cooked beans, or 2 tablespoons peanut butter as 1 ounce of meat.
Milk, Yogurt, Cheese	2 servings (3 servings for women who are pregnant or breast-feeding, teenagers, and young adults to age 24.)	1 cup milk 8 ounces yogurt 1-1/2 ounces natural cheese 2 ounces process cheese

TABLE 2-5 *(continued)*		
Food Group	Suggested Daily Servings	What Counts as a Serving
Fats, Sweets, and Alcoholic Beverages	Use fats and sweets sparingly. If you drink alcoholic beverages, do so in moderation.	

Note: The guide to daily food choices described here was developed for Americans who regularly eat foods from all the major food groups listed. Some people such as vegetarians and others may not eat one or more of these types of foods. These people may wish to contact a dietitian or nutritionist for help in planning food choices.

Source: "Using the Food Guide Pyramid: A Resource for Nutrition Educators" by A. Shaw, L. Fulton, C. Davis, and M. Hogbin. U.S. Department of Agriculture: Food, Nutrition, and Consumer Services; Center For Nutrition Policy and Promotion.

two servings from the bread group. A smaller portion counts as part of a serving—the 1/2 cup of skim milk at breakfast counts as half a serving from the milk group.

■ Desserts and snacks contribute to food group servings. In this menu, plain cookies (gingersnaps), fruit (pineapple chunks for dessert at dinner), crackers, cheese, vegetable juice, and a half sandwich contribute substantially to food group servings and nutrient intake for the day.

■ The relatively high-fat entrée at lunch (Taco Salad) and the cheese for a snack are balanced by a low-fat breakfast, a low-fat entrée for dinner (Pork and Vegetable Stir-fry), and selection of fruit and lower-fat cookies for desserts.

■ Reduced-fat and reduced-salt/sodium products can also help keep the fat and sodium levels in check. This menu uses low-fat, low-sodium cheese and unsalted tortilla chips in the Taco Salad, low-calorie mayonnaise-type dressing in the turkey sandwich, and no-salt-added tomato juice.

In order to keep calories to the target level (2,200), sources of added sugars in this menu are limited to the cookies at lunch.

The Pyramid recommends a number of servings of a certain size daily. However, the number and size of servings a person eats from each food group usually varies from day to day. Therefore, the Pyramid actually applies to the amount of food a person eats over several days, not just one day.

Many foods that Americans eat are mixtures of foods from several food groups—pizza, beef stew, and macaroni and cheese, for example. These mixed dishes or combination foods contain items from more than one group, and often provide a wide variety of nutrients. For example, macaroni and cheese counts as servings from both the bread group (macaroni) and the milk group (cheese).

TABLE 2-6 Counting Food Group Servings in One Day's Menu at 2,200 Calories

Recipe	Bread	Vegetable	Fruit	Milk	Meat oz.	Fat grams	Calories
BREAKFAST							
Medium grapefruit, 1/2			1			trace	41
Medium banana			1			1	108
Ready-to-eat cereal flakes, 1 ounce	1					trace	111
Toasted raisin English muffin	2					1	138
Soft margarine, 2 teaspoons						8	68
Skim milk, 1/2 cup				1/2		trace	43
LUNCH							
Taco salad, 1 serving						19	455
unsalted tortilla chips	3/4						
tomato purée and greens		1-1/2					
low-fat, low-sodium cheddar cheese				1/2			
beef and beans					2-1/2		
Medium gingersnaps, 2	1					2	101
DINNER							
Pork and vegetable stirfry, 1 serving						9	370
rice	1-1/2						
vegetables		1					
pork					3		
Cooked broccoli, 1/2 cup		1				trace	26
Small white rolls, 2	2					3	167
Soft margarine, 2 teaspoons						8	68
Minted pineapple chunks, juice-pack, 1/2 cup			1			trace	75
SNACKS							
Wheat crackers, 6	1					4	86
Cheddar cheese, 1-1/2 ounces				1		14	171
Turkey sandwich, 1/2						4	137
rye bread	1						
turkey					1		
lettuce leaf							
mayonnaise-type salad dressing, fat-free, 1/2 tablespoon							
No-salt-added tomato juice, 3/4 cup		1				trace	31
TOTAL	10-1/4	4-1/2	3	2	6-1/2	73	2,196

Source: "Using the Food Guide Pyramid: A Resource for Nutrition Educators" by A. Shaw, L. Fulton, C. Davis, and M. Hogbin. U.S. Department of Agriculture: Food, Nutrition, and Consumer Services; Center For Nutrition Policy and Promotion.

Pork and Vegetable Stir-fry

Category: Entrée   Yield: 4 servings, 1 cup meat mixture, 1/4 cup sauce and 3/4 cup rice each

INGREDIENTS

Boneless pork loin, lean	1 pound	Lemon juice	1/4 cup
Tarragon leaves	1/2 teaspoon	Carrots, sliced	1 cup
Pepper	1/4 teaspoon	Fresh mushrooms, sliced	1 cup
Garlic powder	1/4 teaspoon	Celery, sliced	1 cup
Salt	1/4 teaspoon	Onions, chopped	1/2 cup
Cornstarch	2 teaspoons	Rice, cooked	3 cups
Water	1 cup		

STEPS

1. Partially freeze meat. Trim fat and slice meat across the grain into 1/4-inch-thick slices.
2. Combine seasonings. Sprinkle mixture over meat.
3. Combine cornstarch, water, and lemon juice. Set aside.
4. Heat nonstick frying pan. Add meat and stir-fry until brown, about 5 minutes. Drain meat, remove to another container, and cover to keep warm.
5. In same pan, stir-fry carrots 5 minutes or until tender-crisp. Add remaining vegetables and stir-fry 2 minutes. Add meat, and cornstarch mixture. Bring to a boil. Cook, stirring constantly, until thickened.
6. Serve over rice.

NUTRITIONAL ANALYSIS:

Calories:	Protein (gm):	Fat (gm):	Carbo (gm):	Sodium (mg):
370	29	9	42	240

EACH SERVING PROVIDES:

3 ounces from meat group
1 serving from vegetable group
1-1/2 servings from bread group

Figure 2-2

Recipe for Pork and Vegetable Stir-fry

Source: "Using the Food Guide Pyramid: A Resource for Nutrition Educators" by A. Shaw, L. Fulton, C. Davis, and M. Hogbin. U.S. Department of Agriculture: Food, Nutrition, and Consumer Services; Center for Nutrition Policy and Promotion.

Most recipes contain foods from more than one food group. For example, Pork and Vegetable Stir-fry (Figure 2-2) uses 1 pound of boneless pork loin, which is expected to yield 12 ounces of meat, or four 3-ounce servings. The recipe also uses 3 1/2 cups of fresh vegetables, which, after cooking, will yield about four 1/2-cup servings. Finally, the recipe includes 3 cups of cooked rice, or four 3/4-cup servings. Each serving will therefore provide 3 ounces from the meat group, 1/2 cup from the vegetable group, and 3/4 cup from the bread group.

In developing the food guide, the typical use of foods by Americans was an important factor in establishing food groups and in developing nutrient profiles for each food group. The Pyramid has been adapted to reflect the customs of numerous ethnic and cultural groups within the United States, including Mexican (Figure 2-3), Asian, and Mediterranean diets (see Appendix D). It has also been adapted to meet the needs of young children and older adults (see Chapter 13), as well as vegetarians (see Chapter 11).

Planning Menus Using the Food Guide Pyramid. Planning menus gives you the opportunity to include a variety of foods from each food group, especially foods from subgroups that provide nutrients often low in American diets. It also provides the chance to balance fat and sodium to maintain healthful levels over time.

Table 2-7 shows a five-day menu designed to meet the Food Guide Pyramid guidelines and provide 2,200 calories.

■ **Bread, Cereal, Rice, and Pasta**—While there are many products to choose from, most people eat less than the minimum of 6 servings per day, and choices tend to be enriched, rather than whole grains. The Whole-Wheat Cornmeal Muffins and the Whole-Wheat Pancakes illustrate some whole-grain products. Rice-Pasta Pilaf illustrates the use of a grain mixture as an attractive side dish, and provides part of a vegetable serving as well. Recipes for Chocolate Mint Pie, Peach Crisp, and Lemon Pound Cake show that desserts can contribute to grain servings too.

■ **Vegetables**—Although most people report having some vegetables each day, they are often potatoes, especially French fries. The Food Guide Pyramid encourages consumption of a variety of different vegetables, with special emphasis on dark-green leafy vegetables and cooked dry beans and peas, and urges preparation in lower-fat ways. The Corn and Zucchini Combo, Spinach-Orange Salad, and Confetti Coleslaw illustrate the use of vegetables in attractive, lower-fat ways. Other recipes—Chili-Baked Potato, Pork and Vegetable Stir-fry, Creole Fish Fillet—suggest ways to increase the use of vegetables as part of main dishes. In some recipes, vegetables add flavor or serve as extenders to make larger por-

TABLE 2-7 Five Days' Menus at 2,200 Calories

	Day 1	Day 2	Day 3	Day 4	Day 5
BREAKFAST					
	Orange juice ......3/4 c	Grapefruit juice ....3/4 c	Grapefruit............1/2	Fresh sliced strawberries ....1/2 c	Cantaloupe ...1/4 melon
	Oatmeal ....1/2 c	Breakfast pita ....1 sandwich	Banana........1 medium	Whole-grain cereal flakes ....1 oz	Turkey patty.... 1-1/2 oz
	White toast ......2 slices	2% fat milk.........1 c	Ready-to-eat cereal flakes........1 oz	Toasted plain bagel ...... 1 medium	Whole-wheat pancakes.. 2
	Margarine.........2 tsp		Toasted English muffin with raisins......1	Cream cheese....1 tbsp	Blueberry sauce ... 1/4 c
	Jelly..............1 tsp		Margarine.........2 tsp	2% fat milk.........1 c	Margarine.........1 tsp
	2% fat milk.........1/2 c		Skim milk........1/2 c		Skim milk.........1 c
LUNCH					
	Split pea soup......1 c	Turkey pasta salad ...... 1-1/4 c	Taco salad greens ....1 c	Broiled chicken fillet sandwich 1	Chili-stuffed baked potato 1
	Quick tuna and sprouts sandwich 1	Tomato wedges on lettuce leaf .....1 serving	chill .... 3/4 c	Mayonnaise ....1 pkt	Lowfat, low-sodium cheddar cheese ..3 tbsp
	Mixed green salad ...1 c	Hard rolls.............2	Gingersnaps 2	Confetti coleslaw....1/2 c	Spinach-orange salad.... 1 c
	Reduced-calorie Italian dressing.........1 tbsp	Margarine.........2 tsp		Fresh orange 1	Wheat crackers........6
	Chocolate mint pie ...... 1 serving	Oatmeal cookies......4		2% fat milk.........1 c	Skim milk.........1 c
		2% fat milk.........1 c			

DINNER

Savory sirloin . . . . . 3 oz
Corn and zucchini combo . . . . . 3/4 c
Tomato and lettuce salad . . . . . 1 serv
French dressing . . . 1 tbsp
Whole-wheat rolls 2
Margarine . . . . . 1 tsp
Yogurt-strawberry parfait . . . . . 1 c

Creole fish fillets . . . 4 oz
Small new potatoes with skin 2
Cooked green peas . . . . . 1/2 c
with margarine . . . 1 tsp
Whole-wheat cornmeal muffins 2
Margarine . . . . . 2 tsp
Peach crisp . . . . . 1/2 c

Pork and vegetable stir-fry mixture . . . . . 1 c
rice . . . . . 3/4 c
Cooked broccoli . . . 1/2 c
White rolls 2
Margarine . . . . . 2 tsp
Minted pineapple chunks . . . . . 1/2 c

Lentil stroganoff mixture . . . . . 1-1/2 c
noodles . . . . . 3/4 c
Cooked whole green beans . . . . . 1/2 c
with margarine . . . 1 tsp
Tomato and cucumber salad . . . . . 1 serv
Reduced-calorie vinaigrette dressing . . . . . 1 tbsp
Pumpernickel roll 1
Margarine . . . . . 1 tsp
Honeydew . . . . 1/8 melon

Apricot-glazed chicken . . . . . 3 oz
Rice-pasta pilaf . . . . . 3/4 c
Tossed salad . . . . . 1 c
Reduced-calorie Italian dressing . . . . . 1 tbsp
Hard rolls 2
Margarine . . . . . 2 tsp
Vanilla ice milk . . . . 1/2 c

SNACKS

Graham crackers . . . . . 6 squares
2% fat milk . . . . . 1 c
Peanut butter . . . . 2 tbsp
Fresh peach 1
Carrot sticks . . . . . 7–8 medium

Bagel . . . . . 1 medium
Margarine . . . . . 2 tsp
Fresh pear 1

Wheat crackers 6
Cheddar cheese . . . . . 1-1/2 oz
Turkey sandwich . . . 1/2
No-salt-added tomato juice . . . . . 3/4 c

No-salt-added vegetable juice . . . . . 3/4 c
Roast beef sandwich . . 1
2% fat milk . . . . . 1 c

Soft pretzel . . . . . 1 large
Fresh apple 1/2

Source: "Using the Food Guide Pyramid: A Resource for Nutrition Educators" by A. Shaw, L. Fulton, C. Davis, and M. Hogbin. U.S. Department of Agriculture: Food, Nutrition, and Consumer Services; Center For Nutrition Policy and Promotion.

Pirámide

Del día con el sabor popular mexicano
A Food Guide Pyramid with a Mexican Flavor

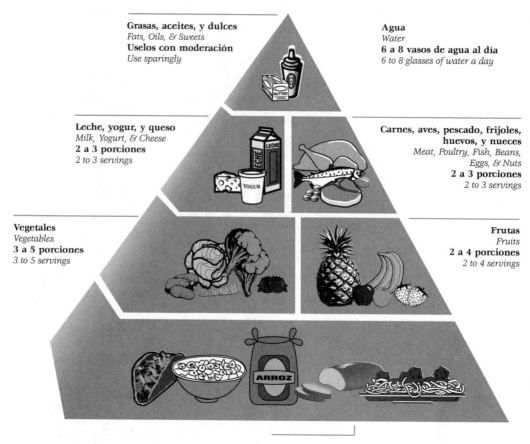

Grasas, aceites, y dulces
Fats, Oils, & Sweets
Uselos con moderación
Use sparingly

Agua
Water
6 a 8 vasos de agua al día
6 to 8 glasses of water a day

Leche, yogur, y queso
Milk, Yogurt, & Cheese
2 a 3 porciones
2 to 3 servings

Carnes, aves, pescado, frijoles, huevos, y nueces
Meat, Poultry, Fish, Beans, Eggs, & Nuts
2 a 3 porciones
2 to 3 servings

Vegetales
Vegetables
3 a 5 porciones
3 to 5 servings

Frutas
Fruits
2 a 4 porciones
2 to 4 servings

Tortillas, panes, cereales, arroz, y pastas
Tortillas, Bread, Cereal, Rice, & Pasta
6 a 11 porciones
6 to 11 servings

Al preparar nutritivos alimentos mexicanos con sus niños o nietos, usted les ayuda a:
—mejorar su alimentación, y
—lograr que esos alimentos sigan siendo parte de su cultura.
Preparing healthy and nutritious Mexican foods with your children and grandchildren will improve their diet and help ensure that these foods remain part of the culture.

Figure 2-3

A Food Guide Pyramid with a Mexican Flavor

Source: U.S. Department of Agriculture/U.S. Department of Health and Human Services.

tions—the Breakfast Pita Sandwich or Tuna Sprouts Sandwich. Fresh vegetables add crunch to the Turkey Pasta Salad. Versatile legumes can count as vegetables or as meat alternates, as in Split Pea Soup or Lentil Stroganoff.

■ **Fruit**—Fruit is particularly underconsumed by Americans. In recent USDA food-consumption surveys, only a little over half the adults reported having fruit or fruit juice on any given day. Even fewer low-income people reported eating any fruit. The recipes included here illustrate the use of fruit in a variety of ways. The Blueberry Sauce makes a tasty, nutritious substitute for syrup; fruit can flavor and enhance meat in a main dish, as in the Apricot-Glazed Chicken. It can be a colorful part of a main-dish salad, as in the Turkey Pasta Salad, or in the Spinach-Orange Salad. It also makes a great low-fat dessert, as in the Strawberry Yogurt Parfait or Peach Crisp. The menus also include a variety of whole fruits, fruit juices, and canned fruit as part of meals and snacks.

■ **Milk, Yogurt, and Cheese**—Milk products are often underconsumed by adults, especially fluid milk. The menus show the use of a variety of milk products in addition to fluid milk that contribute to servings from this group: cheese, ice milk, yogurt, frozen yogurt. Recipes for Strawberry Yogurt Parfait and Chocolate Mint Pie illustrate milk's use in attractive low-fat desserts.

■ **Meat, Poultry, Fish, and Alternates**—The main-dish and sandwich recipes illustrate the use of a variety of meats and alternates. The recipes use lean meats, low-fat preparation techniques, and herbs and spices for flavoring to reduce sodium. Servings of meat, poultry, or fish average 3 ounces in main-dish recipes; addition of vegetables and grains make larger portions. The lentils in Lentil Stroganoff provide meat equivalents for a meatless main dish.

Use the following questions to ensure that your menu follows the Food Guide Pyramid.

1. Does a day's menu on the average provide at least the lower number of servings from each of the major food groups?
2. Does the menu have several servings of whole-grain breads and cereals each day?
3. Does the menu include several servings of each of the vegetable subgroups: dark green leafy (such as spinach, broccoli, romaine lettuce), deep yellow (carrots, sweet potatoes), dry beans and peas (kidney beans, lentils), starchy (potatoes, corn), and other vegetables?
4. Does the menu include some vegetables and fruits with skins and seeds (baked potatoes with skin, berries, or apples or pears with peels)?

5. Are foods high in fat, sugar, and/or sodium balanced with choices lower in these nutrients?

Exchange Systems

Exchange system—A tool to plan diets that groups foods by their nutrient and caloric content. Foods within each group have about the same amount of calories, carbohydrate, protein, and fat, so that any food can be substituted for any other food in the same group.

Whereas the food group system groups foods by their protein, vitamin, and mineral content, the **exchange system** groups foods by their calorie, carbohydrate, fat, and protein content. Each food on a list has approximately the same amount of calories, carbohydrate, fat, and protein as another in the portions listed, so that any food on a list can be exchanged for any other food on the same list.

The **Exchange Lists for Meal Planning** have been developed by the American Diabetes and American Diabetic associations for use primarily by people with diabetes, who need to regulate what and how much they eat. They are also often used in weight control because they are relatively easy to learn and master, and they afford a good deal of control over calorie intake. There are seven exchange lists of like foods. Each food on a list has approximately the same amount of calories, carbohydrate, fat, and protein as another in the portions listed, so that any food on a list can be exchanged, or traded, for any other food on the same list (see Table 2-8). The seven

---

TABLE 2-8 Nutrient Content of Exchange Lists

Groups/Lists	Typical Item	Carbohydrate (grams)	Protein (grams)	Fat (grams)	Calories
Carbohydrate Group					
Starch	1 slice bread	15	3	1 or less	80
Fruit	1 small apple	15	—	—	60
Milk					
Skim	1 cup	12	8	0–3	90
Low-fat	1 cup	12	8	5	120
Whole	1 cup	12	8	8	150
Other carbohydrates	2 small cookies	15	varies	varies	varies
Vegetables	1/2 cup cooked carrots	5	2	—	25
Meat and Meat Substitute Group					
Very lean	1 oz. chicken (no skin, white)	—	7	0–1	35
Lean	1 oz. lean beef	—	7	3	55
Medium-fat	1 oz. ground beef	—	7	5	75
High-fat	1 oz. pork sausage	—	7	8	100
Fat Group		—	—	5	45

Adapted from: *Exchange Lists for Meal Planning*, 1995, American Diabetes Association and American Diabetic Association.

exchange lists are starch, fruit, milk, other carbohydrates, vegetables, meat and meat substitutes, and fat. People with diabetes can exchange starch, fruit, or milk choices within their meal plans because they all have about the same amount of carbohydrate per serving.

Each exchange list has a typical item with an easy-to-remember portion size:

Starch—1 slice bread, 80 calories

Meat—1 ounce lean meat, 55 calories

Vegetable—1/2 cup cooked vegetable, 25 calories

Fruit—1 small apple, 60 calories

Milk—1 cup skim milk, 90 calories

Other carbohydrates—2 small cookies, calories vary

Fat—1 teaspoon margarine, 45 calories

The meat exchange is broken down into very lean, lean, medium-fat, and high-fat meat and meat alternates. Very lean and lean meats are encouraged. The milk exchange contains skim, low-fat, and whole milk exchanges. Fats are divided into three groups, based on the main type of fat they contain: monounsaturated, polyunsaturated, or saturated. There is also a listing of free foods that contain negligible calories.

Both systems have their good and bad points. While the Food Guide Pyramid plan encourages variety, nutrient adequacy, and balance—no group of foods is overemphasized—the Exchange Lists also promote moderation and variety. The Food Guide Pyramid is easier to use, but the Exchange Lists are more accurate in terms of calories and nutrients consumed.

■ **MINI-SUMMARY**

Dietary recommendations have been published for the healthy American public for almost 100 years. The Food Guide Pyramid, with its five food groups (Figure 2-1), is based on the Dietary Guidelines for Americans and nutrient recommendations currently found in the Dietary Reference Intakes. The key concepts of the Pyramid include variety, proportionality, and moderation. By using the number of servings and portion sizes in the Pyramid as well as additional Pyramid guidelines, you can plan healthy menus.

Food Labels

Since 1938, the federal government has required basic information on food labels (Figure 2-4). The Food and Drug Administration (FDA) regulates labels on all packaged foods except for meat, poultry, and egg products—

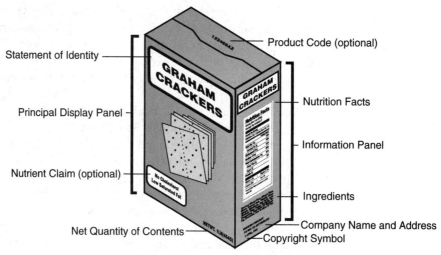

Figure 2-4

Location of nutrition facts

foods regulated by the U.S. Department of Agriculture (USDA). The amount of information on food labels varies, but all food labels must contain at least:

■ The name of the food
■ A list of ingredients
■ The net contents or net weight—the quantity of the food itself without the packaging (in English and metric units)
■ The name and place of business of the manufacturer, packer, or distributor

Nutrition information is also required for most foods, our next topic.

For most foods, all ingredients must be listed on the label and identified by their common names to help consumers identify ingredients that they are allergic to or want to avoid for other reasons. The ingredient that is present in the largest amount, by weight, must be listed first. Other ingredients follow in descending order according to weight (Figure 2-5).

Nutrition Facts

Figure 2-6 is a sample "Nutrition Facts" panel that you see on food labels. Serving size is the first stop when you read the Nutrition Facts, because it tells the calorie and nutrient content per serving. Just how big is a serving?

Figure 2-5
Food label

Serving sizes are designed to reflect the amounts people actually eat. Of course, if you eat more or less than the serving size, you will need to calculate the increase or decrease in calories and nutrients.

If you check serving sizes on similar foods, you'll see that the sizes are similar. That means you don't need to be a math whiz to compare two foods. It's easy to see the calorie and nutrient differences between similar

Figure 2-6
Nutrition label

Nutrition Facts

Serving Size 1 cup (228g)
Servings Per Container 2

Amount Per Serving

Calories 250 Calories from Fat 110

	% Daily Value*
Total Fat 12g	**18%**
Saturated Fat 3g	**15%**
Cholesterol 30mg	**10%**
Sodium 470mg	**20%**
Total Carbohydrate 31g	**10%**
Dietary Fiber 0g	**0%**
Sugars 5g	
Protein 5g	

Vitamin A 4%	•	Vitamin C 2%
Calcium 20%	•	Iron 4%

* Percent Daily Values are based on a 2,000 calorie diet.

servings of canned fruit packed in syrup versus the same fruit in natural juices. The same is true for two brands of packaged macaroni and cheese.

Look for servings in two measurements—common household and metric measures. A serving of applesauce would read 1/2 cup (114 g). The household measure is easier to understand, but the metric measure gives a more precise idea of the amount. For example, 114 g means 114 grams, a measure of weight. There are 28 grams in 1 ounce. The label helps you get familiar with metrics, too.

The next stop on the Nutrition Facts panel is the Calories per Serving category, which lists the total calories in one serving, as well as the calories from fat.

Nutrients are listed next. Information about some nutrients is required. These nutrients are total fat, saturated fat, cholesterol, sodium, total carbohydrate, dietary fiber, sugars, protein, vitamin A, vitamin C, calcium, and iron. Others are listed voluntarily. If a food contains an insignificant amount of a required nutrient, it might be omitted from the label.

Information about other nutrients is required in two cases: if a claim is made about the nutrients on the label, or if the nutrients are added to the food. For example, fortified breakfast cereals must give Nutrition Facts for any added vitamins and minerals.

Daily Value—Nutrient standards used on food labels to allow nutrient comparisons among foods.

Nutrient amounts actually are listed in two ways: in metric amounts (in grams) or as a percentage of the **Daily Value.** Daily Values are nutrient standards used on food labels to allow nutrient comparisons among foods. They are based on a 2,000-calorie diet. Therefore, the Daily Value may be a little high, a little low, or right on target for you. Percent Daily Values show you how much of the Daily Value is in one serving. For example, in Figure 2-6, the Percent Daily Value for total fat is 18 percent and for dietary fiber is 0 percent. The Daily Value for fat (and also carbohydrate and protein) is based on a 2,000-calorie-per-day diet. The Daily Value for dietary fiber is 25 grams.

The Daily Values for vitamins and most minerals are based on the DRI. The DRI can't be used on nutrition labels, because it is set for specific age and gender categories. The Daily Values are generally set at or near the highest DRI value (Table 2-9). Daily values are also set for several nutrients not part of the DRI (carbohydrate, fiber, fat, saturated fat, cholesterol, protein, sodium, potassium). Some of these nutrients may eventually become part of the DRI.

The values listed for total carbohydrate include all carbohydrates, including dietary fiber and sugars listed below it. The sugar values include naturally present sugars, such as lactose in milk and fructose in fruits, as well as those added to the food, such as table sugar and corn syrup. The label can claim no sugar added but still have naturally occurring sugar. An example is

TABLE 2-9   Daily Values*

Nutrient	Daily Value
Carbohydrate	300 grams
Fiber	25 grams
Cholesterol	300 milligrams
Fat	65 grams
Saturated fat	20 grams
Protein	50 grams
Vitamin A	5,000 International Units
Vitamin D	400 International Units
Vitamin E	30 International Units
Vitamin K	80 micrograms
Vitamin C	60 milligrams
Thiamin	1.5 milligrams
Riboflavin	1.7 milligrams
Niacin	20 milligrams
Vitamin B-6	2 milligrams
Folate	400 micrograms
Vitamin B-12	6 micrograms
Biotin	0.3 milligrams
Pantothenic acid	10 milligrams
Calcium	1,000 milligrams
Chloride	3,400 milligrams
Chromium	120 micrograms
Copper	2 milligrams
Iodine	150 micrograms
Iron	18 milligrams
Magnesium	400 milligrams
Manganese	2 milligrams
Molybdenum	75 micrograms
Phosphorus	1,000 milligrams
Potassium	3,500 milligrams
Selenium	70 micrograms
Sodium	2,400 milligrams
Zinc	15 milligrams

* Designed for adults and children over 4 years of age and based on 2,000 calorie diet.

fruit juice. The nutrition label does not list a Percent Daily Value for sugars because there is not enough scientific evidence to establish one.

The values listed for total fat refer to all the fat in the food: saturated, polyunsaturated, and monounsaturated. Only total fat and saturated fat information is required on the label, because high intakes of both are linked

TABLE 2-10 Nutrient Content Claims—A Dictionary

Nutrient (Content Claim)	Definition (Per Serving)
Calories	
Calorie free	less than 5 calories
Low calorie	40 calories or less
Reduced or fewer calories	at least 25% fewer calories*
Sugar	
Sugar free	less than 0.5 gram sugars
Reduced sugar or less sugar	at least 25% less sugars*
No added sugar	no sugars added during processing or packing, including ingredients that contain sugars, such as juice or dry fruit
Fat	
Fat free	less than 0.5 gram fat
Low fat	3 grams or less of fat
Reduced or less fat	at least 25% less fat*
Trans fat free[+]	less than 0.5 gram of trans fat and less than 0.5 gram saturated fat
Saturated fat free	less than 0.5 gram saturated fat and less than 0.5 gram of trans fat
Low saturated fat	1 gram or less saturated fat (and less than 0.5 gram trans fat)[+]
Reduced saturated fat	at least 25% less saturated fat* (and at least 25% less saturated fat and trans fat combined)[+]
Cholesterol	
Cholesterol free	less than 2 milligrams cholesterol and 2 grams or less of saturated fat (and trans fat combined)[+]
Sodium	
Sodium free	less than 5 milligrams sodium
Very low sodium	35 milligrams or less sodium
Low sodium	140 milligrams or less sodium
Reduced or less sodium	at least 25% less sodium*
Light in sodium	50% less*
Fiber	
High fiber	5 grams or more
Good source of fiber	2.5 to 4.9 grams
More or added fiber	at least 2.5 grams more
Other Claims	
High, rich in, excellent source of	20% or more of Daily Value*
Good source	10% to 19% of Daily Value*
More	10% or more of Daily Value*
Fresh	raw, unprocessed, or minimally processed, with no added preservatives

TABLE 2-10 *(continued)*	
Nutrient (Content Claim)	Definition (Per Serving)
Healthy	Low in fat and saturated fat, contains no more than 20% of the Daily Value for sodium and cholesterol, contains at least 10% of the Daily Value for one of the following: vitamin A or C, calcium, iron, protein, fiber (fresh, canned, or frozen fruits and vegetables and enriched breads and cereals are exempt from 10% rule)
Light	One of the following: one-third fewer calories or 50% less fat*; low-calorie, low-fat, containing 50% less sodium than normally present; or light in color and texture (such as light brown sugar)
Lean (meat and poultry only)	Less than 10 grams fat, 4.5 grams saturated fat (and trans fat combined),[†] and 95 milligrams cholesterol
Extra lean (meat and poultry only)	Less than 5 grams fat, 2 grams saturated fat (and trans fat combined),[†] and 95 milligrams cholesterol

* Compared with a standard serving size of the traditional food.
[†] Proposed in 1999.

to high blood cholesterol, which is linked to increased risk of coronary heart disease. Listing the amount of polyunsaturated and monounsaturated fats in the food is voluntary. In 1999, the Food and Drug Administration announced plans to require food manufacturers to include the amount of trans fatty acids on nutrition labels. Trans fatty acids are liquid fats that have been turned into solid fats. They are often found in margarine and baked goods, and they raise the risk of heart disease. Trans fatty acid information will be on food labels about 2002.

Like sugars, you will notice that there is no Percent Daily Value for protein. Because most Americans get more than enough protein, no Daily Value is necessary.

Nutrient Claims

Nutrient content claims, such as "good source of calcium" or "fat free," can appear on food packages only if they follow legal definitions (Table 2-10). For example, a food that is a good source of calcium must provide

10 to 19 percent of the Daily Value for calcium in one serving. Phrases such as "sugar free" describe the amount of a nutrient in a food, but don't tell exactly how much. These nutrient content claims differ from Nutrition Facts, which do list specific nutrient amounts.

If a food label contains a descriptor for a certain nutrient but the food contains other nutrients at levels known to be less healthy, the label would have to bring that to consumers' attention. For example, if a food making a low-sodium claim is also high in fat, the label must state "see back panel for information about fat and other nutrients."

Health Claims

The Nutrition Labeling and Education Act of 1990 provided, for the first time, the authority to allow food labels to carry claims about the relationship between the food and specific diseases or health conditions. This was a major shift in labeling philosophy. These **health claims** state that certain foods or food substances—as part of an overall healthy diet—may reduce the risk of certain diseases. Examples include calcium and osteoporosis, and dietary saturated fat and cholesterol and the risk of coronary heart disease. Although food manufacturers may use health claims to market their products, the intended purpose of health claims is to benefit consumers by providing information on healthful eating patterns that may help reduce the risk of heart disease, cancer, osteoporosis, high blood pressure, dental cavities, or certain birth defects.

Health claims may show links between the following nutrients and conditions (see Table 2-11 for sample claims):

1. Calcium and osteoporosis
2. Sodium and hypertension (high blood pressure)
3. Dietary fat and cancer
4. Dietary saturated fat and cholesterol, and risk of coronary heart disease
5. Fiber-containing grain products, fruits, and vegetables, and cancer
6. Fruits, vegetables, and grain products that contain fiber, particularly soluble fiber, and risk of coronary heart disease
7. Fruits and vegetables, and cancer
8. Folate and neural-tube birth defects
9. Dietary-sugar alcohol and dental caries (cavities)
10. Dietary soluble fiber, such as that found in whole oats and psyllium seed husk, and coronary heart disease
11. Wholegrain foods, and risk of heart disease and certain cancers
12. Soy protein and risk of coronary heart disease

TABLE 2-11 Health Claims

1. *Calcium and osteoporosis.* Low calcium intake is one risk factor for osteoporosis, a condition of lowered bone mass, or density. Lifelong adequate calcium intake helps maintain bone health by increasing as much as genetically possible the amount of bone formed in the teens and early adult life, and by helping to slow the rate of bone loss that occurs later in life.

    Typical Foods: Low-fat and skim milks, yogurts, calcium-fortified citrus drinks.

    Sample Claim: "Regular exercise and a healthy diet with enough calcium helps teen and young adult white and Asian women maintain good bone health and may reduce their high risk of osteoporosis later in life."

    Requirement: Food or supplement must be "high" in calcium and must not contain more phosphorus than calcium.

2. *Sodium and hypertension (high blood pressure).* Hypertension is a risk factor for coronary heart disease and stroke deaths. The most common source of sodium is table salt. Diets low in sodium may help lower blood pressure and related risk in many people.

    Typical Foods: Fruits, vegetables, unsalted tuna, low-fat milk and yogurt, sherbet, ice milk, cereal.

    Sample Claim: "Diets low in sodium may reduce the risk of high blood pressure, a disease associated with many factors."

    Requirements: Foods must meet criteria for "low sodium."

3. *Dietary fat and cancer.* Diets high in fat increase the risk of some types of cancer, such as cancers of the breast and colon. While scientists don't know how total fat intake affects cancer development, low-fat diets reduce the risk.

    Typical Foods: Fruits, vegetables, reduced-fat milk products, cereals, pasta.

    Sample Claim: "Development of cancer depends on many factors. A diet low in total fat may reduce the risk of some cancers."

    Requirements: Foods must meet criteria for "low fat." Fish and game meats must meet criteria for "extra lean."

4. *Dietary saturated fat and cholesterol, and risk of coronary heart disease.* Diets high in saturated fat and cholesterol increase total and low-density (bad) blood cholesterol levels, and thus the risk of coronary heart disease. Diets low in saturated fat and cholesterol decrease the risk.

    Typical Foods: Fruits, vegetables, skim and low-fat milks, cereals, whole-grain products and pastas.

    Sample Claim: "While many factors affect heart disease, diets low in saturated fat and cholesterol may reduce the risk of this disease."

    Foods must meet criteria for "low saturated fat," "low cholesterol," and "low fat." Fish and game meats must meet criteria for "extra lean."

5. *Fiber-containing grain products, fruits, and vegetables, and cancer.* Diets low in fat and rich in fiber-containing grain products, fruits, and vegetables may reduce the risk of some types of cancer. The exact role of total dietary fiber, fiber components, and other nutrients and substances in these foods is not fully understood.

    Typical Foods: Whole-grain breads and cereals, fruits, and vegetables.

    Sample Claim: "Low-fat diets rich in fiber-containing grain products, fruits, and vegetables may reduce the risk of some types of cancer, a disease associated with many factors."

TABLE 2-11 *(continued)*

Requirements: Foods must meet criteria for "low fat" and, without fortification, be a "good source" of dietary fiber.

6. *Fruits, vegetables, and grain products that contain fiber, particularly soluble fiber, and risk of coronary heart disease.* Diets low in saturated fat and cholesterol and rich in fruits, vegetables, and grain products that contain fiber, particularly soluble fiber, may reduce the risk of coronary heart disease.

Typical Foods: Fruits, vegetables, and whole-grain breads and cereals.

Sample Claim: "Diets low in saturated fat and cholesterol and rich in fruits, vegetables, and grain products that contain some types of dietary fiber, particularly soluble fiber, may reduce the risk of heart disease, a disease associated with many factors."

Requirements: Foods must meet criteria for "low saturated fat," "low fat," and "low cholesterol." They must contain, without fortification, at least 0.6 gram of soluble fiber per reference amount, and the soluble fiber content must be listed.

7. *Fruits and vegetables, and cancer.* Diets low in fat and rich in fruits and vegetables may reduce the risk of some cancers. Fruits and vegetables are low-fat foods, and may contain fiber or vitamin A and vitamin C.

Typical Foods: Fruits and vegetables.

Sample Claim: "Low-fat diets rich in fruits and vegetables (foods that are low in fat and may contain dietary fiber, vitamin A, or vitamin C) may reduce the risk of some types of cancer, a disease associated with many factors. Broccoli is high in vitamins A and C, and it is a good source of dietary fiber."

Requirements: Foods must meet criteria for "low fat" and, without fortification, be a "good source" of fiber, vitamin A, or vitamin C.

8. *Folate and neural-tube birth defects.* Defects of the neural tube (a structure that develops into the brain and spinal cord) occur within the first six weeks after conception, often before the pregnancy is known. The U.S. Public Health Service recommends that all women of child-bearing age in the United States consume 0.4 mg of folate daily to reduce their risk of having a baby affected with spina bifida or other neural-tube defects.

Typical Foods: Enriched cereal grain products, some legumes (dried beans and peas), fresh leafy green vegetables, oranges, grapefruit, many berries, some dietary supplements, and fortified breakfast cereals.

Sample Claim: "Healthful diets with adequate folate may reduce a woman's risk of having a child with a brain or spinal-cord birth defect."

Requirements: Foods must meet or exceed criteria for "good source" of folate—that is, at least 40 micrograms of folate per serving (at least 10 percent of the Daily Value). A serving of food cannot contain more than 100 percent of the Daily Value for vitamin A and vitamin D, because of their potential risk to fetuses.

9. *Dietary-sugar alcohol and dental caries (cavities).* Between-meal eating of foods high in sugar and starches may promote tooth decay. Sugarless candies made with certain sugar alcohols do not.

Typical Foods: Sugarless candy and gum.

Sample Claim: "Frequent between-meal consumption of foods high in sugar and starches promotes tooth decay. The sugar alcohols in this food do not promote tooth decay."

Requirements: Foods must meet the criteria for "sugar free."

TABLE 2-11 *(continued)*

10. *Dietary soluble fiber, such as that found in whole oats and psyllium seed husk, and coronary heart disease.* When included in a diet low in saturated fat and cholesterol, soluble fiber may affect blood lipid levels, such as cholesterol, and thus lower the risk of heart disease. However, because soluble dietary fibers constitute a family of very heterogeneous substances that vary greatly in their effect on the risk of heart disease, the FDA has determined that sources of soluble fiber for this health claim need to be considered case by case. To date, the FDA has reviewed and authorized two sources of soluble fiber eligible for this claim: whole oats and psyllium seed husk.

     Typical Foods: Oatmeal cookies, muffins, breads, and other foods made with rolled oats, oat bran, or whole oat flour, hot and cold breakfast cereals containing whole oats or psyllium seed husk; and dietary supplements containing psyllium seed husk.

     Sample Claim: "Diets low in saturated fat and cholesterol that include 3 grams of soluble fiber from whole oats per day may reduce the risk of heart disease. One serving of this whole-oats product provides ___ grams of this soluble fiber."

     Requirements: Foods must meet criteria for "low saturated fat," "low cholesterol," and "low fat." Foods that contain whole oats must contain at least 0.75 gram of soluble fiber per serving. Foods that contain psyllium seed husk must contain at least 1.7 grams of soluble fiber per serving. The claim must specify the daily dietary intake of the soluble fiber source necessary to reduce the risk of heart disease and the contribution that one serving of the product makes toward that intake level. Soluble fiber content must be stated in the nutrition label.

11. *Wholegrain food, and risk of heart disease and certain cancers.*

     Typical Foods: Whole-wheat bread, whole-wheat pasta, wholegrain cereals.

     Sample Claim: "Diets rich in wholegrain foods and other plant foods and low in total fat, saturated fat, and cholesterol may reduce the risk of heart disease and certain cancers."

     Requirements: To qualify, a food must contain 51 percent or more of its weight as wholegrain ingredients and not more than 3 grams of fat per serving.

12. *Soy protein and risk of coronary heart disease.* Soy protein (about 25 grams a day) included in a diet low in saturated fat and cholesterol may reduce the risk of coronary heart disease by lowering blood cholesterol levels.

     Typical Foods: Soy beverages, tofu, tempeh, soy-based meat alternatives, soy flour.

     Sample Claim: "Diets low in saturated fat and cholesterol that include 25 grams of soy protein a day may reduce the risk of heart disease. One serving of this food provides ___ grams of soy protein."

     Requirements: Only foods that contain at least 6.25 grams of soy protein per serving can use this claim.

Food and food substances can qualify for health claims only if they meet Food and Drug Administration requirements (Table 2-11). For example, a food with a health claim about sodium and hypertension must meet criteria for "low sodium."

■ **MINI-SUMMARY**

All food labels must contain the name of the product; the net contents or net weight; the name and place of business of the manufacturer, packer, or distributor; a list of ingredients in order of predominance by weight; and nutrition information (Figure 2-6). Daily Values are nutrient standards used on food labels to allow nutrient comparisons among foods. The Daily Values are generally set at or near the highest DRI value. Any nutrient or health claims on food labels must comply with Food and Drug Administration regulations and definitions as outlined in this chapter.

Portion Size Comparisons

Portion size is an important concept for anyone involved in preparing and serving foods. Serving sizes vary from kitchen to kitchen, but American serving sizes have been steadily increasing. In comparison with the Food Guide Pyramid portion sizes, as well as those served in many European countries, our portion sizes are huge. It wasn't that long ago when a "large" soft drink was typically 16 fluid ounces. Now, that's often the "small" size.

What you may consider to be one serving of the bread group may actually be three or four servings. For example, the Food Guide Pyramid considers one ounce of bread, about one slice, to be one serving. A typical New York style bagel is about 4 ounces, or about four servings using the Food Guide Pyramid.

You may have noticed that the portion sizes in the Food Guide Pyramid do not always match the serving sizes found on food labels. This is because the purpose of the Food Guide Pyramid is not the same as nutrition labeling. The Food Pyramid was designed to be very simple to use. Therefore, the USDA specified only a few serving sizes for each food group so that they could be remembered easily. Food labels have a different purpose: to allow the consumer to compare the nutrients in equal amounts of foods. To compare the nutrient amounts in equal amounts of pasta, the portion size on the label is 2 ounces of uncooked pasta (about 56 grams), which will cook up to about 1 cup of spaghetti or up to 2 cups of a large shaped pasta, such as ziti. Using the Pyramid, the portion size for cooked pasta is only half a cup.

In many cases, the portion sizes are similar on labels and in the food guide, especially when expressed as household measures. For foods falling into only one major group, such as fruit juices, the household measures provided on the label (1 cup or 8 fluid ounces) can help you relate the label serving size to the Pyramid serving size. For mixed dishes, food-guide serving sizes may be used to visually estimate the food item's contribution to each food group as the food is eaten—for example, the amounts of bread, vegetable, and cheese contributed by a portion of pizza.

■ **MINI-SUMMARY**

The portion sizes in the Food Guide Pyramid do not always match the serving sizes found on food labels. This is because the purpose of the Food Guide Pyramid is not the same as nutrition labeling. Portion sizes in the United States have been increasing.

Check-Out Quiz

1. What are the Food Guide Pyramid's daily nutritional goals? Draw a line from the name of the item to the Pyramid's daily goal for that item.

Item	Pyramid's Daily Goal
Energy (calories)	30 percent or less of calories
Added sugar	100 percent of RDA/DRIs
Fiber	2,400 mg or less
Total fat	Increase intake
Saturated fat	Don't exceed caloric needs
Cholesterol	Less than 10 percent of calories
Protein/vitamins/minerals	1,300–3,000
Sodium	300 mg or less

2. What are the Food Guide Pyramid's serving sizes? Draw a line from the food to the Pyramid's serving size. Each serving size may be used more than once.

Food	Serving Size
Apple	2–3 ounces
Fruit juice	1/2 cup
Bread	3/4 cup
Cold cereal	1 cup
Cooked vegetables	2
Raw vegetables	4 tablespoons
Milk or yogurt	1 medium
Chicken	
Cooked kidney beans	
Eggs	
Peanut butter	
Raw leafy vegetables	
Cooked rice or pasta	

3. Which food group(s) is a good source of protein?
    a. bread, cereal, rice, and pasta
    b. vegetable
    c. milk, yogurt, and cheese
    d. meat, poultry, fish, dry beans, eggs, and nuts
4. Which food group(s) provides one or more B vitamins (thiamin, riboflavin, niacin, folate, B_{12})?
    a. bread, cereal, rice, and pasta
    b. vegetable
    c. milk, yogurt, and cheese
    d. meat, poultry, fish, dry beans, eggs, and nuts
5. Upside-down triangles on the Food Guide Pyramid represent:
    a. added fat
    b. added sugar
    c. Fats, Oils, and Sweets group
    d. added sodium
6. Food Guide Pyramid serving sizes are always the same as food-label serving sizes.
    a. True   b. False
7. Daily Values are based on a 1,800-calorie diet.
    a. True   b. False
8. Claims such as "good source of calcium" or "fat free" on food labels are examples of health claims.
    a. True   b. False
9. Health claims on labels are regulated by the federal government.
    a. True   b. False

10. Portion sizes in the United States have been decreasing.
   a. True   b. False

Activities and Applications

1. Checking Out Nutrient Claims

Look at foods from two of the following sections of the supermarket. Write down nutrient claims (such as low fat) given on at least two different foods from each section. Don't forget: fresh fruits and vegetables, meat, poultry, and seafood don't have labels—look for nutrition information nearby. Also look at the label to see which nutrition facts support this claim.

Produce
Frozen Foods
Fresh Meats, Poultry, and Fish
Dairy
Cereals
Cookies

During your search, also find one food item with a health claim and write it down.

2. Label Reading at Breakfast

Look closely at the "Nutrition Facts" for each food you normally eat for breakfast, such as cereal, milk, and juice. Add up the % Daily Values for fat, saturated fat, cholesterol, sodium, total carbohydrate, protein, vitamin A, vitamin C, calcium, and iron. How nutritious is your breakfast?

3. Label Comparison

Below are labels from regular mayonnaise and low-fat mayonnaise dressing. Which is which? Compare and contrast the labels.

Nutrition Facts	Nutrition Facts
Serving size: 1 tablespoon	Serving size: 1 tablespoon
Amount per serving	Amount per serving
Calories 25	Calories 100
Calories from Fat 10	Calories from Fat 99
Total Fat 1 g	Total Fat 11 g
Saturated 0 g	Saturated 2 g
Polyunsaturated 0.5 g	Polyunsaturated 6 g
Monounsaturated 0 g	Monounsaturated 3 g
Cholesterol 0 mg	Cholesterol 5 mg
Sodium 140 mg	Sodium 180 mg

Total Carbohydrate 4 g   Total Carbohydrate 0 g
        Sugars 3 g   Sugars 0 g
        Protein 0 g   Protein 0 g

Not a significant source of dietary fiber, vitamin A, vitamin C, calcium, and iron.

4. Menu Evaluation

Obtain a cycle menu (a menu rotated at specific time intervals, such as two or four weeks) from a college dining hall, school foodservice, or other foodservice. Evaluate the menu using the Food Guide Pyramid and the questions on pages 53–54.

Nutrition Web Explorer

Food Guide Pyramid: www.nal.usda.gov/fnic
On the home page for the Food and Nutrition Information Center of the USDA, click on "Food Guide Pyramid." Then click on "The Interactive Food Pyramid." Click on each food group and write down the eating tips given for each group.

American Dietetic Association: www.eatright.org
Visit the ADA website and get their "Tip of the Day." Also use the dietitian locator to find a list of dietitians in your area.

Quackwatch: www.quackwatch.com
Visit this website and click on "25 Ways to Spot It" under Quackery. What are 10 ways to spot quackery?

Food Facts *Computerized Nutrient Analysis*

Computer software is the standard for analyzing the amount of nutrients in a recipe. Little wonder, when a task that used to take a half hour or more is done in a matter of minutes. Before computers, it was necessary to look up the nutritional content of each recipe ingredient and record it on a piece of paper. If the amount of the ingredient was not the same as was listed in the reference book, you would have to do some multiplication or division on all the nutrient values to come up with the right numbers. Then, after looking up all the ingredients, you would add up all your columns to get totals. In the final step, you would divide the totals by the yield of the recipe to get the amount of nutrients per serving. Sounds complicated! It sure is, and very time-consuming too.

The computer has done a lot to speed up this process and increase its accuracy. Computerized nutrient-analysis programs contain nutrient information of many different resources. When the name of an ingredient is typed in, the computer lists similar ingredients so that you can choose exactly which one is appropriate. Then you type in the amount of that ingredient you want to be used in the analysis, such as 1 cup. After inputting all the ingredients, you can ask the computer to divide the results by the yield, such as 12 portions. Then the computer will tell you exactly how much of each nutrient (and what percent of the RDA or AI) is contained in one portion. Most computer analysis programs can also give you a percentage breakdown of calories from protein, fat, carbohydrate, and alcohol. Of course, these figures can be printed out on a printer and/or stored in the computer's memory. A sample recipe analysis is given in Figure 2-7.

When considering a computer analysis program, consider the following:

- What different functions can the program perform, and how many of these functions do you need?
- What kind of computer system do you have available to run this software? Be sure you have enough hard-disk space, random access memory (RAM), and an appropriate-speed microprocessor.
- How large is the nutrient database? The database may contain from several thousand to more than 30,000 foods.
- Can foods be added to the database? It's also a good idea to check how many foods can be added.
- How is output presented (graphs, tables, pie charts, etc.), and how easily can it be printed?
- How easy is it to use this program?
- What's the price?
- What service and support is available once you purchase the program? Is on-line help available?

Many companies offer demonstration software at no cost. This is a real benefit, because you can try out the program before buying it. At the government website www.nal.usda.gov/fnic, click on "Food and Nutrition Software Programs" to examine descriptions of many different nutrient analysis software programs.

Breakfast-3 pancakes, 1 oz. sausage, 2 T. syrup, 1 cup orange juice

Recipe Nutrient Analysis
Recipe Food ID: 29 Source: Custom

Yield: 1.00 (1.00 SERVING) Category: No Category
Goal: DAILY VALUES/RDI - ADULT/CHILD

Nutrient	Value	Goal	% Goal
Weight (gm)	430.350		
Kilocalories (kcal)	566.180	2000.000	28%
Protein (gm)	12.503	50.000	25%
Carbohydrate (gm)	88.334	300.000	29%
Fat, total (gm)	19.654	65.000	30%
Alcohol (gm)	0.000		
Cholesterol (mg)	84.617	300.000	28%
Saturated Fat (gm)	5.372	20.000	27%
Monounsaturated Fat (gm)	4.888		
Polyunsaturated Fat (gm)	6.702		
MFA 18:1, Oleic (gm)	2.786		
PFA 18:2, Linoleic (gm)	4.533		
PFA 18:3, Linolenic (gm)	0.608		
PFA 20:5, EPA (gm)	0.000		
PFA 22:6, DHA (gm)	0.006		
Sodium (mg)	686.569	2400.000	29%
Potassium (mg)	647.280	3500.000	18%
Vitamin A (RE)	116.946	1000.000	12%
Vitamin A (IU)	738.724	5000.000	15%
Beta-Carotene (ug)	0.000		
Vitamin C (mg)	124.689	60.000	208%
Calcium (mg)	283.126	1000.000	28%
Iron (mg)	2.792	18.000	16%
Vitamin D (ug)	0.000	10.000	0%
Vitamin D (IU)	0.000	400.000	0%
Vitamin E (ATE)	0.223	20.000	1%
Vitamin E (IU)		30.000	
Alpha-Tocopherol (mg)	0.099		
Thiamin (mg)	0.456	1.500	30%
Riboflavin (mg)	0.398	1.700	23%
Niacin (mg)	2.786	20.000	14%
Pyridoxine/Vit B6 (mg)	0.151	2.000	8%
Folate (ug)	118.464	400.000	30%
Cobalamin/Vit B12 (ug)	0.251	6.000	4%
Biotin (ug)	0.800	300.000	0%
Pantothenic Acid (mg)	0.942	10.000	9%
Vitamin K (ug)	0.248	80.000	0%
Phosphorus (mg)	227.020	800.000	23%
Iodine (ug)		150.000	
Magnesium (mg)	46.320	400.000	12%

Nutrient	Value	Goal	% Goal
Zinc (mg)	0.778	15.000	5%
Copper (mg)	0.250	2.000	13%
Manganese (mg)	0.298	2.000	15%
Selenium (mg)	0.017	0.070	24%
Fluoride (ug)			
Chromium (ug)	0.000	0.120	
Molybdenum (ug)		75.000	
Dietary Fiber, total (gm)	1.624	25.000	6%
Soluble Fiber (gm)	0.000		
Insoluble Fiber (gm)	0.000		
Crude Fiber (gm)	0.250		
Sugar, total (gm)	52.544		
Glucose (gm)	14.826		
Galactose (gm)	0.000		
Fructose (gm)	9.440		
Sucrose (gm)	15.002		
Lactose (gm)	0.000		
Maltose (gm)	4.522		
Tryptophan (mg)	96.160		
Threonine (mg)	290.020		
Isoleucine (mg)	358.420		
Leucine (mg)	617.060		
Lysine (mg)	388.260		
Methionine (mg)	175.020		
Cystine (mg)	144.640		
Phenylalanine (mg)	385.980		
Tyrosine (mg)	283.520		
Valine (mg)	409.180		
Arginine (mg)	435.760		
Histidine (mg)	180.720		
Alanine (mg)	316.500		
Aspartic Acid (mg)	664.800		
Glutamic Acid (mg)	1871.640		
Glycine (mg)	248.040		
Proline (mg)	765.760		
Serine (mg)	446.060		
Moisture (gm)	288.930		
Ash (gm)	4.036		
Caffeine (mg)	0.000		

1 SERVING

% of Kcals

Protein	9%
Carbohydrate	62%
Fat, total	31%
Alcohol	0%

Exchanges

Bread/Starch
Fruit
Other Carbohydrate
Milk - Skim
Milk - Low Fat
Milk - Whole
Vegetable
Meat - Very Lean
Meat - Lean
Meat - Medium Fat
Meat - High Fat
Fat

3/19/2000

Figure 2-7

Sample computerized nutrient analysis output

Hot Topic — Quack! Quack!

The U.S. Surgeon General's Report on Nutrition and Health defines food quackery as "the promotion for profit of special foods, products, processes, or appliances with false or misleading health or therapeutic claims." Have you ever seen advertisements for supplements that are guaranteed to help you lose weight, or herbal remedies to prevent serious disease? If a product's claim seems just too good to be true, it probably *is* too good to be true. The problem with quackery is not just loss of money—you can be harmed as well. When you listen to a quack, you usually stop your regular medical treatment and don't receive, or even seek, further care from a legitimate medical professional.

Nutrition is brimming with quackery, in part because nutrition is such a young science. Research on many fundamental nutrition issues, such as the relationship between sodium and hypertension, is far from being resolved, yet research scientists publicize their results long before those results can be said to really prove a scientific theory. Unfortunately, because much research is only in its early stages, the public has been bombarded with conflicting ideas about issues that relate directly to two very important parts of their lives: their health and their eating habits. This conflict leaves the public confused about the truth, and vulnerable to dubious health products (most often nutrition products) and practices—on which people spend $10 billion to $30 billion annually.

Much misinformation proliferates also because, in some states, anyone can call him- or herself a dietitian or nutritionist. In about half the states, a person is required to obtain a license from the state to use the title dietitian. Licensed dietitians are allowed to offer certain services and may use the initials L.D. (licensed dietitian) after their names. In addition, one may even buy mail-order B.S., M.S., or Ph.D. degrees in nutrition from "schools" in the United States. In all states, nutrition books that are entirely bogus can be published and sold in bookstores, dressed up to look like legitimate health books.

A quack is someone who makes excessive promises and guarantees for a nutrition product or practice that is said to enhance your physical and mental health by, for example, preventing or curing a disease, extending your life, or improving some facet of performance. Health schemes and misinformation proliferate because they thrive on wishful thinking. Many people want easy answers to their medical concerns, such as a quick and easy way to lose weight. Often, claims appear to be grounded in science. Here's how to recognize quacks:

1. Their products make claims such as:
   - quick, painless, and/or effortless,
   - contains special, secret, foreign, ancient, or natural ingredients,
   - effective cure-all for a wide variety of conditions,
   - exclusive product not available through any other source.
2. They use dubious diagnostic tests, such as hair analysis, to detect supposed nutritional deficiencies and illnesses. Then they offer you a variety of nutritional supplements, such as bee pollen or coenzymes, as remedies against deficiencies and disease.
3. They rely on personal stories of success (testimonials) rather than on scientific data for proof of effectiveness.
4. They use food essentially as medicine.
5. They often lack any valid medical or health-care credentials.
6. They come across more as salespeople than as medical professionals.
7. They offer simple answers to complex problems.
8. They claim that the medical community or government agencies refuse to acknowledge the effectiveness of their products or treatments.
9. They make dramatic statements that are refuted by reputable scientific organizations.
10. Their theories and promises are not written in medical journals using a peer review process, but appear in books written only for the lay public.

Keep in mind that there are few, if any, sudden scientific breakthroughs. Science is evolutionary, even downright slow, not revolutionary.

But where can you find accurate nutritional information? In the United States, over 50,000 registered dietitians (RDs) comprise the largest and most visible group of professionals in the nutrition field. Registered dietitians are recognized by the medical profession as the legitimate providers of nutrition care. They have specialized education in human anatomy and physiology, chemistry, medical nutrition therapy, foods and food science, the behavioral sciences, and foodservice management. Registered dietitians must complete at least a bachelor's degree from an accredited college or university, a program of college-level dietetics courses, a supervised practice experience, and a qualifying examination. Continuing education is required to maintain RD status. Registered dietitians work in private practice, hospitals, nursing homes, wellness centers, business and industry, and many other settings. Most are members of the American Dietetic Association.

In addition to using the expertise of an RD, you can ask some simple questions that will help you judge the validity of nutrition information seen in the media or heard from friends.

1. What are the credentials of the source? Does the person have academic degrees in a scientific or nutrition-related field?
2. Does the source rely on emotions rather than scientific evidence, or use sensationalism to get a message across?
3. Are the promises of results for a certain dietary program reasonable or exaggerated? Is the program based on hard scientific information?
4. Is the nutrition information presented in a reliable magazine or newspaper, or is the information published in an advertisement or a well-known publication of questionable reputation?
5. Is the information someone's opinion or the result of years of valid scientific studies, with possible practical nutrition implications?

Much nutrition information that we see or read is based on scientific research. It is helpful to first understand how research studies are designed, as well as pitfalls in each design, so that you can evaluate the study's results. The following three types of studies are commonly used in research.

1. *Laboratory studies* use animals, such as mice or guinea pigs, or tissue samples in test tubes, to find out more about a process that occurs in people (that must also occur in animals), to determine if a substance might be beneficial or hazardous in humans, or to test the effect of a treatment. A major advantage of using laboratory animals is that researchers can control many factors that they can't control in human studies. For instance, researchers can make sure that comparison groups are genetically identical and the conditions to which they are exposed are the same as well. However, mice and other animals are not the same as humans, so results from these studies can't automatically be generalized to humans. For example, laboratory studies have indicated that the artificial sweetener saccharin caused cancer in mice, but this has never been proven for humans.
2. Another type of research, called *epidemiological research,* looks at how disease rates vary among different populations and also factors associated with disease. Epidemiological studies rely on observational data from human populations, so they can only suggest a relationship between two factors; they cannot establish that a particular factor causes a disease. These types of observational studies may compare

factors found among people with a disease, such as cancer, to factors among a comparable group without that disease, or studies may try to identify factors associated with diseases that develop over time within a population group. Researchers may find, for example, fewer cases of osteoporosis in women who take estrogen after menopause.

3. A third type of research goes beyond using animals or observational data, and uses humans as subjects. *Clinical trials* refer to studies that assign similar participants randomly to two groups. One group receives the experimental treatment; the other does not. Neither the researchers nor the participants know who is in which group. For example, a clinical trial to test the effects of estrogen after menopause would randomly assign each participant to one of two groups. Both groups would take a pill, but for one group this pill would be a dummy pill, called a placebo. Clinical studies are used to assess the effects of nutrition-education programs and medical nutrition therapy. Unlike epidemiological studies, clinical studies can observe cause-and-effect relationships.

When reading or listening to a news account of a particular study, it is helpful to have a few key questions in the back of your mind to help evaluate the merits of the study, but also whether it is applicable to you. Look to news reports to address the following:

1. How does this work fit with the body of existing research on the subject? Even the most well-written article does not have enough space to discuss all relevant research on an issue. Yet it is extremely important for the article to address whether a study is confirming previous research and therefore adding more weight to scientific beliefs, or whether the study's results and conclusions take a wild departure from current thinking on the subject.

2. Could the study be interpreted to say something else? Scientists often reach different conclusions when commenting on the same or similar data. Look for varying conclusions from experts, because certain issues they address may be important when putting the findings into context.

3. Are there any flaws in how the study was undertaken that should be considered when making conclusions? The more experts are quoted, or provide background, in a news story, the more likely potential flaws will be described.

4. Are the study's results generalizable to other groups? Not all research incorporates all types of people: men, women, older adults, or people of various ethnicities. Also, a study may have been conducted on animals and not humans. If study results are applicable only to a narrow group of people, it should be reported as such.

With permission, this "Hot Topic" used sections of "If It Sounds Too Good to Be True . . . It Probably Needs a Second Look" from *Food Insight*, published by the International Food Information Council Foundation, March/April 1999.

Chapter 3
Carbohydrates

Carbohydrate literally means hydrate (water) of carbon, a name derived from the investigations of early chemists who found that heating sugars for a long period of time in an open test tube produced droplets of water on the sides of the tube and a black substance, carbon. Later chemical analysis of sugars and other carbohydrates indicated that they all contain at least carbon, hydrogen, and oxygen.

Carbohydrates are the major components of most plants, making up from 60 to 90 percent of their dry weight. In contrast, animals and humans contain a comparatively small amount of carbohydrates. Plants are able to make their own carbohydrates from the carbon dioxide in air and water taken from the soil in a process known as **photosynthesis.** Photosynthesis converts energy from sunlight into energy stored in carbohydrates, which the plant uses to grow and be healthy. Animals are incapable of photosynthesis and, therefore, depend on plants as a source of carbohydrates. Green plants, such as wheat or broccoli, supply the carbohydrates in our diets.

Carbohydrates are separated into two categories: simple and complex. **Simple carbohydrates,** also called **sugars,** include both natural and refined sugars. Carbohydrates are much more than just sugars, though, and include the **complex carbohydrates** starch and fiber. Another name for complex carbohydrate is **polysaccharide** (*poly* means many), a good name for starch and most fibers because both are long chains of many sugars.

After completing this chapter, you should be able to:

- Distinguish between simple and complex carbohydrates.
- Identify foods high in added sugars, natural sugars, starch, and fiber.
- Discuss the health benefits of increased consumption of complex carbohydrates and decreased consumption of added sugar.
- Describe how carbohydrates are digested, absorbed, and metabolized by the body.
- Identify foods as being whole grains or refined grains.
- State the dietary recommendations for carbohydrates.
- Discuss the purchasing, storage, cooking, and menuing of grains, legumes, and pasta.

Carbohydrate—A nutrient group containing carbon, hydrogen, and oxygen that includes sugars, starch, and fibers.

Photosynthesis—A process during which plants convert energy from sunlight into energy stored in carbohydrate.

Simple carbohydrates—Sugars including monosaccharides and disaccharides.

Complex carbohydrates—Long chains of many sugars that include starches and fibers.

Functions of Carbohydrates

Carbohydrates are the primary source of the body's energy. Protein and fat can be burned for energy by other cells, but the body uses carbohydrates first, in part because carbohydrate is the most efficient energy source. In fact, the central nervous system, including the brain and nerve cells, relies almost exclusively on glucose, a sugar, for energy.

If there are not enough carbohydrates for energy, the body can burn either fat or protein, but this is not desirable. When fat is burned for energy without any carbohydrates present, the process is incomplete and results in the production of **ketone bodies,** which start to accumulate in the blood. An excessive level of ketone bodies can cause the blood to become too acidic (called **ketosis**), which then interferes with the transport of oxygen in the blood. Ketosis can cause dehydration and may even lead to a fatal coma. Carbohydrates are important to help the body use fat efficiently.

Carbohydrates also spare protein from being burned for energy so protein can be better used to build and repair the body. About 100 grams of carbohydrates are needed daily to spare protein from being burned for fuel, to prevent ketosis, and to provide glucose to the central nervous system. This amount represents what you minimally need, not what is desirable (about two to three times more).

Carbohydrates are part of various materials found in the body, such as connective tissues, some hormones and enzymes, and genetic material.

Fiber promotes the normal functioning of the intestinal tract, and is associated with a reduced risk of diabetes, certain cancers, and heart disease.

Ketone bodies—A group of organic compounds that cause the blood to become too acidic as a result of fat being burned for energy without any carbohydrates present. **Ketosis**—Excessive level of ketone bodies in the blood and urine.

> ### ■ MINI-SUMMARY
>
> Carbohydrates are the primary source of the body's energy. The central nervous system relies almost exclusively on glucose and other simple carbohydrates for energy. Carbohydrates are also important to help the body use fat efficiently. When fat is burned for energy without any carbohydrates present, the process is incomplete and could result in ketosis. Carbohydrates are part of various materials found in the body. Fiber promotes the normal functioning of the intestinal tract, and may reduce the risk of certain diseases.

Sugars

Simple carbohydrates include **monosaccharides,** or single sugars, and **disaccharides,** or double sugars (*mono* means one and *di* means two). The term sugar refers to both monosaccharides and disaccharides collectively. The chemical names of the six sugars to be discussed all end in -ose, which means sugar.

Monosaccharides include the simple sugars glucose, fructose, and galactose, which are the building blocks of other carbohydrates, such as disaccharides and starch.

In photosynthesis, plants make glucose, which provides energy for growth and other plant activities. **Glucose,** also called dextrose, is the most

Monosaccharide—Single sugars such as glucose or fructose.

Disaccharide—Double sugars such as sucrose.

Glucose—The most significant monosaccharide, the body's primary source of energy.

Fructose—A monosaccharide found in fruits and honey.

Galactose—A monosaccharide found linked to glucose to form lactose or milk sugar.

Sucrose—A disaccharide commonly called cane sugar, table sugar, granulated sugar, or simply sugar.

Maltose—A disaccharide made of two glucose units bonded together.

Lactose—A disaccharide found in milk and milk products that is made of glucose and galactose.

significant monosaccharide because, as in plants, it is the human body's number-one source of energy. Most of the carbohydrates you eat are converted to glucose in the body. The concentration of glucose in the blood, referred to as the **blood glucose level** or **blood sugar level,** is vital to the proper functioning of the human body. Glucose is found in fruits such as grapes, in honey, and, in trace amounts, in many plant foods.

Fructose, the sweetest natural sugar, is also found in honey as well as in fruits. Fructose is about 1.5 times as sweet as sucrose. Fructose and glucose are the most common monosaccharides in nature.

The last single sugar, **galactose,** does not occur alone in nature but is linked to glucose to make milk sugar, also called lactose, a disaccharide.

Most naturally occurring carbohydrates contain two or more monosaccharide units linked together. Disaccharides, the double sugars, include sucrose, maltose, and lactose. They each contain glucose (Figure 3-1). **Sucrose** is the chemical name for what is commonly called cane sugar, table sugar, granulated sugar, or simply, sugar. It is refined from sugarcane or

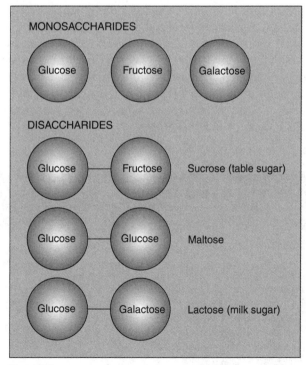

Figure 3-1
Monosaccharides and disaccharides

sugar beet juice and used mainly to sweeten foods. As Figure 3-1 indicates, sucrose is simply two common single sugars—glucose and fructose—linked together. Although the primary source of sucrose in the American diet is refined sugar, sucrose does occur naturally in small amounts in many fruits and vegetables. Table sugar is more than 99 percent pure sugar and provides virtually no nutrients for its 16 calories per teaspoon.

Maltose, which consists of two bonded glucose units, does not occur in nature to any appreciable extent. It is fairly abundant in germinating (sprouting) seeds and is produced in the manufacture of beer.

The last disaccharide, **lactose,** is commonly called milk sugar. It is found naturally only in milk, where it occurs to the extent of about 5 percent, and in certain other dairy products. Unlike most carbohydrates, which are in plant products, lactose is one of the few carbohydrates associated exclusively with animal products. Milk is not thought of as sweet because lactose is one of the lowest-ranking sugars in terms of sweetness (Table 3-1).

Sugars in Food

Sugar occurs naturally in some foods, such as fruits and milk. Fruits are an excellent source of natural sugar, but be aware that some canned fruits contain much added sugar. Canned fruits are packed in one of three styles: in fruit juice, light syrup, or heavy syrup. Both light syrup and heavy syrup have added sugar. Heavy syrup contains the most added sugar (about 4 teaspoons of sugar to 1/2 cup of fruit). Dried fruits, such as raisins, are more concentrated sources of natural sugar than fresh fruits because dried fruits contain much less water.

TABLE 3-1 Relative Sweetness of Sugars and Alternative Sweeteners	
Name	Sweetness Compared to Sucrose
Sugars	
Fructose	1.5
Sucrose	1.0
Glucose	0.7
Lactose	0.2
Alternative Sweetners	
Saccharin (Sweet N'Low)	300
Aspartame (Nutrasweet, Equal)	180
Acesulfame-K (Sunette)	200
Sucralose (Splenda)	600

Although a natural sugar, honey (made by bees) is primarily fructose and glucose, the same two components of table sugar. Therefore, by the time they are absorbed, honey and table sugar are the same thing. Although they are different in flavor and texture, the body can't tell the difference between natural and refined sugars. Honey and sugar contribute only energy and no other nutrients in significant amounts. Because honey is more concentrated, it has 50 percent more calories as an equal volume of sugar.

Lactose, or milk sugar, is present in large amounts in milk, ice cream, ice milk, sherbet, cottage cheese, cheese spreads and other soft cheeses, eggnog, and cream. Hard cheeses contain only traces of lactose.

Added sugars, such as table sugar or corn syrup, are added to foods as sweeteners (Table 3-2). Besides sweetening, they prevent spoilage in jams and jellies and perform several functions in baking, such as browning the crust and retaining moisture in baked goods so they stay fresh. Sugar also

TABLE 3-2 Common Forms of Refined Sugars	
Form of Sugar	Description
Granulated sugar (sucrose)	Most important and most used sugar product on the market. Made from beet sugar or cane sugar, which are identical in chemical composition.
Powdered or confectioners' sugar	Granulated sugar that has been pulverized. Available in several degrees of fineness, designated by the number of X's following the name. 6X is the standard confectioners' sugar and is used in icing and toppings.
Brown sugar	Sugar crystals contained in a molasses syrup with natural flavor and color—91 to 96 percent sucrose. Sold in 4 grades—the higher the grade, the darker the brown sugar, the more flavor.
Turbinado sugar	Raw sugar that has been partially refined and washed.
Syrups	
Corn syrup	Made from cornstarch. Mostly glucose with some maltose. Only 75 percent as sweet as sucrose. Less expensive than sucrose. Used in baked goods and canned goods.
High-fructose corn syrup	Corn syrup treated with an enzyme that converts glucose to fructose, which results in a sweeter product. Used in soft drinks, baked goods, jelly, syrups, fruits, and desserts.
Maple syrup	A concentrated sucrose solution made from mature sugar-maple-tree sap that flows in the spring. Mostly replaced by pancake syrup—a mixture of sucrose and artificial maple flavorings.
Molasses	Thick syrup left over after making sugar from sugarcane. Brown in color with a high sugar concentration.
Honey	Sweet syrupy fluid made by bees from the nectar collected from flowers and stored in nests or hives as food. Made of fructose and glucose.

acts as a food for yeast in breads and other baked goods that use yeast for leavening.

High-fructose corn syrup is corn syrup that has been treated with an enzyme to convert part of the glucose it contains to fructose. The reason for changing the glucose to fructose lies in the fact that fructose is twice as sweet as glucose. High-fructose corn syrup is therefore sweeter, ounce for ounce, than corn syrup, so smaller amounts can be used (making it cheaper). It is used to sweeten almost all nondiet soft drinks and is frequently used in canned juices, fruit drinks, sweetened teas, cookies, jams and jellies, syrups, and sweet pickles.

The sugar content of various foods is listed in Table 3-3. Keep in mind, when you look at "Sugars" on the Nutrition Facts panel, that this number includes naturally occurring sugars and added sugars.

If you chew sugarless gums, you may know that they often contain xylitol, sorbitol, or another sugar alcohol. **Sugar alcohols** are sugarlike compounds that occur naturally in fruits and vegetables. They contain fewer calories per gram than sucrose (from 1.5 to 3 calories per gram) and are used mainly to sweeten sugar-free candies, cookies, and chewing gums. Sugar alcohols are metabolized and absorbed slowly. They are not as sweet as sucrose, so a greater quantity is used in products such as chewing gum, breath mints, and hard candies. The main advantage of these sugarless products is that they do not promote tooth decay, as does sugar. A disadvantage is that large amounts of sugar alcohols can cause diarrhea, and foods using sugar alcohols must be labeled "Excess consumption may have a laxative effect." In addition, mannitol and xylitol have safety issues, which are being investigated by the Food and Drug Administration.

Some of the sugar alcohols on the market are sorbitol, xylitol, mannnitol, and isomalt. Sorbitol is 60 percent as sweet as sucrose and is used in such products as sugarless hard and soft candies, chewing gums, jams, and jellies. Xylitol is almost as sweet as table sugar and is popular in chewing gum and candies. Mannitol is 70 percent as sweet as sucrose and is used in chewing gum. Isomalt is about half as sweet as sucrose and is used in candies and chewing gums, as well as in baked goods and frostings because it does not break down when heated. Lactitol and maltitol are two additional sugar alcohols.

Added sugars—Sugars added to a food for sweetening or other purposes.

High-fructose corn syrup—Corn syrup that has been treated with an enzyme that converts part of the glucose it contains to fructose to make it sweeter.

Added Sugars and Health

In 1998, Americans ate 156 pounds of added sugars per year, up from 127 pounds in 1986. This is an increase of 20 percent. As Americans have been trying to eat less fat, their consumption of added sugars has skyrocketed,

TABLE 3-3 Sugar Content of Foods

Food/Portion	Teaspoons of Sugar
Dairy	
Skim milk, 1 cup	3
Swiss cheese, 1 ounce	Less than 1
Vanilla ice cream, 1/2 cup	4
Meat, Poultry, and Fish	
Meat, poultry, or fish, 3 ounces	0
Eggs	
Egg, 1	0
Grains	
White bread, 1 slice	Less than 1
English muffin, 1	Less than 1
White rice, cooked, 1/2 cup	Less than 1
Cheerios cereal, 1 cup	Less than 1
Honey Nut Cheerios, 1 cup	3
Quaker Oatmeal Squares, 1 cup	2
Fruits	
Apple, 1 medium	4.5
Banana, 1 medium	7
Orange, 1 medium	3
Raisins, 14 grams	2.5
Vegetables	
Broccoli, 1/2 cup raw chopped	Less than 1 gram
Mixed vegetables, 1/3 cup	Less than 1 gram
Beverages	
Cola soft drink, 12 fluid ounces	10
Cakes, Cookies, Candies, and Pudding	
Brownie, 1 average	6
Chocolate graham crackers, 8	2
Chocolate chip cookies, 3	3
Lemon drops, 4 pieces	2.5
M & M candies, 70 pieces	7
Vanilla pudding, 1/2 cup	6
Sweeteners	
White sugar, 1 tablespoon	3
Honey, 1 tablespoon	4
High-fructose corn syrup, 1 tablespoon	4

fueled by soda consumption (accounting for 33 percent of all added sugars) and more and/or larger servings of high-sugar foods such as cookies, ice cream, and other sweets. Is all this sugar good for you? Let's take a look at how it might affect a number of health problems.

Obesity. Although there is no research stating that added sugars alone cause obesity, added sugars are undoubtedly a major factor in rising obesity rates among adults and children. High-sugar foods, such as cookies and candy, are almost always teamed up with fat and high in calories. These foods are also typically low in nutrients and are therefore referred to as **empty calories.** For example, the cupcake pictured in Figure 3-2 supplies 170 calories with virtually no nutrients. If you look at foods with natural sugars, such as fruits and milk, you will notice that they are packaged with many essential nutrients. The foods highest in added sugars are, unfortunately, not nearly so rich in other nutrients.

Empty-calorie foods, such as many snack foods, are very appealing because they taste good and you can eat them quickly, often when you are on the go. Compare, for instance, how much time you need to eat an apple to how long it takes to eat a chocolate-chip cookie. Of course it takes longer to eat the apple, but the apple is a much better snack choice that is just as portable as packaged snack foods.

Empty calories—Foods that provide few nutrients for the number of calories they contain.

Diabetes—A disorder of carbohydrate metabolism characterized by high blood sugar levels and inadequate or ineffective insulin.

Insulin—A hormone that increases the movement of glucose from the bloodstream into the body's cells.

Homemade Cupcake with Icing

Information per serving

Serving size	=	1 Cupcake
Calories		170
Protein		2 grams
Carbohydrates		30 grams
Fat		5 grams
Sodium		110 milligrams

Percentage of U.S. Recommended Daily Allowances

Protein	2
Niacin	*
Thiamin	*
Vitamin A	*
Vitamin C	*
Riboflavin	2
Calcium	*
Iron	*

*Less than 2% of USRDA

Figure 3-2

Example of empty calories

Diabetes. Obesity is more closely linked to diabetes than to any other health problem. **Diabetes** is a disorder in which the body does not metabolize carbohydrates properly. It results from having inadequate or ineffective insulin. **Insulin** is a hormone that increases the movement of glucose from the bloodstream into the body's cells, where it is used to produce energy. People with untreated diabetes have high blood-sugar levels. Treatment for diabetes is individualized to the patient and includes a balanced diet that supports a healthy weight and physical activity, as well as insulin if needed. When overweight people with diabetes lose weight, the disease is usually more controllable.

Heart Disease. A high sugar intake increases heart-disease risk by raising blood triglyceride levels. (Most fat is in a form called triglycerides.) This is especially true for individuals who are "insulin resistant." When an insulin-resistant person eats sugars and refined starches, their bodies make more than the normal amount of insulin, which promotes excess body fat. Up to 25 percent of Americans are insulin resistant, and most of these people are obese. A moderate intake of sugars does not increase heart disease risk.

Hypoglycemia—A symptom in which blood sugar levels are low.

Postpranial hypoglycemia—Low blood sugar that occurs generally two to four hours after meals and includes symptoms such as shakiness, sweating, and dizziness.

Fasting hypoglycemia— Low blood sugar that occurs after not eating for eight or more hours.

Attention Deficit Hyperactivity Disorder— A developmental disorder of children characterized by impulsiveness, distractibility, and hyperactivity.

Hypoglycemia. **Hypoglycemia** is the term used to describe an abnormally low blood glucose level. Two disease conditions may cause hypoglycemia. The most common condition is **postprandial hypoglycemia.** It occurs generally one to four hours after meals and has symptoms such as quickened heartbeat, shakiness, weakness, anxiety, sweating, and dizziness, mimicking anxiety or stress symptoms. It may be caused when a rapid rise in blood glucose after a meal causes a temporary overproduction of insulin, which pulls too much sugar out of the bloodstream.

A second type of hypoglycemia, **fasting hypoglycemia,** is rare. It has numerous causes, such as drugs, and can be serious. Its symptoms occur after not eating for eight or more hours, so it usually occurs during the night or before breakfast.

Hypoglycemia may also be due to cancer, pancreatic diseases, or other reasons. Hypoglycemia is also seen sometimes in people with diabetes. A diet for people with hypoglycemia includes regular, well-balanced meals with moderate amounts of refined sugars and sweets. Protein, fat, and fiber can moderate swings in blood glucose levels.

Hyperactivity in Children. During the 1980s, reports came out linking sugar intake with hyperactivity in children. Unfortunately, the study results were unclear as to whether sugar caused the hyperactivity or the hyperactivity caused the children to eat sugar. Extensive research since then has failed to show that high sugar intake causes hyperactivity or attention deficit hyperactive disorder (ADHD).

Lactose Intolerance. Lactose (milk sugar) is a problem for certain people who lack or, more commonly, don't have enough of the enzyme **lactase.** Lactase is needed to split lactose into its components in the small intestine. If lactose is not split, it travels to the colon, where bacteria ferment it and produce short-chain fatty acids and gas. These by-products do not normally cause any problems or discomfort in small amounts. However, if a lot of lactose travels to the colon, symptoms such as gas, abdominal distention, and diarrhea often occur within about thirty minutes to two hours after ingesting milk products. The symptoms are normally cleared up within two to five hours. This problem, called **lactose intolerance,** seems to be an inherited problem especially prevalent among Asian Americans, Native Americans, African Americans, and Latinos, as well as some other population groups.

> **Lactase**—An enzyme needed to split lactose into its components in the intestines.
>
> **Lactose intolerance**—An intolerance to milk and most milk products due to a deficiency of the enzyme lactase. Symptoms often include flatulence and diarrhea.

Treatment for lactose intolerance requires a diet that is limited in lactose, which is present in large amounts in milk, ice cream, ice milk, sherbet, cottage cheese, eggnog, and cream. Most individuals can drink small amounts of milk without any symptoms, especially if it is taken with food. Lactose-reduced milk and some other lactose-reduced dairy products are available in supermarkets, as is the enzyme lactase (which is also sold in pharmacies). Lactase can be added to milk to reduce the lactose content. Eight fluid ounces of lactose-reduced milk contains only 3 grams of lactose, compared with 12 grams in regular milk. Reducing the lactose content of milk by 50 percent is often adequate to prevent symptoms of lactose intolerance. Although lactose-reduced milk and other lactose-digestive aids are available, they may not be necessary when lactose intake is limited to one cup of milk (or equivalent) or less a day.

Yogurt is usually well tolerated because it is cultured with live bacteria that digest lactose. This is not always the case with frozen yogurt, because most brands do not contain nearly the number of bacteria found in fresh yogurt (there are no federal standards for frozen yogurt at this time). Also, some yogurts have milk solids added to them that can cause problems. Many hard cheeses contain very little lactose, and usually do not cause symptoms because most of the lactose is removed during processing or digested by the bacteria used in making cheese.

People who have difficulty digesting lactose report tremendous variation in which lactose-containing foods they can eat and even the time of day they can eat them. For example, one individual may not tolerate milk at all, whereas another can tolerate milk as part of a big meal. The ability to tolerate lactose is not an all-or-nothing phenomenon. As people with lactase deficiency usually decrease their intake of dairy products and thus their calcium intake, they should try different dairy products to see what they can tolerate.

Dental Caries. The only negative health effect of sugar that most health experts agree on is that sugar (and starches too) do contribute to the development of **dental caries,** or cavities. The more often sugars and starches—even small amounts—are eaten and the longer they are in the mouth before teeth are brushed, the greater the risk for tooth decay. Dental caries are a major cause of tooth loss. This is so because every time you eat something sweet, the bacteria living on your teeth ferment the carbohydrate, which produces acid. This acid eats away at the teeth, and cavities eventually develop. The fermentation may continue for hours. The deposit of bacteria, protein, and polysaccharides that forms on the teeth in the absence of tooth-brushing during a period of 12 to 24 hours is called **plaque.** Without good tooth-brushing habits, plaque may cover all surfaces of the teeth.

Dental caries—Tooth decay.
Plaque—Deposits of bacteria, protein, and polysaccharides found on teeth that contribute to tooth decay.

Other factors influence how much impact foods will have on the development of dental caries. The sequence of eating foods in a meal, the foods' form (liquid or solid and sticky), and combinations of foods also influence dental caries. At meals, if an unsweetened food, such as cheese, is eaten after a sugared food, the plaque will be less acidic, so less acid eats away at the teeth. Cheese also stimulates more saliva, which helps to wash away acids. This is why eating sugary or starchy foods as frequent between-meal snacks is more harmful to teeth than having them at meals. Sticky carbohydrate foods, such as raisins or caramels, cause more problems than liquid carbohydrate foods because they stick to the teeth and provide a constant source of fermentable carbohydrates for the bacteria until washed away. Liquids containing sugars have been considered less harmful to teeth than solid sweets because they clear the mouth quickly.

Food such as dried fruits, breads, cereals, cookies, crackers, and potato chips increase chances of dental caries when eaten frequently. Foods that do not seem to cause cavities include cheese, peanuts, sugarfree gum, some vegetables, meats, and fish. To prevent dental caries, brush your teeth often, floss your teeth once a day, try to limit sweets to mealtime, and see your dentist regularly.

■ **MINI-SUMMARY**

Carbohydrates are separated into two categories: simple and complex. Simple carbohydrates are sugars and include both natural and refined sugars. Complex carbohydrates, such as starch and fiber, are long chains of many sugars. Monosaccharides are the building blocks of other carbohydrates and include glucose (the main source of the body's energy), fructose (found in fruits), and galactose (found in milk sugar). Disaccharides include sucrose (table sugar), maltose, and lactose (milk sugar). Refined sugars, such as high-fructose corn syrup, are used to sweeten soft drinks, breakfast cereals, candy, baked goods such as cakes and pies, syrups, and jams and jellies. Added sugar consumption has risen dramatically. Added sugars are a major factor in rising obesity rates. Obesity is closely linked to the development of diabetes. A high sugar intake increases heart disease risk, especially in individuals who are insulin-resistant, by raising blood triglyceride levels. Sugar contributes to dental decay. Lactose (milk sugar) is a problem for people with lactose intolerance. They experience abdominal cramps, bloating, and diarrhea about thirty minutes to two hours after ingesting milk products. Lactose-reduced milk, yogurt, and hard cheeses are usually well tolerated.

Starch

Plants, such as peas, store glucose in the form of **starch.** Starch is made of many chains of hundreds to thousands of glucoses linked together. The chains may be straight or have tree-like branches (Figure 3-3). When you eat starchy foods, digestive enzymes separate the glucose molecules.

Just as plants store glucose in the form of starch, your body stores glucose in a form called **glycogen.** Like starch, glycogen is a polysaccharide. It is a chain of glucose units, but the chains are longer and have more branches than starch (Figure 3-3). Glycogen is stored in two places in the body: the liver and the muscles. An active 150-pound man has about 400 calories stored in his liver glycogen and about 1,400 calories stored in his muscle glycogen. When the blood sugar level starts to dip and more energy is needed, the liver converts glycogen into glucose, which is then delivered by the bloodstream. Muscle glycogen does not supply glucose to the bloodstream but is used strictly to supply energy for exercise.

Starch—A complex carbohydrate made up of a long chain of glucoses linked together; found in grains, legumes, vegetables, and some fruits.

Glycogen—The storage form of glucose in the body, found in the liver and muscles.

Gelatinization—A process in which starches, when heated in liquid, absorb water and swell in size.

Starch in Food

Starch is found only in plant foods. Cereal grains, the fruits or seeds of cultivated grasses, are rich sources of starch and include wheat, corn, rice, rye, barley, and oats. Cereal grains are used to make breads, baked goods,

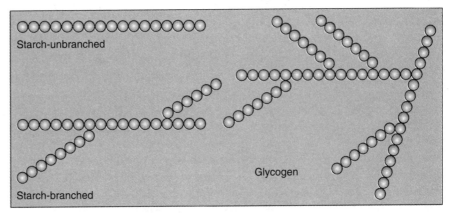

Figure 3-3
The structures of starch and glycogen

breakfast cereals, and pastas. Starches are also found in root vegetables such as potatoes, and dried beans and peas such as navy beans.

Starchy foods in general are not flavorful if eaten raw, so most are cooked to make them taste better and be more digestible. Starch, such as cornstarch, is used extensively as a thickener in cooking, because starch undergoes a process called **gelatinization** when heated in liquid. When starches gelatinize, granules absorb water and swell, making the liquid thicken. When this occurs, the liquid becomes thicker because there is less water. Gelatinization is a process unique to starches, so you find them frequently used as thickeners in soups, sauces, gravies, puddings, and other foods. Other thickeners include arrowroot and purées of starchy vegetables or legumes.

Starch and Health

Starch creates the same problem as sugar in the mouth and therefore contributes to tooth decay and dental caries. Starch from whole-grain sources is preferable to starch found in refined grains such as white flour. This is discussed in detail in the following section.

> ■ **MINI-SUMMARY**
>
> Starch is a storage form of glucose in plants. The body stores glucose as glycogen in the liver and muscles. Glycogen is important to maintain normal blood sugar levels. Starches are found in cereal grains, breads, baked goods, cereals, pastas, root vegetables, and dried beans and peas. They are commonly used as thickeners. Like sugar, starch contributes to dental caries.

Fiber

Dietary fiber is mostly material from plants that resists digestion by our digestive enzymes. Like starch, most fibers are chains of bonded glucose units, but what's different is that the units are linked with a chemical bond that our digestive enzymes can't break down. In other words, most fiber passes through the stomach and intestines unchanged and is excreted in the feces. Fiber was called *roughage* a few generations ago.

Fiber is found only in plant foods; it does not appear in animal foods. There are two major types of fiber, soluble and insoluble. **Soluble fiber** swells in water, like a sponge, into a gel-like substance. **Insoluble fiber** also swells in water, but not nearly to the extent of soluble fiber.

The soluble fibers include gums, mucilages, pectin, and some hemicelluloses. They are generally found around and inside plant cells. The insoluble fibers include cellulose, lignin, and the remaining hemicelluloses. They generally form the structural parts of plants. The amount of fiber in a plant varies among plants and may vary within a species or variety, depending on growing conditions and the plant's maturity at harvest.

Dietary fiber—Material from plant cells that mostly resists digestion by our digestive enzymes.

Soluble fiber—A classification of fiber that includes gums, mucilages, pectin, and some hemicelluloses. They are generally found around and inside plant cells.

Insoluble fiber—A classification of fiber that includes cellulose, lignin, and the remaining hemicelluloses. They generally form the structural parts of plants.

Fiber in Food

Fiber is abundant in plants, so legumes (dried beans, peas, and lentils), fruits, vegetables, whole grains and whole-grain products, nuts, and seeds provide fiber (Table 3-4). Fiber is not found in meat, poultry, fish, dairy products, and eggs.

Foods containing soluble fibers are as follows.

■ Many fruits and vegetables, such as apples, grapes, citrus fruits, and carrots (fruit juices are not good sources of fiber)
■ Some cereal grains, such as oats and barley
■ Beans and peas, such as kidney beans, pinto beans, chickpeas, split peas, and lentils

Insoluble fiber includes the structural parts of plants, such as skins and the outer layer of the wheat kernel. You have seen insoluble fiber in the skin of whole-kernel corn and in celery strings. It is found in the following foods:

■ Wheat bran
■ Whole grains, such as whole wheat and brown rice, and products made with whole grains, such as whole-wheat bread
■ Many fruits and vegetables

TABLE 3-4 Fiber Content of Selected Foods

Food/Portion	Grams Fiber	Food/Portion	Grams Fiber
Dairy Group		*Vegetables*	
Milk, 1 cup	0	Broccoli spears, 1/2 cup	2.4
Plain yogurt, 1 cup	0	Carrots, 1/2 cup	2.0
Cheddar cheese, 1 once	0	Green beans, canned, 1/2	2.0
Ice cream, soft serve, 1 cone	0	Lettuce, iceberg, 1 cup	0.5
Meat, Poultry, Fish		Potato, 1/2 cup	1.5
Meat, poultry, and fish	0	*Legumes*	
Eggs		Chick peas, 1/2 cup	4.3
Eggs	0	Kidney beans, 1/2 cup	6.9
Grains		Lentils, 1/2 cup	5.2
White bread, 1 slice	0.6	Split peas, 1/2 cup	3.1
Whole-wheat bread, 1 slice	1.5	*Nuts and seeds*	
Hamburger bun, 1/2	0.7	Almonds, 6 whole	0.6
Saltine crackers, 6	0.5	Peanuts, 10 large	0.6
Fruits		Peanut butter, smooth,	
Apple, red, 1 small	2.8	1 tablespoon	1.0
Applesauce, 1/2 cup	2.0		
Banana 1/2 small	1.1		
Orange, 1 small	2.9		
Orange juice, 1/2 cup	0.1		
Raisins, 2 tablespoons	0.4		

Source: Anderson, James W. 1990. *Plant Fiber in Foods.* Lexington: HCF Nutrition Research Foundation, Inc. Reprinted with permission.

- Beans and peas
- Seeds

Most foods contain both soluble and insoluble fibers.

Whole Grains

The full name for grains is cereal grains. Cereal grains are the seeds of cultivated grasses such as wheat, corn, rice, rye, barley, and oats, among others. All cereal grains have a large center area high in starch known as the **endosperm.** The endosperm also contains some protein. At one end of the endosperm is the **germ,** the area of the kernel that sprouts when allowed to germinate. The germ is rich in vitamins and minerals, and contains some oil. The **bran,** containing much fiber and other nutrients, covers both the endosperm and the germ. The seed contains everything needed to reproduce the plant: the germ is the embryo, the endosperm contains the nutrients for growth, and the bran protects the entire seed (Figure 3-4).

Most grains undergo some type of processing or milling after harvesting to allow them to cook more quickly and easily, to make them less chewy, and to lengthen their shelf life. Grains such as oats and rice have an outer husk or hull that is tough and inedible, so it is removed. Other processing steps might include polishing the grain to remove the bran and germ (as in making white flour), cracking the grain (as in cracked wheat), or steaming the grain (as in bulgur) to shorten the cooking time. The processes of rolling or grinding a grain, such as oatmeal, also shorten the cooking time.

Whenever the fiber-rich bran and the vitamin-rich germ are left on the endosperm of a grain, the grain is called a **whole grain.** Examples of whole grains include whole wheat, whole rye, oatmeal, whole cornmeal, whole hulled barley, and brown rice. Read the Food Facts in this chapter for more information on a variety of grains.

If the bran and germ are separated (or mostly separated) from the endosperm, the grain is called **refined** or **milled.** Whereas whole-wheat flour is made from the whole grain, white flour (sometimes called wheat flour) is made only from the endosperm of the wheat kernel. Whole-wheat flour does not stay fresh as long as white flours. This is due to the presence of the germ, which contains oil. When the oil turns rancid, or deteriorates, the flour will turn out a poor-quality product.

In baking, wholegrain flours produce breads that are denser and chewier. For example, breads made with only whole-wheat flour are more compact and heavier than breads made with only white flour. This is because the strands of gluten in the whole-wheat bread are cut by the sharp edges of the bran flakes. Some bakers prefer to use some white flour to strengthen the bread.

Endosperm—In cereal grains, a large center area high in starch.

Germ—In cereal grains, the area of the kernel rich in vitamins and minerals that sprouts when allowed to germinate.

Bran—In cereal grains, the part that covers the grain and contains much fiber and other nutrients.

Whole grain—A grain that contains the endosperm, germ, and bran.

Refined or milled—A grain in which the bran and germ are separated (or mostly separated) from the endosperm.

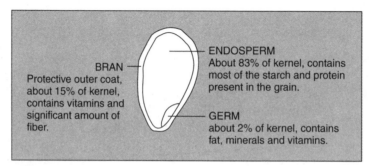

Figure 3-4

A grain of wheat

Many consumers prefer white bread because it lacks the dark color and crunchy texture of whole-wheat bread, but there are also quite a few health-conscious consumers who appreciate a good-quality wholegrain bread.

When you compare the nutrients in whole grains and refined grain, *whole grains are always a far more nutritious choice.* They surpass refined grains in their fiber, vitamin, and mineral content. When wheat is refined, 22 nutrients and most of the fiber are removed. With whole wheat you get more vitamin E, vitamin B$_6$, pantothenic acid, magnesium, zinc, potassium, copper, and, of course, fiber. By federal law, refined grains are enriched with five nutrients that are lost in processing: thiamin, riboflavin, niacin, folate (folic acid), and iron.

It can be quite difficult to determine from the name of a product whether it is indeed a whole grain. A multigrain bagel, for instance, is probably made with refined grains. On the other hand, whole-wheat bread (but *not* wheat, stoned wheat, or seven-grain bread) is a wholegrain product. The only way to be sure the product is wholegrain is to check the ingredient list. The first ingredient should be a whole grain, such as whole wheat or oatmeal.

Fiber and Health

What can fiber do for you? Numerous epidemiologic (population-based) studies have found that diets low in saturated fat and cholesterol and high in fiber are associated with a reduced risk of certain cancers, diabetes, digestive disorders, and heart disease. However, since high-fiber foods may also contain antioxidant vitamins, **phytochemicals** (substances in plants that may reduce risk of cancer and heart disease when eaten often) and other substances that may offer protection against these diseases, researchers can't say for certain that fiber alone is responsible for the reduced health risks. Findings on the health effects of fiber show that it may play a role in the following:

Phytochemicals—Minute substances in plants that may reduce risk of cancer and heart disease when eaten often.

Diverticulosis—A disease of the large intestine in which the intestinal walls become weakened, bulge out into pockets, and at times become inflamed.

Hemorrhoids—Enlarged veins in the lower rectum.

Carcinogen—Cancer-causing substance.

- **Digestive disorders.** Because insoluble fiber aids digestion and adds bulk to stool, it hastens passage of fecal material through the gut, thus helping to prevent or alleviate constipation. Fiber also may help reduce the risk of **diverticulosis,** a condition in which small pouches form in the colon wall (usually from the pressure of straining during bowel movements). People who already have diverticulosis often find that increased fiber consumption can alleviate symptoms, which include constipation and/or diarrhea, abdominal pain, and flatulence. A diet high in insoluble fiber is also used to prevent or treat **hemorrhoids,** enlarged veins in the lower rectum.

■ **Cancer.** Epidemiologic studies have generally noted an association between a combination of low total fat and high fiber intakes, and reduced incidence of colon cancer. The exact mechanism for reducing the risk is not known, but scientists theorize that insoluble fiber adds bulk to stool, which in turn dilutes **carcinogens** (cancer-causing substances) and speeds their transit through the lower intestines and out of the body.

■ **Heart disease.** Clinical studies show that a diet low in saturated fat and cholesterol, and high in fruits, vegetables, and grain products that contain soluble fiber, can lower blood cholesterol levels. As it passes through the gastrointestinal tract, soluble fiber binds to dietary cholesterol, helping the body to eliminate it. This reduces blood cholesterol levels, a major risk factor for heart disease. In particular, the fiber of whole oats or beans can lower total and LDL (low-density lipoprotein), or "bad," blood cholesterol in diets that include these foods at appropriate levels.

■ **Diabetes.** As with cholesterol, soluble fiber traps carbohydrates to slow their digestion and absorption. In theory, this may help prevent wide swings in blood sugar level throughout the day. Additionally, it appears that a high-sugar, low-fiber diet more than doubles a woman's risk of diabetes.

■ **Obesity.** Because insoluble fiber is indigestible and passes through the body virtually intact, it provides few calories. And since the digestive tract can handle only so much bulk at a time, fiber-rich foods are more filling than other foods—so people tend to eat less.

■ **MINI-SUMMARY**

Dietary fiber is mostly material from plants that resists digestion. Most fibers are chains of bonded glucose units. Soluble fiber is found in foods such as fruits, vegetables, oats, barley, and beans. Insoluble fiber is found in foods such as wheat bran and whole grains. Diets low in saturated fat and cholesterol and high in fiber are associated with a reduced risk of certain cancers, diabetes, digestive disorders, and heart disease. Fiber-rich foods are filling, so people tend to eat less.

Digestion, Absorption, and Metabolism of Carbohydrates

Cooking carbohydrate foods makes them easier to digest. As mentioned previously, starches gelatinize, making them easier to chew, swallow, and digest. Cooking usually breaks down fiber in fruits and vegetables, also making them easier to chew, swallow, and digest.

TABLE 3-5 Carbohydrate Digestion			
Site	Name of Enzyme	Carbohydrate Acted Upon	Products Formed
Mouth	Salivary amylase	Starch	Small polysaccharides, maltose
Small intestine	Pancreatic amylase	Starch	Small polysaccharides, maltose
	Sucrase	Sucrose	Glucose, fructose
	Lactase	Lactose	Glucose, galactose
	Maltase	Maltose	Glucose

Before carbohydrates can be absorbed through the villi of the small intestines, they must be broken down into monosaccharides, or one-sugar units. Starch digestion begins in the mouth, where an enzyme, salivary amylase, starts to break down some starch into small polysaccharides and maltose. In the stomach, salivary amylase is inactivated by the stomach acid. Next, the intestine completes the breakdown of starch into maltose, which is then split by an enzyme (maltase) into two glucose units. Glucose can now be absorbed.

Through the work of three enzymes made in the intestinal wall (see Table 3-5), all sugars are broken down into the single sugars: glucose, fructose, and galactose. They are then absorbed and enter the bloodstream, which carries them to the liver. In the liver, fructose and galactose are converted to glucose or further metabolized to make glycogen or fat. The hormone insulin makes it possible for glucose to enter body cells, where it is used for energy or stored as glycogen.

Fiber cannot be digested, or broken down into its components by enzymes, so it continues down to the large intestine to be excreted. Although human enzymes can't digest most fibers, some bacteria in the large intestine can digest soluble fibers. As they digest the soluble fibers, the bacteria produce gas and small fat particles that are absorbed. The fats contribute some calories.

■ **MINI-SUMMARY**

During digestion, various enzymes break down starch and sugars into monosaccharides, which are then absorbed. In the liver, fructose and galactose are converted into glucose or further metabolized. Fiber can't be digested by human enzymes, but some soluble fibers are digested by intestinal bacteria, producing gas and fat fragments that are absorbed.

Dietary Recommendations for Carbohydrates

The Food Guide Pyramid recommends that adults eat at least three servings of vegetables and two servings of fruits daily, and at least six servings of grain products, such as breads, cereals, pasta, and rice, with an emphasis on whole grains. Unfortunately, three-quarters of Americans don't eat a wholegrain food each day. The Daily Value for carbohydrate is 300 grams, or 60 percent of total calories.

Added sugars now account for 16 percent of calories. For teenagers, added sugars account for 20 percent of calories. The Dietary Guidelines for Americans suggests using sugars in moderation. The U.S. Department of Agriculture suggests that you try to limit your added sugars to six teaspoons a day if you eat about 1,600 calories, 12 teaspoons at 2,200 calories, or 18 teaspoons at 2,800 calories. That works out to 6 to 10 percent of calories from added sugars. The World Health Organization suggests limiting added sugars to 10 percent of total calories.

With respect to fiber, current research indicates that daily consumption of 20 to 35 grams of dietary fiber from a variety of foods may be helpful in promoting health and managing certain diseases. The Daily Value for fiber is 20 grams for a 1,600-calorie diet, 25 grams for a 2,000-calorie diet, and 30 grams for a 2,500-calorie diet. The World Health Organization recommends at least 27 grams of dietary fiber daily.

■ **MINI-SUMMARY**

Carbohydrates should be 60 percent of total daily calories, with 10 percent or less of total calories from added sugars. Whole grains are to be emphasized. About 20 to 35 grams of fiber should be consumed daily.

Ingredient Focus: Grains, Legumes, and Pasta

Carbohydrates are found, to varying degrees, in all the groups of the Food Guide Pyramid. This section will discuss grains, legumes, and pasta. Fruits and vegetables are discussed in Chapter 6, and milk and milk products in Chapter 4.

Grains

Nutritionally, grains such as rice have much to offer (Table 3-6). They are low or moderate in calories, high in starch and fiber (if whole grain), low in fat, moderate in protein, and full of vitamins and minerals. In addition, they are inexpensive and can be quite profitable. Both traditional grains, such as rice, and newer grains, such as quinoa, are being featured more often on the menu, and not simply as a side dish, but also in main dishes. The Food Facts section in this chapter talks about many grains, except rice, which is discussed next.

Grains should be stored in their original packaging or in an airtight container. Store them in a cool, dry area. All grains can be refrigerated, which is a good idea if the kitchen is particularly hot and humid. Refrigeration is very important for whole grains, which do contain some oil that can go rancid. Store whole grains in the refrigerator for up to six months. Grains should be rinsed well before cooking to remove dust and dirt. Quinoa, in particular, must be thoroughly rinsed, because its natural coating has a bitter taste if not removed.

Rice, perhaps the first grain ever cultivated by man, is a semiaquatic member of the grass family. Its edible seed is the staple grain for over half the world's population. Brown rice is the whole grain and, of course, more nutritious than white rice. Among white rices, the most nutritious is parboiled or converted rice. Converted rice is a specially processed long-grain rice that has been partially cooked under steam pressure, dried, and then milled to remove the outer hull and bran. The parboiling process results in a grain that is more nutritious and a rice that is particularly separate and fluffy.

TABLE 3-6 Nutrition Information for Cooked Long-Grain Rice (per 1/2-cup serving)

	Brown	Regular— Milled White (enriched)	Parboiled (enriched)	Precooked White (enriched)
Kilocalories	111	131	100	80
Carbohydrate (g)	23	28	22	17
Protein (g)	3	3	2	2
Total fat (g)	1	0.3	0.2	0.1
Saturated fatty acids (g)	0	0	0	0
Cholesterol (mg)	0	0	0	0
Fiber (g)	1.7	0.5	0.5	0.6
Sodium (mg)	3.0	2.0	2	2

Brown rice has a nutty flavor and chewy texture compared with white rice, which has been milled to remove the bran that distinguishes brown rice. Either rice is classified by its shape: short-grain, medium-grain, or long-grain. Long-grain white rice is four to five times as long as it is wide. Its cooked grains are separate and fluffy, and are used for side dishes, entrees, salads, pilaf, and so on. Medium-grain rice is a little shorter and plumper than long-grain rice. After cooking, the rice is more moist, tender, and has a greater tendency to cling together than long-grains. Both medium-grain and short-grain white rice are good choices for making creamy dishes, such as rice pudding, risotto, molds, or croquettes. The shorter the grain, the more tender and clinging it becomes as it cooks. The boiled rice used in Japanese cooking is short-grain.

There are also numerous specialty rices.

- **Basmati rice** is an aromatic variety of extra-long-grain rice with a nutty flavor that is a very important ingredient in India. It is available as both a brown and a white rice, and has a firm consistency.
- From Thailand and other parts of southeast Asia comes **jasmine rice,** a long-grain white rice that somewhat resembles basmati rice. Jasmine rice is excellent in cold salads because it stays fluffy after cooling.
- **Texmati rice** is a cross between basmati and long-grain white rice. It is grown in the United States. It has a nutty flavor and is available both white and brown.
- In Italy, the rice of choice is **arborio rice,** a short-grain rice that is a must for making the classic Italian dish, risotto. Arborio rice is very starchy and can absorb a great deal of liquid without becoming soggy, so it is ideal for dishes such as risotto, paella, and jambalaya, that need slow, gentle cooking. There are a number of varieties available.
- **Glutinous rice** is a short-grain rice used in some Chinese and Japanese desserts. It is very starchy and sticks together when cooked. It comes black (unhulled) or white (polished). This rice is used in sushi and many other Asian dishes.
- **Japonica rice** is a Japanese rice that sticks together when cooked. It is a good choice for Asian dishes.
- **Purple rice,** also called black Thai rice, cooks up purple. It is cooked risotto style.

Wild rice is not a true rice, but the seed of a grass that grows wild in the marshes of the Great Lakes region. It is dark brown to almost black in color, and has a nutty flavor. Because true wild rice is so expensive, some kitchens use cultivated wild rice. The cultivated variety is coarser in texture and not quite as flavorful.

Some popular rice dishes include the following:

■ **Jambalaya**—a traditional Louisiana rice dish, highly seasoned, and flavored with sausage, ham, seafood, pork, chicken, or other meat
■ **Paella**—a traditional Spanish dish of saffron-flavored rice, shellfish, chicken, chorizo, vegetables, and seasonings
■ **Pilaf**—a light and fluffy rice dish originating in the Middle East; the rice is often sautéed in oil and then cooked in a broth with onions, raisins, various spices, and sometimes meat
■ **Risotto**—a rich and creamy Italian rice dish in which the rice is browned in fat or oil with onions and then cooked in broth, often flavored with Parmesan cheese
■ **Arroz con pollo**—a Spanish dish mixing rice with chicken.

These dishes are all good examples of how to use other ingredients, or a certain cooking method, to add taste and flavor to grains, which often have a subtle taste alone. Like most grains, rice is prepared by cooking in liquid until tender and the liquid is absorbed. Cooking times for rice, as well as other grains, appear in Table 3–7.

CHEF'S TIPS
■ Grains work very well as main dishes when mixed with each other, or with lentils. They have a good appearance and fit together. For example, couscous and wheat berries are attractive, as is barley with quinoa. To either dish you could add lentils, vegetables, and seasonings.
■ Orzo works well with barley or quinoa.
■ Rice and beans is a very popular and versatile dish using grains and legumes. For appearance, mix purple rice with white beans, or wild rice with cranberry beans.

Legumes

Legumes include all sorts of dried beans, peas, and lentils. Dried beans are among the oldest of foods and are an important staple for millions of people in other parts of the world. Beans were once considered to be worth their weight in gold—the jeweler's "carat" owes its origin to a pealike bean on the east coast of Africa.

From a nutritional point of view, legumes are a hit. They are:

■ High in complex carbohydrates
■ High in fiber
■ Low in fat (only a trace, except in a couple of cases)
■ Cholesterol free
■ A good source of vitamins and minerals
■ Low in sodium

TABLE 3-7 Cooking Information for Grains

Grain	Appearance	Flavor	Soaking Required	Cooking Time	Cups Liquid for Cooking	Cups Yield	Uses	Storage
Amaranth	Golden	Sweet, nutlike	No	25 minutes	2-1/2	3-1/2	Hot cereal, pilaf, in baking, can be popped as a snack	Airtight container, in cool place for many months, otherwise refrigerate for 5 months
Barley, pearl	White-tan	Mild, nutty	No (but will reduce cooking time)	35–40 minutes	3	3-1/2	Soups, casseroles, stews, cooked cereals, side dishes, pilafs	Airtight container—6–9 months at room temperature
Barley, whole hulled	Brownish-gray	Nutty, chewy	Yes	60–90 minutes	3	4	Same as above	Airtight container 1 month at room temperature, 4–5 months in refrigerator
Buckwheat, whole white	Brown-white	Mild	No	20 minutes	2	2-1/2	Side dishes	Airtight container—1–2 months, better stored in refrigerator
Buckwheat, roasted (kasha)	Brown	Distinct, nutty, chewy	No	10–15 minutes	2	2-1/2	Soups, side dishes, salads, pilaf, stuffing, hot cereal	Same as above

Grain	Color	Flavor		Cooking time			Uses	Storage
Corn, whole hominy	Yellow or white	Sweet, creamy texture	No	2-1/2–3 hours	2-1/2	3	Soups, stews, casseroles, hot cereal, puddings, baked goods	Airtight container—1 month at room temperature, 5 months in refrigerator
Corn, hominy grits	Whitish-gray	Distinct	No	20–25 minutes	4	3	Hot breakfast cereal	Airtight container, many months at room temperature
Millet	Bright gold color, small	Like corn, crunchy	No	30–35 minutes	2	3	Soups, casseroles, meat loaves, porridge, croquettes, pilaf, salads, stuffing, side dishes	Airtight container, 6 months at room temperature
Oats, steel-cut	Off-white	Mild, pleasant	No	45–60 minutes	2	2	Hot cereal	Airtight container—1 month at room temperature, 6 months in refrigerator
Quinoa	Pale yellow	Nutty	No	12–15 minutes	2	2-1/2	In place of rice	Airtight container in cool place for 1 month, otherwise refrigerate for 5 months
Rice, regular-milled long grain	White	Mild	No	15–20 minutes	2	3	Side dishes, casseroles, stews, soups, stuffing, salads	Airtight container—many months at room temperature

TABLE 3-7 *(continued)*

Grain	Appearance	Flavor	Soaking Required	Cooking Time	Cups Liquid for Cooking	Cups Yield	Uses	Storage
Rice, regular-milled, medium or short grain	White	Mild	No	20–25 minutes	1-1/2	3	Same as above	Same as above
Rice, parboiled	White	Mild	No	20–25 minutes	2–2-1/2	3–4	Same as above	Same as above
Rice, brown	Tan–brown	Nutty	No	40–50 minutes	2-1/2	4	Same as above	Airtight container—1 month at room temperature, 6 months in refrigerator
Rice, wild	Dark brown	Nutty	No (but rinse it)	30–45 minutes	3	3-1/2–4	Side dishes, stuffing, casseroles	Airtight container–many months at room temperature
Rice, basmati	White	Nutty, spicy	Yes (rinse also)	25 minutes	1-1/2	3	Side dishes, casseroles	Airtight container–1 month at room temperature, 6 months in refrigerator

Grain	Color	Flavor	Soak	Cooking Time			Uses	Storage
Jasmine Rice	White	Aromatic	No	15–20 minutes	2	3	Side dishes, casseroles, stews, soups	Same as above
Texmati Rice	White	Nutty	No	15–20 minutes	2	3	Same as above	Same as above
Rye, whole berries	Brown, oval	Distinct rye flavor	No	1-1/2 hours	3	3	Hot cereal, side dishes	Airtight container—1 month at room temperature, 5 months in refrigerator
Wheat, bulgur	Dark brown	Nutty	No	20–25 minutes	2-1/2	2	Salads, soups, breads, desserts, with rice, meat dishes, in place of rice pilaf, stuffing	Airtight container in cool place, or refrigerator for 5–6 months
Wheat, whole berries	Deep brown	Nutty, crunchy	Yes (1 cup to 3-1/2 cups cold water)	1 hour	3	2	Salads, meat loaves, croquettes, breads, side dishes	Airtight container in cool place up to 1 month, up to 5 months in refrigerator

Besides being so nutritious, they are very cost-effective.

Many varieties of beans may be found on the grocery shelf. Here are popular varieties, and their uses.

■ **Adzuki beans**—These beans are small and reddish-brown in color, with a mild nutty flavor and soft texture. They originally came from China and Japan. They are colorful with rice and in many other dishes as well.

■ **Anasazi beans**—Originally grown by Native Americans in the Southwest, these white beans are kidney-shaped and spotted with maroon. They are used in Mexican and southwestern dishes, such as refried beans.

■ **Black beans** (also called turtle beans)—These beans have an oval shape and a black skin. They are used in thick soups, chili, and spicy dishes, and in Central American, South American, and Caribbean cuisine.

■ **Black-eye peas** (also called black-eye beans or cowpeas)—These beans are small, oval-shaped, and creamish white with a black spot on one side. They are often used as a main dish vegetable, such as in the classic southern dish Hoppin' John. Black-eye peas are excellent in salads, stews, and soups.

■ **Cannellini beans**—These white kidney beans are popular in many Italian dishes. They work well in soups or salad and can be pureed, due to their creamy texture, with herbs, garlic, and other flavorings.

■ **Cranberry beans**—These light pink beans with beige spots lose their pink color during cooking. They are popular for baked beans and Italian dishes.

■ **Fava beans**—These large, flat brown beans have an earthy flavor and are used in Mediterranean dishes such as falafel.

■ **Garbanzo beans** (also called *chickpeas*)—These beans are nut-flavored and commonly pickled in vinegar and oil for salads. They are shaped like acorns, beige to yellow in color, and crunchy in texture. They can be used in soups or pasta salads, or puréed with tahini and lemon to make hummus or to make an Indian curry.

■ **Great Northern beans**—Larger than but similar to pea beans, these beans are used in soups, salads, casserole dishes, and especially in baked beans. They have a mild flavor.

■ **Kidney beans**—These beans are large and have a red color and kidney shape. They are popular for chili con carne and red beans and rice, and add zest to salads, soups, and Mexican dishes.

■ **Lima beans**—Although not widely known as dry beans, lima beans make an excellent main-dish vegetable and can be used in casseroles. They are broad and flat and come in three sizes: large, regular, and baby.

■ **Mung beans**—These small, ground beans are green or yellow and popular in Asian cuisine.

■ **Navy beans**—Also called pea beans, navy beans are white beans that are smaller than Great Northern beans. They are often used in soups and baked beans.

- **Pinto beans**—These beans are of the same species as the kidney and red beans. Beige-colored and speckled, they turn a uniform pink when cooked. They are used often in salads, refried beans, and chili.
- **Red and pink beans**—Pink beans have a more delicate flavor than red beans. Both are used in many Mexican dishes and chili.

Beans should be of similar color and size. When beans get old, they lose their color. Look for beans and other legumes that have no obvious defects, such as cracks or pinholes that may indicate insect damage, and make sure field debris (such as stones or twigs) are not in the bag. Once a package of beans has been opened, transfer them to an airtight container and store in a cool, dry spot—but not in the refrigerator. They can be stored at room temperature for one year.

Before soaking beans to rehydrate them, wash them carefully and pick over to remove any foreign particles. After soaking beans, be sure to discard the soaking water. Substances in the soaking water contribute to indigestion and can cause flatulence. Beans, like other legumes, are simmered in liquid (see Table 3-8 for cooking times). Casseroles, stews, stuffings, sandwich spreads, salads, and soups can all be prepared with beans as the central ingredient.

Dry peas are an interesting and versatile food that adds variety to meals. Dry peas may be green or yellow and may be bought either split or whole. Whole dry peas are available, but many cooks prefer to start with the half-circles of split peas. Green dry peas have a more distinct flavor than yellow dry peas. Split peas are popular in soups and are also great in side dishes, such as salads and pilafs.

Dry peas are served in many ways—with grains or as side dishes, or they can be puréed and made into dips, patties, croquettes, stuffed peppers, and even souffles. They go well with vegetables, pasta, fish, meat, poultry, and more. They can go in soups, salads, side dishes, main dishes, and casseroles.

The lentil is an old-world legume that is disc-shaped and about the size of a pea. Thousands of years old, lentils were perhaps the first of the convenience foods—they do not require soaking. Lentils come in colors such as green, red, black, or brown. Lentils may come whole or split (split lentils cook faster).

When cooking lentils, you can add seasonings or flavorful ingredients to the cooking water, since lentils absorb flavors well. Do not add acid ingredients, such as tomatoes or lemon juice, until later, since they will slow cooking. Salt should be added at the end of the cooking time for the same reason. Drain, if necessary, and they are ready. Cooked lentils may be stored in liquids such as broth, fruit juice, or salad dressing, to boost flavor.

Lentils are an excellent partner with many foods. They make excellent side dishes and go well in soups, stews, sauces, stuffings, and salads. Con-

TABLE 3-8 Cooking Information for Legumes

Bean, Pea, or Lentil	Size/Shape/Color	Flavor	Soaking Required	Cooking Time	Cups Liquid for Cooking	Yield[a]	Uses
Adzuki beans	Small, reddish brown	Nutty, sweet	Yes	1–1-1/2 hours	3	2	Rich, Asian cooking
Anasazi beans	Kidney shaped, white with maroon	Rich, meaty	Yes	2 hours	3	2	Chili and other Mexican dishes
Black beans (turtle beans)	Small, pea-shaped, black	Full, mellow	Yes	1-1/2 hours	4	2	Mediterranean cuisine, soups (black bean soup), chilis, salads, with rice
Black-eyed peas (cowpeas, black-eyed beans)	Small, oval, creamy white with black spot	Earthy, absorb other flavors	No	50–60 minutes	3	2	Casseroles, with rice, with pork, Southern dishes
Chickpeas (garbanzo beans, ceci beans)	Round, tan, large	Nutty	Yes	2-1/2 hours	4	4	Salads, soups, casseroles, hors d'oeuvres, hummus and other Middle East dishes
Fava beans, whole	Large, round, flat, off white or tan	Full	Yes	3 hours	2-1/2	4	Soups, casseroles, salads
Great Northern beans	Large, oval, white	Mild	Yes	1-1/2 hours	3-1/2	2	Soups, casseroles, baked beans, and mixing with other varieties

Bean	Description	Flavor	Soak	Cooking time	Cups[a]	Yield (cups)	Uses
Kidney beans	Large, kidney-shaped, red or white (red is much more common)	Rich, meaty, sweet	Yes	1–1-1/2 hours	3	2	Chili, casseroles, salads, soups, a favorite in Mexican and Italian cooking
Lentils	Small, flat, disk-shaped, green, red, or brown, split or whole	Mild, earthy	No	30–45 minutes	2	2-1/4	Soups, stews, salads, casseroles, stuffing, sandwiches, spreads, with rice
Lima beans	Flat, oval, cream or greenish, large or baby size	Large—full Baby—mild	Yes	1-1/2 hours (large) 1 hour (baby)	2	1-1/4	Soups, casseroles, side dishes
Navy beans (pea beans)	Small to medium, round to oval, white	Mild	Yes	1-1/2 hours	3	2	Baked beans, soups, salads, side dishes, casseroles
Peas, split	Small, flat on one side, green or yellow	Rich, earthy	No	30 minutes	3	2-1/4	Soups, casseroles
Peas, whole	Small-medium, round, yellow or green	Rich, earthy	Yes	40 minutes	3	2-1/4	Soups, casseroles, Scandinavian dishes
Pinto beans	Medium, kidney-shaped, pinkish brown	Rich, meaty	Yes	1-1/2 hours	3	2	A favorite for chili, refried beans, and in other Mexican cooking
Pink beans	Medium, oval, pinkish brown	Rich, meaty	Yes	1 hour	3	2	Popular in barbecue-style dishes
Soybeans	Medium, oval-round, creamy yellow	Distinctive	Yes	3-1/2 hours or more	3	2	Soups, stews, casseroles

[a] From 1 cup of uncooked bean, pea, or lentil.

109

sider them as you would potatoes or rice. Lentil purée, not unlike peanut butter, can be used on bread or muffins, and in sandwiches, dips, spreads, Mexican dishes, and vegetable fillings.

CHEF'S TIPS
- When choosing legumes for a dish, think color and flavor. Make sure the colors you pick will look good when the dish is complete. Also think of other ingredients you will use for flavor. In a salad, for example, black-eyed peas (black and white) go well with flageolet beans and red adzuki beans. To add a little more color and develop the flavor, you might add chopped tomatoes, fresh cilantro (Chinese parsley), and haricots vert (green beans) or fresh corn.
- Bigger beans, such as gigante white beans, hold their shape well and lend a hearty flavor to stews, regouts, and salads.
- Chickpeas can be puréed, as in hummus, and used as a dip, a spread, a sandwich filling layered with grilled vegetables, or a filling for pasta, crêpes, or twice-baked potatoes.
- You can cook together several types of beans, such as cranberry, turtle, and white beans, in stock flavored with herbs, vinegar, and carrots.
- A number of dried beans are also available fresh: cannellini, cranberry, fava, black-eyed peas, flagelot, lima, mung, and soybeans. If you can afford them, they are excellent products. They are plumper in size and have a fresher flavor than dry beans that you rehydrate.
- Use whole lentils, such as black or French green lentils, in grain dishes or salads because whole lentils hold their shape better. Use split lentils, such as brown, red, or yellow lentils, in soups, where they help thicken the liquid and shape is not as important.

Pasta

When we hear the word *pasta,* we relate it quickly to Italian cuisine. Pasta has been eaten for over 5,000 years and is very closely associated with Italian cooking. Pasta, from the Italian word for paste, is an edible dough made from flour and water that is rolled and cut into one of over 150 pasta shapes found in the United States (Figure 3-5).

Pasta may be dried or fresh. Dried pasta includes both *macaroni* and *noodles.* **Macaroni** products are pastas made from flour and water. These include spaghetti, elbow macaroni, lasagne, ziti, and other shapes. Many are available using whole wheat, or half whole wheat and half white flour. Noodles are also made from flour and water, but, by law, must contain 5.5 percent egg solids. Noodles are usually flat, like a ribbon, and come in different widths. Noodles can also contain flour made from products such

Macaroni—Pastas made from flour and water.

Noodles—Pastas made from flour, water, and egg solids.

Semolina—The roughly milled endosperm of a type of wheat called durum wheat.

Couscous—A granular form of semolina, like a tiny pasta.

as legumes. For example, cellophane noodles are made with flour made from mung beans.

Semolina is preferred for making dried pasta. Semolina is the roughly milled endosperm of a type of wheat called durum wheat. Durum wheat is known as a very hard wheat, meaning that it has a high protein content. Semolina is used almost exclusively for making pasta. Whole-wheat dried pasta is also available. Less expensive pasta products are made from a softer flour. High-quality pasta should be brittle and yellow in color, and holds its shape well when cooked. Poor-quality pasta is often a whitish-gray color, and it becomes soft and loses its shape when it is cooked. Dried pasta needs to be stored in a cool, dry place.

Couscous is a granular form of semolina, and is essentially a tiny pasta. It is cooked by soaking and then steaming. Couscous is a staple of North African and some Middle Eastern cuisines. It is also available precooked, like instant rice. Couscous is often used in place of rice.

Fresh pasta is more perishable and expensive than dried pasta. It is available as dough or in shapes. For example, ravioli is a soft dough, stuffed with a filling. Spaetzle and gnocchi are also soft pasta doughs. Flavored pastas usually contain vegetables, such as red tomato, artichoke, beet, carrot, or spinach (only a little is used). They are very colorful products, and the vegetables can add a subtle flavor.

Most pasta is high in starch, low in fat, and moderate in protein content. Pasta is cooked in a generous amount of boiling water. Dried pasta requires a much longer cooking time than fresh pasta, which cooks in a few minutes at most. Cook pasta until it is *al dente,* or firm to the bite. Cooked pasta is over 60 percent water in weight.

Each shape of pasta is appropriate for certain types of dishes.

■ Elbow macaroni is good in salads and soups because it retains its shape.
■ Tube pastas (such as elbow macaroni) or pastas with a hollow space (such as shells) work well with meat or vegetable sauces, as they trap the sauce in their spaces.
■ Fresh pasta, because it is softer in texture and absorbs sauce more readily than dried, is better with a smooth, light sauce that coats the pasta evenly.
■ Flat noodles are also better with smooth, light sauces.
■ Delicate pasta should be served with delicate sauces, and hearty pasta is best with hearty sauces.

Many shapes of pasta are suitable for filling, and all kinds of fillings may be used. Here is another challenge to the cook's ingenuity. Pasta sauces are discussed in Chapter 8.

Asian noodles are another type of pasta that is also excellent in main dishes and side dishes. *Bean thread vermicelli,* also called cellophane noodles,

Figure 3-5

Forms of Pasta

(Courtesy National Pasta Association.)

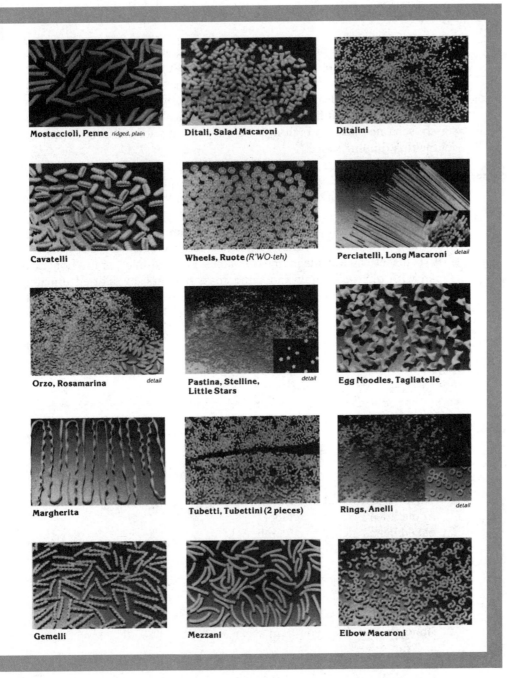

Mostaccioli, Penne *ridged, plain*

Ditali, Salad Macaroni

Ditalini

Cavatelli

Wheels, Ruote *(R'WO-teh)*

Perciatelli, Long Macaroni *detail*

Orzo, Rosamarina *detail*

Pastina, Stelline, Little Stars *detail*

Egg Noodles, Tagliatelle

Margherita

Tubetti, Tubettini (2 pieces)

Rings, Anelli *detail*

Gemelli

Mezzani

Elbow Macaroni

are very starchy, thin noodles made from mung beans. They work well in soups, stews, and ragouts. *Rice vermicelli* are also very starchy and can be used much like bean thread vermicelli. When soaked in water, they can also be used in stir-fries. *Soba noodles* are flat, thin, brownish-gray noodles made from buckwheat flour. Whereas the vermicelli noodles are almost all starch, soba noodles are high in protein. They are served in hot broth or cold with a dipping sauce.

CHEF'S TIPS

■ When buying dried pasta, always buy a high-quality product.

■ Pasta and beans work well together, such as in Pasta e Fagioli (see recipe on page 303).

■ Pasta is easy to prepare, comes in many shapes and sizes, and can serve as the base for a wide variety of entrées and side dishes.

■ Pasta dishes can include many vegetables. For example, spray a nonstick pan with olive oil spray and sauté roast peppers, oven-dried tomato, lightly steamed broccoli florets, and roast fresh garlic. Add a reduced broth with herbs, then add cooked rigatoni and toss the pasta with the vegetables. This can serve as the base for a 4 ounce serving of fish, chicken, veal, or pork paillard (a cutlet that has been pounded flat and then grilled or sautéed).

Check-Out Quiz

1. Match the food below with the nutrient(s) it is rich in.

Food	*Nutrient*
White bread	Added sugars
Whole-wheat bread	Natural sugars
Apple juice	Fiber
Baked beans	Starch
Milk	
Bran flakes	
Sugar-frosted oats	
Cola drink	
Broccoli	

2. Honey is better for you than sugar.
   a. True
   b. False

3. Carbohydrates spare protein from being burned for energy so that protein can be used to build and repair the body.
   a. True
   b. False

4. Maltose is made up of galactose and glucose.
   a. True
   b. False
5. Mannitol is an example of a sugar alcohol.
   a. True
   b. False
6. Obesity is more closely linked to diabetes than any other health problem.
   a. True
   b. False
7. All types of carbohydrates contribute to dental decay.
   a. True
   b. False
8. Seven-grain bread is an example of a whole grain.
   a. True
   b. False
9. Because insoluble fiber aids digestion and adds bulk to stool, it hastens passage of fecal material through the gut, thus helping to prevent or alleviate constipation.
   a. True
   b. False
10. Legumes are low in fiber and sodium.
    a. True
    b. False

Activities and Applications

1. Self-Assessment
List how many servings of the following foods you normally eat daily.

Refined Sugars	Number of Servings
Sugar in coffee or tea	_____
Sweetened beverages	_____
Sweetened breakfast cereals	_____
Candy	_____
Commercially made baked goods, including cakes, pies, cookies, and doughnuts	_____
Jam, jelly, pancake syrup	_____

Complex Carbohydrates	Number of Servings
Breads and rolls	_____
Ready-to-eat and cooked cereals	_____
Pasta, rice, other grains	_____
Dried beans and peas	_____
Potatoes	_____
Fruits	_____
Vegetables	_____

How do you rate? Do you get at least six servings per day of breads, rolls, cereals, pasta, rice, and other grains? Is at least one serving a whole grain? Do you get a daily serving of dried beans and peas? Do you get at least five servings per day of fruits and vegetables combined? If not, you should not be choosing foods from the "refined sugar" column of the chart until your more important nutritional needs are met. Compare the number of servings you have daily of foods high in refined sugars and those high in complex carbohydrates. The idea is to push complex-carbohydrate intake and to minimize refined sugars.

2. Carbohydrate Basics

Check off under the appropriate column(s) when you think the food contains a significant amount of sugar, starch, and/or fiber.

Food	Sugar	Starch	Fiber
1. Hamburger	_____	_____	_____
2. Chicken wing	_____	_____	_____
3. Flounder	_____	_____	_____
4. Boiled egg	_____	_____	_____
5. American cheese	_____	_____	_____
6. Sour cream	_____	_____	_____
7. White bread	_____	_____	_____
8. Whole-wheat bread	_____	_____	_____
9. Chocolate cake	_____	_____	_____
10. Macaroni	_____	_____	_____
11. Brown rice	_____	_____	_____
12. Split peas	_____	_____	_____
13. Peanuts	_____	_____	_____
14. Fresh orange	_____	_____	_____
15. Broccoli	_____	_____	_____

3. Whole Grain or Refined Grain?

Read the following ingredient labels for breads. Which one is white bread and which one is whole grain bread?

#1 Made from: Unbromated unbleached enriched wheat flour, corn syrup, partially hydrogenated soybean oil, molasses, salt, yeast, raisin juice concentrate, potato flour, wheat gluten, honey, vinegar, mono and diglycerides, cultured corn syrup, unbleached wheat flour, xanthan gum, and soy lecithin.

#2 Made from: Stoneground whole-wheat flour, water, high-fructose corn syrup, wheat gluten, yeast, honey salt, molasses, partially hydrogenated soybean oil, raisin syrup, soy lecithin, mono and diglycerides.

4. How Many Teaspoons of Sugar?

One teaspoon of sugar weighs 4 grams. Determine how many teaspoons of sugar are in each of the following foods, as described on their nutrition labels. Which food contains more sugar? Which food contains more fiber?

NUTRITION FACTS	NUTRITION FACTS
Amount per serving	Amount per serving
Calories 230	Calories 140
Calories from Fat 140	Calories from Fat 20
Total Fat 16 g	Total Fat 2.5 g
Saturated 6 g	Saturated 0.5 g
Polyunsaturated 1 g	Polyunsaturated 1.0 g
Monounsaturated 7 g	Monounsaturated 0.5 g
Cholesterol 74 g	Cholesterol 0 g
Sodium 180 mg	Sodium 180 mg
Total Carbohydrate 28 g	Total Carbohydrate 21 g
Dietary Fiber 0 g	Dietary Fiber 5 g
Sugar 24 g	Sugar 4 g
Protein 21 g	Protein 8 g

5. Nonnutritive Sweetener Sleuth

Check your refrigerator and cupboards to see what kind of foods, and how many, contain nonnutritive sweeteners. Look for the words Equal, aspartame, saccharin, acesulfame potassium, Sunette, or Sweet One.

Nutrition Web Explorer

International Food Information Council www.ificinfo.health.org
On IFIC's home page, click on "Food Safety and Nutrition Information," then click on "Sugars and Sweeteners." Read about the two newest artificial sweeteners, acesulfame potassium and sucralose.

Joslin Diabetes Center www.joslin.harvard.edu/education/library
Joslin Diabetes Center is an excellent site to learn almost anything about diabetes. At their library's homepage, click on "Meal Planning Using Carbohydrate Counting." Read through it completely, then complete the exercise "Meal Planning Practice."

Food Facts *Amber Waves of Grain*

The variety of grains is astonishing. Read on to learn more about these high-carbohydrate foods.

Barley. When the term *barley* is used in a recipe, it usually refers to pearl barley a white variety that has had the inedible husks or hulls, germ, and bran removed. Pearl barley can be boiled and used in soups, casseroles, stews, stuffings, cooked cereals, as a side dish, or as pilaf. Cooked barley can be sautéed with vegetables such as onions and mushrooms. Its taste is mild and nutty, and its texture is chewy.

Buckwheat. Although buckwheat has many grain-like characteristics, it is from an entirely different family and is actually a fruit. Two forms of buckwheat are available for purchase: whole white buckwheat and roasted hulled buckwheat, also called kasha. Kasha can be purchased whole or ground. Roasting gives the buckwheat kernels a distinct, nutty flavor. Kasha is best mixed with less flavorful grains. Kasha can be used in soups, stuffings, side dishes, and salads. Whole white buckwheat has a mild flavor and can be used to replace rice or pasta.

Millet. Millet, a golden grain, is grown and used in the Far East and China. Although the millet raised in the United States is used mostly to feed animals and birds, millet has been eaten by people in other parts of the world since Old Testament times. Millet is an important dietary component for many Africans, Indians, and Chinese. No wonder; it contains a high-quality protein. Cooking millet has had the inedible hull (that the birds just love!) removed, as well as an outer bran layer that is also inedible. It can be boiled and used in soups, casseroles, meat loaves, porridge, croquettes, pilaf, salads, stuffings, or as a side dish with chopped onions and fresh herbs, such as basil.

Oats. Oats are a unique grain: When they are milled, only the inedible hull is removed, and the bran and germ are left with the kernel. Whichever form of oats you buy, you are getting whole-grain nutrition. Rolled oats, or "old-fashioned oats," are made by cutting up raw oats into a product that is then steamed, shaped into flakes, and dried. Rolled oats require only five minutes' cooking time.

Quick oats start out as rolled oats but are sliced finer and slightly precooked, so they cook quicker—in about one minute. Instant oats are cut even smaller and result in a product that needs only to be mixed with boiling water.

Rye. Although rye is known most for its flour, you can buy whole rye berries that can be cooked in liquid into a hot cereal or side dish. Rye flakes, the equivalent of rolled oats, are also available and can be used as a hot cereal or side dish. They have a tangy taste.

Wheat. Wheat, the most important food grown in the world, is a rich source of nourishment. It is, of course, used to make flours, breads, cereals, and pastas.

Whole-wheat grains that have been steamed, dried, and ground into small pieces are called *bulgur*. Depending on how it is processed, some or all of the bran may be removed. Check the bulgur you want to purchase to see if the dark brown bran is still on. Bulgur has a nutty flavor that is excellent by itself or mixed with rice. Its uses are numerous—from salads to soups, from breads to desserts. It is also a nutritious extender and thickener for meat dishes and soups.

Wheat bran is available either processed or unprocessed, and is used as an ingredient in cooking or baking. It makes a high-fiber addition to baked goods, such as breads and muffins, and can be substituted for bread crumbs in most recipes.

Wheat germ is separated from the wheat grain and can be purchased at the supermarket either toasted or raw. Wheat germ has a nutty, crunchy texture and can be added to cereal, pancakes, baked goods, casseroles, salads, and breading.

Wheat berries, the actual whole-wheat kernels, are also available cracked, as cracked wheat, in coarse, medium, or fine qualities. Cracked wheat is particularly popular in breads.

Amaranth. Amaranth is one of the newer grains to arrive on the market, yet it has been around for at least 5,000 years! An important part of the Aztec diet, it contains a high-quality protein and is rich in calcium. Amaranth seeds are tiny and yellow-brown in color. A spicy grain with a slightly peppery taste, amaranth seeds can be cooked with other grains or to make pilaf. Amaranth cooks up soupy instead of fluffy.

Quinoa. Quinoa is a tiny, pale yellow seed that is technically not a cereal grain but a dried fruit. Whereas amaranth was popular with the Aztecs, quinoa was a staple of the Incas in Peru. Like amaranth, quinoa is rich is complete protein (unlike other plant foods) and calcium. Also, quinoa, unlike grains in general, contains an appreciable amount of oil (7 grams per 8 ounces). The best-quality quinoa is altiplano quinoa from Bolivia or Peru. It can be used in any recipe to replace rice.

Hot Topic ## Nonnutritive Sweeteners

The introduction of diet soft drinks in the 1950s sparked the widespread use of nonnutritive sweeteners. **Nonnutritive sweeteners** contain either no or very few calories. Four different ones will be discussed here: saccharin, cyclamate, aspartame, and acesulfame-K.

Saccharin. Sacchrain, discovered in 1879, has been consumed by Americans for more than 100 years. Its use in foods increased slowly until the two world wars, when its use increased dramatically due to sugar shortages. Saccharin is 300 times sweeter than sucrose and is excreted unchanged directly into the urine. It is used in a number of foods and beverages, and when combined with aspartame, its sweetness is intensified. Saccharin by itself has a bitter aftertaste. It is sold in liquid, tablet, packet, and bulk form.

In 1977, the Food and Drug Administration (FDA), which regulates the use of food additives, proposed a ban on its use in foods and allowed its sale as a tabletop sweetener only as an over-the-counter drug. This proposal was based on studies that showed the development of urinary bladder cancer in second-generation rats fed the equivalent of 800 cans of diet soft drinks a day. The surge of public protest against this proposal (there were no other alternative sweeteners available at that time) led Congress to postpone the ban, and the postponement is now extended to 2002. Products containing saccharin must have a warning label that states, "Use of this product may be hazardous to your health. This product contains saccharin, which has been determined to cause cancer in laboratory animals."

Cyclamate. Discovered accidentally in 1937, cyclamate was introduced into beverages and foods in the early 1950s. By the 1960s it dominated the noncaloric sweetener market. It is 30 times sweeter than sucrose and is not metabolized by most people. Cyclamate was banned in 1970 after studies showed that large doses of it, given with saccharin, were associated with increased risk of bladder cancer. Cyclamate is still banned in the United States but is approved and used in more than 40 other countries worldwide. Cyclamate is again under consideration for use in specific products, such as tabletop sweeteners and nonalcoholic beverages. It is stable at hot and cold temperatures and has no aftertaste.

Aspartame. In 1965, aspartame, a low-calorie sweetener, was also discovered accidentally. After being tested in more than 100 scientific studies in animals and humans, it was approved by the FDA in 1981. Aspartame is marketed in the United States under the brand name NutraSweet and as Equal tabletop sweetener. It is 200 times sweeter than sucrose and has an acceptable flavor with no bitter aftertaste.

Aspartame is made by joining two protein components, aspartic acid and phenylalanine, and a small amount of methanol. Aspartic acid and phenylalanine are building blocks of protein. Methanol is found naturally in the body and in many foods, such as fruit and vegetable juices. In the intestinal tract, aspartame is broken down into its three components, which are metabolized in the same way as if they had come from food. Aspartame contains 4 calories per gram, but so little of it is needed that the calorie content is negligible.

Aspartame is used as a tabletop sweetener, to sweeten many prepared foods, and in simple recipes that do not require lengthy heating or baking. Aspartame's components separate when heated over time, resulting in a loss of sweetness. It is best used at the end of cooking. Aspartame can be found in diet soft drinks, powdered drink mixes, cocoa mixes, pudding and gelatin mixes, frozen desserts, and fruit spreads and toppings. If you drink canned diet soft drinks, chances are they are sweetened with aspartame. Fountain-made diet soft drinks are more commonly sweetened with a blend of aspartame and saccharin, because saccharin helps maintain the right amount of sweetness.

The stability of aspartame in liquid in storage, the safety of the products of its metabolism, and symptoms possibly related to its use have provoked concerns and much research. Available evidence suggests that aspartame consumption is safe over the long term and is not associated with serious health effects.

The FDA uses the concept of an **Acceptable Daily Intake (ADI)** for many food additives, including aspartame. The ADI represents an intake level that, if maintained each day throughout a person's lifetime, would be considered safe by a wide margin. The ADI for aspartame has been set at 50 milligrams per kilogram of body weight. To take in the ADI for a 150-pound adult, someone would have to drink twenty 12-ounce cans of diet soft drinks daily.

The only individuals for whom aspartame is a known health hazard are those who have the disease phenylketonuria (PKU), because they are unable to metabolize phenylalanine. For this reason, any product containing aspartame carries a warning label. Some other people may also be sensitive to aspartame and need to limit their intake.

Acesulfame–K. In 1988, the FDA approved a new noncaloric sweetener, acesulfame potassium, or Acesulfame–K, for use in dry food products and as a powder or tablet to be used as a tabletop sweetener. It is marketed under the brand names Sunette and Sweet One tabletop sweetener. It is about as sweet as aspartame but is more stable, and can be used in baking. Acesulfame–K is approved for use in soft drinks, chewing gums, dry beverage mixes, gelatins, puddings, and baked goods. One major beverage maker mixes Acesulfame–K with aspartame to sweeten one of its diet sodas. Its taste is reportedly clean and sweet, with no aftertaste in most products.

Acesulfame–K passes through the digestive tract unchanged. The sweetener is used in over 50 countries, including France, Britain, and Russia.

Sucralose. After reviewing more than 110 animal and human safety studies conducted over 20 years, the FDA approved sucralose in 1998 as a tabletop sweetener and for use in a number of food products. In 1999, sucralose was approved as a general-purpose sweetener for all foods.

Known by its trade name, Splenda, sucralose is 600 times sweeter than sugar. It tastes like sugar because it is made from table sugar. But it cannot be digested, so it adds no calories to food. Because sucralose is so much sweeter than sugar, it is bulked up with maltodextrin, a starchy powder, so it will measure more like sugar. It has a good shelf life and doesn't degrade when exposed to heat. Numerous studies have shown that it does not affect blood glucose levels, making it an option for people with diabetes.

It is important to have a variety of nonnutritive sweeteners in the marketplace, especially for people with diabetes. Though artificially sweetened products are not magic foods that will melt pounds away, they can be a helpful part of an overall weight control program that includes exercise and a moderate diet.

Chapter 4
Lipids: Fats and Oils

The word *fat* is truly an all-purpose word. We use it to refer to the excess pounds we carry, the blood component that seems to cause heart disease, and the greasy foods in our diet that we feel we ought to cut out. To be more precise about the nature of fat, we need to look at fat in more depth.

To begin, **lipid** is the chemical name for a group of compounds that includes fats, oils, cholesterol, and lecithin. Fats and oils are the most abundant lipids in nature and are found in both plants and animals. A lipid is customarily called a **fat** if it is a solid at room temperature, and it is called an **oil** if it is a liquid at the same temperature. Lipids obtained from animal sources are usually solids, such as butter or beef fat, whereas oils are generally of plant origin. Therefore, we commonly speak of animal fats and vegetable oils, but we also use the word fat to refer to both fats and oils, which is what we will do in this chapter.

Like carbohydrates, lipids are made of carbon, hydrogen, and oxygen. Unlike most carbohydrates, lipids are not long chains of repeating units. Most of the lipids in foods (over 90 percent), and also in the human body, are in the form of **triglycerides.** Therefore, when we talk about fat in food or in the body, we are really talking about triglycerides. This chapter will help you to:

Lipid—The chemical name for a group of fatty substances, including fats, oils, cholesterol, and lecithin, that is present in blood and body tissues.
Fat—A nutrient that provides 9 calories per gram; a lipid that is solid at room temperature.
Oil—A form of lipid that is usually liquid at room temperature.
Triglyceride—The major form of lipid in food and in the body, made of three fatty acids attached to a glycerol backbone.

- Compare and contrast fats and oils.
- List the functions of lipids in foods and in the body.
- Identify the predominant form of lipid.
- Define saturated, monounsaturated, and polyunsaturated fats and list foods in which each is found.
- Define trans fatty acids and give examples of foods they are found in.
- Define cholesterol and lecithin, explain their functions, and tell where they are found in the body and in foods.
- Discuss how fats are digested, absorbed, and metabolized.
- Discuss the relationship between lipids and conditions such as heart disease and cancer.
- State recommendations for dietary intake of fat, saturated fat, trans fat, monounsaturated fat, polyunsaturated fat, and cholesterol.
- Discuss the purchasing, storage, cooking, and menuing of meat, poultry, fish, and shellfish.
- Describe rancidity.
- Distinguish between the percentage of fat by weight and the percentage of calories from fat.

Functions of Lipids

Fats have many vital purposes in the body, where they account for 15 to 25 percent or more of your weight. Fat is an essential part of all cells. At least 50 percent of your fat stores are located under the skin, where fat provides insulation, a cushion (like shock absorbers) around critical organs, and optimum body temperature in cold weather.

Most cells store only small amounts of fat, but specific cells, called fat cells or adipose cells, can store loads of fat and actually increase 50 times in weight! If your fat cells are completely filled with fat and you need to store more fat, your body can even produce new fat cells. Dietary fat can be stored in fat cells more easily than glucose can be stored as glycogen. Unlike fat, glycogen contains much water, so that it is bulky. Fat cells are a compact way to store lots of energy. Remember that one gram of fat yields 9 calories, compared to 4 calories for carbohydrate or protein. Fats provide much of the energy to do the work in your body, especially work involving your muscles.

Fat is an important part of all cell membranes. Fat also transports the fat-soluble vitamins (A, D, E, and K) throughout the body.

In foods, fats enhance taste, flavor, aroma, crispness (especially in fried foods), juiciness (especially of meat), and tenderness (especially in baked goods). Fats such as cooking oils do a wonderful job of carrying many flavors, such as the flavor of an Indian curry. Fats also provide a smooth texture and a creamy feeling in the mouth. The love of fatty foods cuts across all ages (just watch a preschooler devour French fries or an elderly adult eat a piece of chocolate cake) and cultures (where fatty foods are available). Eating a meal with fat makes people feel full, because fat delays the emptying of the stomach. This lasting feeling of fullness is called **satiety.**

Satiety—A feeling of being full after eating.

Essential fatty acids—Fatty acids that the body cannot produce, making them necessary in the diet: linoleic acid and linolenic acid.

A group of vitamins called the fat-soluble vitamins appear mainly in foods that contain some fat. Fat helps these vitamins get absorbed into the body. Certain fat-containing foods also provide the body with two fatty acids that are considered **essential fatty acids** because the body can't make them. Fatty acids are a component of triglycerides and are discussed next. The essential fatty acids are needed for normal growth in infants and children. They are a part of cell membranes, and also of lipids in the brain and nervous system. From the essential fatty acids, the body makes hormonelike substances (like messengers) that control a number of body functions, such as blood pressure.

Lipids also include cholesterol and lecithin. Their functions will be discussed later in this chapter.

■ **MINI-SUMMARY**

Many fat cells are located just under the skin, where fat provides insulation for the body, a cushion around critical organs, and optimum body temperature in the cold. Fat stores are a compact way to store lots of energy at 9 calories per gram. In foods, fats enhance taste, flavor, aroma, crispness, juiciness, tenderness, and texture. Fats have satiety value.

Triglycerides

A triglyceride (Figure 4-1) is made of three **fatty acids** (*tri-* means three) attached to **glycerol,** a derivative of carbohydrate. Glycerol contains three carbon atoms, each attached to one fatty acid. You can think of glycerol as the backbone of the triglyceride.

Fatty acids in triglycerides are made of carbon atoms joined like links in a straight chain. Interestingly, the number of carbons is always an even number. Fatty acids differ from one another in two respects: the length of the carbon chain and the degree of saturation. The length of the chain may be categorized as short chain (6 carbons or less), medium chain (8 to 12 carbons), or long chain (14 to 20 carbons). Most food lipids contain long-chain fatty acids. The length of the chain influences the fat's ability to dissolve in water. Generally, triglycerides do not dissolve in water, but the short- and medium-chain fatty acids have some solubility in water, which will have implications later in our discussion on their digestion, absorption, and metabolism.

Fatty acids are referred to as **saturated** or **unsaturated.** To understand this concept, think of each carbon atom in the fatty-acid chain as having hydrogen atoms attached like charms on a bracelet, as you can see in Figure

Figure 4-1

A triglyceride

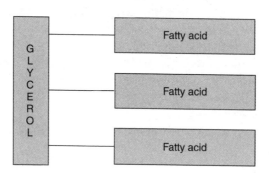

4-2. Each "C" represents a carbon atom, each "H" represents a hydrogen atom, and each "O" represents an oxygen atom. Each carbon atom can have a maximum of four bonds, so it can attach to four other atoms. Typically a carbon atom has one bond each to the two carbon atoms on its sides and one bond each to two hydrogens. If each carbon atom in the chain is filled to capacity with hydrogens, it is considered a saturated fatty acid. That's how saturated fatty acid got its name: It is saturated with hydrogen atoms. When a hydrogen is missing from two neighboring carbons, a double bond forms between the carbon atoms, and this type of fatty acid is considered unsaturated.

If you look at Figure 4-2, the top fatty acid in the illustration is saturated: It is filled to capacity with hydrogens. By comparison, the middle and lower fatty acids are unsaturated. This is evident because there are empty spaces without hydrogens in the picture. Wherever hydrogens are missing, the carbons are joined by two lines, indicating a double bond. The spot where the double bond is located is called the **point of unsaturation.**

Now that you know what saturated and unsaturated fatty acids are, we need to look at the two types of unsaturated fatty acids. Unsaturated fatty acids are either **monounsaturated** or **polyunsaturated.** A fatty acid that contains only one (*mono* means one) point of unsaturation is called monoun-

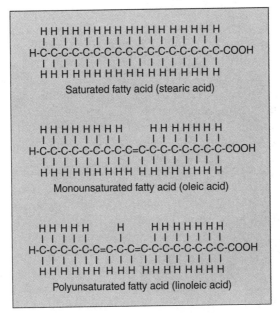

Figure 4-2
Types of fatty acids

saturated; if the chain has two or more points of unsaturation, the fatty acid is called polyunsaturated. Figure 4-2 gives an example of a monounsaturated and a polyunsaturated fatty acid.

Now that you know about the different types of fatty acids, it's time to get back to the concept of triglycerides. From the three types of fatty acids, we get three types of triglycerides, commonly called fats.

1. A saturated triglyercide, also called a **saturated fat,** is made of mostly saturated fatty acids.
2. A monounsaturated triglyceride, also called a **monounsaturated fat,** is made of mostly monounsaturated fatty acids.
3. A polyunsaturated triglyercide, also called a **polyunsaturated fat,** is made of mostly polyunsaturated fatty acids.

Now we are ready to see which types of foods the three different fats appear in.

Triglycerides in Food

All food fats, animal or vegetable, contain a mixture of saturated and unsaturated fats. A fat or oil is classified as saturated, monounsaturated, or polyunsaturated based on which type of fatty acid predominates.

Saturated fats are mostly found in foods of animal origin, and monounsaturated and polyunsaturated fats are mostly found in foods of plant origin and some seafoods. Foods of animal origin include meat, poultry, seafood, milk and dairy products such as butter, and eggs. Foods of plant origin include fruits, vegetables, dried beans and peas, grains, foods made with grains such as breads and cereals, nuts, seeds, and vegetable oils such as corn oil. The more unsaturated a fat is, the more liquid it is at room temperature.

Before going into more detail on foods that contain the three types of fats, let's see what each food group contributes in terms of overall fat.

1. *Fruits and Vegetables.* Whether fresh, canned, or frozen, most fruits and vegetables are practically fat free. The exceptions are avocados, olives, and coconuts, which contain significant amounts of fat. Also, frozen vegetables that have butter, margarine, or sauces added are probably high in fat. Last, fried vegetables, such as French-fried potatoes, are high in fat.
2. *Breads, Cereals, Rice, Pasta, and Grains.* Most breads and cereals in this group are low in fat. Exceptions include granolas, croissants, biscuits, cornbread, and many crackers. Most baked goods such as cakes, pies, cookies, and quick breads are also high in fat, especially when commercially made.

3. *Dry Beans and Peas, Nuts and Seeds.* Dry beans and peas are very low in fat. Most nuts and seeds, however, such as peanuts and peanut butter, are quite high in fat.
4. *Meat, Poultry, and Fish.* Meat and poultry, and to some extent fish, contain a bit of fat. The fat content of meat tends to be higher than that of poultry, and poultry tends to have more fat than seafood does. Of course, within each group there are choices that are quite high in fat and choices that are much more moderate in fat. For example, chicken without the skin is low in fat, because most of its fat is just under the skin.
5. *Dairy Foods.* Most regular dairy foods are high in fat. Luckily, there are plenty of choices with no fat or reduced fat, such as skim milk, nonfat yogurt, and low-fat cheeses.
6. *Fats, Oils, and Condiments.* Fats, such as vegetable shortening, and oils are almost all fat. Table 4-1 lists the total calories and fat in selected fats and oils. Condiments such as regular mayonnaise and salad dressings also contain much fat.

TABLE 4-1   Total Calories and Fat in Selected Fats and Oils

Fat or Oil	Calories/ Tablespoon	Grams Fat/ Tablespoon	Grams Saturated Fat/Tablespoon
Coconut oil	120	13	12
Palm kernel oil	120	13	11
Palm oil	120	13	7
Butter, stick	108	12	7
Lard	115	13	5
Cottonseed oil	120	13	3
Olive oil	119	13	1
Canola oil	120	13	1
Peanut oil	119	13	2
Safflower oil	120	13	1
Corn oil	120	13	2
Soybean oil	120	13	2
Sunflower oil	120	13	1
Shortening	106	12	3
Margarine, stick	100	11	1–3
Margarine, soft tub	100	11	1–2
Margarine, liquid	90	10	1–2
Margarine, whipped	70	8	1–2
Margarine, spread	60	7	1–2
Margarine, diet	50	6	1

Source: U.S. Department of Agriculture.

The main contributors to fat in the American diet are beef, margarine, salad dressings, mayonnaise, cheese, milk, and baked goods. You can't see most of the fat you get in the foods you eat, except of course when you add oils and fats. The fatty streaks in meat, in milk and cheese, and in fried foods are not as obvious as the margarine you spread on bread.

In addition to understanding which foods are high in fat, let's look at which foods contain mostly saturated, monounsaturated, and polyunsaturated fat.

1. *Saturated fat.* The biggest sources of saturated fat in the American diet are cheese, beef (more than half comes from ground beef), milk, the fats in baked goods, margarine, and butter. Saturated fat is also found in other full-fat dairy products such as ice cream, eggs, and poultry skin. Animal fat tends to contain at least 50% saturated fat. Although most vegetable oils are rich in unsaturated fats, several are high in saturated fat. These include the so-called tropical oils: coconut, palm kernel, and palm oils. They are used in some processed foods, such as baked goods and frozen whipped nondairy toppings.

2. *Monounsaturated fat.* Good examples of monounsaturated fats include olive oil, canola oil, and peanut oil. Like other vegetable oils, these are used in cooking and in salad dressings. Canola oil is also used to make some margarines.

3. *Polyunsaturated fat.* Polyunsaturated fats are found in greatest amounts in safflower, corn, soybean, sesame, and sunflower oils. These oils are commonly used in salad dressings and as cooking oils. Nuts and seeds also contain polyunsaturated fats, enough to make them a rather high-calorie snack food depending on serving size.

Figure 4-3 shows the fat composition of common foods.

Trans Fats

Trans fats, short for trans fatty acids, are a result of a process called **hydrogenation.** Hydrogenation, discovered at the turn of the twentieth century, converted liquid vegetable oils into solid fats by using hydrogen, heat, and certain metal catalysts. The partial hydrogenation process was quickly commercialized to make vegetable shortening, which is cheaper to make than butter or lard (pork fat) and has a longer shelf life. Vegetable shortening is simply vegetable oils that have been partially hydrogenated. The hydrogenation process has also been used to produce margarines (hydrogenation makes them easy to spread) and many oils used in deep-fat

Trans fat—Unsaturated fatty acids that lose a natural bend or kink so they become straight (like saturated fatty acids) after being hydrogenated; they act like saturated fats in the body.

Hydrogenation—A process in which liquid vegetable oils are converted into solid fats (such as margarine) by the use of heat, hydrogen, and certain metal catalysts.

Fatty Acid Composition of 1 Tablespoon of Fats and Oils

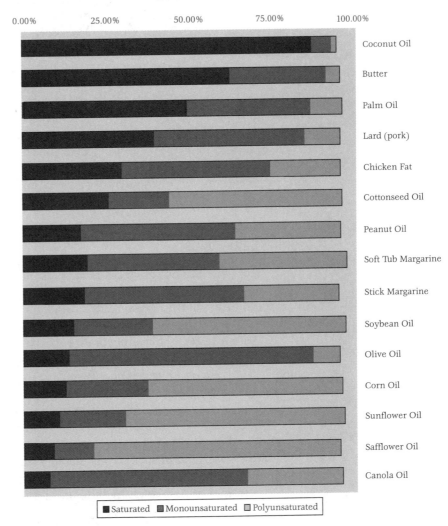

Figure 4-3
Fatty acid composition of 1 tablespoon of fats and oils

frying (hydrogenation gives them a high smoking point). Hydrogenation also helps these products stay fresh longer. If you check food labels, you will find partially hydrogenated oils popping up in many cookies, crackers, peanut butters, and salad dressings.

During hydrogenation, some of the unsaturated fatty acids become saturated. Other unsaturated fatty acids lose their natural bend or kink and become straight, like saturated fatty acids. These are the trans fatty acids. Because they are straight, they can fit closer together, which makes them more solid. This explains why vegetable shortening is a solid. Because they are straight, they also behave like saturated fats in the body. Like saturated fats, trans fatty acids have been shown to raise blood levels of LDL cholesterol, the "bad" cholesterol. This is discussed in more depth in a later section on health and lipids.

On food labels you will soon see the amount of trans fat per serving of the food. The percent Daily Value per serving on the Nutrition Facts panel for saturated fat will be based on the sum of saturated *and* trans fats (see Figure 4-4). The amount of trans fatty acids in selected foods appears in Table 4-2.

Essential Fatty Acids

As mentioned, there are two essential fatty acids that the body can't make. The essential fatty acids are both polyunsaturated fatty acids: **linoleic acid** and **linolenic acid.** Linoleic acid is called an omega-6 fatty acid because its double bonds appear after the sixth carbon in the chain (see Figure 4-2). Linolenic acid is the leading omega-3 fatty acid found in food, and its double bonds appear after the third carbon in the chain.

Linoleic acid is found in vegetable oils such as corn, safflower, soybean, cottonseed, sunflower, and peanut. Most Americans get enough linoleic acid from foods containing vegetable oils, such as margarine, salad dressings, and mayonnaise. Other foods that supply linoleic acid include whole grains, nuts, seeds, leafy vegetables, and the fatty portion of meats.

Linolenic acid is found in several oils, notably canola, soybean, walnut, and wheat germ oil (or margarines made with canola or soybean oil). Linolenic acid is also high in fish such as salmon, mackerel, sardines, halibut, bluefish, trout, and tuna. Notice that these fish tend to be fatty fish, not lean fish. Whereas Americans generally get plenty of linoleic acid, that is not the case with linolenic acid. Two or more servings per week of fish is suggested to increase intake of linolenic acid, and to improve the balance of intake for both essential fatty acids.

Linoleic acid—Omega-6 fatty acid found in vegetable oils such as corn, safflower, soybean, cottonseed, sunflower, and peanut oils; an essential fatty acid that increases blood clotting and inflammatory responses in the body.

Linolenic acid—Omega-3 fatty acid found in several oils, notably canola, soybean, walnut, and wheat germ oil (or margarines made with canola or soybean oil) and fatty fish; this essential fatty acid is important for normal growth and development and seems to protect against heart disease.

Nutrition Facts

Serving Size 1 Tbsp (14 g)
Servings per Container 32

Amount Per Serving

Calories 100 Calories from Fat 100

	% Daily Value*
Total Fat 11 g	17%
Saturated Fat** 4 g	20%
Polyunsaturated Fat 3.5 g	
Monounsaturated Fat 3.5 g	
Cholesterol 0 mg	0%
Sodium 115 mg	5%
Total Carbohydrate 0 g	0%
Protein 0 g	

Vitamin A 6%

Not a significant source of dietary fiber, sugars, vitamin C, calcium, and iron.

*Percent Daily Values are based on a 2000 calorie diet.

**Includes 2 g trans fat.

Figure 4-4
Nutrition facts

Linoleic and linolenic acids sometimes have opposite roles in the body. Whereas linoleic acid tends to increase blood clotting and inflammatory responses in the body, linolenic does the reverse. Diets rich in fish, meaning high in linolenic acid, may protect against heart disease by decreasing blood clotting, a factor in heart attacks. Diets rich in linolenic acid may also protect against certain forms of cancer. Both essential fatty acids also have additional roles. Linolenic acid is essential for normal growth and development, and linoleic acid is converted to other substances that allow cell membranes to function properly.

TABLE 4-2 Trans Fatty Acids in One Serving of Selected Foods	
Food	Trans Fatty Acids (grams/serving)
Vegetable shortening	1.4–4.2
Margarine, stick, 83% fat	2.4
Margarine, stick, 68% fat	1.8
Margarine, tub, 80% fat	1.1
Margarine, tub, 40% fat	0.6
Salad dressings (regular)	0.06–1.1
Vegetable oils	0.01–0.06
Pound cake	4.3
Doughnuts	0.3–3.8
Microwave popcorn (regular)	2.2
Chocolate-chip cookies	1.2–2.7
Vanilla wafers	1.3
French fries (fast food)	0.7–3.6
Snack crackers	1.8–2.5
Snack chips	0–1.2
Chocolate candies	0.04–2.8
White bread	0.06–0.7
Ready-to-eat breakfast cereals	0.05–0.5

Source: USDA food composition data, 1995.

Cholesterol

Cholesterol—The most abundant sterol (a category of lipids); a soft, waxy substance present only in foods of animal origin, it is found in all parts of the body and has important functions.

Bile acids—A component of bile that aids in the digestion and absorption of fats in the duodenum of the small intestine.

Cholesterol is the most abundant *sterol,* a category of lipids. Pure cholesterol is an odorless, white, waxy, powdery substance. You cannot taste it or see it in the foods you eat.

Your body needs cholesterol to function normally. It is present in every cell in your body, including the brain and nervous system, muscle, skin, liver, intestines, heart, and skeleton. The body uses cholesterol to make **bile acids,** which allow us to digest fat, and to make cell membranes, many hormones, and vitamin D. Unfortunately, high blood cholesterol is a risk factor for heart disease and is found in the walls of clogged arteries. This will be discussed in more detail later on.

So which foods contain cholesterol? Cholesterol is found only in foods of animal origin: egg yolks (it's not in the whites), meat, poultry, fish, milk, and milk products (Table 4-3). It is not found in foods of plant origin. Egg yolk and organ meats (liver, kidney, sweetbreads, brain) contain the most cholesterol—one egg yolk contains 213 milligrams of cholesterol. About

TABLE 4-3 Cholesterol in Foods	
Food and Portion	Cholesterol (milligrams)
Liver, braised, 3 ounces	333
Egg, whole, 1	213
Beef, short ribs, braised, 3 ounces	80
Beef ground, lean, broiled medium, 3 ounces	74
Beef, top round, broiled, 3 ounces	73
Chicken, roasted, without skin, light meat, 3 1/2 ounces	75
Shrimp, moist heat cooked, 3 ounces	167
Scallops, broiled, 3 ounces	47
Lobster, moist heat cooked, 3 ounces	61
Haddock, baked, 3 ounces	63
Mackerel, baked, 3 ounces	64
Swordfish, baked, 3 ounces	43
Milk, whole, 8 ounces	33
Milk, 2% fat, 8 ounces	18
Milk, 1% fat, 8 ounces	10
Skim milk, 8 ounces	4
Cheddar cheese, 1 ounce	30
American processed cheese, 1 ounce	27
Cottage cheese, low-fat, 1%, 1/2 cup	5

Source: National Institutes of Health. 1994. *Step by Step: Eating to Lower Your High Blood Cholesterol.* NIH Publication No. 94–2920.

four ounces of meat, poultry, or fish (trimmed or untrimmed) contain 100 milligrams of cholesterol, with the exception of shrimp, which is higher in cholesterol.

In milk products, cholesterol is mostly in the fat, so lower-fat products contain less cholesterol. For example, 1 cup of whole milk contains 33 milligrams of cholesterol, whereas a cup of skim milk contains only 4 milligrams (Table 4-3).

Egg whites and foods that come from plants, such as fruits, nuts, vegetables, grains, cereals, and seeds, have *no* cholesterol.

We take in about 200 to 400 milligrams of cholesterol daily, and the liver also makes a significant amount of cholesterol—about 1,000 milligrams. Because the body produces cholesterol, it is not considered an essential nutrient.

■ **MINI-SUMMARY**

Cholesterol is present in every cell and is used by the body to make bile acids, which allow us to digest fats, and to make cell membranes, many hormones, and vitamin D. Cholesterol is found only in foods of animal origin, such as egg yolks, meat, poultry, fish, milk, and milk products. It is not found in foods of plant origin. The liver produces cholesterol, so it is not an essential nutrient.

Lecithin

Lecithin—A phospholipid and a vital component of cell membranes that acts as an emulsifier (a substance that keeps fats in solution).

Lecithin is considered a *phospholipid,* a class of lipids that are like triglycerides except that one fatty acid is replaced by a phosphorus-containing substance. Lecithin functions as a vital component of cell membranes. It also acts as an *emulsifier.* As you may know, fats and water do not normally stay mixed together, but separate into layers. An emulsifier is capable of breaking up the fat globules into small droplets, resulting in a uniform mixture that won't separate. Lecithin keeps fats in solution in the blood, and elsewhere in the body, a most important function. Lecithin is used commercially as an emulsifier in foods such as salad dressings and bakery products.

Although the media have featured lecithin as a wonder nutrient that can burn fat, improve memory, and other similar feats, none of these is true. Since lecithin is made in the liver, it is not considered an essential nutrient.

■ **MINI-SUMMARY**

Lecithin, a phospholipid, is a vital component of cell membranes and acts as an emulsifier (keeps fats in solution). It is not an essential nutrient.

Digestion, Absorption, and Metabolism

Fats are difficult for the body to digest, absorb, and metabolize. The problem is simple: Fat and water do not mix. Minimal digestion of fats occurs before they reach the upper part of the small intestine. Once they reach this area, the gallbladder is stimulated to release **bile** into the intestine. Bile is made by the liver, stored in the gallbladder, and squirted into the intestinal tract when fat is present. Bile contains bile acids that emulsify fat, meaning that they split fats into small globules or pieces. In this manner, fat-splitting

Bile—A yellow-green liver secretion that is stored in the gallbladder and released when fat enters the small intestine because it emulsifies fat.

Monoglycerides— Triglycerides with only one fatty acid.

Lipoprotein—Protein-coated packages that carry fat and cholesterol through the bloodstream, classified according to their density.

Chylomicron—The lipoprotein responsible for carrying mostly triglycerides, and some cholesterol, from the intestines through the lymph system to the bloodstream.

Lipoprotein lipase—An enzyme that breaks down triglycerides in the blood into fatty acids and glycerol so they can be absorbed in the body's cells.

Very low-density lipoproteins (VLDL)—Lipoproteins made by the liver to carry triglycerides and some cholesterol through the body.

Low-density lipoproteins (LDL)—Lipoproteins that contain most of the cholesterol in the blood; they carry cholesterol to body tissues, including the arteries.

enzymes (such as pancreatic lipase) can then do their work. The enzymes break down the triglycerides in food to their component parts—fatty acids and glycerol—so they can be absorbed across the intestinal wall. **Monoglycerides,** triglycerides with only one fatty acid, are also produced.

Once absorbed into the cells of the small intestine, triglycerides are re-formed. Now it is time for them to travel in the blood. Both shorter-chain fatty acids and glycerol can travel freely in the blood because they are water soluble. However, triglycerides, monoglycerides, cholesterol, and longer-chain fatty acids would float in clumps and wreak havoc in either the blood or lymph. Because of this, the body wraps them with protein to make them water soluble. The resulting substance is called a **lipoprotein,** a combination of fat (*lipo-*) and protein. Lipoproteins contain triglycerides, protein, cholesterol, and phospholipid, another type of lipid. There are four types of lipoproteins.

Chylomicron is the name of the lipoprotein responsible for carrying mostly triglycerides, and some cholesterol, from the intestines through the lymph system to the bloodstream. In the bloodstream, an enzyme—**lipoprotein lipase**—breaks down the triglycerides into fatty acids and glycerol so they can be absorbed into the body's cells. Once the triglycerides are disposed of, most of what remains of the chylomicron is some protein and cholesterol that is metabolized by the liver.

The primary sites of lipid metabolism are the liver and the fat cells. The liver manufactures triglycerides and cholesterol too. Triglycerides, and some cholesterol, are carried through the body by the liver's version of chylomicrons: **very low-density lipoproteins (VLDLs).** The VLDLs release triglycerides, with the help of lipoprotein lipase, throughout the body. Once the majority of triglycerides are removed, VLDLs are converted in the blood into another type of lipoprotein called **low-density lipoprotein (LDL).**

The LDLs, mostly made of cholesterol, transport much of the cholesterol found in the blood to the body's cells. Certain cells (especially in the liver) have the ability to absorb the entire LDL particle. The LDLs not absorbed by cells are somehow involved in depositing cholesterol on the inner blood vessel wall, causing hardening and narrowing of the arteries.

A last type of lipoprotein, **high-density lipoprotein (HDL),** contains much protein and travels throughout the body picking up cholesterol. It is thought that the HDLs carry cholesterol back to the liver for disposal. Thus HDLs help remove cholesterol from the blood, preventing the buildup of cholesterol in the arterial walls.

Most body cells can store only small amounts of fat, but fat cells can become greatly enlarged with fat. The fat cells of obese people may be many times larger than those of a normal-weight or underweight individual.

High-density lipoproteins (HDL)—Lipoproteins that contain much protein and carry cholesterol away from body cells and tissues to the liver for excretion from the body.

However, if not enough calories are being consumed to meet the body's needs, the fat cells release fat and start to shrink.

■ MINI-SUMMARY

With the help of bile and various enzymes, fats are broken down into monoglycerides, fatty acids, and glycerol so they can be absorbed across the intestinal wall. Once they are absorbed into the cells of the small intestine, triglycerides are reformed. Chylomicrons, a type of lipoprotein, carry triglycerides and cholesterol from the intestines through the lymph system to the bloodstream. In the bloodstream, the enzyme lipoprotein lipase breaks down the triglycerides into fatty acids and glycerol so they can be absorbed into the body's cells. Very low-density lipoproteins are the liver's equivalent of chylomicrons.

Low-density lipoproteins are mostly made of cholesterol and transport much of the cholesterol found in the blood. High-density lipoprotein travels throughout the body picking up cholesterol.

Lipids and Health

Plaque—Deposits on arterial walls that contain cholesterol, fat, fibrous scar tissue, calcium, and other biological debris.
Atherosclerosis—The most common form of artery disease, characterized by plaque buildup along artery walls.
Angina—Symptoms of pressing, intense pain in the heart area, often due to stress or exertion when the heart muscle gets insufficient blood.
Myocardial infarction—heart attack.

Heart disease is the number-one killer of both men and women in the United States. More than 90 million American adults, or about 50 percent, have elevated blood cholesterol levels, one of the key risk factors for heart disease. Anyone can develop high blood cholesterol, regardless of age, gender, race, or ethnic background.

Too much circulating cholesterol can injure arteries, especially the heart's arteries (called the coronary arteries) that supply the heart with what it needs to keep pumping. This leads to accumulation of cholesterol-laden **plaque** in blood vessel linings, a condition called **atherosclerosis.**

When blood flow to the heart is impeded, the heart muscle becomes starved for oxygen, causing chest pain (called **angina**). If a blood clot completely obstructs a coronary artery affected by atherosclerosis, a heart attack (called a **myocardial infarction**) can occur.

One of the primary ways in which LDL cholesterol levels can become too high in blood is through eating too much saturated fat, trans fat, and cholesterol. Several other factors also affect blood cholesterol levels.

- **Heredity** — High cholesterol often runs in families. Even though specific genetic causes have been identified in only a minority of cases, genes still play a role in influencing blood cholesterol levels.

■ **Weight** — Excess weight tends to increase blood cholesterol levels. Losing weight may help to lower levels.

■ **Exercise** — Regular physical activity may not only lower LDL cholesterol, but it may increase levels of desirable HDL.

■ **Age and gender** — Before menopause, women tend to have lower total cholesterol levels than men's at the same age. Cholesterol levels naturally rise as men and women age. Menopause is often associated with increases in LDL cholesterol in women.

Though high total cholesterol and LDL cholesterol levels, along with low HDL cholesterol, can increase heart-disease risk, there are several other risk factors. These include cigarette smoking, high blood pressure, diabetes, obesity, and physical inactivity. If any of these is present in addition to high blood cholesterol, the risk of heart disease is even greater.

The good news is that these factors can be brought under control either by changes in lifestyle — such as a different diet, losing weight, or an exercise program — or quitting a tobacco habit. Drugs also may be necessary for some people. Sometimes one change can help bring several risk factors under control. For example, weight loss can reduce blood cholesterol levels, help control diabetes, and lower high blood pressure.

But some risk factors can't be controlled. These include age (45 or older for men, and 55 or older for women) and family history of early heart disease (father or brother stricken before age 55; mother or sister stricken before age 65).

When a patient without heart disease is first diagnosed with elevated blood cholesterol, doctors often prescribe a program of diet, exercise, and weight loss to bring levels down. National Cholesterol Education Program guidelines suggest at least a six-month program of reduced dietary saturated fat and cholesterol, together with physical activity and weight control, as the primary treatment before resorting to drug therapy. Typically, doctors prescribe the Step I/Step II diet to lower dietary fat, especially saturated fat. Many patients respond well to this diet and end up sufficiently reducing blood cholesterol levels.

On the Step I diet, the patient should eat 8 to 10 percent of the day's total calories from saturated fat (include trans fat here), 30 percent or less of total calories from fat, less than 300 milligrams of dietary cholesterol daily, and just enough calories to achieve and maintain a healthy weight. If the Step I diet doesn't result in a desirable cholesterol level, doctors may try the Step II diet, which changes the daily saturated fat limits to below 7 percent of daily calories and dietary cholesterol to below 200 milligrams. Step II also is the diet for people with heart disease.

In many patients, blood cholesterol levels should begin to drop a few weeks after starting on a cholesterol-lowering diet. Just how much they drop depends on factors such as how high the cholesterol level is and how the person's body responds to changes made. With time, cholesterol levels may be reduced 10 to 50 milligrams per deciliter or more, a clinically significant amount.

But sometimes diet and exercise alone are not enough to reduce cholesterol to goal levels. Perhaps a patient is genetically predisposed to high blood cholesterol. In these cases, doctors often prescribe drugs. The National Cholesterol Education Program estimates that as many as 9 million Americans take some form of cholesterol-lowering drug therapy. The most prominent cholesterol drugs are in the statin family, an array of powerful treatments that includes drugs such as Mevacor (lovastatin) and Pravachol (pravastatin).

Cancer is the second leading cause of death in the United States following heart disease. Diets high in fat and cholesterol are associated with many types of cancer. Although dietary fat does not seem to initiate the cancer process, it may promote its development once started. Fat may influence cancer because it contains so many calories, and energy itself seems to promote cancer. It may be that certain forms of fat promote cancer more than others. Monounsaturated fatty acids (as in olive oil) and omega-3 fatty acids (as in fatty fish) may be protective, whereas linoleic acid, an omega-6 fatty acid, and saturated fatty acids may be promoters.

Dietary Recommendations

The Dietary Guidelines for Americans and the American Heart Association recommend a diet that provides no more than 30 percent of total calories from fat (Table 4-4). The American Heart Association goes on to recommend

TABLE 4-4 Recommended Fat and Saturated Fat Intake

If Your Total Daily Calories Are:	Total Fat (grams)	Saturated Fat (grams)
1,200	40	13
1,500	50	17
1,800	60	20
2,000	67	22
2,200	73	24
2,400	80	27
2,600	86	29
2,800	93	31
3,000	100	33

that saturated fat intake be limited to 8 to 10 percent, polyunsaturated fat to 10 percent, and monounsaturated fat to 15 percent of total calorie intake. There seems to be agreement that if fat intake goes higher than 30 percent of total calories, the diet should emphasize monounsaturated fats such as olive oil (as in the Mediterranean diet—see Appendix D) or canola oil. Many Americans eat about 33 percent of calories from fat.

The American Heart Association also recommends 300 milligrams or less of cholesterol daily. Recommendations for omega-3 and omega-6 fatty acids are expected at a later date.

These recommendations do not apply to children age 2 and under. They need fat in order to grow and develop properly.

Ingredient Focus: Meats, Poultry, and Fish

Before going into meats, poultry, and fish, two concepts need to be explained: percent fat and rancidity. When looking at fat in food, it is important to distinguish between two different concepts: the percentage of fat by weight and the percentage of calories from fat. To explain these two concepts, let's look at an example. In a supermarket, you find sliced turkey breast that is advertised as being "96 percent fat free." What this means is that if you weighed out a 3-ounce serving, 96 percent of the weight is lean or without fat. In other words, only 4 percent of its weight is actually fat. The statement "96 percent fat free" does not tell you anything about how many calories come from fat.

Now, if you look at the Nutrition Facts on the label, you read that a 3-ounce serving contains 3 grams of fat, 27 calories from fat, and 140 total calories. The label also states that the percentage of calories from fat in a serving is 19 percent. To find out the percentage of calories from fat in any serving of food, simply divide the number of calories from fat by the number of total calories, then multiply the answer by 100, as follows.

$$\frac{\text{Calories from fat} \times 100}{\text{Total calories}} = \text{Percent of calories from fat}$$

$$\frac{27 \text{ calories from fat} \times 100}{140 \text{ calories}} = 19 \text{ percent}$$

This percentage has become more important as recommendations on fat consumption target 30 percent or less of total calories as a desirable daily total from fat. This does not mean, however, that every food you eat needs to derive only 30 percent of its calories from fat. If this were the case, you could not even have a teaspoon of margarine because all of its calories come from fat. It is your total fat intake over a few days that is important, not the percentage of fat in just one food or just one meal.

Another important concept, rancidity, is related to the storing and use of fats in the kitchen. **Rancidity** is the deterioration of fat, resulting in undesirable flavors and odors. In the presence of air, fat can lose a hydrogen at the point of unsaturation and take on an oxygen atom. This change creates unstable compounds that start a chain reaction, quickly turning a fat rancid. You can tell whether a fat is rancid by its odd odor and taste. The greater the number of points of unsaturation, the greater the possibility that rancidity will develop. This explains why saturated fats are more resistant to rancidity than unsaturated fats. Rancidity is also quickened by heat and ultraviolet light rays. Luckily, vitamin E is present in plant oils, and it naturally resists deterioration of the oil. Food additives such as BHA (butylated hydroxyanisole) and BHT (butylated hydroxytolulene) are added to packaged foods such as salad dressings to maintain freshness.

To prevent rancidity, store fat and oils tightly sealed in cool, dark places. For butter and margarine, check the date on the packaging. When oils are refrigerated, they sometimes become cloudy and thicker. This usually clears up after they are left at room temperature again or put under warm water.

Food Facts in this chapter discusses the wide variety of oils, butter, and margarine. This section will discuss meats, poultry, and fish. Milk and dairy products, which are also often high in fat, are discussed in the next chapter.

> **Rancidity**—The deterioration of fat, resulting in undesirable flavors and odors.

Nutrition

Purchasing lean, fresh cuts of meat, poultry, and fish are important, but first let's compare these items nutritionally (see Table 4-5).

- Most fish is lower in fat, saturated fat, and cholesterol than are meat and poultry.
- Chicken is twice as fatty as turkey.
- Chicken and turkey breast (meaning white meat) without skin are low in fat—only about 3 grams of fat in 3 ounces of chicken, and 1 gram of fat in 3 ounces of turkey. By comparison, white meat with skin and dark meat (such as thighs and drumsticks) are much higher in fat. Also, chicken wings may be considered white meat, but they are fattier than the drumstick.
- If buying ground turkey or chicken, make sure it is made from only skinless breast meat for the least amount of fat. If the product includes skin and dark meat, it will be *much* higher in fat.

TABLE 4-5 Meat, Poultry, and Fish: A Comparison				
Food Type (3 ounces, cooked)	Saturated Fat (grams)	Dietary Cholesterol (milligrams)	Total Fat (grams)	Calories
Beef, top round, broiled	3	73	8	185
Beef, whole rib, broiled	10	72	26	313
Chicken, light meat without skin, roasted	1	64	4	130
Chicken, light meat with skin, roasted	3	71	19	189
Ground turkey—breast meat only	<1	35	<2	130
Ground turkey (meat and skin), cooked	3	87	11	200
Cod, baked	<1	47	<1	89

Source: National Institutes of Health, 1994. *Step by Step: Eating to Lower Your High Blood Cholesterol.* NIH Publication No. 94-2920.

■ Trimmed veal is leaner than skinless chicken.
■ When choosing beef, you will get the least fat from eye of round, followed by top round and bottom round.

Meat is a good source of many important nutrients, including protein, iron, copper, zinc, and some of the B vitamins, such as B_6 and B_{12}. Meat is also a significant source of fat, saturated fat, and cholesterol.

In comparison to red meats, skinless white-meat chicken and turkey are comparable in cholesterol, but lower in total fat and saturated fat. The skin of chicken and turkey contains much of the bird's fat. The skin should be left on during cooking to keep in moisture, but can be removed before serving. Chicken and turkey are rich in protein, niacin, and vitamin B_6. They are also good sources of vitamin B_{12}, riboflavin, iron, zinc, and magnesium. Duck and goose are quite fatty in comparison, because they contain all dark meat.

Fish and shellfish are excellent sources of protein, and are relatively low in calories. Most are also low to moderate in cholesterol content and a good source of certain vitamins, such as vitamins E and K, and minerals, such as iodine and potassium. Certain fish (Tables 4-6 and 4-7) are fattier than others, such as mackerel and herring, but fatty fish are an important source of omega-3 fatty acids.

TABLE 4-6 Fat Content of Fish		
Low-Fat Fish (fat content less than 2.5 percent)	Medium-Fat Fish (fat content 2.5–5 percent)	High-Fat Fish (fat content over 5 percent)
Cod	Bluefish	Albacore tuna
Croaker	Swordfish	Bluefin tuna
Flounder	Yellowfin tuna	Herring
Grouper		Mackerel
Haddock		Sablefish
Pacific halibut		Salmon
Pollock		Sardines
Red snapper		Shad
Rockfish		Trout
Sea bass		Whitefish
Shark		
Sole		
Whiting		

Purchasing

The *first step* is to select a lean cut, such as one of these:

◼ **Beef:** Eye of round, inside (top) round, outside (bottom) round, sirloin tip, flank steak, top sirloin butt.
◼ **Veal:** Any trimmed cut except commercially ground and veal patties.
◼ **Pork:** Pork tenderloin, pork chop (sirloin), pork chop (top loin), pork chop (loin).
◼ **Lamb:** Shank, sirloin.
◼ **Poultry:** Breast (skin removed after cooking).
◼ **Fish:** All fish and shellfish.

The *second step* is to order a quality product, such as USDA Choice for beef (avoid Prime—it contains more fat), from a reputable purveyor. Check the product for quality when it is received, and reject it if necessary. Put it immediately into the refrigerator or freezer. Fresh meat should be wrapped. Fresh poultry and whole fish should be stored on shaved or crushed ice. The ice should be put in drip pans and changed daily. Whole fish should be drawn (remove entrails) as soon as possible. Fish steaks (cross-cut section), fillets (boneless lengthwise section), or other cut fish should be wrapped in moisture-proof packaging before placing on ice. If icing is not possible, the refrigerator should maintain a temperature from 30 to 34 °F. Fresh poultry and fish should be used preferably within 24 hours.

TABLE 4-7　Seafood Nutrition Chart (based on 3-1/2-ounce portions)

Species	Calories	Fat (grams)	Saturated Fat (grams)*	Cholesterol (milligrams)
Finfish				
Carp, cooked, dry heat	162	7	1	84
Cod, Atlantic, cooked, dry heat	105	1	—	55
Grouper, cooked, dry heat	118	1	—	47
Haddock, cooked, dry heat	112	1	—	74
Halibut, cooked, dry heat	140	3	1	41
Herring, Atlantic, cooked, dry heat	203	12	3	77
Mackerel, Atlantic, cooked, dry heat	262	18	4	75
Perch, cooked, dry heat	117	1	—	115
Pike, Northern, cooked, dry heat	113	1	—	50
Pollock, Walleye, cooked, dry heat	113	1	—	96
Pompano, cooked, dry heat	211	12	5	64
Salmon, Coho, cooked, moist heat	185	8	1	49
Salmon, Sockeye, canned, drained solids with bone	153	7	2	44
Sea bass, cooked, dry heat	124	3	1	53
Smelt, Rainbow, cooked, dry heat	124	3	1	90
Snapper, cooked, dry heat	128	2	—	47
Swordfish, cooked, dry heat	155	5	1	50
Trout, Rainbow, cooked, dry heat	151	4	1	73
Tuna, Bluefish, fresh, cooked, dry heat	184	6	2	49
Whiting, cooked, dry heat	115	2	—	84
Shellfish				
Clam, cooked, moist heat	148	2	—	67
Crab, Alaska King, cooked, moist heat	97	2	—	53
Crayfish, cooked, moist heat	114	1	—	178
Lobster, Northern, cooked, moist heat	98	1	—	72
Oyster, Eastern, cooked, moist heat	137	5	1	109
Scallops, raw	88	1	—	33
Shrimp, cooked, moist heat	99	1	—	195

* A dash (—) means less than 1 gram of saturated fat.

Source: United States Department of Agriculture.

Preparation

The *third step* is to trim the meat of all visible fat and perform other techniques that decrease calories, fat, and cholesterol. Of course, even after fat is trimmed, the meat still contains invisible fat. Studies performed on cooked poultry have shown that poultry cooked with the skin on (where most of the fat lurks) does not significantly add fat to the poultry meat itself and does help prevent the meat from drying out. So it's a good idea to cook poultry with the skin, then remove the skin before serving. Select a 4- to 5-ounce raw portion to produce a 3- to 3 1/2-ounce cooked portion.

The *fourth step* is to use flavorful rubs and marinades, when appropriate, to allow new and creative flavor options. **Rubs** combine dry ground spices, such as cinnamon, and finely cut herbs, such as cilantro. Rubs may be dry or wet. Wet rubs, also called pastes, use liquid ingredients such as mustard or vinegar. Pastes produce a crust on the food. Wet or dry seasoning rubs work particularly well with beef, and can range from a mesquite barbecue seasoning rub to a Jamaican jerk rub. To make a rub, various seasonings are mixed together and spread or patted evenly on the meat just before cooking for delicate items, or up to 24 hours in advance for heartier meat cuts (see Table 4-8 for ingredients in 13 Cajun Spice Rub). The larger the piece of meat or poultry, the longer the rub can stay on. The rub flavors the exterior of the meat as it cooks.

Marinades, seasoned liquids used for soaking a food before cooking, are useful for adding flavor as well as for tenderizing meat and poultry. Marinades bring out the strongest flavors naturally so you don't need to drown the food in fat, cream, or sauces. Marinades allow a food to stand on its own with a light dressing, chutney, sauce, or relish (discussed more in Chapter 8). Fish can also be marinated. Although fish is already tender, a short marinating time (about 30 minutes) can develop a unique flavor. A marinade usually contains an acidic ingredient, such as wine, beer, vinegar, citrus juice, or plain yogurt, to break down the tough meat or poultry. The other ingredients add flavor. Without the acidic ingredient, you can marinate fish for a few hours to instill flavor. Oil is often used in marinades to carry flavor, but it isn't essential. Fat-free salad dressings such as Italian work

Rubs—A dry marinade made of herbs and spices (and other seasonings), sometimes moistened with a little oil, and rubbed or patted on the surface of meat, poultry, or fish (which is then refrigerated and cooked at a later time).

Marinades—A seasoned liquid used before cooking to flavor and moisten foods; usually based on an acidic ingredient.

TABLE 4-8

13 Cajun Spice Rub

INGREDIENTS

4 cups paprika

2 cups chili powder

4 tablespoons cayenne pepper

4 tablespoons black and white pepper

4 tablespoons garlic powder

4 tablespoons onion powder

2 cups cumin

4 tablespoons thyme

4 tablespoons oregano

4 tablespoons marjoram

4 tablespoons basil

4 tablespoons gumbo filet

4 tablespoons fennel powder

well in marinades. To give marinated foods flavor, try minced fruits and vegetables, low-sodium soy sauce, mustard, fresh herbs, and spices. For example, fruit-juice marinades can be flavored with Asian seasonings such as ginger and lemongrass. Even a simple fat-free Italian dressing can serve as a marinade.

Cooking

The *fifth step* is to choose a cooking method that will produce a flavorful, moist product, and that adds little or no fat to the food. Possibilities include roasting, grilling, broiling, sautéing, poaching, and braising (discussed in detail in Chapter 8). The *sixth step* is to think of how you want to flavor the dish (discussed in detail in Chapter 8). For example, smoking can be used to complement the taste of meat, poultry, or fish. Hardwoods or fruitwoods, such as the following, are best for producing tasty foods:

- **Fruit (apple, cherry, peach):** These woods are too strong for fish but work well with pork, chicken, or turkey.
- **Hickory and maple:** These are great for beef or pork.
- **Mesquite:** Mesquite produces an aromatic smoke that works well with beef and pork.

The *seventh step* is to make the portion look large and attractive on the plate. Keep in mind that a cooked 3-ounce serving of meat, poultry, or fish is about the size of a deck of cards. Slice the meat or poultry thin and fan it across the plate so it looks like more. More information on presenting foods is found in Chapter 8.

CHEF'S TIPS
- Marinate top sirloin butt or sirloin tip with tomato juice, herbs, and spices. Cut into strips and use in fajitas or stir-fries. Cut into cubes and grill them as kabobs.
- Grilling is a wonderful way to add flavor to meats, poultry, and fish without adding fat. Since it must be done just prior to serving, the food is always fresh. Marinate your foods ahead of time to add flavor and moisture.
- Remember food safety when marinating foods. Do not reuse liquid marinades. Marinate meat, poultry, and fish in covered, clean, sanitized pans in the coldest part of the refrigerator on the bottom shelf, to avoid cross-contamination of other foods. If you make a marinade and part of it is to be used in the sauce, put that portion aside *before* you marinate the meat, poultry, or fish.
- Organic chicken is very much worth the extra money for its superb sweet taste. When you butcher the whole chicken, there is also less fat under the skin.

■ Fish is a very versatile and nutritious food. Anything, such as rice or beans or pasta, goes with fish. Serve fish on top of a vegetable ragout, or salmon with couscous.

■ You can marinate fish without any citrus, which ruins the texture of the fish if it is marinated for very long. By eliminating citrus, you can marinate the fish longer, for two hours or even overnight. Try a marinade that includes fish stock, chives, tarragon, thyme, and black pepper. The fish will absorb some liquid, which keeps it moister during cooking.

■ Cedar-planked fish is another way to add flavor. Soak an untreated cedar plank, then put marinated fish on it and bake in the oven. The cedar plank will impart a unique flavor.

■ Fish must be cooked very carefully. Fish is done when it *just* separates into flakes and turns opaque. Once cooking is completed, serve immediately for the best flavor and texture.

Check-Out Quiz

Directions: In the following columns, check off each food that is a significant source of fat and/or cholesterol.

Food	Fat	Cholesterol
1. Butter	_____	_____
2. Margarine	_____	_____
3. Split peas	_____	_____
4. Peanut butter	_____	_____
5. Porterhouse steak	_____	_____
6. Flounder	_____	_____
7. Skim milk	_____	_____
8. Cheddar cheese	_____	_____
9. Chocolate chip cookie made with vegetable shortening	_____	_____
10. Green beans	_____	_____

Match each statement on the left with the term on the right that it describes. The terms will be used more than once.

_____ 1. Present in every cell in the body   a. Lecithin
_____ 2. Emulsifies fats   b. Cholesterol
_____ 3. Found only in animal foods
_____ 4. Vital component of cell membranes
_____ 5. Used to make bile

Match each numbered statement with the lettered term it describes.

_____ 1. Cottonseed oil is a good source of this fat   a. Rancidity
    b. Chylomicron
_____ 2. Lessens possibility of blood clots   c. Monounsaturated fat

_____ 3. Liquid at room temperature, solid when refrigerated
_____ 4. Deterioration of fat in air and heat
_____ 5. Olive oil is a good source of this fat
_____ 6. Breaks up fat globules into small droplets
_____ 7. Carries triglycerides and cholesterol

d. Saturated fat
e. Emulsifier
f. Polyunsaturated fat
g. Long-chain omega-3 fatty acid

Activities and Applications

1. Self-Assessment

To find out if your diet is high in fat, saturated fat, and cholesterol, check _yes_ or _no_ to the following questions.

Do You Usually:	YES	NO
1. Put butter on popcorn?	_____	_____
2. Eat more red meats (beef, pork, lamb) than chicken and fish?	_____	_____
3. Leave the skin on chicken?	_____	_____
4. Eat whole-milk cheeses, such as Cheddar, American, and Swiss, more than three times a week?	_____	_____
5. Sauté or fry foods more than once or twice a week?	_____	_____
6. Eat regular lunch meats, hot dogs, and bacon more than three times a week?	_____	_____
7. Leave visible fat on meat?	_____	_____
8. Use regular creamy salad dressings such as Russian, blue cheese, thousand island, and creamy French?	_____	_____
9. Eat potato chips, nacho chips, and/or cream dips more than twice a week?	_____	_____
10. Drink whole milk?	_____	_____
11. Eat more than four eggs a week?	_____	_____
12. Eat organ meats (liver, kidney, etc.) more than once a week?	_____	_____
13. Use mayonnaise, margarine, and/or butter often on your sandwiches?	_____	_____
14. Use vegetable shortening in baking or cooking?	_____	_____
15. Eat commercially baked goods, including cakes, pies, and cookies, more than twice a week?	_____	_____

Ratings: If you answered yes to:

1–3 questions: You are probably eating a diet not too high in fat, saturated fat, and cholesterol.

4–7 questions: You could afford to make some food substitutions, such as skim for regular milk, to reduce your fat and saturated fat intake.

8–15 questions: Your diet is very likely to be high in fat, saturated fat, and cholesterol.

2. Changing Eating Habits

If you are eating too much fat, you can make changes a little at a time! Check off one of these things to try (if you are not already doing it) or make up your own.

- The next time, I eat chicken, I will take the skin off.
- I will limit my daily meat and poultry servings to two 3-ounce servings a day. A 3-ounce serving looks about the size of a deck of cards.
- This week, I will try a new type of fresh or plain frozen fish.
- I will try a low-fat cheese, like low-fat Swiss.
- I will switch to 1% or skim milk.
- I will try sherbet or ice milk for dessert instead of ice cream.
- I will count the number of eggs I eat a week and see whether I meet the recommendations.
- To cut back on fat, I will try to use a lower-in-fat margarine, salad dressing, or mayonnaise.
- I will keep more fruit and vegetables in the refrigerator, so they will be handy for a snack instead of cookies or chips.
- I will buy pretzels instead of potato chips.
- For breakfast, instead of doughnuts, I will try a hot or cold cereal with skim milk and toast with jelly.
- I will top my spaghetti with stir-fried vegetables instead of a creamy sauce.

3. Reading Food Labels

Following are food labels from two brands of lasagne. One is heavy on cheese and ground beef. The other is a vegetable lasagne made with moderate amounts of cheese. Using the Nutrition Facts given, can you tell which is which? How did you tell?

Lasagne#1	Lasagne#2
NUTRITION FACTS	NUTRITION FACTS
Amount per serving	Amount per serving
Calories 230	Calories 140
Calories from Fat 140	Calories from Fat 20
Total Fat 16 g	Total Fat 2.5 g
Saturated 6 g	Saturated 0.5 g
Polyunsaturated 1 g	Polyunsaturated 1.0 g
Monounsaturated 7 g	Monounsaturated 0.5 g
Cholesterol 74 g	Cholesterol 0 g

Sodium 180 mg
Total Carbohydrate 0 g
  Dietary Fiber 0 g
    Sugar 0 g
Protein 21 g

Sodium 180 mg
Total Carbohydrate 21 g
  Dietary Fiber 5 g
    Sugar 0 g
Protein 8 g

4. Meat, Poultry, and Seafood Comparison

Pick out three meat items you eat, three poultry items you eat, and three fish/shellfish items you eat. Make a chart listing the calories, fat, saturated fat, and cholesterol of all these foods. Once the chart is done, ask yourself the following questions.

■ Which food has the least/most fat?
■ Which food has the least/most saturated fat?
■ Which food has the least/most cholesterol?

5. Name That Fat Substitute!

Following are ingredient listings from four products made with fat substitutes. Using "Hot Topics: Fat Substitutes" on page 157 as a guide, identify the fat substitutes in these foods.

Creme-Filled Chocolate Cupcakes—0 grams fat/cupcake
Sugar, water, corn syrup, bleached flour, egg whites, nonfat milk, defatted cocoa, invert sugar, modified food starch (corn, tapioca), glycerine, fructose, calcium carbonate, natural and artificial flavors, leavening, salt, dextrose, calcium sulfate, oat fiber, soy fiber, preservatives, agar, sorbitan monostearate, mono- and dislycerides, carob bean gum, polysorbate 60, sodium stearoyl lactylate, xanthan gum, sodium phosphate, maltodextrin, guar gum, pectin, cream of tartar, sodium aluminum sulfate, artificial color.

Low-Fat Mayonnaise Dressing—1 gram fat per tablespoon
Water, corn syrup, liquid soybean oil, modified food starch, egg whites, vinegar, maltodextrin, salt, natural flavors, gums (cellulose gel and gum, xanthan), artificial colors, sodium benzoate and calcium disodium EDTA.

Lite Italian Dressing—0.5 grams fat/2 tablespoons
Water, distilled vinegar, salt, sugar, contains less than 2% of garlic, onion, red bell pepper, spice, natural flavors, soybean oil, xanthan gum, sodium benzoate, potassium sorbate and calcium disodium EDTA, yellow 5 and red 40.

Light Cream Cheese—5 grams fat per 2 tablespoons
Pasteurized skim milk, milk, cream, contains less than 2% of cheese culture, sodium citrate, lactic acid, salt, stabilizers (xanthan and/or carob bean and/or guar gums), sorbic acid, natural flavor, vitamin A palmitate.

Nutrition Web Explorer

Nutrition Site of the American Heart Association www.deliciousdecisions.org
Visit the nutrition site for the American Heart Association and click on "Cookbook." Take the Heart Healthy Chef's Tour. Write down three cooking methods, three seasonings, and three cooking substitutions that are heart-healthy.

California Olive Oil www.olive-oil.com
Visit this website of an olive-oil producer and find out what an oil mister is.

Food Facts *Oils and Margarines*

There is an ever-widening variety of oils and margarines on the market. They can differ markedly in their color, flavor, uses, and nutrient makeup.

When choosing vegetable oils, pick those high in monounsaturated fats, such as olive oil, canola oil, and peanut oil. Olive oil contains from 73 to 77 percent monounsaturated fat. The color of olive oil varies from pale yellow to dark green and its flavor from subtle to a full, fruity taste. The color and flavor of olive oil depend on the olive variety, level of ripeness, and processing method. When buying olive oil, make sure you are buying the right product for your intended use.

1. **Extra virgin olive oil,** the most expensive form, has a rich, fruity taste that is ideal for flavoring finished dishes and in salads, vegetable dishes, marinades, and sauces. It is not usually used for cooking because it loses some flavor. It is made by putting mechanical pressure on the olives, a more expensive process than using heat and chemicals.

2. **Olive oil,** also called pure olive oil, is golden and has a mild, classic flavor. It is an ideal, all-purpose product that is great for sautéing, stir-frying, salad dressings, pasta sauces, and marinades.

3. **Light olive oil** refers only to color or taste. These olive oils lack the color and much of the flavor found in the other products. Light olive oil is good for sautéing, stir-frying, or baking because the oil is used mainly to transfer heat rather than to enhance flavor.

Polyunsaturated fats, such as corn oil, safflower oil, sunflower oil, and soybean oil are also good choices, but not as good as monounsaturated fats such as olive oil, canola oil, and peanut oil.

Table 4-9 gives information on various oils. Be prepared to spend more for exotic oils such as almond, hazelnut, sesame, and walnut oils. Because these oils tend to be cold pressed (meaning they are processed without heat), they are not as stable as the all-purpose oils and should be purchased in small quantities. They are strong, so you don't need to use much of them. Don't purchase these oils to cook with—they burn easily.

Vegetable oils are also available in convenient sprays that can be used as nonstick

Oil	Characteristics/Uses	Smoke Point	Oil	Characteristics/Uses	Smoke Point
Canola oil (Monounsaturated)	Light yellow color Bland flavor Good for frying, sautéing, and in baked goods Good oil for salad dressings	420°F	Olive Oil	Extra virgin or virgin olive oil—good for flavoring finished dishes and in salad dressings, strong olive taste	*
Corn oil	Golden color Bland flavor Good for frying, sautéing, and in baked goods Too heavy for salad dressings	420°F		Pure olive oil—can be used for sautéing and in salad oils, not as strong an olive taste as extra virgin or virgin	280°F
Cottonseed oil	Pale yellow color Bland flavor Good for frying, sautéing, and in baked goods Good oil for salad dressings	420°F		Light olive oil—the least flavorful, good for sautéing, stir-frying, or baking.	
			Peanut oil	Pale yellow color Mild nutty flavor Good for frying and sautéing Good oil for salad dressings	420°F
Hazelnut oil	Dark amber color Nutty and smoky flavor Not for frying or sautéing as it burns easily Good for flavoring finished dishes and salad dressings Use in small amounts Expensive	*	Safflower oil	Golden color Bland flavor Has a higher concentration of polyunsaturated fatty acids than any other oil Good for frying, sautéing, and in baked goods Good oil for salad dressings	470°F
Olive oil	Varies from pale yellow with sweet flavor to greenish color and fuller flavor to full, fruity taste (color and flavor depend on olive variety, level of ripeness, and how oil was processed)		Sesame oil	Light gold flavor Distinctive, strong flavor Good for sautéing Good oil for flavoring dishes and in salad dressings Use in small amounts Expensive	440°F

TABLE 4-9 Vegetable Oils

TABLE 4-9 *(continued)*

Oil	Characteristics/Uses	Smoke Point	Oil	Characteristics/Uses	Smoke Point
Soybean oil	More soybean oil is produced than any other type; used in most blended vegetable oils and margarines Light color Bland flavor Good for frying, sautéing, and in baked goods Good oil for salad dressings	420°F	Sunflower oil	Pale golden color Bland flavor Good for frying, sautéing, and in baked goods Good oil for salad dressings	340°F
			Walnut oil	Medium yellow to brown color Rich, nutty flavor For flavoring finished dishes and in salad dressings Use in small amounts Expensive	*

* Not recommended for cooking.

spray coatings for cooking and baking with a minimal amount of fat. Vegetable-oil cooking sprays come in a variety of flavors (butter, olive, Italian, mesquite), and a quick two-second spray adds about 1 gram of fat to the product. To use, spray the pan first away from any open flames (the spray is flammable), heat up the pan, then add the food.

Margarine was first made in France in the late 1800s to provide an economical fat for Napoleon's army. It didn't become popular in the United States until World War II, when it was introduced as a low-cost replacement for butter. Margarine must contain vegetable oil and water and/or milk or milk solids. Flavorings, coloring, salt, emulsifiers, preservatives, and vitamins are usually added. The mixture is heated and blended, then firmed by exposure to hydrogen gas at very high temperatures (see hydrogenation, page 130). The firmer the margarine, the greater the degree of hydrogenation and the longer its shelf life.

Standards set by the U.S. Department of Agriculture and the Food and Drug Administration require margarine and butter to contain at least 80 percent fat by weight and to be fortified with vitamin A. One tablespoon of either has approximately 11 grams of fat and 100 calories. You can compare the fat profile of butter and margarine in Table 4-10. Butter contains primarily saturated fat and no polyunsaturated fat, whereas margarine is low in saturated fat and rich in monounsaturated and polyunsaturated fat. Although some margarines contain more trans fat than butter does, the total of trans and saturated fat is always less than the total for butter. The total for butter is much higher because of all the saturated fat it contains.

Butter must be made from cream and milk. Salt and/or colorings may be added.

		Grams	Grams Saturated	Grams Trans	Grams Monounsaturated	Grams Polyunsaturated
	Calories/	Fat/	Fat/	Fat/		
Fat or Oil	Tablespoon	Tablespoon	Tablespoon	Tablespoon	Fat/Tablespoon	Fat/Tablespoon
Coconut oil	120	14	12	0	1	0
Palm kernel oil	120	14	11	0	2	0
Palm oil	120	14	7	0	5	1
Butter, stick	108	12	8	0.3	4	0
Lard	115	13	5	N/A	6	1
Cottonseed oil	120	14	4	0	2	7
Olive oil	119	14	2	0	10	1
Canola oil	120	14	1	0	8	4
Peanut oil	119	14	2	0	6	4
Safflower oil	120	14	1	0	2	10
Corn oil	120	14	2	0	3	8
Soybean oil	120	14	2	0	3	8
Sunflower oil	120	14	1	0	6	6
Shortening	106	12	3	1.4–4.2	6	3
Margarine, stick	102	11	2	2.4	5	3
Margarine, soft tub	102	11	2	1	4	5
Margarine, liquid	102	11	2	varies	3	6
Margarine, whipped	70	7	2	varies	2	3
Margarine spread	78	9	1	varies	5	3
Margarine, diet	51	6	1	varies	2	2
Margarine, fat-free	0	0	0	0	0	0

TABLE 4-10 Total Calories and Fat in Selected Fats and Oils

Source: U.S. Department of Agriculture Food and Drug Administration, and manufacturers.

Margarine must contain vegetable oil and water and/or milk. Salt, food coloring, other vitamins, emulsifying agents such as lecithin, and preservatives may be added to margarine.

If a butter or margarine product does not contain at least 80 percent fat by weight, it can't be called "butter" or "margarine," but instead is classified as a *spread*. The percent of fat (by weight) must appear on the label. Water, gums, gelatins, and various starches are used in spreads to replace some or all of the fat, or air may be whipped into the product.

Margarines basically vary along these dimensions:

■ *Their physical form.* Margarine comes in either sticks or in tubs. Tub margarines contain more polyunsaturated fatty acids than stick margarines, so they melt at lower temperatures and are easier to spread. Spreads come in sticks, tubs, liquids, and pumps. Liquid spreads are packaged in squeeze bottles or pump dispensers in which the margarine spread is really liquid, even in the refrigerator. They work well when drizzled on hot vegetables and other cooked dishes. Fat-free sprays also can be used to coat cooking pans.

■ *Type of vegetable oil(s) used.* The vegetable oil may be mostly corn oil, safflower oil, canola oil, or others. Check the ingredients label to compare how much liquid oil and/or partially hydrogenated oil are used. If the first ingredient is liquid corn oil, then more liquid oil is used, meaning that there will be less saturated fat than there is in a product with hydrogenated corn oil as the first ingredient.

■ *Percent fat by weight and nutrient profile.* Margarine and spreads are available with amounts of fat that vary from 0 percent to 80 percent by weight. Look on the label for the percentage of fat by weight. Also look for terms such as light, diet, or fat-free. Light margarine means that the product contains one-third fewer calories or half the fat of the regular product. Diet margarine, also called reduced-calorie margarine, has at least 25 percent fewer calories than the regular product. Fat-free margarine has less than 0.5 grams of fat per serving. If you are wondering how they make a margarine fat-free, part of the answer is gelatin. Water, rice starch, and other fillers are used to make it taste like fat. As mentioned in this chapter, information on the trans fat content of products will appear soon on labels. A number of margarines without any trans fats are available.

Not all margarines and spreads can be used in the same way. Spreads with lots of water can make bread or toast soggy, and may spatter and evaporate quickly in hot pans, causing foods to stick. In baking, low-fat spreads are not recommended because product quality suffers.

In addition to butter and margarine, blends and butter-flavored buds are also available. Blends are part margarine and part butter (about 15 to 40 percent). They are made of vegetable oils, milk fat, and other dairy ingredients added to make the product taste like butter. Blends may have as much fat as regular margarine or butter (in other words, at least 80 percent fat) or they may be reduced in fat. Butter-flavored buds are made from carbohydrates and a small amount of dehydrated butter. They are virtually fat free and cholesterol free, and are designed to melt on hot, moist foods such as a baked potato. When mixed with water, they can make butter-flavored sauces.

Two new margarine-like spreads, Take Control made by Lipton and Benecol made by McNeil Consumer Products, appeared on the market in 1999 after FDA approval. Both use ingredients that reportedly lower blood cholesterol levels. Take Control uses a stanol-like ingredient from soybeans. Benecol contains a plant stanol ester that comes from pine trees.

Lipton recommends one to two servings (1–2 tablespoons) of Take Control every day as part of a diet low in saturated fat and cholesterol. Take Control is low in saturated fat and free of trans fats, and contains 6 grams of fat and 1.1 grams soybean extract per serving. Benecol is recommended in three daily servings (1-1/2 tablespoons total) of its regular or light spread. Both products cost considerably more than regular margarine. These spreads are examples of functional foods, a topic discussed in detail in Chapter 6.

Hot Topic ## Fat Substitutes

Food manufacturers are making it easier for fat-conscious consumers to have their cake and eat it too—and their cheeses, chips, chocolate, cookies, ice cream, salad dressings, and various other foods that are now available in lower-fat versions. A host of fat substitutes that replace some, or sometimes all, of the fat in a food, makes these lower-fat foods possible. Most of these fat replacers are ingredients already approved by the Food and Drug Administration for other uses in food. For instance, starches and gums, two popular fat replacers, have long since been approved as thickeners and stabilizers. Newer fat replacers, such as olestra, have undergone or will undergo close scrutiny by FDA to assess their safety.

Fat is a difficult substance to replace because it gives taste, consistency, stability, and palatability to foods. Fat replacers can help reduce a food's fat and calorie levels while maintaining some of the desirable qualities fat brings to food, such as mouth feel, texture, and flavor.

Under FDA regulations, fat replacers usually fall into one of two categories: food additives or "generally recognized as safe" (GRAS) substances. Each has its own set of regulatory requirements.

Food additives must be evaluated for safety and approved by FDA before they can be marketed. They include substances with no proven track record of safety; scientists just don't know that much about their use in food. Examples of food additives are olestra, polydextrose, and carrageenan, which are used as fat replacers.

GRAS substances, on the other hand, do not have to undergo rigorous testing before they are used in foods because they are generally recognized as safe by knowledgeable scientists, usually because of the substances' long history of safe use in foods. Many GRAS substances are similar to substances already in food. Examples of GRAS substances used as fat replacers are cellulose gel, dextrins, and guar gum.

In addition to water, the simplest fat substitute, fat replacers may be carbohydrate-, protein-, or fat-based. The first to hit the market used carbohydrate as the main ingredient. Avicel, for example, is a cellulose gel introduced in the mid-1960s as a food stabilizer. Carrageenan, a seaweed derivative, was approved for use as an emulsifier, stabilizer, and thickener in food in 1961. Its use as a fat replacer became popular in the early 1990s. Polydextrose (made from dextrose and small amounts of sorbitol and citric acid) came on the market in 1981 as a humectant, meaning that it helps retain moisture in a food. Other carbohydrate-based fat replacers include starch, modified food starch, dextrins and maltodextrins (both made from starch), fiber, and gums. Gums are made from seeds, seaweed extracts, and plants. Examples include xantham gum, guar gum, alginates, and cellulose gum.

Fruit purées, such as prune purée, are also useful to replace some of the fat in a product. Fruit purées have been used successfully in baked products, where they add tenderness and moisture.

Although their original purpose was to perform certain technical functions in food that would improve overall quality, many carbohydrate-based fat replacers are now used to replace fat and reduce calories. They provide from 0 to 4 calories per gram and are used in a variety of foods including dairy-type products, sauces, frozen desserts, salad dressings, processed meat, baked goods, spreads, chewing gum, and sweets.

Protein-based fat substitutes came along in the 1990s. These (and fat-based replacers) were designed specifically to replace fat in foods. One protein-based fat substitute, known as Simplesse, is made from egg white and milk protein that are blended and heated using a process called microparticulation. The protein is shaped into microscopic round particles that roll easily over one another. The aim of the process is to create the feel of a creamy liquid and the texture of fat. Simplesse cannot be used to fry foods, but can be used in some cooking and baking. It contains 1 to 2 calories per gram. Simplesse has been used in frozen dessert–type foods. Another type of protein-based fat replacers, called protein blends, combine animal or vegetable protein, gums, food starch, and water. They are made with FDA-approved ingredients and are used in frozen desserts and baked goods.

Olestra is an example of a fat-based fat replacer. FDA approved olestra (its brand name is Olean, made by Procter & Gamble) in 1996 for use in preparing potato chips, crackers, tortilla chips, and other savory snacks. Olestra has properties similar to those of naturally occuring fat, but it provides zero calories and no fat. That's because it is indigestible. It passes through the digestive tract but is not absorbed into the body. This is due to its unique configuration: a center unit of sucrose (sugar) with six, seven, or eight fatty acids attached. Olestra's configuration also makes it possible for the substance to be exposed to high temperatures, such as frying—a quality most other fat replacers lack.

As promising as olestra sounds, it does have some drawbacks. Studies show that it may cause intestinal cramps and loose stool in some individuals. Also, according to clinical tests, olestra reduces the absorption of fat-soluble nutrients, such as vitamins A, D, E, and K and carotenoids, from foods eaten at the same time as olestra-containing products.

To address these concerns, FDA required that the four vitamins be added to olestra-containing foods and that the following statement appear on products made with olestra:

> This product contains Olestra. Olestra may cause abdominal cramping and loose stools. Olestra inhibits the absorption of some vitamins and other nutrients. Vitamins A, D, E and K have been added.

Another fat-based replacer, Salatrim, is the generic name for a family of reduced-calorie fats that are only partially absorbed in the body. Salatrim provides 5 calories per gram. It is used in a brand of reduced-fat baking chips. Caprenin is a 5-calorie-per-gram fat substitute for cocoa butter that is used in candy bars.

In using reduced-fat foods, be aware that fat-free does *not* mean calorie-free. The calories lost in removing regular fat from a food can be regained through sugars added for palatability, as well as fat replacers, many of which provide calories too. Use the Nutrition Facts panel on the product to compare calories and other nutrition information between fat-reduced and regular-fat foods. Many nutrition experts agree that, used properly, fat replacers can play an important role in improving adult Americans' diets. But, as with any diet or food, variety and moderation are emphasized to ensure a healthy intake.

Chapter 5
Protein

Have you ever wondered why meat, poultry, and seafood are often considered entrées, or main dishes, whereas vegetables and potatoes are side dishes? As recently as the 1950s and 1960s, the abundant protein found in meat, poultry, and seafood was considered the mainstay of a nutritious diet. You could say that these foods took center stage, or more accurately, center plate. As a child, I can remember going to visit my grandparents on Sundays and eating a roast beef dinner during our visit. Yes, we had vegetables too, but the big deal at dinner was the roast that was carefully cooked, sliced, and served (with brown gravy, of course).

Today, protein foods continue to be an important component of a nutritious diet; however, we are much more likely to see foods such as lentils or pasta occupying the center of the plate. For adults who grew up when beef was king (and not nearly as expensive as it is today) and full-fat bologna sandwiches filled many lunchboxes, making spaghetti without meatballs takes a little getting used to, but more and more meatless meals are being served.

So just what are **proteins?** They are an essential part of all living cells found in animals and plants. The protein found in animal and plant foods is such an important substance that the term *protein* is derived from the Greek word meaning "first." About 16 percent of your body weight (if you're not overweight) is protein. Proteins reside in your skin, hair, nails, muscles, and tendons, to name just a few places. They function in a very broad sense to build and maintain the body. This chapter discusses protein's structure, functions, metabolism, and relationship to diet, and will help you to:

- Identify the building blocks of protein.
- List the functions of protein in the body.
- Explain how protein is digested, absorbed, and metabolized.
- Distinguish between complete and incomplete protein.
- Explain the consequences of eating too much or too little protein.
- State the dietary recommendations for protein.
- Discuss the purchasing, storage, cooking, and menuing of milk, dairy products, and eggs.

Structure of Protein

Like carbohydrates and fats, proteins contain carbon, hydrogen, and oxygen. Unlike carbohydrates and fats, proteins contain nitrogen and provide much of the body's nitrogen. Nitrogen is necessary for bodily function; life as we know it wouldn't exist without nitrogen.

Figure 5-1

An amino acid

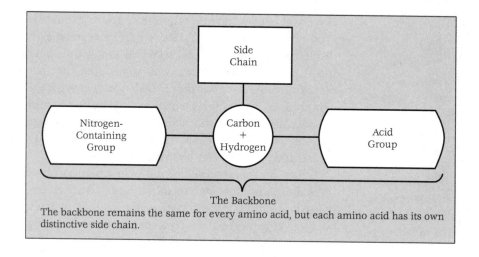

The Backbone
The backbone remains the same for every amino acid, but each amino acid has its own distinctive side chain.

Proteins are long chains of **amino acids** strung together much like rail-road cars. Amino acids are the building blocks of protein. There are 20 different ones, each consisting of a backbone to which a side chain is attached (Figure 5-1). The amino acid backbone is the same for all amino acids, but the side chain varies. It is the side chain that makes each amino acid unique.

TABLE 5-1 Amino Acids	
Essential Amino Acids	Nonessential Amino Acids*
Histidine	Alanine
Isoleucine	Arginine
Leucine	Asparagine
Lysine	Aspartic acid
Methionine	Cysteine
Phenylalanine	Glutamic acid
Threonine	Glutamine
Tryptophan	Glycine
Valine	Proline
	Serine
	Tyrosine

* Under some circumstances, one or more of these may become essential.

Amino acids—The building blocks of protein.
Essential or indispensable amino acids—Amino acids that either cannot be made in the body or cannot be made in the quantities needed by the body; must be obtained in foods.
Peptide bonds—The bonds that form between adjoining amino acids.
Polypeptides—Protein fragments with ten or more amino acids.
Primary structure—The number and sequence of the amino acids in the protein chain.
Secondary structure—The bending and coiling of the protein chain.

Of the 20 amino acids in proteins (see Table 5-1), nine either cannot be made in the body or cannot be made in the quantities needed. They must therefore be obtained in foods for the body to function properly. This is why we call these amino acids **essential** or **indispensable amino acids.** The remaining 11, called **nonessential amino acids,** can be made in the body. However, under certain circumstances, one or more of these amino acids may become essential.

When the amino-acid backbones join end to end, a protein forms (Figure 5-2). The bonds that form between adjoining amino acids are called **peptide bonds.** Proteins often contain from 35 to several hundred or more amino acids. Protein fragments with 10 or more amino acids are called **polypeptides.**

Each of the over 100,000 different proteins in the body contains its own unique number and sequence of amino acids. In other words, each protein differs in terms of what amino acids it contains, how many it contains, and the order in which they are contained. The number and sequence of the amino acids in the protein chain is called the **primary structure.** The number of possible arrangements is as amazing as the fact that all the words in the English language are made of different sequences of 26 letters. Also, some proteins are made of more than one chain of amino acids. For example, hemoglobin contains four chains of linked amino acids.

After a protein chain has been made in the body, it does not remain a straight chain. In the instant after a new protein is created, the side chain of each amino acid in the strand either attracts or repels other side chains, resulting in the protein either bending or coiling (Figure 5-3). This bending and coiling is called the protein's **secondary structure.**

One more step must take place before the protein can do any work in the body. Due in part to the interaction of amino acids at some distance

Figure 5-2
A part of a protein

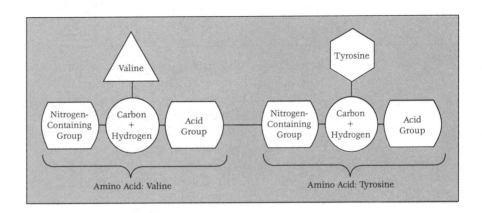

Figure 5-3
Primary structure—the number and sequence of amino acids; secondary structure—bends or coils; tertiary structure—folds and loops

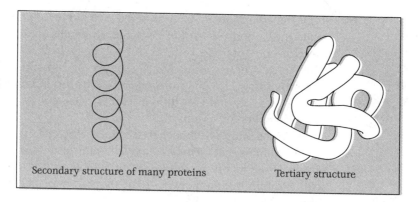

Secondary structure of many proteins   Tertiary structure

Tertiary structure—The folding of the protein chain.

from each other in the chain (Figure 5-3), the protein folds and loops. This process of folding results in the protein's **tertiary structure.** In case you are wondering whether a protein's tertiary structure has any real importance, it does! A protein's tertiary structure—how it bends and folds—makes the protein able to perform its functions in the body. Up to this point, the protein wasn't functional. So what exactly does a protein do? That's our next topic.

◼ **MINI-SUMMARY**

Proteins contain nitrogen. They are long chains of 20 different amino acids, some of which are essential, joined end to end by peptide bonds. Each protein has its own characteristic primary structure (number and sequence of amino acids), secondary structure (bending or coiling), and tertiary structure (folding and looping), which make it functional.

Functions of Protein

After reviewing all of the jobs proteins perform, you will have a greater appreciation of this nutrient. In brief, protein is part of most body structures; builds and maintains the body; is a part of many enzymes, hormones, and antibodies; transports substances around the body; maintains fluid and acid-base balance, and can provide energy for the body (Table 5-2). Now let's take a look at each function separately.

Proteins function as part of the body's structure. For example, protein can be found in skin, bones, hair, fingernails, muscles, blood vessels, the digestive tract, and blood. Protein appears in every cell.

TABLE 5-2 Functions of Protein

■ Acts as a structural component of the body
■ Builds and maintains the body
■ Found in many enzymes and hormones, and all antibodies
■ Transports iron, fats, minerals, and oxygen
■ Maintains fluid and acid-base balance
■ Provides energy as last resort

Proteins are used for building and maintaining body tissues. Worn-out cells are replaced throughout the body at regular intervals. For instance, your skin today will not be the same skin in a few months. It is constantly being broken down and rebuilt or remodeled, as are most body cells, including the protein within the cells. The greatest amount of protein is needed when the body is building new tissues rapidly, such as during pregnancy or infancy. Additional protein is also needed when body protein is either lost or destroyed, as in burns, surgery, or infections.

Proteins are found in many **enzymes,** some hormones, and all antibodies. Thousands of enzymes have been identified. Almost all the reactions that occur in the body, such as food digestion, involve enzymes. Enzymes are catalysts, meaning that they increase the rate of these reactions, sometimes more than a million times. They do this without being changed in the overall process. Enzymes contain a special pocket called the *active site.* You can think of the active site as a lock into which only the correct key will fit. Various substances will fit into the pocket, undergo a chemical reaction, and then exit the enzyme in a new form, leaving the enzyme to speed up other reactions.

Hormones are chemical messengers secreted into the bloodstream by various organs, such as the liver, to regulate certain body activities so a constant internal environment (called **homeostasis**) is maintained. For example, the hormone **insulin** is released from the pancreas when your blood sugar level goes up, such as after eating lunch. Insulin pushes sugar from the blood into your cells, resulting in lower, more normal blood sugar levels. Amino acids are components of insulin as well as other hormones.

Antibodies are blood proteins whose job is to bind with foreign bodies or invaders (scientific name is **antigens**) that do not belong in the body. Invaders could be viruses, bacteria, or toxins. Each antibody fights a specific invader. For example, there are many different viruses that cause the common cold. An antibody that binds with a certain cold virus is of no use to

Deoxyribonucleic acid (DNA)—The protein carrier of the genetic code in the cells.

Enzymes—Catalysts in the body.

Hormones—Chemical messengers in the body.

Homeostasis—A constant internal environment in the body.

Antibodies—Proteins in the blood that bind with foreign bodies or invaders.

Antigens—Foreign invaders in the body.

you if you have a different strain of the cold virus. However, exposure to a cold virus results in increased amounts of the specific type of antibody that can attack it. Next time that particular cold virus comes around, your body remembers and makes the right antibodies. This time the virus is destroyed faster, and your body's response (called the **immune response**) is enough to combat the disease.

Proteins also act as taxicabs in the body, transporting iron and other minerals, fats, and oxygen through the blood.

Protein also plays a role in body fluid balance (to be discussed in chapter 7), and the **acid-base balance** of the blood. Normal bodily processes produce acids and bases that can cause major problems, even death, if not buffered or neutralized. The blood must remain neutral; otherwise, dangerous conditions known as **acidosis** (above normal acidity) and **alkalosis** (above normal alkalinity) can occur. Some blood proteins have the chemical ability to buffer, or neutralize, both acids and bases.

In addition, amino acids can be burned to supply energy (4 calories per gram) if absolutely needed. Of course, burning amino acids for energy takes them away from their vital functions. Some amino acids can also be converted to glucose when necessary to maintain normal blood glucose levels. On the other hand, excess calories from protein will result in the same thing that always results from too many calories: fat.

Immune response—The body's response to a foreign substance, such as a virus, in the body.

Acid-base balance—The process by which the body buffers the acids and bases normally produced in the body so that the blood is neither too acidic nor too basic.

Acidosis—A dangerous condition in which the blood is too acidic.

■ **MINI-SUMMARY**

Protein is part of most body structures; builds and maintains the body; is a part of many enzymes, hormones, and antibodies; transports substances around the body; maintains fluid and acid-base balance; and provides energy for the body.

Denaturation

Under certain circumstances, a protein's shape is distorted, causing it to lose its ability to function. This process is called **denaturation.** In most cases, the damage cannot be reversed. Denaturation can occur both to proteins in food and to proteins in our bodies.

Denaturation can be caused by high temperatures (as in cooking), ultraviolet radiation, acids and bases, agitation or whipping, and high salt concentration. For example, when you fry an egg, the proteins in the egg white become denatured and turn from clear to white. Gluten, the protein in flour, denatures during baking to give bread and other baked goods their structure. Denaturation of protein can also occur in the body whenever the blood becomes too acidic or too basic.

> ■ **MINI-SUMMARY**
>
> When a protein is denatured, its shape gets distorted so it can no longer function. Protein foods denature during cooking, and proteins in the body can denature if the blood becomes too acidic or basic.

Digestion, Absorption, and Metabolism

Like lipids, proteins cannot be absorbed across the intestinal membranes until they are broken down into their amino-acid units. Protein digestion starts in the stomach, where stomach acid uncoils the proteins (denaturation) enough to allow enzymes to enter them to do their work. One stomach enzyme, **pepsin,** is the principal digestive enzyme. It splits peptide bonds, making proteins shorter in length.

Digestion is then completed in the small intestine, where pancreatic and intestinal enzymes work on releasing amino acids to be absorbed across the intestinal wall. Because they are water-soluble, amino acids easily travel in the blood to the liver and then to the cells that require them.

An **amino acid pool** in the body provides the cells with a supply of amino acids for making protein. The amino acid pool refers to the overall amount of amino acids distributed in the blood, the organs (such as the liver), and the body's cells. Amino acids from foods, as well as amino acids from body proteins that have been dismantled, stock these pools. In this manner, the body recycles its own proteins. If the body is making a protein and can't find an essential amino acid for it, the protein can't be completed, and the partially completed protein is disassembled or taken apart. This is important to consider for the next section on protein quality.

Pepsin—The principal digestive enzyme of the stomach.

Amino acid pool—The overall amount of amino acids distributed in the blood, organs, and body cells.

> ■ **MINI-SUMMARY**
>
> Protein digestion takes place in the stomach and small intestine, where stomach acid and enzymes help break up food proteins into amino acids to be absorbed across the wall of the small intestine. Because they are water-soluble, amino acids easily travel in the blood (as part of the amino acid pool) to the liver and cells that require them.

Protein in Food

Protein is found in animal and plant foods (Table 5-3). Protein is highest in animal foods, such as beef, chicken, fish, and dairy products. Of the plant foods, grains, legumes, and nuts usually contribute more protein than vegetables and fruits. Protein-rich foods are usually higher in fat and saturated fat, and always higher in cholesterol, than plant foods (plant foods have no cholesterol). Protein-rich foods also tend to be the most expensive foods on the menu.

To understand the concept of protein quality, you need to recall that nine of the 20 amino acids either can't be made in the body or can't be

TABLE 5-3 Fat, Saturated Fat, Protein, Cholesterol, and Fiber in Animal and Plant Foods

Animal Foods	Fat (grams)	Saturated Fat (grams)	Protein (grams)	Cholesterol (milligrams)	Fiber (grams)
Beef, ground, broiled, 3 oz.	16	6	21	74	0
Chicken breast, roasted, 3 oz.	3	1	27	73	0
Cod, baked, 3 oz.	1	0	19	47	0
Milk, 2%, 8 fl. oz.	5	3	8	18	0
Cheese, American, 1 oz.	9	6	6	27	0
Egg, 1	6	2	6	274	0
Plant Foods					
Lentils, cooked, 1/2 cup	0	0	8	0	5
Peanut butter, 2 tablespoons	16	3	10	0	2
Brown rice, cooked, 1/2 cup	1	0	2	0	1
Spaghetti, whole wheat, 1 cup	1	0	7	0	3
Whole-wheat bread, 2 slices	2	0	6	0	3
Broccoli, chopped 1 cup	0	0	6	0	3
Apple, 1 medium	0.5	0	0.3	0	3

Sources: United States Department of Agriculture Handbook Number 72, 8-1, 8-5, 8-13, 8-15, 8-20.

made in sufficient quantity. Food proteins that provide all of the essential amino acids in the proportions needed are called high-quality or **complete proteins.** Examples of complete proteins include the animal proteins, such as meats, poultry, fish, eggs, milk, and other dairy products.

An essential amino acid in lowest concentration in a protein is referred to as a **limiting amino acid** because it limits the protein's usefulness unless another food in the diet contains it. Lower-quality or **incomplete protein** contains at least one limiting amino acid. Plant proteins, including dried beans and peas, grains, vegetables, nuts, and seeds, are incomplete. When certain plant foods, such as peanut butter and whole-wheat bread, are eaten over the course of a day, the limiting amino acid in each of these groups is supplied by the other group. Such combinations are called **complementary proteins.**

Although plant proteins are incomplete and score lower than animal proteins, they are not low quality. When plant proteins are eaten with other foods, the food combinations usually result in complete protein. This is the case, for example, when grains are consumed with legumes. Some plant proteins, such as the grains amaranth and quinoa and protein made from soybeans (called "isolated soy protein"), are complete proteins. In adequate amounts and combinations, plant foods can supply the essential nutrients needed for growth and development and overall health. Many cultures around the world use plant proteins extensively. Plant protein foods contribute 65 percent of the protein for each person in the world. For North America alone, plant protein foods contribute only about 32 percent of the protein for each person.

Researchers have developed various ways to score the quality of food proteins. They judge them on how much of their nitrogen the body retains, or how well the proteins support growth or maintenance of body tissue. Animal proteins tend to score higher than vegetable proteins, and animal protein is also more digestible. Protein scores have little use in countries where protein consumption is adequate, but are useful to scientists working in countries where protein intakes are low.

■ MINI-SUMMARY

Animal proteins are examples of complete proteins, and most plant proteins are examples of incomplete proteins. By eating complementary plant proteins, you can overcome the problem presented by limiting amino acids and eat a nutritionally adequate diet. In the right amounts and combinations, plant proteins can support growth and maintenance.

Protein and Health

If you eat too much or too little protein, your health may be affected. First let's look at a high-protein diet. Eating too much protein has no benefits. It will not result in bigger muscles, stronger bones, or increased immunity. In fact, eating more protein than you need may add excessive calories beyond what you require. Extra protein is not stored as protein but is stored as fat if too many calories are being taken in.

Diets high in protein can also be a concern if you are eating a lot of high-fat animal proteins, such as bacon and hamburger, and few vegetable proteins. Comparison of the fat and fiber content of animal and vegetable proteins (Table 5-3) makes clear that plant sources of protein contain less fat and more fiber. They also contain no cholesterol and are rich in vitamins and minerals. Eating too much high-fat animal protein can raise your blood cholesterol levels, which in turn increases your risk of cardiovascular disease. By eating fewer plant proteins, you are also missing out on good sources of fiber and antioxidant nutrients (see Chapter 6), which may protect against cancer.

Published studies show that increased protein intake leads to increased calcium loss. This does not necessarily mean that everyone who takes in too much protein is calcium deficient, since the body will make up for this loss by absorbing more calcium in the intestine. However, if an individual has a high protein intake and a low calcium intake, the increased calcium absorption won't compensate enough for its loss. Published studies also show that high protein intakes tax the kidneys and can worsen kidney problems in patients with renal (kidney) disease.

Protein-energy malnutrition (PEM)—A broad spectrum of malnutrition from mild to serious cases.

Kwashiorkor—A type of PEM associated with children with insufficient protein intake and who have a preexisting disease.

Marasmus—A type of PEM characterized by gross underweight and severe food shortage.

On the other hand, eating too little protein can cause problems too, such as slowing down the protein rebuilding and repairing process and weakening the immune system. Developing countries have the most problems with **protein-energy malnutrition (PEM).** PEM refers to a broad spectrum of malnutrition, from mild to serious cases. PEM can occur in infants, children, adolescents, and adults, although it is seen most often in infants and children. PEM develops gradually over weeks or months. In mild cases of PEM, there is weight loss, stunted or slowed growth, and more sedentary behavior.

In severe cases of PEM, physicians often see the clinical syndromes called kwashiorkor and marasmus. **Kwashiorkor** is usually seen in children with an existing disease who are getting totally inadequate amounts of protein and only marginal amounts of calories. Characterized by retarded growth and development, the child has a protruding abdomen due to edema (swelling), peeling skin, a loss of normal hair color, irritability, and sadness.

Marasmus is characterized by severe insufficiency of calories and protein, which accounts for the child's gross underweight, lack of fat stores, and wasting away of muscles. There is no edema. Whereas marasmus is usually associated with severe food shortage and prolonged semistarvation, kwashiorkor is associated with poor protein intake and early weaning from mother's milk due to arrival of a new baby.

■ **MINI-SUMMARY**

There is no benefit to eating too much protein. Eating too much high-fat animal protein can increase your blood cholesterol levels, increase calcium loss from the body (a concern when calcium intake is low), and worsen kidney problems in people with renal disease. Eating too little protein is associated with protein-energy malnutrition.

Dietary Recommendations for Protein

New Dietary Reference Intakes have not yet been established. The 1989 RDA for protein will be used until new DRIs are announced. The RDA for protein for a 174-pound man is 63 grams. For a 138-pound woman, the RDA is 50 grams. For healthy adults, the RDA works out to be 0.36 grams of protein per pound of body weight. This allows for adequate protein to make up for daily losses in urine, feces, hair, and so on. In other words, taking in enough protein each day to balance losses results in a state of protein balance, called nitrogen balance. The RDA for protein is generous and is based on the recommendation that proteins come from both animal and plant foods.

Positive nitrogen balance—A condition in which the body excretes less protein than is taken in; this can occur during growth and pregnancy.

Negative nitrogen balance—A condition in which the body excretes more protein than is taken in; this can occur during starvation and certain illnesses.

The amount of protein needed daily is proportionally higher during periods of growth, such as during pregnancy and infancy. During these periods, a person needs to eat more protein than is lost, a condition known as **positive nitrogen balance. Negative nitrogen balance** occurs during starvation and some illnesses when the body excretes more protein than is taken in.

In the United States, meeting the RDA for protein is rarely a problem. According to U.S. Department of Agriculture (USDA) surveys, 14 to 18 percent of calories in the American diet come from protein, with animal proteins contributing about 65 percent (USDA, 1983, 1986, and 1987). Using the RDA for protein and energy as a guide, the percent of calories from protein should be lower, from 10 to 12 percent.

■ **MINI-SUMMARY**

Most Americans eat more than the RDA for protein. More protein is needed during periods of growth and positive nitrogen balance. Negative nitrogen balance occurs during starvation and some illnesses.

Ingredient Focus: Milk, Dairy Products, and Eggs

Milk

The age of the long-necked glass bottle containing milk that required shaking prior to consumption was not so long ago. In the early stages of milk processing, milk was **pasteurized** (heated to kill harmful germs) but not homogenized. When this milk was allowed to stand, the lighter fat would float to the top and the thinner milk would settle below. In other words, the top contained cream, and the bottom clear liquid was skim milk. By shaking the bottle prior to use, the consumer created a "temporary emulsion" (a combination of fat and liquid) and the milk would become smooth and blended. Since fat and liquid do not combine permanently under normal circumstances, milk would start to separate within a few minutes after shaking. Today's milk products are **homogenized** to create a permanent bond between the milk and cream. In addition, the cardboard containers help to better preserve some nutrients.

Milk is a good source of:

- high-quality protein
- carbohydrate
- riboflavin
- vitamins A and D (if fortified)
- calcium and other minerals such as phosphorus, magnesium, and zinc

Different types of fluid milk (see Table 5-4) contain varying amounts of fat. In addition to whole, reduced fat, low-fat and fat-free, there are also the following.

- **Cultured Buttermilk.** Buttermilk is made most often by adding a bacterial culture to fresh, pasteurized skim milk. The bacteria convert the sugar in milk (lactose) into lactic acid, thereby giving buttermilk a thick consis-

TABLE 5-4	Nutrient Composition of 1 Cup of Milk			
Name	Calories	Saturated Fat	Fat	Calcium
Whole milk	150	8 g	5 g	300 mg
Reduced-fat milk, also called 2% reduced-fat milk	120	5 g	3 g	300 mg
Lowfat milk, also called 1% lowfat milk	100	2.5 g	1.5 g	300 mg
Fat-free milk, also called skim or nonfat milk	80	0 g	0 g	300 mg

tency and a tart and buttery taste. Buttermilk is enjoyed as a beverage and is used in baking when sour milk is needed. It is also used in making salad dressings and cold soups.

■ **Eggnog.** Eggnog is a mixture of dairy ingredients (cream, milk, partially skimmed milk or skim milk), eggs, and sweeteners. It may be flavored with rum extract, nutmeg, vanilla, or other flavorings. Commercial eggnog is pasteurized, so there shouldn't be any concern about the safety of the egg yolks in this product.

■ **Lactase-treated milk.** Milk that has been treated with the enzyme lactase is particularly helpful for individuals who have lactose intolerance. Lactose intolerance, or lactase deficiency, is a disease caused by a lack of the enzyme lactase, which is needed by the intestine to split lactose into its two components for absorption. After drinking milk or eating a dairy product with lactose, the lactose-intolerant individual experiences symptoms that include abdominal cramps, bloating, and diarrhea. Lactose-treated milk is otherwise nutritionally identical to regular milk.

When purchasing milk, decide on the fat content you want, and specify U.S. Grade A milk that has been pasteurized, homogenized, and fortified with vitamins A and D. Fresh milk is very perishable and should be stored in the refrigerator for up to four days after the pull date (whole milk stays fresher longer than either skim or low-fat milks).

Milk is also available with some or all of its water content removed.

■ **Evaporated milk and evaporated skim milk.** These products are made by heating milk to stabilize the milk protein, then removing about 60 percent of the water. Both products are sold in cans. Evaporated milk has 7.5 percent milkfat; evaporated skim milk has no more than 0.5 percent milkfat. They are used mostly in cooking and baking. One-half cup of either product can be reconstituted with one-half cup of water to make 1 cup of milk.

■ **Sweetened condensed milk and skimmed milk.** These products also have about 60 percent of their water removed. What makes them different is that they are heavily sweetened, usually with sugar. Sweetened condensed milk has 8 percent milkfat; sweetened condensed skimmed milk has no more than 0.5 percent milkfat. They are used mostly in cooking and baking.

■ **Nonfat dry milk (powdered milk).** This product is made by removing the water from pasteurized skim milk. It can be easily reconstituted with water—3 tablespoons of the dry milk to 8 ounces of water. Although it doesn't have the taste of fluid milk, it is acceptable for use in baking and cooking.

Evaporated, condensed, and nonfat dry milk can all be stored unopened at room temperature for at least one year or, if available, you can use the dating code marked on the container. Once opened, evaporated and condensed milk should be poured into another container, refrigerated, and used within five days. Once reconstituted, nonfat dry milk should be refrigerated and used within five days.

Milk products are a very important part of many recipes. The addition of milk products can add a creamy consistency to many sauces, help dissolve dry ingredients in baked products, or simply add good nutrition and flavor to a food. Substituting low-fat or skim milk for whole milk works well in many cooking and baking recipes to reduce the fat content. Evaporated skim milk can be used in place of half-and-half in some recipes such as soups.

When cooking with milk, remember a very important rule—use a moderate heat and heat the milk slowly (but not too long) to avoid "curdling"—a grainy appearance with a lumpy texture. From a scientific point of view, milk curdles when the casein (protein in milk) separates out of the milk. Add other food products to hot milk products slowly, either with a spoon or a wire whisk, if preparing a sauce, to avoid lumps. Be especially careful when adding foods high in acid—milk has a tendency to curdle if not beaten quickly.

Another problem with milk is the creation of a top layer of "skin"—a film that forms on top of the milk during cooking. It can be prevented by keeping the pot of milk covered. A final problem with milk is scorching—when milk sticks to the bottom of the pot during cooking and eventually starts to burn, making the food you are cooking taste terrible. Because milk is very sensitive to direct heat, heat it slowly; in some cases you will need to heat milk in a double boiler.

Cheese

Cheese is a versatile food and can be used in many different applications. Cheese flavors range from mild to very sharp, and cheese can be not only an important ingredient in an end product, but can be consumed as a snack.

Cheese is made from various types of milks—cow's and goat's being the most popular. In many parts of the world cheese is produced from various sources of milk: sheep, reindeer, the yak, buffalo, camel mares, and donkeys. Cheese is produced when bacteria or rennet (or both) are added to milk and the milk then curdles. The liquid, known as the whey, is separated from the solid, known as the curd, which is the cheese. It is thought that cheese was discovered by accident thousands of years ago in the ancient Far East.

Cheese is an excellent source of nutrients such as protein and calcium. However, because most cheeses are prepared from whole milk or cream, they are also high in saturated fat and cholesterol. Ounce for ounce, meat, poultry, and most cheeses have about the same amount of cholesterol. But cheeses tend to have much more saturated fat.

Determining which cheeses are high or low in saturated fat and cholesterol can be confusing, because there are so many different kinds on the market: part-skim, low-fat, processed, and so on. Not all reduced-fat or part-skim cheeses are always low in fat; they are only lower in fat than similar natural cheeses. For instance, one reduced-fat Cheddar gets 56 percent of its calories from fat—considerably less than the 71 percent of regular Cheddar, but not super lean either. The trick is to read the label. Table 5-5 is a guide to fat in cheeses.

Cheese can be divided into different categories, depending on texture or whether it has been ripened (also called cured). The texture of cheese varies from soft to semisoft, firm, and hard. *Ripened cheeses* are those that have been fermented with bacteria or molds. By contrast, *unripened cheeses* are fresh and untreated. Unripened cheeses are generally more perishable than ripened cheeses. The processing time is very short and no time is needed for ripening. When the cheese is manufactured, it is packaged and sold with a relatively short expiration date compared with ripened cheeses. Popular types of unripened cheeses include ricotta cheese and cottage cheese.

Ripened cheeses need many months to age and develop into a marketable cheese. The longer they age, the sharper their flavor—and the higher their cost. Cheddar and Swiss are popular types of ripened cheese. Cheddar cheese comes in many varieties, ranging in flavor from very mild to very sharp. Cheddar is used as an appetizer, in sandwiches, and as a dessert cheese. Two kinds of imported Swiss are frequently used in the kitchen: Emmenthaler and Gruyère. Domestic Swiss is usually used in sandwiches.

TABLE 5-5 Guide to Fat in Cheeses†

Lowfat 0–3 g fat/oz	Medium Fat 4–5 g fat/oz	High Fat 6–8 g fat/oz	Very High Fat 9–10 g fat/oz
Natural Cheeses			
*Cottage Cheese (1/4 c) Dry curd	*Mozzarella Part skim	Blue Cheese	Cheddar
Cottage Cheese (1/4 c) Lowfat 1%	*Ricotta (1/4 c) Part skim	*Brick	Colby
Cottage Cheese (1/4 c) Lowfat 2%	String cheese Part skim	Brie	*Cream Cheese (1 oz = 2 Tbsp)
Cottage Cheese (1/4 c) Creamed 4%		Camembert	Fontina
Sap Sago		Edam	*Gruyere
		Feta	Longhorn
		Gjetost	*Monterey Jack
		Gouda	Muenster
		*Light Cream Cheese (1 oz = 2 Tbsp)	Roquefort
Look for special low-fat brands of mozzarella, ricotta, Cheddar, and Monterey jack.	Look for reduced-fat brands of Cheddar, colby, Monterey jack, muenster, and Swiss.	Limburger	
		Mozzarella, whole milk	
		Parmesan (1 oz = 3 Tbsp)	
		*Port du Salut	
		Provolone	
		*Ricotta (1/4 c), whole milk	
		Romano (1 oz = 3 Tbsp)	
		*Swiss	
		Tilsit, whole milk	
Modified Cheeses			
Pasteurized process, imitation, and substitute cheeses with 3 g fat/oz or less.	Pasteurized process, imitation, and substitute cheeses with 4–5 g fat/oz.	Pasteurized Process Swiss cheese	Some pasteurized process cheeses are found in this category—check the labels.
		Pasteurized Process Swiss cheese food	
		Pasteurized Process American cheese	
		Pasteurized Process American cheese food	
		American cheese food cold pack	
		Imitation and substitute cheeses with 6–8 g fat/oz.	

† Check the labels for fat and sodium content. 1 serving = 1 oz. unless otherwise stated.
* These cheeses contain 160 mg or less of sodium per 1 oz.

Source: Reprinted by permission of the American Heart Association, Alameda County Chapter, 11200 Golf Links Road, Oakland, CA 94605.

When a cheese or combination of cheeses is manufactured into another type of cheese, this is referred to as a *process cheese.* Whereas natural cheeses are made directly from milk or whey, process cheese is a modified form of natural cheeses that have been ground or shredded from a variety of natural cheeses. Processed cheese takes many forms.

- **Pasteurized process cheese**—a combination of Cheddar and other cheeses with emulsifiers to make it smooth (for this reason, these cheeses melt better); includes American cheese
- **Pasteurized process cheese foods and spreads**—made like process cheese except for the addition of optional ingredients such as cream, milk, buttermilk, nonfat dry solids, or whey; these contain more moisture and less fat than process cheese
- **Cold pack cheese**—made by mixing ripened cheeses without heat

American cheese is probably the most well-known process cheese. It is excellent in sandwiches and melted on cheeseburgers. Table 5-6 is a guide to many of the different types of cheeses available.

Buy the best quality cheese that you can. Some cheeses are graded by the USDA. The four grades are AA, A, B, or C. All Cheddar cheese that is graded must also show the cure category.

Mild: Cured for two to three months.
Mellow aged: Cured for four to seven months.
Sharp: Full ripened—cured for eight to twelve months.
Very Sharp: Aged over twelve months.

When receiving cheese, check for mold—even when it is in airtight packaging. Mold commonly appears as white, blue, or green fuzzy spots. Also check for dryness—the cheese will have darker edges if this is the case. Many processed cheeses are dated with sell-by or use-by dates, so check for one. Make sure the cheese has been kept cold. Only four types of cheese do not require refrigeration—cold pack cheese, cold pack cheese food, pasteurized process cheese food, and pasteurized process cheese spread—until opened.

Because of their fat content, many cheeses readily absorb refrigerator odors, resulting in a poor-tasting product. Therefore, store all cheeses in tight plastic wrap or foil (except blue cheese, which should be wrapped loosely). Cheeses also need to be wrapped tightly to prevent them from drying out. Hard and firm cheeses keep from a week to several months in the refrigerator, while semisoft and soft cheeses are much more perishable and keep only one to two weeks. For unripened cheeses, such as cottage cheese and cream cheese, check the "sell by" or "use by" date. These products should last at least several days beyond the sell-by date.

TABLE 5-6 A Guide to Cheeses

Cheese	Characteristics	Uses
UNRIPENED		
Cottage	Mild, slightly acid flavor; soft, open texture with tender curds of varying size; white to creamy white	Appetizers, salads, cheesecakes, dips
Cream	Delicate, slightly acid flavor; soft, smooth texture; white	Appetizers, salads, sandwiches, desserts, and snacks
Neufchâtel	Mild, acidic flavor; soft, smooth texture similar to cream cheese but lower in fat; white	Salads, sandwiches, desserts, snacks, dips
Ricotta	Mild, sweet, nutlike flavor; soft, moist texture with loose curds (fresh ricotta) or dry and suitable for grating; white	Salads, main dishes such as lasagne and ravioli, and desserts, mostly in cooked dishes
SOFT, RIPENED		
Bel Paese	Mild, sweet flavor; light, creamy-yellow interior; slate-gray surface; soft to medium-firm, creamy texture	Appetizers, sandwiches, desserts, and snacks
Brie	Mild to pungent flavor; soft, smooth texture; creamy-yellow interior; edible thin brown and white crust	Appetizers, sandwiches, desserts, snacks, salads
Camembert	Distinctive mild to pungent flavor; soft, smooth texture—almost fluid when fully ripened; creamy-yellow interior; edible thin white or gray-white crust	Appetizers, desserts, and snacks
Limburger	Highly pungent, very strong flavor and aroma; soft, smooth texture that usually contains small irregular openings; creamy-white interior; reddish-yellow surface	Appetizers, desserts, snacks, sandwiches
SEMISOFT, RIPENED		
Blue	Tangy, piquant flavor; semisoft, pasty, sometimes crumbly texture; white interior marbled or streaked with blue veins of mold; resembles Roquefort	Appetizers, salads and salad dressings, desserts, and snacks
Brick	Mild, pungent, sweet flavor; semisoft to medium-firm, elastic texture; creamy white-to-yellow interior; brownish exterior	Appetizers, sandwiches, desserts, and snacks
Gorgonzola	Tangy, rich, spicy flavor; semisoft, pasty, sometimes crumbly texture; creamy-white interior, mottled or streaked with blue-green veins of mold; clay-colored surface	Appetizers, salads, desserts, and snacks
Mozzarella (also called Scamorza)	Delicate, mild flavor; slightly firm, plastic texture; creamy white	Main dishes such as pizza or lasagne, sandwiches, snacks, and salads

178

Cheese	Description	Uses
Muenster	Mild to mellow flavor; semisoft texture with numerous small openings; creamy-white interior; yellowish-tan or white surface	Appetizers, sandwiches, desserts, and snacks
Port du Salut	Mellow to robust flavor similar to Gouda; semisoft, smooth elastic texture; creamy white or yellow	Appetizers, desserts, and snacks
Roquefort	Sharp, peppery, piquant flavor; semisoft, pasty, sometimes crumbly texture; white interior streaked with blue-green veins of mold	Appetizers, salads and salad dressings, desserts, and snacks
Sapsago	Sharp, pungent, clover-like flavor; very hard texture suitable for grating; light green or sage green	Grated for seasoning
Stilton	Piquant flavor, milder than Gorgonzola or Roquefort; open, flaky texture; creamy-white interior streaked with blue-green veins of mold; wrinkled, melon-like rind	Appetizers, salads, desserts, snacks, in cooked foods
HARD, RIPENED		
Cheddar (often called American)	Mild to very sharp flavor; smooth texture, firm to crumbly; light cream to orange	Appetizers, main dishes, sauces, soups, sandwiches, salads, desserts, and snacks
Colby	Mild to mellow flavor, similar to Cheddar; softer body and more open texture than Cheddar; light cream to orange	Sandwiches, snacks, cooked foods
Edam	Mellow, nutlike, sometimes salty flavor; rather firm, rubbery texture; creamy-yellow or medium yellow–orange interior; surface coated with red wax; usually shaped like a flattened ball	Appetizers, salads, sandwiches, sauces, desserts, and snacks
Gouda	Mellow, nutlike, often slightly acid flavor; semisoft to firm, smooth texture, often containing small holes; creamy-yellow or medium yellow–orange interior; usually has red wax coating; usually shaped like a flattened ball	Appetizers, salads, sandwiches, sauces, desserts, and snacks
Gruyère	Nutlike, salty flavor, similar to Swiss, but sharper; firm, smooth texture with small holes or eyes light yellow	Appetizers, desserts, snacks, fondue and other cooked dishes
Monterey (Jack)	Semisoft; smooth, open texture, mild flavor; Cheddar-like; hard when aged	Appetizers, sandwiches, salads
Parmesan	Sharp, distinctive flavor; very hard, granular texture; yellowish white	Grated on cooked (Italian) dishes, salads, as seasoning
Provolone	Mellow to sharp flavor, smoky and salty; firm, smooth texture; cuts without crumbling; light creamy yellow; light-brown or golden-yellow surface	Appetizers, main dishes, sandwiches, desserts, and snacks
Romano	Very sharp, piquant flavor; very hard, granular texture; yellowish-white interior; greenish–black surface	Seasoning and general table use; when cured a year, it is suitable for grating
Swiss (also called Emmenthaler)	Mild, sweet, nutlike flavor; firm, smooth, elastic body with large round eyes; light yellow	Sandwiches, salads, snacks, fondue and other cooked dishes

There are several ways to use cheese in light and healthy cooking.

- Use a regular cheese with a strong flavor, and use less of it than called for in the recipe.
- Use less cheese.
- Substitute low-fat cheeses for regular ones.
- Use a mixture of half regular cheese and half low-fat cheese.

When cooking with cheese, observe some simple guidelines to come out with a tasty dish.

1. *Use low heat.* It is best to use as low a heat as possible when cooking with cheese. Cheese has a tendency to toughen when subjected to high heat, due to its high protein content. Avoid "boiling" at all costs.
2. *Use short cooking times.* Most recipes will require the addition of cheese at the end of the recipe to avoid overcooking. Remember to stir often to enhance the blend of flavors and establish a good, smooth consistency.
3. *Grate the cheese.* The best way to add a cheese to a recipe is to grate it. Grating will break the cheese into small, thin pieces that will melt and blend quickly and evenly into the end product.

Cream

Cream is used in many ways in the kitchen. It is used in sauces, soups, hot and cold beverages, baked goods, and, when whipped, as a topping for desserts and hot beverages.

There are several types of cream. Most have been pasteurized. In some cases you may find ultrapasteurized cream at the supermarket. In comparison to pasteurized cream, ultrapasteurized cream has a much longer shelf life but does not whip as well.

- **Heavy whipping cream or heavy cream.** Heavy cream is a very thick, semi-fluid liquid that is at least 36 percent fat (by weight). It is often used as a topping after a short period of whipping and can be used in sauces. Remember not to overbeat heavy cream, or it will turn into butter!
- **Light whipping cream or whipping cream.** Light whipping cream contains 30 to 35 percent milkfat. Whipping cream labeled as ultrapasteurized does not whip as well as regular whipping cream, but does have a longer shelf life.
- **Light cream.** Light cream is also called table cream or coffee cream. It has a fat content between 18 and 30 percent (usually it's 18 percent). It, too, has a strong, rich flavor and is often used as an ingredient in both sauces and baked products. It can't be successfully whipped.

- **Half-and-half.** Half-and-half (half milk, half cream) normally has a fat content of 10 to 12 percent. Its consistency is heavy and the flavor is still rich. Like light cream, it can't be successfully whipped.
- **Sour cream.** Sour cream is made from pasteurized cream (with 18 percent milkfat content) to which bacteria are added. The bacteria convert the milk sugar, lactose, into lactic acid, thereby giving the product its thick consistency and tangy taste.

Although cream does contain some nutrients, it mostly supplies fat (see Table 5-7).

Always purchase high-quality, very fresh cream. Be sure the container is clean, tightly sealed, and cold. Check for a date code on it as well. Cream is very perishable and should be kept refrigerated in its closed carton. Do not leave cream out on the kitchen table while you are having your morning coffee—keep it cold! Observe the use-by date on the carton. If sell-by dates are used, allow three to four days beyond the date. For sour cream, allow about ten days past the sell-by date. If sour cream gets moldy, even if it is just a few dots, throw it out. Cream that has been ultrapasteurized can be kept refrigerated (unopened) for up to six weeks. Once opened, it can keep one week.

If you want *real* whipped cream (in moderation anything is fine), use light whipping cream, or drained yogurt and whipped egg whites (described under Chef's Tips). Make sure the cream is cold, as cold cream whips better than warm cream. For best results, place the cream, bowl, and beaters into the freezer for about ten minutes before whipping. Use a bowl that is deep enough to accommodate the beaters and small enough for the beaters to

	TABLE 5-7 Calories and Fat in Milk and Cream		
Product	Serving Size	Calories	Fat (grams)
Skim milk	1 cup	86	0.5
1% milk	1 cup	102	3
2% milk	1 cup	121	5
Whole milk	1 cup	150	8
Half-and-half	1 tablespoon	20	2
Light cream (coffee cream)	1 tablespoon	29	3
Light whipping cream, fluid	1 tablespoon	44	5
Heavy whipping cream, fluid	1 tablespoon	52	6
Sour cream	1 tablespoon	26	3

Source: U.S.D.A.

maintain contact with the cream. Beat rapidly, scraping the bowl frequently for two to three minutes, until you get stiff peaks. Don't overbeat—the product will be granular and turns into butter. That is all it usually takes. If you are adding sugar to the cream, do so after whipping, because it makes the cream harder to whip and less stable. Also, use confectioner's rather than granulated sugar for a smoother product. One cup of whipping cream yields two cups whipped. It is best to use whipped cream right away, but it can be put into the refrigerator, covered, for a few hours.

When cooking with sour cream, don't let it boil, or it may curdle due to the high heat. To prevent separation, you can mix in 1 tablespoon of flour per 1/2 cup of sour cream before cooking.

Ice Cream and Ice Milk

In order to be labeled as ice cream, a product must have at least 10 percent fat by weight. Premium ice creams have much more than the minimum required: about 16 percent fat. By comparison, ice milk, which is prepared from the same ingredients as ice cream, must have at least 3 percent fat. Ice milk usually has more sugar added.

Ice cream and ice milk are a source of calcium, riboflavin, and protein. Ice cream is also a significant source of fat, as seen in Table 5-8. Ice milk has much less fat.

Ice cream quality varies in three ways:

■ **The amount of fat.** The more fat an ice cream contains, the richer it tastes and the more expensive it is.
■ **The kind of flavoring.** Ice cream contains natural flavorings, artificial flavorings, or a combination of both. Natural flavorings make a better-quality ice cream that is then more expensive. If only natural flavorings are used, for example, the vanilla product is called vanilla ice cream; if

TABLE 5-8 Calories and Fat in Ice Cream, Ice Milk, and Frozen Yogurt

Dessert	Serving Size	Calories	Fat (grams)
Premium vanilla ice cream (16% fat)	1 cup	349	24
Vanilla ice cream (10% fat)	1 cup	269	14
Vanilla ice milk	1 cup	184	6
Vanilla soft-serve ice milk	1 cup	223	5
Frozen yogurt, vanilla	1 cup	240	varies

both natural and artificial flavorings are used, it is called vanilla-flavored ice cream; if only artificial flavorings are used, it is called artificially flavored ice cream.

■ **The amount of air in the ice cream.** All ice cream has air in it—less than half of the product can be air if it is to be labeled as ice cream. Air cells act as a cushion that keep the ingredients from forming into a solid, icy mass. The cranking of an ice-cream freezer whips air into the mixture. Premium ice creams have less air in them than lesser brands. That's why premium ice creams always weigh more than equivalent volumes of cheaper brands.

Make sure any ice cream or ice milk you receive is rock solid and the container is clean. If it feels sticky or looks frosty, it has probably thawed and refrozen. Ice cream and ice milk last about two months in the freezer. For best quality, it is best to use them within two weeks.

Yogurt

Yogurt is one of the oldest fermented milks known. Yogurt is cultured with the live active cultures lactobaccillus bulgaricus and streptococcus thermophilus. In the modern commercial production of yogurt, the milk base for the product is pasteurized to condition it for fermentation. Only then are the culture organisms added. After approximately three hours of incubation at about 110 degrees F., the milk acquires its custardlike texture. After cooling, the product is ready for distribution. Yogurt is not required to contain live bacterial cultures. Yogurt with live, active bacterial cultures will state this fact on the label.

Yogurt is a good source of:

■ Protein
■ Calcium, phosphorus, and potassium
■ Riboflavin and vitamin B_{12}

Yogurt contains the same fats that are found in the milk product it was made from. Whole-milk yogurts contain at least 3.5 grams fat per 100 grams (3-1/2 ounces) of yogurt, low-fat contains between .5 and 2 grams fat, and nonfat yogurts contain less than .5 grams fat.

There are three main types of yogurt to choose from.

1. Unflavored, plain yogurt
2. Flavored, containing no fruit (such as vanilla, lemon, coffee)
3. Flavored and containing fruit, which may be of two styles:
   ■ **Sundae style**—fruit is at the bottom of the container with plain or flavored yogurt on top; this product is normally stirred before eating

▪ **Blended style (also called Swiss style)** — fruit is blended throughout plain or flavored yogurt

The calorie content of fresh yogurt (see Table 5-9) varies dramatically, due to its fat content and whether fruit is added. Many yogurts with fruit also have a lot of sugar or other sweetener added to help preserve the fruit. Plain yogurt with fresh fruit has much less sugar and fewer calories.

Like fresh yogurt, frozen yogurt varies in the amount of fat it contains. On the high end are yogurts with about 6 grams of fat per 6 ounces (still lower than an equivalent amount of vanilla ice cream, with 10 grams of fat). On the low end are nonfat yogurts. Most frozen yogurt does not contain nearly the number of live bacteria (if they're alive at all) that fresh yogurt does. There are no federal standards for frozen yogurt, and many brands do not use live culture. Even when live bacteria are used, their numbers are far below those found in fresh yogurt. With or without active cultures, frozen yogurt still is a healthier option than ice cream (that is, of course, without the addition of pieces of candy bars, chocolate chips, etc.).

When yogurt is received, see that it is fresh and the containers are clean. Store fresh yogurt in the refrigerator up to the use-by date, or seven to ten days past the sell-by date. If you open the container and find some liquid sitting on top of the yogurt, don't worry, it's still safe to eat. Just drain off the liquid or stir it into the yogurt.

Plain yogurt, either nonfat or low-fat, can be substituted for many higher-fat ingredients in salad, salad dressings, soups, sauces, and desserts. Here are some examples.

▪ Substitute plain yogurt for mayonnaise. In situations where the taste of mayonnaise is desired, use half reduced-calorie mayonnaise and half yogurt. This works well for dishes such as potato salad, coleslaw, tuna salad, cold pasta salads, and appetizers.

TABLE 5-9 Calories and Fat in Yogurt (1-cup portion)		
Type of Yogurt	Calories	Fat (grams)
Whole milk, plain	139	7
Low-fat, plain	144	4
Low-fat, vanilla or coffee flavored	194	3
Low-fat, fruit flavored	225	3
Nonfat, plain	127	0

- Substitute plain yogurt for sour cream in dips and salad-dressing recipes.
- In baking, substitute yogurt for sour cream in recipes for pancakes, waffles, loaf breads, and muffins.
- In cooking, substitute 1 cup yogurt for 1 cup sour cream.
- On baked potatoes, offer plain yogurt mixed with fresh herbs instead of sour cream.
- Mix plain yogurt with ricotta cheese (made from skim milk) and use as a spread on toast, bagels, and crackers.

Where you want a creamy texture, drain the yogurt first to remove some of the liquid. Because of yogurt's acidity, you may want to decrease the amount of other acidic ingredients in your recipe, such as lemon juice.

When cooking with yogurt, use only low heat. High temperatures may cause separation, evaporation of liquid, and a curdled appearance. To help keep yogurt from separating during cooking, blend one tablespoon of cornstarch (unless the recipe calls for flour to be mixed with it) with a few tablespoons of yogurt, then stir the mixture into the remaining yogurt to be used, and proceed according to the recipe. Yogurt might also become thin if it is overmixed, so do not overstir.

Eggs

Eggs are truly a unique food product—they provide versatility and can be used for any meal. Eggs not only are excellent traditional breakfast foods, but are used in many breads, pies, cakes, custards, beverages, and entrées, to name a few.

Eggs are very nutritious and full of high-quality protein, as well as varying amounts of many vitamins and minerals. The concern with overconsumption of eggs stems from the fact that they are very high in cholesterol—215 milligrams per egg (compare that to the suggested maximum of 300 milligrams to eat daily). One egg also contributes 5 grams of fat, of which 2 grams are saturated fat.

Eggs are sold according to their grade and size. Standards for both grade and size are established by the U.S. Department of Agriculture. "Grade" refers to the quality of the egg and the shell when it is packed. Grades are AA, A, and B. The grade of an egg has nothing to do with whether it is nutritious or wholesome. Instead, grade is based on freshness, as well as the interior quality of the egg white and yolk. Lower-grade eggs have a thinner white and flatter yolk than do higher-grade eggs. U.S. Grade A is the most common grade available and has a 30-day shelf life. Grade AA is not often seen because that grade must have a 10-day expiration date.

Eggs come in various sizes: jumbo, extra large, large, medium, small, and peewee. Eggs are sized by weight, so in a given box, some eggs may be below size and others above; one dozen eggs must meet the minimum weight per dozen set for the marked size.

When eggs are received, check the carton to make sure it contains only clean, uncracked eggs. Make sure the eggs have been refrigerated.

There is no difference between brown- and white-shelled eggs. Shell color is determined by the breed of the hen, and it does not affect the grade, nutritive value, flavor, or cooking performance of the egg. Brown eggs are often more expensive because they come from larger hens that require more food.

Store eggs in the carton they came in, because the carton will help keep out odors the eggs might absorb and it also helps prevent the loss of carbon dioxide and moisture from the eggs, which causes them to age quicker. Keep eggs refrigerated, as they maintain freshness better this way. Eggs should be stored with the large end up to keep the yolk centered. As an egg ages, the yolk flattens and the thick section of the white becomes watery and thins out. Stale eggs also usually have a "rotten egg" odor that smells like sulfur.

Eggs can be cooked in many ways: baked, cooked in the shell, poached, fried, and scrambled. When making fried eggs, scrambed eggs, and omelets, the use of nonstick pans and vegetable cooking sprays are important to keep down the amount of fat. When cooking eggs, always use low to medium temperatures to prevent overcooking. Overcooked eggs are tough and rubbery.

Egg substitutes are available that are low in cholesterol, but not always low in fat. They are often made from egg whites and vegetable oil.

Besides using egg substitutes, there are other ways to make egg dishes with less cholesterol.

1. To make scrambled eggs and omelets, use one whole egg, and add two egg whites for each additional egg.
2. In baking, replace one whole egg with two egg whites, and two whole eggs with one whole egg and two egg whites.
3. Use 1/2 cup egg substitute to equal one whole egg.

CHEF'S TIPS
- ■ To make whipped cream, drain plain fat-free yogurt in cheesecloth to remove as much liquid as possible. Fold whipped egg whites into the yogurt and add a little honey for flavor. Use frozen pasteurized egg whites to avoid any food safety (salmonella) problem.
- ■ To make an excellent omelet without cholesterol, whip egg whites until they foam. Add a touch of white wine, fresh mustard, and chives. Spray a nonstick pan with oil and add your eggs. Cook like a traditional omelet. When the omelet is close to done, put the pan under the broiler to finish.

The omelet will puff up. Stuff the omelet, if desired, with vegetables or other low-fat filling, then fold over and serve.

■ For color and flavor, serve an omelet with spicy salsa poured on top of it, or serve with salsa or black bean relish and blue corn tortilla chips.

Check-Out Quiz

1. Proteins contain nitrogen.
   a. True   **b.** False
2. Essential amino acids cannot be made in the body.
   a. True   **b.** False
3. Every protein has a unique primary structure.
   a. True   **b.** False
4. In denaturation, the protein's shape is distorted but the protein can still function.
   a. True   **b.** False
5. During digestion and before absorption, proteins are broken down into their amino acid units, which can then be absorbed and transported in the blood.
   a. True   **b.** False
6. Americans tend to eat just enough protein.
   a. True   **b.** False
7. The stomach enzyme pepsin aids in the digestion of protein.
   a. True   **b.** False
8. You should try to balance your intake of protein from animal and plant sources.
   a. True   **b.** False
9. Kwashiorkor is associated with children with insufficient protein intake and who have a preexisting disease.
   a. True   **b.** False
10. Most plant foods are examples of incomplete proteins.
    a. True   **b.** False

Activities and Applications

1. Self-Assessment

Write the number of times per week that you eat the foods listed below in the space provided. Is your protein coming mostly from animal or plant sources, or is it somewhat evenly balanced between the two? Think about the serving sizes of the animal proteins versus the plant proteins. Are the meats, poultry, and fish usually the entrées, and the pasta, rice, vegetables, and dried beans or peas served as side dishes in smaller quantities? What can you do to balance the two sides better, if necessary?

Animal Protein Sources		Plant Protein Sources	
Red meats	_____	Dried beans	_____
Poultry	_____	Dried peas	_____
Fish	_____	Bread	_____
Milk	_____	Cereals	_____
Cheese	_____	Pasta	_____
Yogurt	_____	Rice	_____
Eggs	_____	Nuts and seeds	_____
		Vegetables (including potatoes)	_____
Total Number of Servings	_____	Total Number of Servings	_____

2. Reading Food Labels

Following are food labels from a beef burger and a vegetable burger. Compare and contrast their nutritional content.

Beef Burger (3 oz.)	Vegetable Burger (2.5 oz.)
NUTRITION FACTS	NUTRITION FACTS
Amount per serving	Amount per serving
Calories 230	Calories 140
Calories from Fat 140	Calories from Fat 20
Total Fat 16 g	Total Fat 2.5 g
Saturated 6 g	Saturated 0.5 g
Polyunsaturated 1 g	Polyunsaturated 1.0 g
Monounsaturated 7 g	Monounsaturated 0.5 g
Cholesterol 74 g	Cholesterol 0 g
Sodium 180 mg	Sodium 180 mg
Total Carbohydrate 0 g	Total Carbohydrate 21 g
Dietary Fiber 0 g	Dietary Fiber 5 g
Sugar 0 g	Sugar 0 g
Protein 21 g	Protein 8 g

3. How Much Protein Do You Need?

Calculate how many grams of protein you need by multiplying your weight (in pounds) times 0.36.

Example: 150 pounds × .36 grams protein per pound = 54 grams protein

4. How Much Protein Do You Eat?

Write down everything you ate yesterday, including approximate portion sizes. If yesterday was not a typical day, write down what you normally eat during the course of a day. Using Appendix A, find the amount of protein in each food and total up your protein intake for the day.

Now you can compare how much protein you ate on one day to the RDA. Do you consume too much, too little, or just about the right amount of protein daily? If you are eating too much, what foods would you cut down on and what foods would you replace them with?

5. Meat Diet vs. Mostly Plant Diet

Using the tables in this chapter, Appendix A, and/or food labels, find the amount of protein in each food listed below and total up each list. Each list represents one day's intake.

Meat-Based Diet		Plant-Based Diet with Dairy	
2 eggs	_____	1 cup oatmeal	_____
2 slices white toast	_____	1/2 cup raisins	_____
1/2 cup orange juice	_____	1 corn muffin	_____
1/2 cup milk	_____	1 cup milk	_____
1 doughnut	_____	1 apple	_____
3 ounces roast beef	_____	1 vegetarian burger	_____
1 ounce American cheese	_____	1 wholegrain bun	_____
2 slices white bread	_____	Lettuce and tomato slices	_____
1 tablespoon mayonnaise	_____	1 banana	_____
1 oz. package corn chips	_____	iced tea	_____
2 cupcakes	_____	granola bar	_____
2 slices pizza	_____	1 cup vegetable soup	_____
1 cup vegetable salad	_____	1 cup meatless chili	_____
1 tablespoon dressing	_____	1 cup vegetable salad	_____
12 oz. soft drink	_____	1 tablespoon dressing	_____
1 cup vanilla ice cream	_____	1/2 cup milk	_____
		Peach cobbler	_____
Total Protein:	_____	Total Protein:	_____

Which diet contained more protein? Do either or both of these diets meet your protein RDA? Is it possible for you to get the protein you need without eating meat, poultry or seafood?

Nutrition Web Explorer

Vegetarian Eating   www.veg.org
On this organization's home page, click on "Frequently Asked Questions," then click on "Glossary." Write down the different styles of vegetarian eating.

Eat Ethnic   www.eatethnic.com
Use this website to find out what types of high-protein foods another ethnic group eats.

Food Facts *Soybeans*

The soybean plant was first domesticated in China 3,000 years ago. The Chinese call it the *yellow jewel* or the *great treasure,* for several reasons. Soybeans are easy to farm, and the plants do not deplete the soil. They are inexpensive to buy, contain the most protein of all legumes (with no cholesterol), and are a very versatile food, although, when merely boiled, they have a strong taste with a metallic aftertaste. Perhaps due to this problem, the soybean has been used to make a tremendous variety of products.

Soybeans are grown in abundance in this country, but most are sold as animal feed after being processed for their oil. Soybean oil is used extensively in salad dressings, in margarine, and as salad/cooking oil.

In addition to soy oil, another important soybean product, particularly for vegetarians, is tofu, or bean curd. Tofu was invented by a Chinese scholar in 164 B.C. and is the most important of the foods prepared from soybeans in the East. Tofu is made in a process similar to making cheese. Soybeans are crushed to produce soy milk, which is then coagulated, causing solid curds (the tofu) and liquid whey to form. Tofu is white in color, soft in consistency, and bland in taste. It readily picks up other flavors, making it a great choice for mixed dishes such as lasagne.

Tofu is available shaped in cakes of varying textures and packed in water, which must be changed daily to keep it fresh. Firm tofu is compressed into blocks, and holds its shape during preparation and cooking. Firm tofu can be used for stir-frying, grilling, or marinating. Soft tofu contains much more water and is more delicate. Soft tofu is good to use in blenderized recipes to make dips, sauces, salad dressings, spreads, puddings, cream pies, pasta filling, and cream soups. Silken tofu is even softer and more delicate, and works well in creamy desserts. Tofu should be kept refrigerated and used within one week.

CHEF'S TIPS FOR USING TOFU

- Marinade tofu with ginger lime sauce.
- Crumble firm tofu and sauté with chopped onions, bell peppers, and other vegetables, herbs, and seasonings to make tacos and other Mexican dishes.
- Grill tofu and serve as the "meat" in a sandwich with Portabello mushrooms.
- Replace part of the cream in creamed soups with blended silken tofu.
- Use blended silken or soft tofu instead of ricotta cheese in Italian dishes and other mixed dishes, such as Indian curry or a hot Thai dish.
- Use soft tofu in place of mayonnaise in salad dressings such as green goddess.

Other soybean products include the following:

- **Soy sauce** combines fermented soy and wheat. The wheat is first roasted, and contributes both the soy sauce's brown color and its sharp, distinctive flavor.
- **Miso** is similar to soy sauce but pasty in consistency. It is made by fermenting soybeans with or without rice or other grains. A number of varieties are available, from light-colored and sweet to dark and robust.

It is used in soups and gravies, as a marinade for tofu, as a seasoning, and as a spread on sandwiches and fried tofu.

- **Tempeh** is a white cake made from fermenting soybeans. It is a pleasant-tasting, high-protein food that can be cooked quickly to make dishes such as barbecued or fried tempeh, or cut into pieces to add to soups. Tempeh is cultured like cheese and yogurt, and therefore must be used when fresh or it will spoil.
- **Textured vegetable protein (TVP)** is made of granules of isolated soy protein that must be rehydrated before using in recipes. TVP is actually a brand name. The generic name is textured soy protein, or TSP. It can replace up to one-quarter of the meat in a recipe without tasting unac-

ceptable. It is a very concentrated source of protein and is almost fat-free. Because of its strong flavor, TSP is most successfully used in highly flavored dishes such as chili, spaghetti sauce, and curries.

- **Meat analogs** are imitation meat products made from soy protein without any animal products. They are offered in forms resembling meat, such as hamburgers, hot dogs, bacon, ham, and chicken patties and nuggets. They contain little or no fat and no cholesterol, but are often high in sodium. Some are fortified with vitamin B_{12} and iron.

Additional soy products include soy milk, soy yogurt, soy ice cream, and soy nuts (great for salads and snacks).

Hot Topic Irradiation

Beef is one of the U.S. food industry's hottest sellers—to the tune of about 8 billion pounds a year. In recent years, though, beef, especially ground beef, has shown a dark side: It can harbor the bacterium *E. coli* 0157:H7, a pathogen that threatens the safety of the domestic food supply. If not properly prepared, beef tainted with *E. coli* 0157:H7 can make people ill, and can, in the case of children or the elderly, kill them. In 1993, *E. coli*-0157:H7-contaminated hamburgers sold by a fast-food chain were linked to the deaths of four children and hundreds of illnesses in the Pacific Northwest.

In 1997, the potential extent of *E. coli* 0157:H7 contamination came to light when Arkansas-based Hudson Foods Inc. voluntarily recalled 25 million pounds of hamburger suspected of contained *E. coli* 0157:H7. It was the largest recall of meat products in U.S. history.

Nationally, *E. coli* 0157:H7 causes about 20,000 illnesses and 500 deaths a year, according to the federal Centers for Disease Control and Prevention. Scientists have known only since 1982 that this form of *E. coli* causes human illness.

To help combat this public health problem, the Food and Drug Administration (FDA) approved in 1997 the treatment of red-meat products with a measured dose of radiation. This process, commonly called *irradiation,* has drawn praise from many food-industry and health organizations because it can control *E. coli* 0157:H7 and several other disease-causing microorganisms. Since 1963, the FDA has been allowing the irradiation of a number of foods, such as poultry, fresh fruits and vegetables, dry spices, and seasonings.

The process is similar to sending luggage through a radiation field — typically gamma rays produced from radioactive cobalt-60. That amount of energy is not strong enough to add any radioactive material to the food. The same irradiation process is used to sterilize medical products such as bandages, contact lens solutions, and hospital supplies such as gloves and gowns. Many spices solid in this country also are irradiated, which eliminates the need for chemical fumigation to control pests. American astronauts have eaten irradiated foods since 1972.

Irradiation is a "cold" process that gives off little heat, so foods can be irradiated within their packaging and remain protected against contamination until opened by users. Because a few bacteria can survive the process in poultry and meats, it's important to keep products refrigerated and to cook them properly.

Irradiation interferes with bacterial genetics, so the contaminating organism can no longer survive or multiply. Although chemicals called radiolytic products are created when food is irradiated, the FDA has found them to pose no health hazard. In fact, the same kinds of products are formed when food is cooked.

As part of its approval, the FDA requires that irradiated foods include labeling with either the statement "treated with radiation" or "treated by irradiation" and the international symbol for irradiation, the radura (Figure 5-4). Irradiation labeling requirements apply only to foods sold in stores. Irradiation labeling does not apply to restaurant foods.

The FDA has evaluated irradiation safety for 40 years and found the process safe and effective for many foods. Before approving red-meat irradiation, the agency reviewed numerous scientific studies conducted worldwide. These include research on the chemical effects of radiation on meat, the impact that the process has on nutrient content, and potential toxicity.

In reviews of the irradiation process, FDA scientists concluded that irradiation reduces or eliminates pathogenic bacteria, insects, and parasites. It reduces spoilage; in certain fruits and vegetables, it inhibits

Figure 5-4

sprouting and delays the ripening process. Also, it does not make food radioactive, compromise nutritional quality, or noticeably change food taste, texture, or appearance, as long as it's applied properly to a suitable product.

Health experts say that in addition to reducing *E. coli* 0157:H7 contamination, irradiation can help control the potentially harmful bacteria *Salmonella* and *Campylobacter,* two chief causes of foodborne illness. *Salmonella*—commonly found in poultry, eggs, meat, and milk—sickens as many as 4 million and kills 1,000 per year nationwide. *Campylobacter,* found mostly in poultry, is responsible for 6 million illnesses and 75 deaths per year in the United States. FDA officials emphasize that though irradiation is a useful tool for reducing foodborne illness risk, it complements, but doesn't replace, proper food-handling practices by producers, processors, and consumers.

Chapter 6
Vitamins

In the early 1900s, scientists thought they had found the compounds needed to prevent **scurvy** and **pellagra,** two diseases caused by vitamin deficiencies. These compounds originally were believed to belong to a class of chemical compounds called amines and were named from the Latin *vita,* or life, plus *amine—vitamine.* Later, the "e" was dropped when it was found that not all of the substances were amines. At first, no one knew what they were chemically, so vitamins were identified by letters. Later, what was thought to be one vitamin turned out to be many, and numbers were added, such as the vitamin B complex (for example, vitamin B_6). Later on, some vitamins were found unnecessary for human needs and were removed from the list, which accounts for some of the numbering gaps. For example, vitamin B_8, adenylic acid, was later found not to be a vitamin.

This chapter will help you to:

- State the general characteristics of vitamins.
- Identify the functions and food sources of each of the 13 vitamins.
- List which vitamins are more likely to be deficient in the American diet, and the possible side effects of vitamin toxicity.
- Describe ways to conserve vitamins when handling and cooking foods.
- Discuss the purchasing, storage, cooking, and menuing of fruits and vegetables.
- Define functional foods and give examples of phytochemicals and the foods in which they are found.

Characteristics of Vitamins

Let's start with some basic facts about vitamins.

1. Very small amounts of vitamins are needed by the human body and very small amounts are present in foods. Some vitamins are measured in IU's (international units), a measure of biological activity; others are measured by weight, in micrograms or milligrams. To illustrate how small these amounts are, remember that 1 ounce is 28.3 grams. A milligram is 1/1000 of a gram, and a microgram is 1/1000 of a milligram.
2. Although vitamins are needed in small quantities, the roles they play in the body are enormously important, as you will see in a moment.
3. Most vitamins are obtained through food. Some are also produced by bacteria in the intestine (and are absorbed into the body), and one (vitamin D) can be produced by the skin when it is exposed to sunlight.
4. There is no perfect food that contains all the vitamins in just the right amounts. The best way to assure an adequate vitamin intake is to eat a varied and balanced diet.

5. Vitamins do not contain calories, so they do not directly provide energy to the body. Vitamins provide energy indirectly because they are involved in energy metabolism.
6. Some vitamins in foods are not the actual vitamin but are **precursors.** The body chemically changes the precursor to the active form of the vitamin.
7. A **megadose** of a vitamin is defined as more than 10 times the RDA or AI. It often has toxic effects. Vitamin D, for example, can be toxic when taken at only 5 to 10 times the AI.
8. The body can't detect whether a vitamin is synthetic or natural.

Vitamins are classified according to how soluble they are in either fat or water. **Fat-soluble vitamins** (A, D, E, and K) generally occur in foods containing fat, and they can be stored in the body. **Water-soluble vitamins** (vitamin C and the B-complex vitamins) are not stored appreciably in the body (except vitamins B_6 and B_{12}) and don't often reach toxic levels. Now let's take a closer look at the 13 different vitamins.

Precursors—Forms of vitamins that the body changes chemically to active vitamin forms.

Megadose—A supplement intake of 10 times the RDA of a vitamin or mineral.

Fat-soluble vitamins—A group of vitamins that generally occur in foods containing fats; these include vitamins A, D, E, and K.

Water-soluble vitamins—A group of vitamins that are soluble in water and are not stored appreciably in the body; these include vitamin C, thiamin, riboflavin, niacin, vitamin B_6, folate, vitamin B_{12}, pantothenic acid, biotin.

■ MINI-SUMMARY

Very small amounts of vitamins are needed by the human body, and very small amounts are present in foods. Although vitamins are needed in small quantities, the roles they play in the body are enormously important. Vitamins must be obtained through foods, because vitamins are either not made in the body or not made in sufficient quantities. There is no perfect food that contains all the vitamins in just the right amounts. The best way to assure an adequate intake of vitamins is to eat a varied and balanced diet. Vitamins have no calories, so they do not directly provide energy to the body. Some vitamins in foods are not the actual vitamin, but rather are precursors. The body chemically changes the precursor to the active form of the vitamin. Megadoses are often toxic. Vitamins are classified according to how soluble they are in either fat or water.

Fat-Soluble Vitamins

Fat-soluble vitamins include vitamins A, D, E, and K. They generally occur in foods containing fats and are stored in the body either in the liver or in adipose (fatty) tissue until they are needed. Fat-soluble vitamins are absorbed and transported around the body like other fats. If anything interferes with normal fat digestion and absorption, these vitamins may not be absorbed.

Although it is convenient to be able to store these vitamins so you can survive periods of poor intake, excessive vitamin intake (such as large doses of vitamin pills) causes large amounts of vitamins A, D, and K to be stored and may lead to undesirable symptoms.

Vitamin A

Xerosis—A condition in which the cornea of the eye becomes dry and cloudy; often due to a deficiency of vitamin A.

Xerophthalmia—Hardening and thickening of the cornea that can lead to blindness; usually caused by a deficiency of vitamin A.

Night blindness—A condition caused by insufficient vitamin A in which it takes longer to adjust to dim lights after seeing a bright light at night; this is an early sign of vitamin A deficiency.

Preformed vitamin A—The form of vitamin A called retinol.

Provitamin A—Precursors of vitamin A, such as beta carotene.

Carotenoid—A class of pigments that contribute red, orange, or yellow colors to fruits and vegetables.

During World War I, many children in Denmark developed eye problems. Their eyes became dry and eyelids became swollen, and eventually blindness resulted. A Danish physician read that an American scientist gave milkfat to laboratory animals to cure similar eye problems in animals. At the time Danish children were drinking skim milk, because all the milkfat was being made into butter and sold to England. When the Danish doctor gave whole milk and butter to the children, they got better. The Danish government later restricted the amount of exported dairy foods. Dr. E. V. McCollum, the American scientist, eventually found vitamin A (the first vitamin to be discovered) to be the curative substance in milkfat.

Vitamin A has two roles involving the eyes. First, it is essential for the health of the cornea, the clear membrane covering your eye. Without enough vitamin A, the cornea becomes cloudy. Eventually it dries (called **xerosis**) and thickens and can result in permanent blindness (**xerophthalmia**).

Vitamin A is well known for its role in night vision. When there is insufficient vitamin A, you may experience symptoms of night blindness. In **night blindness,** it takes longer to adjust to dim lights after seeing a bright flash of light (such as oncoming car headlights) at night. This is an early sign of vitamin A deficiency. If the deficiency continues, xerosis and xerophthalmia can occur.

Vitamin A is involved in many other functions. It plays a role in cell growth and development and in healthy skin and hair, as well as in proper bone growth and tooth development in children. Vitamin A is also needed for proper immune-system functioning (so you can fight infections) and to maintain the protective linings of your lungs, intestines, urinary tract, and other organs (this also helps to fight infection and disease). Vitamin A is essential for normal reproduction.

Vitamin A is found in foods in two forms: **preformed vitamin A** or **retinol,** and **provitamin A.** Provitamin A refers to certain members of a class of pigments called **carotenoids** that contribute red, orange, or yellow color to fruits and vegetables. Provitamin A is converted to vitamin A in the body.

Beta-carotene—A precursor of vitamin A that functions as an antioxidant in the body; the most abundant carotenoid.

Antioxidant—A compound that combines with oxygen to prevent oxygen from oxidizing or destroying important substances; antioxidants prevent the oxidation of unsaturated fatty acids in the cell membrane, DNA, and other cell parts that substances called free radicals try to destroy.

Free radical—An unstable compound that reacts quickly with other molecules in the body.

Retinol equivalent—The unit of measure for vitamin A activity in both its forms (retinol and beta-carotene). 1 RE = 1 microgram retinol or 6 micrograms beta-carotene.

Beta-carotene, the most abundant carotenoid, is an **antioxidant.** Antioxidants combine with oxygen so that the oxygen is not available to oxidize, or destroy, important substances in the cell. Antioxidants prevent the oxidation of unsaturated fatty acids in the cell membrane, DNA (the genetic code), and other cell parts that substances called **free radicals** destroy. Free radicals are highly reactive compounds that normally result from cell metabolism and functioning of the immune system. In the absence of antioxidants, free radicals destroy cells (possibly accelerating the aging process) and alter DNA (possibly increasing the risk for cancerous cells to develop). Free radicals may also contribute to the development of cardiovascular disease. In the process of functioning as an antioxidant, beta-carotene is itself oxidized or destroyed. The role of beta-carotene as an anticancer agent is being studied, but it appears that beta-carotene supplements do not protect against cancer (or heart disease).

Certain plant foods are excellent sources of carotenoids. These include dark green vegetables, such as spinach, and deep orange fruits and vegetables, such as apricots, carrots, and sweet potatoes. Beta-carotene has an orange color seen in many vitamin-A-rich fruits and vegetables, but in some cases its orange color is masked by dark green chlorophyll found in vegetables such as broccoli or spinach.

Sources of preformed vitamin A include animal products such as liver (a very rich source), vitamin-A-fortified milk, eggs, and fortified cereals. Most-ready-to-eat and instant cereals are fortified with vitamin A. Fortified ready-to-eat cereals usually contain at least 25 percent of the U.S. RDA for vitamin A. Butter and margarine are also fortified with vitamin A. Retinol, the active form of vitamin A found in animal foods, is used in fortification. Preformed retinol makes up about two-thirds of our total vitamin A intake.

The RDA for vitamin A is expressed in **retinol equivalents.** Retinol equivalents measure the amount of retinol derived from eating foods that contain preformed vitamin A or provitamin A. To get enough vitamin A, it is recommended that you eat a dark green vegetable or deep orange fruit or vegetable at least every other day. Because the body stores vitamin A in the liver, it is not absolutely necessary to eat a good source every day.

Prolonged use of high doses of preformed vitamin A (even just three times one's RDA) may cause hair loss, bone pain and damage, soreness, liver damage, nausea, and diarrhea. High doses are particularly dangerous for pregnant women (they may cause birth defects) and the elderly (they can cause joint pain, nausea, muscle soreness, itching, hair loss, and liver and bone damage). Only preformed vitamin A, not carotenoids such as beta-carotene, has such effects. When supplements of beta-carotene are taken, the beta-carotene can't be converted fast enough to be toxic, and less is absorbed as larger doses are taken.

Vitamin D

Vitamin D differs from all the other nutrients in that it can be made in the body. It also is unique in that it functions at times as a hormone. When ultraviolet rays shine on your skin, a cholesterol-like compound is converted into a precursor of vitamin D and absorbed into the blood. Of course, if you are not in the sun much or if the ultraviolet rays are cut off by heavy clothing, clouds, smog, fog, sunblock, and window glass, there will be less vitamin D produced. On the positive side, a light-skinned person needs only about 15 minutes of sun on the face, hands, and arms two to three times per week to make enough vitamin D (a dark-skinned person needs more). Several months' supply of vitamin D can be stored in the body, which is helpful during winter months when the sun is not as strong in northern climates and you need to wear more clothing.

Vitamin D is a member of a team of nutrients and hormones that maintain blood calcium levels and ensure an adequate supply of calcium (and phosphorus) for building bones and teeth. Calcium is also used to contract muscles and transmit nerve impulses, so vitamin D ensures that calcium is available for these functions. Vitamin D increases blood calcium levels in three ways: it increases calcium absorption in the intestine, decreases the amount of calcium excreted by the kidney, and pulls calcium out of the bones.

Vitamin D also helps cells develop in the skin and the immune system. Lastly, vitamin D plays an as-yet-undefined role in some cancer cells.

Significant food sources of vitamin D include liver, egg yolks, and fatty fish. Except for these few foods, only small amounts of vitamin D are found in food. For this reason, milk is normally fortified with vitamin D. Some breakfast cereals are also fortified with vitamin D. If you drink two cups of milk each day, you will get about half the RDA of vitamin D (the rest comes from other foods and sun exposure). There is no vitamin D added to milk products such as yogurt or cheese.

Rickets—A childhood disease in which bones do not grow normally, resulting in bowed legs and knock knees; it is generally caused by a vitamin D deficiency.

Osteomalacia—A disease of vitamin D deficiency in adults in which the leg and spinal bones soften and may bend.

Vitamin D deficiency in children causes **rickets,** a disease in which bones do not grow normally, resulting in soft bones and bowed legs. Vitamin D deficiency in adults causes **osteomalacia,** a disease in which leg and spinal bones soften and may bend and break. Rickets is rarely seen, but osteomalacia may be seen in elderly individuals with poor milk intake and little sun exposure, and in individuals with diseases such as renal disease or alcoholism.

Vitamin D, when taken in excess of the RDA, is the most toxic of all the vitamins. All you need is about 4 to 5 times the RDA to start feeling symptoms of nausea, vomiting, diarrhea, fatigue, confusion, and thirst. It can lead to calcium deposits in the heart and kidneys that can cause severe health problems and even death. Young children and infants are especially susceptible to the toxic effects of too much vitamin D, and megadoses can cause growth failure.

Vitamin E

When you talk about vitamin E, you are actually referring to its four different forms, named after the first four letters of the Greek alphabet: alpha, beta, gamma, and delta. Vitamin E has an important function in the body as an antioxidant, especially to red blood cells in the lungs that pick up oxygen and white blood cells that defend the body from disease. Vitamin E helps develop nervous tissues. Vitamin E's possible role in preventing cardiovascular disease and cancer and in delaying aging is still in the research stages.

Vitamin E is widely distributed in plant foods. Rich sources include vegetable oils, margarine and shortening made from vegetable oils, seeds, and nuts. In oils, vitamin E acts like an antioxidant, thereby preventing the oil from going rancid or bad. Other good sources include wholegrain and fortified breads and cereals, and soybeans. Animal foods are poor sources for vitamin E.

Deficiency is rare, except in infants born prematurely, in part because of vitamin E's wide distribution in plant foods and because the body has significant storage capacity.

Vitamin K

Vitamin K has an essential role in the production of several chemicals (such as prothrombin) involved in blood clotting. Blood clotting is vital in the prevention of excessive blood loss when the skin is broken. Vitamin K is also needed to make an important protein used in making bone.

Vitamin K appears in certain foods and is also produced in the body. There are billions of bacteria that normally live in your intestines, and some of them make a form of vitamin K. It is thought that the amount of vitamin K produced by the bacteria is significant and may meet about half of your needs. (An infant is normally given this vitamin after birth to prevent bleeding because the intestine does not yet have the bacteria to produce vitamin K.) Food sources of vitamin K provide the balance needed. Excellent sources of vitamin K include liver and dark green leafy vegetables such as broccoli, collard greens, cauliflower, spinach, and cabbage.

Vitamin K deficiency is rare. Excessive supplementation of a synthetic version of vitamin K can be toxic, so supplements of vitamin K alone are not available unless prescribed by a physician.

■ **MINI-SUMMARY**

Table 6-1 summarizes the functions and sources of the fat-soluble vitamins: A, D, E, and K.

Water-Soluble Vitamins

Water-soluble vitamins include vitamin C and the B-complex vitamins. The B vitamins work in every body cell, where they function as *coenzymes*. A coenzyme combines with an enzyme to make it active. Without the coenzyme, the enzyme is useless. The body stores only limited amounts of water-soluble vitamins (except vitamins B_6 and B_{12}). Due to their limited storage, these vitamins need to be taken in daily. Excesses are excreted in the urine. Even though excesses are excreted, excessive supplementation of certain water-soluble vitamins can cause toxic side effects.

Vitamin C

Scurvy, the name for vitamin C deficiency disease, has been known since biblical times. It was most common on ships, where sailors developed bleeding gums, weakness, loose teeth, broken capillaries (small blood vessels) under the skin, and eventually death. Because sailors' diets included fresh fruits and vegetables only for the first part of a voyage, longer voyages resulted in more cases of scurvy. Once it was discovered that citrus fruits prevented scurvy, British sailors were given daily portions of lemon juice. In those days, lemons were called limes; hence British sailors got the nickname "Limeys."

Vitamin C (its chemical name is ascorbic acid, meaning "no-scurvy acid") is important in forming **collagen,** a protein substance that provides strength and support to bones, teeth, skin, cartilage, and blood vessels. It has been said that vitamin C acts like cement, holding together our cells and tissues. It is also important in healing wounds, absorbing iron into the body, fighting infection, and helping to make the hormone thyroxine, which regulates your basal metabolic rate.

Like beta-carotene and vitamin E, vitamin C is an important antioxidant; it prevents the oxidation of vitamin A and polyunsaturated fatty acids in the intestine. Its antioxidant properties have made vitamin C widely used as a food additive. It may appear on food labels as sodium ascorbate, calcium ascorbate, or simply ascorbic acid.

Scurvy—A vitamin-C-deficiency disease marked by bleeding gums, weakness, loose teeth, and broken capillaries under the skin.

Collagen—The most abundant protein in the body; a fibrous protein that is a component of skin, bone, teeth, ligaments, tendons, and other connective structures.

TABLE 6-1 Vitamins

Vitamin	Recommended Intake	Functions	Sources
Fat-soluble Vitamins			
Vitamin A	Males: 1,000 Retinol Equivalents Females: 800 Retinol Equivalents (1989 RDA)	Cell growth and development Healthy skin and hair Bone and tooth development Maintenance of protective linings of lungs, etc. Proper functioning of immune system Normal reproduction Night vision; healthy cornea	Preformed: Liver, fortified milk, fortified cereals, eggs Provitamin: Dark green vegetables, deep orange fruits and vegetables
Vitamin D	5 micrograms (1998 AI) Upper Level: 50 micrograms	Maintenance of blood calcium levels Building bones	Fortified milk, fatty fish, fortified breakfast cereals, egg yolk, sunshine
Vitamin E	15 milligrams (2000 RDA) Upper Level: 1,000 milligrams	Antioxidant Development of nervous tissue Protects red and white blood cells	Vegetable oils, margarine, shortening, seeds, nuts, wheat germ, wholegrain and fortified breads and cereals, soybeans
Vitamin K	Males: 80 micrograms Females: 65 micrograms (1989 RDA)	Blood clotting Makes protein used in making bones	Dark green leafy vegetables, cabbage Intestinal bacteria
Water-soluble Vitamins			
Vitamin C	75 milligrams (2000 RDA) Upper Level: 2,000 milligrams	Antioxidant Formation of collagen Wound healing Iron absorption Functioning of immune system Synthesis of thyroxine (hormone)	Citrus fruits, bell peppers, kiwifruit, broccoli, strawberries, tomatoes, potatoes, juices and cereals fortified with Vitamin C.

Vitamin	Amount	Functions	Sources
Thiamin	Males: 1.2 milligrams Females: 1.1 milligrams (1998 RDA)	Coenzyme in energy metabolism Functioning of nervous system Normal growth	Pork, sunflower seeds, wheatgerm, peanuts, dry beans, and wholegrain and enriched/fortified breads and cereals
Riboflavin	Males: 1.3 milligrams Females: 1.1 milligrams (1998 RDA)	Coenzyme in energy metabolism Healthy skin Normal vision	Milk and milk products, wholegrain and enriched/fortified breads and cereals, some meats, eggs
Niacin	Males: 16 milligrams equivalent Females: 14 milligrams equivalent (1998 RDA) Upper Level: 35 milligrams	Coenzyme in energy metabolism Healthy skin Normal functioning of nervous system Normal functioning of digestive tract	Meat, poultry, fish, wholegrain and enriched/fortified breads and cereals, milk, eggs
Vitamin B$_6$	Adults: 1.3 milligrams (1998 RDA) Upper Level: 100 milligrams	Coenzyme in carbohydrate, fat, and protein metabolism Synthesis of red blood cells, white blood cells, and neurotransmitters	Meat, poultry, fish, fortified cereals, some leafy green vegetables, potatoes, bananas, watermelon
Folate	Adults: 480 micrograms (1998 RDA) Upper Level: 1000 micrograms	Formation of DNA Formation of new cells	Green leafy vegetables, legumes, orange juice, enriched/fortified breads and cereals
Vitamin B$_{12}$	Adults: 2.4 micrograms (1998 RDA)	Activation of folate Normal functioning of the nervous system	Meat, poultry, seafood, eggs, dairy products, fortified breads and cereals
Pantothenic Acid	5 milligrams (1998 AI)	Energy metabolism	Widespread
Biotin	30 micrograms (1998 AI)	Energy metabolism Carbohydrate, fat, and protein metabolism	Widespread

RDA and AI amounts are for adults.

Foods rich in vitamin C include citrus fruits (oranges, grapefruits, limes, and lemons), bell peppers, kiwifruit, strawberries, tomatoes, broccoli, and potatoes. Only foods from the fruit and vegetable groups contribute vitamin C. There is little or no vitamin C in the meat group (except in liver, of course) or the dairy group. Some juices are fortified with vitamin C (if not already rich in vitamin C), as are most ready-to-eat cereals. Many people meet their needs for vitamin C simply by drinking orange juice. This is a good choice because vitamin C is easily destroyed in food preparation and cooking. Table 6-2 lists the vitamin C content of selected foods.

Certain situations require additional vitamin C. These include pregnancy and nursing, growth, fevers and infections, burns, fractures, surgery, cancer, and cigarette smoking. Smoking interferes with the use of vitamin C. The RDA for smokers, 100 milligrams, is almost double the RDA for nonsmoking adults.

Deficiencies resulting in scurvy are rare. Scurvy might occur in elderly people with poor intake of fruits and vegetables, and in alcoholics.

Doses of vitamin C of over 1,000 milligrams (1 gram) daily are probably hazardous, and diarrhea and stomach inflammation may occur. Doses as low as 250 milligrams can cause certain medical tests, such as for diabetes, to be inaccurate.

Thiamin, Riboflavin, and Niacin

Thiamin, riboflavin, and niacin all play key roles as part of coenzymes in energy metabolism. They are essential in the release of energy from carbohydrates, fats, and proteins. They are also needed for normal growth.

Thiamin also plays a vital role in the normal functioning of the nerves and muscles, since the nerves send messages to the muscles to contract and relax. Riboflavin is important for healthy skin and normal vision. Niacin is needed for the maintenance of healthy skin and the normal functioning of the nervous system and digestive tract.

Because thiamin, riboflavin, and niacin all help to release food energy, the needs for these vitamins increase as calorie needs increase.

Thiamin is widely distributed in foods, but mostly in moderate amounts. Pork is an excellent source of thiamin. Other sources include sunflower seeds, wheat germ, peanuts, acorn squash, dry beans, and wholegrain and enriched/fortified breads and cereals.

Milk and milk products are the major source of riboflavin in the American diet. Other sources include organ meats such as liver (very high in riboflavin), wholegrain and enriched/fortified breads and cereals, eggs, and some meats.

TABLE 6-2 Vitamin C in Foods	
Food	Milligrams Vitamin C
Fruits	
Orange, 1	80
Kiwi, 1 medium	75
Cranberry juice cocktail, 6 ounces	67
Orange juice, from concentrate, 1/2 cup	48
Papaya, 1/2 cup cubes	43
Strawberries, 1/2 cup	42
Grapefruit, 1/2	41
Grapefruit juice, canned, 1/2 cup	36
Cantaloupe, 1/2 cup cubes	34
Tangerine, 1	26
Mango, 1/2 cup, slices	23
Honeydew melon, 1/2 cup cubes	21
Banana, 1	10
Apple, 1	8
Nectarine, 1	7
Vegetables	
Broccoli, chopped, cooked, 1/2 cup	49
Brussels sprouts, cooked, 1/2 cup	48
Cauliflower, cooked, 1/2 cup	34
Sweet potato, baked, 1	28
Kale, cooked, chopped, 1/2 cup	27
White potato, baked, 1	26
Tomato, 1 fresh	22
Tomato juice, 1/2 cup	22
Cereals	
Cornflakes, 1 cup	15

Source: U.S. Department of Agriculture.

The main sources of niacin are meat, poultry, and fish. Organ meats, again, are quite high in niacin. Wholegrain and enriched/fortified breads and cereals, as well as peanut butter, are also important sources of niacin. All foods containing complete protein, such as those just mentioned and also milk and eggs, are good sources of the precursor of niacin, tryptophan. **Tryptophan,** an amino acid present in some of these foods, is converted to niacin. About half the niacin we use is made from tryptophan.

Deficiencies in thiamin, riboflavin, and niacin are rare in the United States, in part because foods such as enriched breads contain all three nutrients. General symptoms for B-vitamin deficiencies include fatigue, decreased appetite, and depression. Alcoholism can create deficiencies in these three vitamins.

Toxicity is not a problem except in the case of niacin. Nicotinic acid, a form of niacin, has been prescribed by physicians to lower elevated blood cholesterol levels. Unfortunately, it has some undesirable side effects. Starting at doses of 100 milligrams, typical symptoms include flushing, tingling, itching, rashes, hives, nausea, diarrhea, and abdominal discomfort. Flushing of the face, neck, and chest lasts for about 20 minutes after taking a large dose. More serious side effects of large doses include liver damage, high blood sugar levels, and abnormal heart rhythm.

Vitamin B$_6$

Vitamin B$_6$ is actually three different compounds that can each be changed to the active coenzyme form. Vitamin B$_6$ plays an important role as part of a coenzyme involved in carbohydrate, fat, and protein metabolism. It is particularly important in protein metabolism, specifically in making amino acids. It is also used to make red blood cells (which transport oxygen around the body), white blood cells (which are crucial for the immune system to function properly), and **neurotransmitters** (substances that allow nerve cells to communicate). Vitamin B$_6$ is also used to break down glycogen to glucose.

Good sources for vitamin B$_6$ include meat, poultry, and fish. Vitamin B$_6$ also appears in plant foods; however, it is not as well absorbed from these sources. Good plant sources include potatoes, some fruits (such as bananas and watermelon), and some leafy green vegetables (such as broccoli and spinach). Fortified ready-to-eat cereals are also good sources of vitamin B$_6$

Deficiency of vitamin B$_6$, which may occur in some women and older adults, can cause symptoms such as fatigue and irritability. Symptoms can become much more serious if the deficiency continues.

Excessive use of vitamin B$_6$ (more than 2 grams daily for two months or more or 200 milligrams daily for a longer period of time) can cause irreversible nerve damage and symptoms such as numbness in hands and feet and difficulty walking. The Upper Level for B$_6$ is 100 milligrams. The problem with B$_6$ is that, unlike other water-soluble vitamins, it is stored in the muscles. Vitamin B$_6$ supplementation became popular when it appeared that the vitamin may relieve some of the symptoms of premenstrual syndrome and carpal tunnel syndrome, a condition in which a compressed nerve in the wrist causes much pain. There is no scientific research showing that B$_6$ helps in either condition.

Neurotransmitters—Chemical messengers released at the end of the nerve cell to stimulate or inhibit a second cell, which may be another nerve cell, a muscle cell, or a gland cell.

Folate and Vitamin B$_{12}$

Folate and vitamin B$_{12}$ often work together in the body. Folate, also called folic acid or folacin, is a component of coenzymes required to form DNA, the genetic material contained in every body cell. Folate is therefore needed to make all new cells. Much folate is used to produce adequate numbers of red blood cells, white blood cells, and digestive-tract cells, since these cells divide frequently.

Excellent sources of folate include green leafy vegetables (the word folate comes from the word *foliage,* meaning leaves), legumes, orange juice, and fortified breads and ready-to-eat cereals. Much folate is lost during food preparation and cooking, so fresh and lightly cooked foods are more likely to contain more folate.

Vitamin B$_{12}$, a group of related compounds that contain the mineral cobalt, is present in all body cells. One of its important functions is to convert folate into its active forms so that it can make DNA. It also helps in the normal functioning of the nervous system by maintaining the protective cover around nerve fibers.

Vitamin B$_{12}$ differs from other vitamins in that it is found only in animal foods such as meat, poultry, fish, shellfish, eggs, milk, and milk products. Plant foods do not naturally contain any vitamin B$_{12}$. Vegetarians who do not eat any animal products need to include vitamin-B$_{12}$-fortified cereals or soy products in their diet or supplements.

Intrinsic factor—A protein secreted by stomach cells that is necessary for the absorption of vitamin B$_{12}$.

Megaloblastic anemia—A form of anemia seen with a deficiency of vitamin B$_{12}$ or folate and characterized by large, immature red blood cells.

Vitamin B$_{12}$ also differs from other vitamins in that it requires an **intrinsic factor** (a protein-like substance produced in the stomach) to be absorbed. Vitamin B$_{12}$ attaches to the intrinsic factor in the stomach and is then carried to the small intestine, where it is absorbed. Vitamin B$_{12}$ is stored in the liver.

A folate deficiency can cause **megaloblastic anemia,** a condition in which the red blood cells are larger than normal and function poorly. Other symptoms may include digestive-tract problems such as diarrhea, and mental confusion and depression. Groups particularly at risk for folate deficiency are pregnant women, low-birthweight infants, and the elderly.

A folate deficiency may have especially serious consequences during the first few weeks of pregnancy, when it may cause **neural tube defects.** The neural tube is the tissue in the embryo that develops into the brain and spinal cord. The neural tube closes within the first month of pregnancy. Neural tube defects are diseases in which the brain and spinal cord form improperly in early pregnancy. They affect one to two of every 1,000 babies born each year. Neural tube defects include anencephaly, in which most of the brain is missing, and spina bifida. In one form of spina bifida, a piece of the spinal cord protrudes from the spinal column, causing paralysis of parts of the lower body.

The need for folate is critical during the earliest weeks of pregnancy—when most women don't know they are pregnant. However, consuming enough folate during the entire pregnancy is also important.

Because of the importance of folate during pregnancy and the difficulty most women encounter trying to get double the RDA for folate in the diet, the Food and Drug Administration requires manufacturers of enriched breads, flour, pasta, cornmeals, rice, and some other foods to fortify their products with a very absorbable form of folate. Women eating folate-fortified foods should not assume that these foods will meet all their folate needs. They should still seek out folate-rich foods.

A number of frequently used medications interfere with the normal use of folate in the body. Frequent users of aspirin, antacids, anticonvulsant medications, or oral contraceptives should speak with their physicians about obtaining adequate folate.

A vitamin B_{12} deficiency in the body is usually not due to poor intake, but rather to a problem with absorption. Absorption may decline with age. When vitamin B_{12} is not properly absorbed, **pernicious anemia** develops. Pernicious means ruinous or harmful, and this type of anemia is marked by a megaloblastic anemia (as with folate deficiency), also called macrocytic anemia, in which there are too many large, immature red blood cells. Symptoms include extreme weakness and fatigue. Nervous-system problems also erupt. The cover surrounding the nerves in the body becomes damaged, making it difficult for impulses to travel along them. This causes a poor sense of balance, numbness and tingling sensations in the arms and legs, and mental confusion.

Because deficiencies in folate or vitamin B_{12} cause macrocytic anemia, a physician may mistakenly administer folate when the problem is really a vitamin B_{12} deficiency. The folate would treat the anemia, but not the deterioriation of the nervous system due to a lack of vitamin B_{12}. If untreated, this damage can be significant and sometimes irreversible, although it takes many years to occur. When vitamin B_{12} is deficient due to an absorption problem, injections of the vitamin must be given. Table 6-3 lists food sources of folate.

Pernicious anemia—A type of anemia caused by a deficiency of vitamin B_{12} and characterized by macrocytic anemia and deterioration in the functioning of the nervous system.

Pantothenic Acid and Biotin

Both pantothenic acid and biotin are parts of coenzymes involved in energy metabolism. Pantothenic acid is needed to release energy from carbohydrates, fats, and protein, and to make fatty acids. Biotin is involved in the metabolism of carbohydrates, fats, and proteins.

Both pantothenic acid and biotin are widespread in foods. Good sources of pantothenic acid include peanuts, eggs, milk, some vegetables, and legumes.

TABLE 6-3   Food Sources of Folate

Foods Especially Rich in Folate	Folate (micrograms)
Fruits	
1 cup orange juice from concentrate	109
1 cup canned pineapple juice	58
1 cup orange sections	55
1 cup papaya	49
Vegetables and legumes	
1/2 cup cooked lentils	179
1/2 cup cooked spinach	131
1/2 cup cooked black beans	128
1 cup raw, chopped spinach	108
1/2 cup cooked turnip greens	86
1/2 cup cooked collard greens	65
Grains and cereals	
1 ounce bran flakes	100
1 cup corn flakes	100
1 ounce shredded wheat	14
Nuts	
1/3 cup peanuts	117
1/3 cup sunflower seeds	109
Meat and poultry	
4 ounces cooked beef liver	162

Good sources of biotin include peanuts, egg yolks, cheese, and soybeans. Intestinal bacteria make considerable amounts of biotin. Deficiency and toxicity concerns are not known.

Vitaminlike Substances

In the DRI/RDA tables in Appendix B, you will see an "Adequate Intake" level for choline. The AI is set well below what a typical intake is, so deficiency is very rare. Choline is used in the body to make cell membranes and also functions in nerve transmission and proper developing and functioning of the brain. Further research is being done to determine whether choline is indeed an essential nutrient, in other words, a real vitamin. Other vitaminlike substances, such as carnitine, lipoic acid, inositol, and taurine, are not essential and are therefore not vitamins.

Ingredient Focus: Fruits and Vegetables

Nutrition

Vegetables and fruits, in general, have the following characteristics:

- Low in calories
- Low or no fat (except avocados)
- No cholesterol
- Good sources of fiber
- Excellent sources of vitamins and minerals, particularly vitamins A and C (see Table 6-4)
- Low in sodium (except for canned vegetables)

Also, dried fruits, such as raisins and apricots, provide some iron. The nutritional content of many fruits and vegetables appear in Tables 6-4 and 6-5.

Purchasing and Receiving

Always purchase quality fruits and vegetables, preferably Fancy or No. 1 grade. Seriously consider organic produce, for its fuller flavor, if your budget will allow it. Figure 6-1 shows some specialty vegetables and fruits that you may want to consider purchasing for variety.

When receiving fresh produce, randomly check for the following.

1. Characteristic color—should be bright and lively.
2. Characteristic shape—misshapen produce is often inferior in taste and texture and harder to prepare.
3. Appropriate size—small or immature vegetables lack full flavor, and overgrown vegetables are often coarse.
4. Freshness, which may be evident by crispness, lack of wilting, etc., depending on the item.
5. Bruises, blemishes, decay, cracks, frostbite, etc.

TABLE 6-4 Produce Nutrition

"Eating five fruits and vegetables a day is one of the most important choices you can make to help maintain your health"—National Cancer Institute

	Household Serving Sizes	Serving Size (g)	Serving Size (oz)	Calories (Kcal)	Protein (g)	Carbohydrate (g)	Fat (g)	Sodium (mg)	Dietary Fiber (g)	Vitamin A	Vitamin C	Calcium	Iron
										(% of U.S. RDA)			
Apple	1 med. apple	154	5.5	80	0	18	1	0	5	*	6	*	*
Asparagus	5 spears	93	3.5	18	2	2	0	0	2	10	10	*	*
Avocado	1/3 med. avocado	55	2	120	1	3	12	5	2	*	5	*	*
Banana	1 med. banana	126	4.5	120	1	28	1	0	3	*	15	*	2
Bell Pepper	1 med. pepper	148	5.5	25	1	5	1	0	2	2	130	*	*
Broccoli	1 med. stalk	148	5.5	40	5	4	1	75	5	10	240	6	4
Cabbage	1/12 med. head	84	3	18	1	3	0	30	2	*	70	4	*
Cantaloupe	1/4 med. melon	134	5	50	1	11	0	35	0	80	90	2	2
Carrot	1 med., 7" long, 1-1/4" diameter	78	3	40	1	8	1	40	1	330	8	2	*
Cauliflower	1/6 med. head	99	3	18	2	3	0	45	2	*	110	2	2
Celery	2 med. stalks	110	4	20	1	4	0	140	2	*	15	4	*
Cherry	21 cherries; 1 cup	140	5	90	1	19	1	0	3	*	10	2	*
Cucumber	1/3 med. cucumber	99	3.5	18	1	3	0	0	0	4	6	2	2
Grape	1-1/2 cups grapes	138	5	85	1	24	0	3	2	3	9	2	2
Grapefruit	1/2 med. grapefruit	154	5.5	50	1	14	0	0	6	6	90	4	*
Green Bean	3/4 cup cut beans	83	3	14	1	2	0	0	3	2	8	4	*
Green Onion	1/4 cup chopped	25	1	7	0	1	0	0	0	3	20	*	5
Honeydew	1/10 med. melon	134	5	50	1	12	0	50	1	*	40	*	2
Iceberg Lettuce	1/6 med. head	89	3	20	1	4	0	10	1	2	4	*	*
Kiwifruit	2 med. kiwifruit	148	5.5	90	1	18	1	0	4	2	230	4	4
Leaf Lettuce	1-1/2 cups shredded	85	3	12	1	1	0	40	1	20	4	4	*
Lemon	1 med. lemon	58	2	18	0	4	0	10	0	*	35	2	*
Lime	1 med. lime	67	2.5	20	0	7	0	1	3	*	35	2	2
Mushrooms	5 med. mushrooms	84	3	25	3	3	0	0	0	*	2	*	*
Nectarine	1 med. nectarine	140	5	70	1	16	1	0	3	20	10	*	*
Onion	1 med. onion	148	5.5	60	1	14	0	10	3	*	20	4	*
Oranges	1 med. orange	154	5.5	50	1	13	0	0	6	*	120	4	*
Peach	2 med. peaches	174	6	70	1	19	0	0	1	20	20	*	*
Pear	1 med. pear	166	6	100	1	25	1	1	4	*	10	2	2

TABLE 6-4 *(continued)*

"Eating five fruits and vegetables a day is one of the most important choices you can make to help maintain your health"—National Cancer Institute

	Household Serving Sizes	Serving Size (g)	Serving Size (oz)	Calories (Kcal)	Protein (g)	Carbohydrate (g)	Fat (g)	Sodium (mg)	Dietary Fiber (g)	Vitamin A	Vitamin C	Calcium	Iron
										(% of U.S. RDA)			
Pineapple	2 slices, 3" diameter 3/4" thick	112	4	90	1	21	1	10	2	*	35	*	*
Plum	2 med. plums	132	4.5	70	1	17	1	0	1	9	20	*	*
Potato	1 med. potato	148	5.5	110	3	23	0	10	3	*	50	*	8
Radishes	7 radishes	85	3	20	0	3	0	35	0	*	30	*	*
Strawberries	8 med. berries	147	5.5	50	1	13	0	0	3	*	140	2	2
Summer Squash	1/2 med. squash	98	3.5	20	1	3	0	0	1	4	25	2	2
Sweet Corn	kernels from 1 med. ear	90	3	75	3	17	1	15	1	5	10	*	3
Sweet Potato	1 med., 5" long, 2" diameter	130	4.5	140	2	32	0	15	3	520	50	3	4
Tangerine	2 med., 2-3/8" diameter	168	6	70	1	19	0	2	2	30	85	2	*
Tomato	1 med. tomato	148	5.5	35	1	6	1	10	1	20	40	*	2
Watermelon	1/18 med. melon; 2 cups diced	280	10	80	1	19	0	10	1	8	25	*	2

* Contains less than 2% of the U.S. RDA of this nutrient.

Source: U.S. Food and Drug Administration.

Frozen vegetables and fruits have much the same flavor as fresh fruit, but the texture often changes during freezing. Vegetables and fruits are frozen at the peak of ripeness and are very convenient to use. Look for frozen fruit that is thoroughly frozen and shows no signs of having thawed and refrozen.

Canned fruits are available in heavy syrup, in light syrup, juice-packed, and water-packed. Syrup-packed fruit contains much sugar. Canned fruits and vegetables are mostly available as Grade A or Grade B—each offers the same nutrition, but the Grade A will have a better appearance, texture, and flavor. Look for clean cans without dents, swelling, or rust.

TABLE 6-5   Vitamin A and C and Fiber in Fruits and Vegetables

In selecting your daily intake of fruits and vegetables, the National Cancer Institute recommends choosing:

- At least one serving of a vitamin A-rich fruit or vegetable a day.
- At least one serving of a vitamin C-rich fruit or vegetable a day.
- At least one serving of a high-fiber fruit or vegetable a day.
- Several servings of cruciferous vegetables a week. Studies suggest that these vegetables may offer additional protection against certain cancers, although further research is needed.

High in Vitamin A*	High in Vitamin C*	High in Fiber or Good Source of Fiber*	Cruciferous Vegetables
apricots	apricots	apple	bok choy
cantaloupe	broccoli	banana	broccoli
carrots	brussels sprouts	blackberries	brussels sprouts
kale, collards	cabbage	blueberries	cabbage
leaf lettuce	cantaloupe	brussels sprouts	cauliflower
mango	cauliflower	carrots	
mustard greens	chili peppers	cherries	
pumpkin	collards	cooked beans and peas	
romaine lettuce	grapefruit	(kidney, navy, lima,	
spinach	honeydew melon	and pinto beans,	
sweet potato	kiwi fruit	lentils, black-eyed	
winter squash (acorn,	mango	peas)	
hubbard)	mustard greens	dates	
	orange	figs	
	orange juice	grapefruit	
	pineapple	kiwi fruit	
	plum	orange	
	potato with skin	pear	
	spinach	prunes	
	strawberries	raspberries	
	bell peppers	spinach	
	tangerine	strawberries	
	tomatoes	sweet potato	
	watermelon		

* Based on FDA's food labeling regulations.
Source: National Cancer Institute.

Storage

All vegetables and fruits require careful handling and storage to conserve quality. Before storing, discard any that are damaged. Date all containers to make sure produce is used on a first-in, first-out basis. Stack containers

ALFALFA SPROUTS are the tender young sprouts produced by alfalfa seeds and harvested only six days after they begin growing. They have a crunchy, nutty flavor.

BEAN SPROUTS grow from mung beans and are harvested and sold within six days from the time they first appear. The shorter the bean sprouts, the crispier and more tender they are.

BITTER MELON (Balsam Pear) is shaped like a cucumber, 6–8 inches long. The outer surface is clear green and wrinkled; inside it contains a layer of white or pink spongy pulp and seeds.

BOK CHOY (Chinese Chard, White Mustard Cabbage) is a combination of chard and celery with long and broad crisp white stalks with shiny, deep green leaves. The flavor is sweet but mild.

CACTUS LEAVES (Nopales) grow on a cactus plant and are light green and crisp, but also tender. Served like a vegetable, they have the texture and flavor of green beans.

CELERIAC (Celery Root) looks like celery, but only the bulb-type root is edible. The small young root knob is more tender and less woody than the larger, mature roots.

CHARD (Swiss Chard) is a lush green, leafy vegetable resembling spinach in appearance, with flavor and texture similar to asparagus. The leaves and ribs are excellent as fresh salad greens.

CHAYOTE (Vegetable Pear) has a very dark green hard surface and can be either pear shaped or round. It is 3–5 inches long, looks like an acorn squash but has a more delicate flavor, and has flesh the color of honeydew melon.

CHINESE LONG BEANS (Dow Kwok) are pencil-thin beans, 12–25 inches long, with light green tender pods. The tiny beans inside resemble immature blackeyed peas.

CORIANDER (Cilantro) resembles parsley with small leafy sprigs, but is more tender. The flavor is stronger than parsley and lingers on the tongue. Therefore, as a parsley substitute, less is used.

SPECIALTY FRUITS & VEGETABLES

CRABAPPLES are tiny apples with a strong, pleasing flavor for cooking and use in jellies and jams. They are too tart to eat out of hand. They can be yellow, red or green.

DAIKON (Japanese White Radish) is a large, tapered white radish about 8–10 inches long. It has a crisp texture and a flavor similar to an ordinary radish, but sharper and hotter.

FENNEL (Anise) has broad leaf stalks which overlap at the base of the stem and form a firm, rounded, white bulb. The leaves are green and featherlike and the plant has a licorice aroma and flavor. The bulb as well as the inside stalks are edible.

GINGER ROOT is the underground stem or root of the tropical ginger plant. It is light reddish brown and knobby, and is used as a tangy flavoring when shredded or mashed.

HORSERADISH ROOT (German Mustard) has a potent flavor and is used in small quantities. The edible part of the plant is the white root, which should be grated or shredded.

JERUSALEM ARTICHOKES (Sunchokes) are root vegetables of a gnarled, knotty appearance with a crisp, crunchy texture and delicious flavor. Eaten cooked or raw, they are very versatile and can be used as a substitute for potatoes.

JICAMA is a large turnip-shaped root vegetable with a light brown peel. Inside, it is white and crisp with a delicate sweet flavor similar to water chestnuts. The peel is not usually eaten.

KIWIFRUIT has an unattractive brown fuzzy surface. The inside, however, is smooth, creamy, bright green flesh with a strawberry-melon flavor and tiny edible black seeds. Kiwifruit is about the size of a large egg, 3 inches long and 2 inches in diameter.

KOHLRABI is a green leafy vegetable with a stem that thickens above ground and looks like a bulb. The whole plant is edible. The leaves are used like spinach and the bulb (about 2–3 inches in diameter) is served cooked or raw.

Figure 6-1

Specialty fruits and vegetables

Courtesy United Fresh Fruit and Vegetable Association.

KUMQUATS are small, orange-gold citrus fruits with sweet-flavored skins and tart, tangy insides. Eaten whole, the flavors mix nicely. Remove seeds before eating.

MANGOS are subtropical fruits of varied size, shape and color. Their smooth skins can be green, yellow or red, and their shapes vary from round to oval. Sizes can be from 2 to 10 inches and 2 to 5 pounds.

NAPPA (Sui Choy, Chow Choy, Won Bok, Chinese Cabbage) has broad-ribbed stalks varying in color from white to light green, resembling romaine, with crinkly leaves forming a long, slender head. The flavor is mild.

PAPAYA is an oblong melonlike tropical fruit, green to yellow on the outside with golden yellow to orange flesh inside. A medium papaya weighs about 1 pound. The small black seeds inside are not to be eaten.

PERSIMMONS are a brilliant orange in color and soft to the touch (like a tomato). They are about the size of an apple. Enjoy their flavorful taste by eating out of hand.

PLANTAINS look like large green bananas with rough, mottled peels. They are used as a vegetable and must be cooked. Never eat them raw. Black skins indicate that they are sweet and fully ripe.

POMEGRANATES are about the size and shape of apples, golden yellow to deep red, with hundreds of kernels or seeds which are juicy and sweet. Inside, the pulp is crimson and too bitter to eat. Only the seeds are edible.

PRICKLY PEARS (Cactus Pear, Indian Fig, Barberry Fig, Tuna) are pear-shaped with yellow to crimson skins covered with spines. Usually the spines have been removed before marketing. The flesh is yellow and sweet with a taste of watermelon; the skin is bright red when ripe. Peel back the skin to eat.

QUINCE looks like an apple with an odd-shaped stem end and skin that is pale yellow when the fruit is fully ripe. Quince has a distinctive biting flavor and should be served cooked in jams, jellies, sauces, puddings, etc., not eaten raw.

SALSIFY (Oyster Plant) looks like parsnip with a full, grassy top, a firm gray-white root, and the juicy flavor of fresh oysters. The edible part is the carrot-shaped root, which should be peeled and then cooked.

SNOW PEAS (Sugar Peas, China Peas) are an oriental vegetable looking like regular peas, except that the pods are much flatter and translucent. They have a delicate flavor and add crunch and texture to a dish. Serve hot or cold after blanching 2–3 minutes.

SPAGHETTI SQUASH (Cucuzzi, Calabash, Suzza Melon) is a large edible gourd that is round to oval, about 2 feet long and 3–4 inches in diameter. The skin is pale yellow and edible; the inside forms translucent strands similar to spaghetti when cooked. Use it like pasta.

SUGAR CANE is woody sugar stalks cut into 4–6 inch lengths. Eat by pulling away the outer covering and chewing the cane. Boil the cane for a sweet syrup topping.

TARO ROOT (Dasheen) is a highly digestible starchy root with a mild flavor, used in many different ways like a potato. It should be peeled and cooked before eating.

TOFU (Soybean Curd) is similar to a very soft cheese or custard, with a bland flavor and the ability to pick up the flavor of the foods with which it is cooked.

TOMATILLOS (Ground Tomatoes, Husk Tomatoes) are small green vegetables that grow on ground vines and look like tiny tomatoes. They turn from bright green to yellow when fully ripe but are often used unripe, raw or cooked.

UGLI FRUIT is an unattractive, juicy citrus hybird, a cross between a grapefruit and a tangerine. It is the size and shape of a grapefruit, with a mottled rough peel that looks loose. The rind has light-green blemishes that turn orange when ripe. The fruit has a mild, orangelike flavor.

WATER CHESTNUTS (Chinese Water Chestnuts, Waternuts) are the small edible roots of the tropical Water Chestnut plant. The deep brown skin of the roots is removed and only the white, crunchy inside is used (mainly in salads or stir-fry dishes) after blanching 2–4 minutes.

YUCCA ROOT (Manioc, Cassava, Casava) is the large, starchy root of the Yucca plant. Used like a potato, it must be peeled before cooking and cooked before eating. Its flavor is mild but very pleasing.

so that air can circulate and produce can continue to breathe. Most produce needs refrigeration — except green bananas, avocados, onions, and potatoes. The enzymes responsible for ripening bananas are more active at warm temperatures.

The saying about a bad apple spoiling the whole bunch is true: one damaged apple (or any fruit) can accelerate the decay of fruit stored with it. The same is true for potatoes. Also, never store tomatoes near lettuce, because tomatoes cause the lettuce to brown.

Most produce need not be washed before storing. In particular, if you wash the following vegetables and fruits, it may cause mold or wilting: strawberries, raspberries, blueberries, mushrooms, plums, lettuce, and grapes.

Some fruits are picked when they are ripe, and therefore don't ripen after being packed. These include grapes, citrus fruit, berries, and apples. Other fruits benefit from further ripening or softening. These include avocados, bananas, kiwifruit, melons, peaches/nectarines, and tomatoes.

Store frozen vegetables and fruits at 0 degrees F. or below for 10 to 12 months. Canned vegetables and fruits can be stored in a cool, dry place for as long as a year. When stored too long or at too high a temperature, canned vegetables and fruits lose quality (but are still safe to eat). Once opened, store in another container and use within two to three days. Dried fruits should be stored in tightly closed plastic bags or containers. They can last up to one month at room temperature or six months in the refrigerator.

Preparation and Cooking

When preparing and cooking vegetables and fruits, an important consideration is doing so without the loss of valuable nutrients. The Food Facts section in this chapter explains how to retain nutrients.

Vegetables can be cooked using many different cooking methods, including (but not limited to) steaming, baking, broiling, stir-frying, grilling, and sautéing. Following are guidelines for cooking and serving vegetables with optimum color, flavor, and texture.

1. Cook vegetables for as short a time as possible. Do not overcook in order to retain an appropriate texture and the most nutrition, color, and flavor.
2. Steam vegetables when possible, because steaming is an excellent cooking method for retaining nutrients, color, flavor, and texture.
3. Be very careful about timing when a vegetable is ready to come out of the steamer to avoid overcooking. Vegetables such as broccoli, cauliflower, brussels sprouts, and cabbage develop an unpleasant taste and smell when overcooked.

4. When boiling vegetables, start with boiling water to decrease the cooking time and use enough water to cover them (to retain color and flavor), except for green and strong-flavored vegetables. Green vegetables actually need more water to keep their color, and strong-flavored vegetables need more water to help release their strong flavors. Boil green and strong-flavored vegetables uncovered for the same reasons. An exception is spinach, which can be cooked covered in a small amount of water.

5. Do not add baking soda to vegetables, because it causes a mushy texture and nutrient loss.

6. Reheat canned vegetables by bringing half their liquid to a boil, then adding the vegetables and bringing to serving temperature. Canned vegetables are already cooked.

7. When microwaving single servings of vegetables, cover them and be careful not to overcook.

Fruits are not cooked as frequently as vegetables, but they can be easily baked (baked apples), broiled (broiled bananas), sautéed (sautéed peaches), and simmered (applesauce).

On a menu, vegetables have endless possibilities—in appetizers, salads, soups, stir-fries, ragouts, pasta, pizza, and even breads (zucchini bread), cakes (carrot cake), and pies (pumpkin pie). Like vegetables, fruits offer wonderful colors and a variety of fresh flavors and textures to be incorporated into any menu category. Fruits are excellent raw or in salads, poached in a sweetened liquid, roasted, puréed in fruit soup or batter for baked products, chopped in salsas and chutneys, cooked in sauces, and baked in cobblers and pies. Salsas are chunky mixtures of fruits and/or vegetables and flavorings that are served cold as sauces for meat, poultry, and fish. Chutney is a sweet-and-sour condiment made of fruits and/or vegetables cooked in vinegar with sugar and spices. Recipes for chutney and salsa appear in Chapter 8.

CHEF'S TIPS
- Fruits work especially well with breakfast. For example, there is fruit compote (fresh or dried fruit, cooked in syrup flavored with spices or liqueur), glazed grapefruit, baked apple wrapped with phyllo dough, or French crêpes with chopped kiwi salsa.
- Fruits are a natural in salads with vegetables: combine pineapple, raisins, and carrots, or grapes with cucumbers. Feature colorful fruits such as kiwi with carrots in a chef's salad. Cooked fruits and vegetables have many possibilities too: sweet potatoes with apples or lemon-glazed baby carrots.
- Roasted fruits are a wonderful base to place entrées on. For example, place chicken paillard (cutlets pounded and grilled or sautéed) on a bed

of roasted peaches and mango, or monkfish on a bed of roasted pears and fennel. Fruits can also be used to make relishes.

■ Fruits have always been a natural for dessert—fresh, roasted, or perhaps baked into a cobbler with a granola crust or a (strawberry-baked) strudel.

■ Fruits such as pineapple, kiwi, mango, and papaya work well in salsa and relishes.

■ Berries, such as blueberries or strawberries, are wonderful when you want vibrant colors.

■ Vegetables allow you to serve what appears to be a sumptuous portion without the dish being high in calories, fat, or cholesterol. What's also wonderful about vegetables is that they are not very expensive when in season.

■ When using vegetables, you need to *think*—about what's in season for maximum flavor, about how the dish will look, and about how the dish will taste. Think flavor and color. For example, halibut works well with beets, haricots verts, and a grilled tomato.

■ Also, think variety. Serving vegetables doesn't mean switching from broccoli to cauliflower and then back to broccoli. There are many, many varieties of vegetables to choose from. Be adventurous.

■ Most of your meal's eye appeal comes from vegetables—use them to your advantage.

Check-Out Quiz

1. Vitamin E deficiency in adults causes osteomalacia.
   a. True   b. False

2. Water-soluble vitamins are not toxic when taken in excess of the RDA, because they are fully excreted.
   a. True   b. False

3. Vitamin E is a fat-soluble vitamin.
   a. True   b. False

4. Vitamins are needed in very small amounts.
   a. True   b. False

5. Vitamins supply calories and energy to the body.
   a. True   b. False

6. Provitamin A is an active form of the vitamin.
   a. True   b. False

7. Water-soluble vitamins are more likely than fat-soluble to be needed in the diet on a daily basis.
   a. True   b. False

8. A megadose of a vitamin is 5 times the RDA or AI.

a. True   **b.** False

9. Thiamin, riboflavin, and niacin all play key roles as coenzymes in energy metabolism.

a. True   **b.** False

10. Good sources of folate include green leafy vegetables, legumes, and orange juice.

a. True   **b.** False

11. Name the vitamins described in the following.

A. Which vitamin(s) is only present in animal foods?

B. Which vitamin(s) is found in high amounts in pork and ham?

C. Which vitamin(s) is found mostly in fruits and vegetables?

D. Which vitamin(s) needs a compound made in the stomach in order to be absorbed?

E. Which vitamin, when deficient, causes osteomalacia?

F. Which vitamin is made from tryptophan?

G. Which vitamin(s) is made by intestinal bacteria?

H. Which vitamin(s) do you need more of if you eat more protein?

I. Which vitamin(s) is needed for clotting?

J. Which vitamin(s) is increased during pregnancy?

K. Which vitamin(s) is known for forming a cellular cement?

L. Which vitamin has a precursor called beta carotene?

M. Which vitamin is made in your skin?

N. Which vitamin(s) is an antioxidant?

O. Which vitamin(s) is needed for bone growth and maintenance?

P. Which vitamin(s), when deficient, causes night blindness?

Q. Which vitamin(s) is purposely put into milk because there are no other good sources of it available?

Activities and Applications

1. Your Eating Style

Using Table 6-1, circle any foods that you do not eat at all, or that you eat infrequently, such as dairy products or green vegetables. Do you eat most of the foods containing vitamins, or do you hate vegetables and maybe fruits too? In terms of frequency, how often do you eat vitamin-rich foods? The answers to these questions should help you assess whether your diet is adequately balanced and varied, which is necessary to ensure adequate vitamin intake.

2. Supermarket Sleuth

Check at your local pharmacy or supermarket to view the selection of supplements available. How many supply only 100 percent of the RDA? How many supply 500 percent and more of the RDA? Are any "nonvitamins" being sold? Name three.

3. Vitamin Salad Bar

You are to set up a salad bar using Figure 6-2. You may use any foods you like in your salad bar as long as you have a good source of each of the 13 vitamins and you fill each of the circles. In each circle, write down the name of the food and which vitamin(s) it is rich in.

Nutrition Web Explorer

Food and Drug Administration   vm.cfsan.fda.gov/~dms/supplmnt.html
Use this special FDA site on supplements to search for information about any supplement you or someone you know is taking.

Vitamins   bookman.com.au/vitamins
Use this site to read about the results of two scientific studies on vitamins.

Figure 6-2
Vitamin salad bar

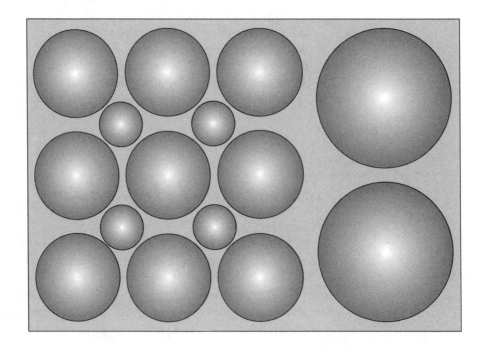

Food Facts *How Food Processing, Storage, Preparation, and Cooking Affect Vitamin and Mineral Content*

Five factors are responsible for most nutrient loss: heat, exposure to air and light, cooking in water, and baking soda. Because the fat-soluble vitamins are insoluble in water, they are fairly stable in cooking. Water-soluble vitamins easily leach out of foods during washing or cooking. Of all the vitamins, vitamin C is the most fragile and the most easily destroyed during preparation, cooking, or storage. Oxygen and high temperatures readily oxidize or destroy vitamin C. Thiamin and folate are also fragile. Here are some tips for retaining food nutrients.

1. Buy fresh, high-quality food.
2. Examine fresh fruits and vegetables thoroughly for appropriate color, size, and shape.
3. Store fruits and vegetables in the refrigerator (except green bananas, avocados, potatoes, and onions) to inhibit enzymes that make fruits and vegetables age and lose nutrients. The enzymes are more active at warm temperatures.
4. Foods should not be stored for too long, as lengthy storage causes nutrient loss. Store canned goods in a cool place. Refrigerated goods should be maintained at a temperature of 41 degrees F. or lower, freezer goods at 0 degrees F. or lower. Thermometers should be kept in the refrigerator and freezer to monitor temperatures.
5. When storing food, close up wrapping tightly to decrease exposure to the air.
6. Wash fruits and vegetables when ready to use, and avoid soaking them.
7. Potatoes and other vegetables that are boiled or baked without being peeled retain many more nutrients than peeled and cut vegetables. In general, the smaller you cut vegetables before cooking, the higher the vitamin loss, because leaching and oxidation are increased with more surface area. Cut vegetables no more than necessary to conserve vitamins A, C, E, and some B vitamins.
8. Keep skins of fruits as much as possible, because more vitamins and minerals reside under the skin than in the center of the fruit.
9. Steaming microwaving, and stir-frying are good choices to retain nutrients when cooking. Each method is fast and uses little or no water. For boiled vegetables, the longer the cooking time and the more water used, the higher the nutrient loss. Cook quickly and use as little water as possible. Also, add the vegetables to the water after it is boiling to conserve vitamin C.
10. Frying temperatures can destroy vitamins such as A, C, E, and K. For instance, French-fried potatoes lose much of their vitamin C.

11. Never use baking soda with green vegetables to improve appearance, as it causes nutrient loss.
12. Broiled or roasted meats retain more B vitamins than meats that are braised or stewed.
13. Use the cooking water from vegetables and the drippings from meats (after skimming off the fat) to prepare soup and gravy.
14. Prepare foods close to the time they will be served.
15. Don't keep milk in clear glass containers, as light destroys the riboflavin. Factors that destroy vitamins often spoil the color, flavor, and texture of food as well.

Hot Topic Functional Foods and Phytochemicals

Foods such as bread and salt were originally enriched and fortified to prevent nutritional-deficiency disease. Today, the fastest-growing category of foods is made up of so-called *functional foods*. These are foods supplemented with ingredients thought to help prevent diseases, such as cancer and heart disease, or to improve health. In other words, functional foods, also called designer foods or nutraceuticals, go beyond basic nutrition. At the supermarket you will find margarine with an ingredient to lower cholesterol, drinks that contain medicinal herbs such as ginseng, and genetically engineered foods like canola oil with beta-carotene, a *carotenoid*. In some stores, you'll even find memory boosters in chewing gum.

Many of the health-promoting ingredients used in functional foods are vitamins, minerals, herbs, or **phytochemicals,** substances such as beta-carotene that are found largely in fruits and vegetables and that seem to be helpful in preventing cancer and/or heart disease. How phytochemicals work and optimal amounts for human consumption are still being investigated. Some act as antioxidants, such as *lycopene* in tomatoes or *phenols* in tea, or they may act in other ways.

For example, broccoli contains the chemical *sulforaphane,* which seems to initiate increased production of cancer-fighting enzymes in the cells. *Isoflavonoids,* found mostly in soy foods, are known as plant estrogens or *phytoestrogens* because they are similar to estrogen and interfere with its actions (estrogen seems to promote breast tumors). Members of the cabbage family (cabbage, broccoli, cauliflower, mustard greens, kale), also called **cruciferous vegetables,** contain phytochemicals such as *indoles* and *dithiolthiones.* They activate enzymes that destroy cancer-causing substances. *Flavonoids,* which are found in citrus fruits, onions, apples, grapes, wine, and tea, are thought to be helpful in preventing heart disease and cancer.

Various phytochemicals reduce heart-disease risk by reducing blood cholesterol levels or preventing blood clotting. Examples of these phytochemicals occur in grapes and garlic.

There are concerns about how effective and safe functional foods are. When manufacturers add herbs to foods, they don't have to disclose how much is added. Is it safe? Do herbs, or other added ingredients really work? Will making tea with St. John's Wort (a herb) really help your depression? Will beta-carotene in a food it normally doesn't appear in still work like the beta-carotene in a carrot? Will any of the added ingredients interact with a medication or dietary supplement you take? Is the fortified food a healthy one, or is it a fortified candy bar?

There are many questions about functional foods, and phytochemicals as well. Like dietary supplements, functional foods will not compensate for poor diets. Whole foods still contain the right amount and balance of nutrients and phytochemicals to promote health.

Chapter 7
Water and Minerals

Major minerals—Minerals needed in relatively large amounts in the diet—over 100 milligrams daily.
Trace minerals—Minerals needed in smaller amounts in the diet—less than 100 milligrams daily.

If you were to weigh all the minerals in your body, they would amount to only 4 or 5 pounds. You need only small amounts of minerals in your diet, but they perform enormously important jobs in your body—building bones and teeth, regulating your heartbeat, and transporting oxygen from the lungs to tissues, to name a few.

Some minerals are needed in relatively large amounts in the diet—over 100 milligrams daily. These minerals, called **major minerals,** include calcium, chloride, magnesium, phosphorus, potassium, sodium, and sulfur. Other minerals, called **trace minerals** or trace elements, are needed in smaller amounts—less than 100 milligrams daily. Iron, fluoride, and zinc are examples of trace minerals. Table 7-1 lists all the major and trace minerals.

Minerals have some distinctive properties not shared by other nutrients. For example, whereas over 90 percent of dietary carbohydrates, fats, and proteins are absorbed into the body, the percentage of minerals that are absorbed varies tremendously. For example, only 5 to 10 percent of the iron in your diet is normally absorbed, about 30 percent of calcium is absorbed, and almost all the sodium you eat is absorbed. Minerals in animal foods tend to be absorbed better than those in plant foods, because plant foods contain fiber and other substances that bind minerals that prevent them from being absorbed. The degree to which a nutrient is absorbed and available to be used in the body is called *bioavailability.* Sometimes minerals even compete with each other for absorption.

Unlike vitamins, minerals are not destroyed in food storage or preparation. They are, however, water-soluble, so there is some loss in cooking liquids. Like vitamins, minerals can be toxic when consumed in excessive amounts and may interfere with the absorption and metabolism of other minerals.

TABLE 7-1	Major and Trace Minerals
Major Minerals	Trace Minerals
Calcium	Chromium
Chloride	Cobalt
Magnesium	Copper
Phosphorus	Fluoride
Potassium	Iodine
Sodium	Iron
Sulfur	Manganese
	Molybdenum
	Selenium
	Zinc

This chapter will help you to:

- List the functions of water.
- Identify the percent of body weight made up of water.
- Recognize the importance of minerals and water to a healthy diet.
- Identify the functions and food sources of the major minerals: calcium, phosphorus, sodium, potassium, chloride, magnesium, and sulfur; and the trace minerals: chromium, copper, fluoride, iodine, iron, selenium, and zinc.
- Discuss how high levels of minerals, such as lead, can be toxic, and how deficiencies can cause diseases such as osteoporosis and iron-deficiency anemia.
- Discuss the purchasing, storage, and use of nuts and seeds in cooking.
- Identify when supplements may be necessary.

Water

Deny someone food and he or she can still live for weeks. But death comes quickly, in a matter of a few days, if you deprive a person of water. Nothing survives without water, and virtually nothing takes place in the body without water playing a vital role.

Although variations may be great, the average adult's body weight is generally 50 to 60 percent water—enough, if it were bottled, to fill 40 to 50 quarts. For example, in a 150-pound man, water accounts for about 90 pounds, fat about 30 pounds, with protein, carbohydrates, vitamins, and minerals making up the balance. Men generally have more water than women, a lean person more than an obese person. Some parts of the body have more water than others. Human blood is about 92 percent water, muscle and brain about 75 percent, and bone 22 percent.

The body uses water for virtually all its functions: for digestion, absorption, circulation, excretion, transporting nutrients, building tissue, and maintaining temperature. Almost all body cells need and depend on water to perform their functions. Water carries nutrients to the cells and carries away waste materials to the kidneys.

Water is needed in each step of the process of converting food into energy and tissue. Water in the digestive secretions softens, dilutes, and liquifies the food to facilitate digestion. It also helps move food along the gastrointestinal tract. Differences in the fluid concentration on either side of the intestinal wall enhance the absorption process.

Water serves as an important part of body lubricants, helping to cushion the joints and internal organs, keeping tissues in the eyes, lungs, and air passages moist, and surrounding and protecting the fetus during pregnancy.

Many adults take in and excrete between 8 and 10 cups of fluid daily. Nearly all foods have some water. Milk, for example, is about 87 percent water, eggs about 75 percent, meat between 40 and 75 percent, vegetables from 70 to 95 percent, cereals from 8 to 20 percent, and bread around 35 percent.

The body gets rid of the water it doesn't need through the kidneys and skin and, to a lesser degree, from the lungs and gastrointestinal tract. Water is also excreted as urine by the kidneys along with waste materials carried from the cells. About 4 to 6 cups a day are excreted as urine. The amount of urine reflects, to some extent, the amount of an individual's fluid intake, although despite the amount consumed, the kidneys will always excrete a certain amount each day to eliminate waste products generated by the body's metabolic actions. In addition to the urine, air released from the lungs contains some water, and evaporation that occurs on the skin (when sweating or not sweating) contains water as well.

If normal and healthy, the body maintains water at a constant level. A number of mechanisms, including the sensation of thirst, operate to keep body-water content within narrow limits. You feel thirsty when the blood starts to become too concentrated. Unfortunately, by the time you feel thirsty, you are already much in need of extra fluid, but this can be easily remedied by drinking promptly. It is therefore very important not to ignore feelings of thirst, a concern that is particularly appropriate for the elderly. For healthy individuals, it is not possible to drink too much water—it will simply be excreted.

There are, of course, conditions in which the various body mechanisms for regulating water balance do not work, such as severe vomiting, diarrhea, excessive bleeding, high fever, burns, and excessive perspiration. In these situations, large amounts of fluids and minerals are lost. These conditions are medical problems to be managed by a physician.

■ **MINI-SUMMARY**

The average adult's body weight is generally 50 to 60 percent water. The body uses water for virtually all its functions: for digestion, absorption, circulation, excretion, transporting nutrients, building tissue, and maintaining temperature. Almost all body cells need and depend on water to perform their functions. Water carries nutrients to the cells and carries away waste materials to the kidneys. A number of mechanisms, including the sensation of thirst, operate to keep body-water content within narrow limits. You feel thirsty when the blood starts to become too concentrated.

Major Minerals

Calcium and Phosphorus

Calcium and phosphorus are used for building bones and teeth. Most of the body's calcium and phosphorus is found in the bones and teeth, where they give rigidity to the structures. Bone is being rebuilt every day, with new bone being formed and old bone being taken apart. There is little turnover of calcium in teeth.

Calcium also circulates in the blood, where a constant level is maintained so it is always available for use. Calcium helps blood to clot, muscles to contract (including the heart muscle), and nerves to transmit impulses, and also helps maintain normal blood pressure. In cases of inadequate dietary intake, calcium is taken out of the bones to maintain adequate blood levels.

Like calcium, phosphorus circulates in the blood and is involved in the metabolic release of energy from fat, protein, and carbohydrates. It is also a part of DNA (genetic material) and many enzymes. Normal body processes produce acids and bases that can cause major blood and body problems, such as coma and death, if not buffered (or neutralized) somehow. Phosphorus has the ability to buffer both acids and bases.

The major sources of calcium are milk and milk products. Not all milk products are as rich in calcium as milk (see Table 7-2). As a matter of fact, butter, cream, and cream cheese contain little calcium. One glass of milk or yogurt or 1-1/2 ounces of cheese each have a little less than one-third of the AI (Adequate Intake) for most adults.

Without milk or milk products in your diet, it may be difficult to get enough calcium. Other good sources of calcium include calcium-set tofu, calcium-fortified foods such as orange juice, and several greens such as broccoli, collards, kale, mustard greens, and turnip greens. Other greens such as spinach, beet greens, Swiss chard, and parsley are calcium-rich but also contain a binder (**oxalic acid**) that seems to prevent some calcium from being absorbed. Dried beans and peas, whole-wheat bread, and certain shellfish contain moderate amounts of calcium, but are usually not eaten in sufficient quantities to make a significant contribution. Meats are poor sources.

About 30 percent of the calcium you eat is absorbed. The body absorbs more calcium (up to 60 percent) during growth and pregnancy, when additional calcium is needed. Absorption is higher for younger people than for older people. Postmenopausal women, who are at high risk of developing osteoporosis if not taking estrogen supplements, often absorb the least

TABLE 7-2 Calcium in Selected Foods	
Food	Calcium Content
Milk, skim, 8 ounces	302
Milk, 2%, 8 ounces	297
Milk, whole, 8 ounces	291
Yogurt, low-fat, 8 ounces	415
Yogurt, low-fat with fruit, 8 ounces	345
Yogurt, frozen, 1 cup	200
Ice cream, vanilla, 1 cup low-fat	176
Cottage cheese, low-fat, 1 cup	155
Swiss cheese, 1 ounces	272
Parmesan, grated, 2 tablespoons	138
Cheddar, 1 ounce	204
Mozzarella, 1 ounce	147
American cheese, 1 ounce	174
Cheese pizza, 1/4 of 14-inch pie	332
Macaroni and cheese, 1/2 cup	181
Orange juice, calcium fortified, 8 ounces	330
Sardines, drained, 2 ounces	175
Oysters, cooked, 3 ounces	76
Shrimp, cooked, 3 ounces	33
Tofu, calcium set, 3-1/2 ounces	128
Dried navy beans, cooked, 1 cup	95
Turnip greens, frozen and cooked, 1 cup	249
Kale, frozen and cooked, 1 cup	179
Mustard greens, 1 cup	104
Broccoli, cooked, 1/2 cup	35
Oatmeal, instant, fortified, 1 packet	160
Pancakes, from mix, one 4-inch pancake	30
Wheat bread, 2 slices	40

Source: U.S. Department of Agriculture Handbook 8, and Home and Garden Bulletin Number 72, and manufacturers.

calcium. Sufficient vitamin D helps calcium absorption and is added to milk. Certain substances interfere with calcium absorption, such as the tannins in tea and large amounts of phytic acid, a binder found in wheat bran and whole grains.

Phosphorus is widely distributed in foods and is rarely lacking in the diet. Milk and milk products are excellent sources of phosphorus, as they are for calcium. Other good sources of phosphorus are meat, poultry, fish, eggs, and legumes. Fruits and vegetables are generally low in this mineral.

Compounds made with phosphorus are used in processed foods, especially soft drinks (phosphoric acid).

Magnesium

Magnesium is found in all body tissues, with about 60 percent in the bones and the remainder in the soft tissues, such as muscles, and in the blood. It is essential to many enzyme systems responsible for energy conversions in the body, and other functions. Magnesium is used in building bones and maintaining teeth. Also like calcium, magnesium is necessary for muscle contraction and nerve transmission. Magnesium also helps the immune system to work properly.

Magnesium is a part of the green pigment **chlorophyll** found in plants, so good sources include green leafy vegetables, potatoes, nuts (especially almonds and cashews), legumes, and wholegrain cereals. Seafood is also a good source. Meat and dairy products supply small amounts. Absorption is only 30 to 40 percent.

Most people probably take in less than the RDA, but deficiency symptoms are rare.

Sodium

Sodium, potassium, and chloride are collectively referred to as **electrolytes** because, when dissolved in body fluids, they separate into positively or negatively charged particles called **ions.** Potassium, which is positively charged, is found mainly within the cells. Sodium (positively charged) and chloride (negatively charged) are found mostly in the fluid outside the cells.

The electrolytes maintain two critical balancing acts in the body: **water balance** and **acid-base balance.** Water balance means maintaining the proper amount of water in each of the body's three "compartments": inside the cells, outside the cells, and in the blood vessels. Electrolytes maintain the water balance by moving the water around in the body. Electrolytes are also able to buffer, or neutralize, various acids and bases in the body. In addition to its roles in water and acid-base balance, sodium is needed for muscle contraction and transmission of nerve impulses.

The major source of sodium in the diet is salt—a compound made of sodium and chlorine. Salt by weight is 39 percent sodium, and 1 teaspoon contains 2,300 milligrams (a little more than 2 grams) of sodium. Many processed foods have high amounts of sodium added during processing and manufacturing, and it is estimated that these foods provide fully 75 percent of the sodium in most people's diets. The following is a list of processed foods high in sodium:

Electrolytes—Chemical elements or compounds that ionize in solution and can carry an electric current; they include sodium, potassium, and chloride.

Water balance—The process of maintaining the proper amount of water in each of three bodily "compartments": inside the cells, outside the cells, and in the blood vessels.

Acid-base balance—The process by which the body buffers the acids and bases normally produced in the body so that the blood is neither too acidic nor basic.

- Canned, cured, and/or smoked meats and fish, such as bacon, salt pork, sausage, scrapple, ham, bologna, corned beef, frankfurters, luncheon meats, canned tuna fish and salmon, and smoked salmon.
- Many cheeses, especially processed cheeses such as processed American cheese
- Salted snack foods, such as potato chips, pretzels, popcorn, nuts, and crackers
- Food prepared in brine, such as pickles, olives, and sauerkraut
- Canned vegetables, tomato products, soups, and vegetable juices
- Prepared mixes for stuffings, rice dishes, and breading
- Dried soup mixes and bouillon cubes
- Certain seasonings such as salt, sea salt, garlic salt, onion salt, celery salt, seasoned salt, soy sauce, Worcestershire sauce, horseradish, ketchup, and mustard

Table 7-3 illustrates the salt content of food in the various food groups. Salt is also used in food preparation and at the table for seasoning.

TABLE 7-3   Where's the Salt?

Food Groups	Sodium, mg	Food Groups	Sodium, mg
Bread, Cereal, Rice, and Pasta		Natural cheeses, 1-1/2 oz.	110–450
Cooked cereal, rice, pasta, unsalted, 1/2 cup	Trace	Process cheeses, 2 oz.	800
Ready-to-eat cereal, 1 oz.	100–360	*Meat, Poultry, Fish, Dry Beans,*	
Bread, 1 slice	110–175	*Eggs, and Nuts*	
		Fresh meat, poultry, fish, 3 oz.	Less than 90
Vegetable		Tuna, canned, water pack, 3 oz.	300
Vegetables, fresh or frozen, cooked without salt, 1/2 cup	Less than 70	Bologna, 2 oz.	580
		Ham, lean, roasted, 3 oz.	1,020
Vegetables, canned or frozen with sauce, 1/2 cup	140–460	*Other*	
Tomato juice, canned, 3/4 cup	660	Salad dressing, 1 tbsp.	75–220
Vegetable soup, canned, 1 cup	820	Ketchup, mustard, steak sauce, 1 tbsp.	130–230
Fruit		Soy sauce, 1 tbsp.	1,030
Fruit, fresh, frozen, canned 1/2 cup	Trace	Salt, 1 tsp.	2,000
		Dill pickle, 1 medium	930
Milk, Yogurt, and Cheese		Potato chips, salted, 1 oz.	130
Milk, 1 cup	120	Corn chips, salted, 1 oz.	235
Yogurt, 8 oz.	160	Peanuts, roasted in oil, salted, 1 oz.	120

Source: U.S. Department of Agriculture.

In addition to the sodium in salt, sodium appears in monosodium glutamate (MSG), baking powder, and baking soda. Other possible sources of dietary sodium include the sodium found in some local water systems and in medications, such as some antacids. Unprocessed foods also contain natural sodium, but in small amounts (with the exception of milk and some milk products).

There is no RDA for sodium. Instead, there is an estimated minimum requirement: 500 milligrams. The sodium intake of Americans is easily 6 times this amount—varying from 3 to 8 grams daily. The Nutrition Facts Label lists a Daily Value of 2,400 mg per day for sodium. Overconsumption of sodium, particularly as salt, is a concern because a high sodium intake aggravates high blood pressure in some people.

Potassium

Potassium, an electrolyte found mainly in the fluid inside the individual body cells, helps maintain water balance and acid-base balance along with sodium. In the blood, potassium assists in muscle contraction, including maintaining a normal heartbeat, and sending nerve impulses.

Potassium is distributed widely in foods, both plant and animal. Unprocessed, whole foods such as fruits and vegetables (especially winter squash, potatoes, oranges, and grapefruits), milk, yogurt, legumes, and meats are excellent sources of potassium.

A potassium deficiency is uncommon in healthy people but may result from dehydration, certain diseases, or drugs. Diuretics, a certain class of blood pressure drugs, cause increased urine output, and some cause an increased excretion of potassium as well. Symptoms of a deficiency include muscle cramps, weakness, nausea, and abnormal heart rhythms that can be very dangerous, even fatal.

Chloride

Chloride, another important electrolyte, helps maintain water balance and acid-base balance. It is also a part of hydrochloric acid, which is quite highly concentrated in the stomach juices. Hydrochloric acid aids in protein digestion, destroys harmful bacteria, and increases the absorption of calcium and iron. The most important source of dietary chloride is sodium chloride, or salt. If sodium intake is adequate, there will be ample chloride as well.

Other Major Minerals

The body doesn't use the mineral sulfur by itself, but uses the nutrients it is found in, such as protein and thiamin. The protein in hair, skin, and nails is particularly rich in sulfur. There is no RDA for sulfur. Protein foods supply plentiful amounts of sulfur, and deficiencies are unknown.

■ **MINI-SUMMARY**

The functions and sources of the major minerals are listed in Table 7-4.

Trace Minerals

Trace minerals (Table 7-1) represent an exciting area for research, because our understanding of many trace minerals is still emerging. Like vitamins, minerals can be toxic at high doses. Unlike vitamins, many trace minerals are toxic at levels only several times higher than recommendations. Also, trace minerals are highly interactive with each other. For example, taking extra zinc can cause a copper deficiency. For this reason, it is advisable to consult a physician for supplementation exceeding one and one-half times the recommended levels.

Iron

Hemoglobin—A protein in red blood cells that carries oxygen through the body.
Myoglobin—An iron-containing protein found in muscles that stores and carries oxygen.
Heme iron—The predominant form of iron in animal foods, absorbed and used more readily than iron in plant foods.
Nonheme iron—A form of iron found in all plant sources of iron and also as part of the iron in animal food sources.

Iron, one of the most abundant metals in the universe and one of the most important in the body, is a key component of **hemoglobin,** a part of red blood cells that carries oxygen to body cells. Cells require oxygen to break down glucose and produce energy. Iron is also part of **myoglobin,** a muscle protein that stores and carries oxygen that the muscles will use to contract. Iron works with many enzymes in energy metabolism and is needed to make amino acids as well as certain hormones and neurotransmitters.

About one-third of iron in the American diet comes from beef and enriched breads, rolls, and crackers. Other sources include shellfish, legumes, fortified cereals, green leafy vegetables, and poultry (see Table 7-5).

Only 5 to 10 percent of dietary iron is absorbed in healthy individuals. Iron-deficient individuals absorb more, up to 20 percent. The ability of the body to absorb and use iron from different foods varies. The predominant form of iron in animal foods, called **heme iron,** is absorbed and used twice as readily as iron in plant foods, called **nonheme iron.** Animal foods also contain some nonheme iron. The presence of vitamin C in a meal increases

TABLE 7-4	Minerals		
Minerals	Recommended Intake[1]	Functions	Sources
Major Minerals			
Calcium	1,000 milligrams (1997 AI) Upper Level: 2,500 milligrams	Mineralization of bones and teeth Blood clotting Muscle contraction Transmission of nerve impulses	Milk and milk products, calcium-set tofu, calcium-fortified foods, broccoli, collards, kale, mustard greens, turnip greens, legumes, whole-wheat bread
Phosphorus	700 milligrams (1997 RDA) Upper Level: 4,000 milligrams	Mineralzation of bones and teeth Energy metabolism Formation of DNA and many enzymes Buffer	Milk and milk products, meat, poultry, fish, eggs, legumes
Magnesium	Males: 420 milligrams Females: 320 milligrams (1997 RDA) Upper Level: 350 milligrams nonfood magnesium	Energy metabolism Formation of bones and maintenance of teeth Muscle contraction, nerve transmission, immune system	Green leafy vegetables, potatoes, nuts, legumes, wholegrain cereals
Sodium	500 milligrams (1989 Estimated Minimum Requirement)	Water balance Acid-base balance Buffer Muscle contraction Transmission of nerve impulses	Salt, processed foods, MSG
Potassium	2,000 milligrams (1989 Estimated Minimum Requirement)	Water balance Acid-base balance Buffer Muscle contraction Transmission of nerve impulses	Many fruits and vegetables (potatoes, oranges, grapefruit, milk, and yogurt, legumes, meats
Chloride	750 milligrams (1989 Estimated Minimum Requirement)	Water balance Acid-base balance Part of hydrochloric acid in stomach	Salt

Mineral	Recommended Intake[1]	Functions	Food Sources
Sulfur	No RDA	Part of some amino acids Part of thiamin	Protein foods
Trace Minerals			
Copper	1.5–3.0 milligrams (1989 Estimated Safe and Adequate Intake)	Iron metabolism Formation of hemoglobin Collagen formation Energy release	Seafood, wholegrain breads and cereals, legumes, nuts, seeds
Fluoride	Men: 3.8 milligrams Women: 3.1 milligrams (1997 AI)	Strengthening of developing teeth	Water (naturally or artificially fluoridated), tea, seafood
Iodine	150 micrograms (1989 RDA)	Normal functioning of thyroid gland Normal metabolic rate Normal growth and development	Iodized salt, saltwater fish
Iron	Males: 10 milligrams Females: 15 milligrams (1989 RDA)	Part of hemoglobin and myoglobin Part of some enzymes Energy metabolism Needed to make amino acids	Red meats, shellfish, legumes, wholegrain and enriched breads and cereals, green leafy vegetables
Selenium	55 micrograms (2000 RDA)	Activation of an antioxidant	Seafood, meat, liver, eggs, whole grains and vegetables (if soil is rich in selenium)
Zinc	Males: 15 milligrams Females: 12 milligrams (1989 RDA)	Cofactor of many enzymes Wound healing DNA and protein synthesis Bone formation Development of sexual organs General growth and maintenance Taste perception and appetite	Protein foods, legumes, dairy products, wholegrain products, fortified cereals
Chromium	50–200 micrograms (1989 Estimated Safe and Adequate Intake)	Works with insulin	Liver, meat, wholegrains, nuts

[1] Recommended intakes are for adults.

TABLE 7-5 Iron in Foods	
Foods	Milligrams Iron
Meat and Poultry	
Beef liver, 3 ounces	5.3
Sirloin steak, 3 ounces	2.6
Hamburger, 3 ounces, lean	1.8
Chicken breast, 3 ounces	0.9
Shellfish	
Oysters, breaded, fried, 1	3.0
Clams, raw, 3 ounces	2.6
Vegetables	
Spinach, 1 cup, frozen, cooked	2.9
Legumes	
Great Northern beans, 1 cup	4.9
Black beans, 1 cup	2.9
Tofu, 2-1/2 inch × 2-3/4 inch × 1 inch cube	2.3
Eggs	
Egg yolk, 1	0.9
Dried Fruits	
Apricots, 1/4 cup	1.5
Raisins, 1/4 cup	0.7
Breads and Cereals	
Corn flakes cereal, 1 ounce	1.8
Whole-wheat bread, 1 slice	1.0
White bread, 1 slice	0.7

Source: Nutritive Value of Foods, U.S. Department of Agriculture Home and Garden Bulletin Number 72.

nonheme iron absorption, as does consuming meat, poultry, and fish. Some foods actually decrease iron consumption: coffee, tea, phytic acid in whole grains, oxalic acid in some vegetables, and calcium supplements. The body increases or decreases iron absorption as needed, but it absorbs iron more efficiently when iron stores are low and during growth spurts or pregnancy. Some iron is stored in the bone marrow and in the liver.

Iron-deficiency
anemia—A condition in
which the size and number
of red blood cells are
reduced, which may result
from inadequate iron intake
or from blood loss;
symptoms include fatigue,
pallor, and irritability.

Iron overload—A
common genetic disease in
which individuals absorb
about twice as much iron
from their food and
supplements as other
people do
(hemochromatosis).

Iron deficiency is the most common nutritional deficiency. If severe enough, it results in **iron-deficiency anemia,** a condition in which the size and number of red blood cells are reduced. The blood hemoglobin concentration also falls. This condition may result from inadequate intake of iron or from blood loss. Symptoms of iron-deficiency anemia include fatigue, pallor, pale skin, and apathy. Iron-deficiency anemia is a real concern in the United States, much more so for women than for men. It is also a concern for infants and children. Children with this condition show symptoms of behavioral problems (short attention span, restlessness) when in reality they need more dietary iron. The RDA for iron is higher for women of childbearing age than for men because women have to replace menstrual blood losses. It is not unusual for menstruating women to regularly consume less than the iron RDA. Iron supplementation authorized by a physician may be helpful.

Although the body generally avoids absorbing huge amounts of iron, some people can absorb large amounts. The problem with iron is that once it is in the body, it is hard to get rid of. For individuals who can absorb much iron, large doses of iron supplements can damage the liver and do other damage, a condition called **iron overload** or *hemochromatosis.* This condition is usually caused by a genetic disorder.

It is especially important to keep iron supplements away from children, because they are so toxic they can kill. Children under age 3 are at particular risk.

Zinc

Zinc is involved in enzymes that catalyze at least 50 different metabolically important bodily reactions. Zinc assists in wound healing, bone formation, DNA and protein synthesis, development of sexual organs, and general tissue growth and maintenance. Zinc is also important for taste perception and appetite.

Protein-containing foods are all good sources of zinc, particularly meat, poultry, and shellfish. At best, only about 40 percent of the zinc we eat is absorbed into the body. Legumes, dairy foods, whole grains, and fortified cereals are good sources as well. Zinc is much more readily available, or absorbed better, from animal foods. Iron and zinc are often found in the same foods. Like iron, zinc is more likely to be absorbed when animal sources are eaten and when the body needs it.

Deficiencies are more likely to show up in pregnant women, the young, and the elderly. Adults deficient in zinc may have symptoms such as poor appetite, diarrhea, skin rash, and hair loss. Signs of severe deficiency in children include growth retardation, delayed sexual maturation, decreased sense of taste, poor appetite, delayed wound healing, and immune deficiencies. Marginal deficiencies do occur in the United States.

When overdoses of zinc (only a few milligrams above the RDA) are taken, it causes a copper deficiency and less iron is absorbed. Since zinc supplements can be fatal at lower levels than many of the other trace minerals, zinc supplements should be avoided unless a physician prescribes them.

Iodine

Thyroid gland—A gland found on either side of the trachea that produces and secretes two important hormones that regulate the level of metabolism.

Iodine, the form in which iodine is found in food, is required in extremely small amounts for normal **thyroid gland** functioning. The thyroid gland, located in the neck, is responsible for producing two important hormones that maintain a normal level of metabolism and that are essential for normal growth and development, and much more. Iodine is necessary for these hormones to be produced.

Iodine is not found in many foods: mostly saltwater fish and grains grown in iodine-rich soil (once covered by the oceans, soil in the central states contains little iodine). Iodized salt was introduced in 1924 to combat iodine deficiencies. Iodine also finds its way accidentally into milk (cows receive iodine-containing drugs, and dairy equipment is sterilized with iodine-containing compounds), into baked goods through iodine-containing compounds used in processing, and foods that have certain food colorings. Because iodine can be toxic at high levels, some industries are trying to cut back on its use. Average intake in the U.S. is more than recommended, but less than toxic.

Selenium

Until 1979, it was not known that selenium is an essential mineral. The first RDA for selenium was announced in 1989. Selenium is a component of enzymes that act like antioxidants along with vitamin E to prevent oxidative tissue damage. Excellent sources include seafood, meat, and liver. Because selenium is found in the soil, vegetables and whole grains may be a good source of selenium as well if the soil is selenium-rich. Selenium deficiency can cause a type of heart disease, but this is rare in the United States. When megadoses of selenium are taken over a long time span, nausea, vomiting, hair loss, and malaise can result. Supplements should be 200 micrograms or less.

Fluoride

Fluoride is the term used for the form of fluorine that appears in drinking water and in the body. The terms **fluoride** and fluorine are used interchangeably. In children, fluoride strengthens the mineral composition of the developing teeth so they resist the formation of dental caries, or cavities. In adults, fluoride helps teeth by decreasing the activity of the bacteria in your mouth that cause dental caries.

The major source of fluoride is drinking water, although tea and seafood also contain fluoride. Some water supplies are naturally fluoridated, and many supplies have fluoride added, usually at a concentration of one part fluoride to a million parts water. Fluoride levels in water are stated in concentrations of parts per million (ppm). About 1 ppm is ideal. Less than 0.7 ppm isn't adequate to protect developing teeth. More than about 1.5 to 2.0 ppm can lead to mild **fluorosis,** a condition that causes small, white, virtually invisible opaque areas on teeth. In its most severe form, fluorosis causes a distinct brownish mottling or discoloring. To avoid these problems, children should be advised to use small amounts of fluoride toothpaste and not to swallow it.

Only fluoride taken internally, whether in drinking water or dietary supplements, can strengthen babies' and children's developing teeth to resist decay. Once the teeth have erupted, they are beyond help from ingested flouride. Supplements are important for the approximately 40 percent of Americans who do not have adequately fluoridated water supplies.

For both children and adults, fluoride applied to the surface of the teeth can nonetheless add protection, at least to the outer layer of enamel, where it plays a role in reducing decay. The most familiar form, of course, is fluoride-containing toothpaste, introduced in the early 1960s. Fluoride rinses are also available, as are applications by dental professionals. All such products are regulated by the Food and Drug Administration (FDA). They are considered effective adjuncts to ingested fluoride, and are the only useful sources of tooth-strengthening fluoride for teenagers and adults.

Some health groups have been concerned that fluoride is a cancer risk. In 1991, based on a review of 50 human studies, scientists from the Public Health Service (which includes the National Institutes of Health, the Center for Disease Control, and the FDA) reported that optimal fluoridation of drinking water does not pose a detectable cancer risk. The report did observe, however, that there are multiple sources of fluoride and commented that, "In accordance with the prudent health practice of using no more than the amount necessary to achieve a desired effect, health professionals and the public should avoid excessive and inappropriate fluoride exposure."

Chromium

Chromium works with insulin to transfer glucose and other nutrients from the bloodstream into the body's cells. Chromium deficiency results in a condition much like diabetes in which the blood glucose level is abnormally high. Chromium also plays an important role in the body's metabolism of **lipids.** Although chromium has been advertised as helping you lose weight and put on muscle, well-designed research studies have not shown these effects. Good sources of chromium are whole, unprocessed foods, such as whole grains and breads and cereals made with whole grains, wheat germ, and lean beef. The richest source is brewer's (nutritional) yeast. Megadoses are not advised.

Copper

Copper works as an important part of many enzymes, such as working with iron to form hemoglobin. It also aids in forming collagen, a protein that gives strength and support to bones, teeth, muscle, cartilage, and blood vessels. As part of many important enzymes, it is involved in the nervous system, energy release, and immune system.

Copper occurs mostly in unprocessed foods. Seafood, legumes, nuts, seeds, and wholegrain breads and cereals are rich sources.

Copper deficiency is rare, but marginal deficiencies do occur. Single doses of copper only 4 times the recommended level can cause vomiting and nervous-system disorders.

Other Trace Minerals

Manganese is needed for bone formation, and as part of many enzymes involved in energy metabolism and the metabolism of carbohydrate, fats, and protein. It is found in many foods, especially whole grains, dried beans, nuts, and leafy vegetables. A deficiency is rare.

Molybdenum is a part of a number of enzymes. It appears in legumes, whole grains, nuts, and milk. Deficiency does not seem to be a problem.

As time goes on, more trace minerals will be recognized as essential to human health. There are currently several trace minerals essential to animals that are likely to be essential to humans as well. Possible candidates for nutrient status include arsenic, boron, nickel, silicon, tin, and vanadium.

> ■ **MINI-SUMMARY**
> The functions and sources of the trace minerals are listed in Table 7-4.

Osteoporosis

Osteoporosis is the most common bone disease. Characterized by loss of bone density and strength, osteoporosis is associated with debilitating fractures, especially in people age 45 and older. Bone loss develops over a span of many years and is largely symptomless, although some women may experience chronic spinal pain or muscle spasms in the back. Often, the first sign of osteoporosis is a wrist or hip fracture or a compression fracture that causes the vertebrae in the upper back to collapse, curving the spine into the "dowager's hump" that has come to symbolize osteoporosis.

As many as 8 million Americans, 80 percent of them women, now suffer from the condition, with more than 1.5 million osteoporosis-related fractures occurring annually. A little less than half of all women over 50 will experience an osteoporosis-related fracture in their lifetime. Another 17 million women are at risk.

Luckily, osteoporosis can be prevented, detected, and treated, and it is never too late to do something about it.

Although bones seem to be as lifeless as rocks, they are in fact composed of living tissue that is continually being broken down and rebuilt in a process called *remodeling*. It takes about 90 days for old bone to be broken down and replaced by new bone; then the cycle begins anew. Bones continue to grow in strength and size until a person's early thirties, when peak bone mass is attained. Men achieve more peak bone mass than women. Optimal bone mass and size will be attained only if there has been enough calcium in the diet. After that, bone is broken down faster than it is deposited, resulting in decreased bone mass (about 1 percent per year). In men, bone loss is slow but constant. For women, bone loss speeds up during the five years following menopause due to decreased production of estrogen, and then slows to about the same rate as before menopause.

Besides being influenced by age, sex, and estrogen levels, bone health is also influenced by diet and exercise. For maximum bone health, adequate amounts of calcium and vitamin D need to be taken in. Unfortunately, most women (especially teenagers) do not consume the AI for calcium. During the five to ten years after the beginning of menopause, optimal calcium intake is important. Although some bone loss in inevitable, it can be kept to its programmed minimum with adequate calcium from the diet. The National Institutes of Health has recommended 1,500 milligrams of calcium for postmenopausal women.

Exercise also influences bone health, and participation in sports and exercise increases bone density in children. For older adults, exercise helps

Osteoporosis—The most common bone disease, characterized by loss of bone density and strength; associated with debilitating fractures, especially in people 45 and older, due to a tremendous loss of bone tissue in midlife.

improve strength and balance, making it less likely for them to have a fall. To benefit bone health, exercise must be weight-bearing or strength training. Also, exercise benefits only the bones used, such as the leg bones when biking or walking.

The best approach to osteoporosis is prevention—the reason why calcium intake is so important. Starting in childhood through young adulthood (when bones are forming), adequate calcium intake is vital to having more bone mass at maturity. Adequate intake of calcium is also important after early adulthood.

Individuals who are aware of the problems of osteoporosis sometimes take calcium supplements. Many calcium supplements provide mixtures of calcium with other compounds such as calcium carbonate, a good source of calcium. There are also powdered forms of calcium-rich substances, such as bonemeal and dolomite (a rock mineral). These are dangerous because they may contain lead and other elements in amounts that constitute a risk. Choose a calcium supplement with the U.S.P. seal of approval. Excessive intake of calcium can cause problems such as possible urinary stone formation, constipation, and decreased absorption of iron and other nutrients.

Other ways to prevent osteoporosis include regular weight-bearing or strength-training exercise (as already mentioned), consumption of adequate milk for vitamin D, exposure to the sun (for more vitamin D), estrogen therapy for women, moderate consumption of alcohol and caffeine, and avoiding smoking.

It has been known for some time that estrogen therapy at and after menopause prevents osteoporosis-related fractures. Estrogen received FDA approval as a treatment for osteoporosis in 1988. Although the hormone decreases the risk of osteoporosis, with long-term use it may increase the risk of breast and endometrial cancers. Estrogen therapy (which is also helpful in treating menopause symptoms) is a poor choice for some groups of women, such as those who have estrogen-sensitive breast cancer.

■ **MINI-SUMMARY**

The best approach to osteoporosis is prevention—taking in the AI for calcium, regular exercise, consuming milk for adequate vitamin D, consuming moderate amounts of alcohol, and avoiding smoking. Estrogen therapy may be advised for some menopausal women.

Ingredient Focus: Nuts and Seeds

Nuts and seeds pack quite a few vitamins (such as folate and vitamin E) and minerals, along with fiber and protein, in their small sizes. Nuts, in particular, also contain quite a bit of fat. Luckily most of the fat (except in walnuts) is monounsaturated. One ounce of many nuts contains from 13 to 18 grams of fat, making them also a relatively high-calorie food, although they may improve success when included in a calorie-controlled diet. By comparison, seeds contain less fat but still quite a few calories.

Nuts usually grow on trees and are characterized by a hard, removable outer shell. Some commonly used nuts include the following:

- **Almonds** were common ingredients in the cuisines of ancient China, Greece, Turkey, and the Middle East. Today much of the world's supply of almonds is grown in California. Almonds are sweet with a delicate butterlike flavor.
- **Brazil nuts** are the firm, but tender, fruit of a South American tree. They have a clean, slightly oily taste. They are high in calories and do not benefit from roasting or toasting.
- **Cashews** form on the bottom of a pear-shaped fruit. Because of the process needed to remove the shell, they are not readily available in the shell.
- **Macadamia nuts** are grown in Hawaii, Australia, and Central America. They are high in cost and very high in fat (store in refrigerator).
- **Peanuts** probably came from Brazil and are grown in the southern United States. Three types of peanuts are most commonly grown: Virginias and Runners, which have red skins, and Spanish, which are smaller and have a more tan skin.
- **Pecans** grow on huge trees native to the Mississippi River Valley. Georgia is the main source of pecans, which is a wonderful all-purpose nut. Kernels are best stored in the refrigerator.
- **Pine nuts,** also known as pignoli, are popular in Italian cuisine, where they are used in rice, sauces, and cakes. They are also used in Turkish, Middle Eastern, and Mexican cooking. There are two varieties: The Mediterranean or Italian pine nut with a light flavor, and the Chinese pine nut with a stronger flavor.
- **Pistachios,** originally from the Middle East and Asia, are now grown in California. The pistachio shell splits naturally as part of the ripening process. Pistachios were originally dyed red by importers to cover stains in imported nuts. Most California pistachios are sold with their natural ivory shell.
- **Walnuts** were introduced to California by the Franciscan Fathers in the 1700s. The mellow flavor of the walnut works well with a variety of

foods. Most walnuts marketed in the United States are the English variety. The black walnut is sweet and has a deeper flavor. Walnuts are a good source of omega-3 fatty acids.

Nuts in their shell can be stored at room temperature in a cool, dry location. Once shelled, most nuts need to be refrigerated.

Nuts are used in all their forms and styles (whole, sliced, pieces, ground, butters, oils) in baked goods, in stews and ragouts, and as toppings for salads, cooked vegetables, and entrées such as fish. Nuts often add eye appeal and an unexpected change in texture. In part due to their high caloric and fat content, nuts are often used in small amounts. By toasting or roasting nuts, you can bring out a more intense flavor and use less. Small amounts of flavorful nuts and seeds can often replace fats such as butter or margarine. Other foods, such as pumpkin seeds or roasted chickpeas, can also be used to replace part or all of the nuts in a dish and still provide a crunchy texture.

Seeds are versatile as well.

- Pumpkin and sunflower seeds are large compared to seeds such as sesame and caraway. They can be used in casseroles, stews, vegetables, stuffings, or salads.
- Sesame seeds and caraway seeds are often used in baking. Toasted sesame seeds can be sprinkled on soups, fish, and cooked vegetables for flavor and texture.

Seeds should be stored in a tightly covered container in a cool, dry, dark area.

CHEF'S TIPS
- To toast nuts such as almonds, spread them in a single layer in a shallow pan. Bake at 325 degrees F for 10 to 15 minutes, or until the almonds are lightly colored. Toss occasionally. They will continue to brown slightly after you remove them from the oven.
- To roast nuts, lightly coat the kernels with oil (about two tablespoons per pound of nuts) and proceed as in toasting.
- Nuts are wonderful in muffins—such as honey-almond muffins or walnut-strawberry muffins.
- Nuts and seeds work well in granolas, give crunch and flavor to casseroles, and add interest to salads, such as fennel, oranges, watercress, and walnut salad.
- Nuts, and seeds turn rancid easily due to their fat content. Store in airtight containers away from heat and light.

Check-Out Quiz

1. Our understanding of many trace minerals is still emerging.
   a. True   **b.** False
2. Two cups of yogurt contain about as much calcium as one cup of milk.
   a. True   **b.** False
3. Sodium and potassium are involved in muscle contraction and transmission of nerve impulses.
   a. True   **b.** False
4. Canned soft drinks are high in sodium.
   a. True   **b.** False
5. Iodine is needed to maintain a normal metabolic rate.
   a. True   **b.** False
6. Few minerals are toxic in excess.
   a. True   **b.** False
7. Nearly all foods contain water.
   a. True   **b.** False
8. The kidneys will always excrete a certain amount each day to eliminate waste products.
   a. True   **b.** False
9. Sodium, potassium, and calcium are referred to as electrolytes.
   a. True   **b.** False
10. Name the mineral(s) referred to by the following descriptions:
    a. Involved in bone formation.
    b. Found mostly in milk and milk products.
    c. Helps maintain water balance and acid-base balance.
    d. Some diuretics deplete the body of this mineral.
    e. Found in the stomach juices.
    f. Important for a healthy heart.
    g. Found in certain water supplies.
    h. Found in salt.
    i. Found in heme.
    j. Causes a form of anemia.
    k. Occurs in the soil.
    l. Part of vitamin B_{12}.

Activities and Applications

1. Your Eating Style

Using Table 7-4, circle any food sources of minerals that you do not eat at all, or eat infrequently, such as dairy products or green vegetables. Do you eat many of the foods containing minerals, or only a fair amount of them? How often do you eat mineral-rich foods? The answers to these questions should help you assess whether your diet is adequately balanced and varied to ensure adequate mineral intake.

2. How Much Do You Drink?

Keep a diary of how many fluid ounces you drink in one day, including beverages such as coffee and soft drinks, but excluding alcoholic beverages. Convert the number of fluid ounces to the number of cups by dividing the total ounces by 8. Compare this number of cups to the 8 to 10 cups you need daily. Did you drink enough fluid? When might you need over 10 cups daily?

3. Mineral Salad Bar

You are to set up a salad bar using Figure 7-1. You may use any foods you like in your salad bar as long as you have a good source of each of the minerals listed in Table 7-4 and you fill each of the circles. In each circle, write down the name of the food and which mineral(s) it is rich in.

4. Sodium Countdown

Using Appendix A, "Nutritive Value of Foods," list the sodium content of 10 of your favorite foods. How much would each contribute to the recommendation of 2,400 milligrams sodium (maximum)?

Figure 7-1
Minerals salad bar

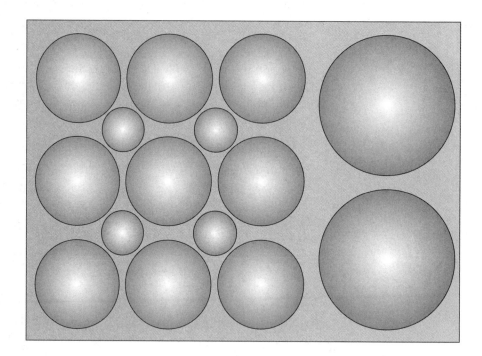

Nutrition Web Explorer

National Institutes of Health Osteoporosis Resource Center www.osteo.org.
On the NIH home page, click on "Bone Health Information." Next click on "Osteoporosis," and then "Fast Facts." Find out why osteoporosis is called the "silent" disease.

Iron Overload Diseases Association www.ironoverload.org
Read the Fact Sheet available at this site and find out if iron overload disease can be controlled by diet.

Food Facts *The Dangers of Lead*

Lead has no known functions or health benefits for humans. In fact, it is a highly toxic metal that can damage the nervous, cardiovascular, renal, immune, and gastrointestinal systems and is particularly dangerous for children. In children, lead has a particularly damaging effect on intellectual development. In addition, lead interferes with the manufacture of heme, the oxygen-carrying part of hemoglobin in red blood cells. Lead consumption in childhood can mean stunted growth and a lower IQ. Damage to the child's nervous system is permanent.

New research on lead shows that it may be dangerous at low levels in adults. Low levels of lead may contribute to hypertension and harm the kidneys. Significant sources of lead include lead-based paint, the overwhelming source, and also drinking water carried in lead pipes, lead-soldered cans, and some ceramic dishes covered with lead glaze. Lead can be in any home's or business's water, so it is advisable to have your water tested. In addition to the problem with lead pipes, some faucets contain lead and leach lead into the water.

Although the number of cans produced in the United States that use lead solder has decreased to under 4 percent, the number of imported cans with lead solder is unknown. The only way to make sure that a can is not soldered with lead is to choose one-piece aluminium cans (like those for soft drinks) or cans with welded seams that have shiny metal around the seam. Cans that are soldered are not so shiny around the seam because some solder is usually obvious.

Most ceramic glaze contains lead, which, if properly fired and sealed, does not cause any concern. However, some ceramic cookware (and dishware), from both outside and inside the United States, has been found to leach lead in dangerous amounts into food. Ceramic items may include fine china, stoneware, earthenware, and ironstone. Unfortunately, there is no way of telling whether a ceramic piece has an unsafe amount of lead unless you use a lead-testing kit. Lead also leaches from lead crystal, and it can leach into any liquid, not just alcohol. Here are precautions to follow to lessen your chances of lead poisoning.

1. Don't store foods in ceramic cookware or dishes unless you are sure they are lead-free.
2. Avoid using ceramic dishes to serve acidic foods and beverages, which cause more lead to be leached out. Examples of acidic foods include citrus juices, apple juice, tomato products, cola-flavored soft drinks, coffee, or tea.
3. Heat also causes more lead to be leached out, so avoid cooking and microwaving with ceramic cookware or dishes that you are not sure of.
4. Do not use, and perhaps dispose of, any china on which the glaze is corroded or has a chalky gray residue when dry.
5. Be cautious about using very old china and highly decorated handcrafted china.
6. If buying new ceramic cookware or dishes, select a manufacturer, such as Corning, that produces lead-free glazes.
7. Do not use lead crystal every day, and never use it to store food or beverages. Also, never let children use it.
8. If in doubt, check it out: buy a lead-testing kit. They cost from $25 and up in hardware stores.

Hot Topic Dietary Supplements

Surveys show that more than half of the U.S. adult population use dietary supplements. In 1996 alone, consumers spent more than $6.5 billion on not only vitamins and minerals, but herbs, amino acids, and a bewildering array of substances. Traditionally, the term "dietary supplements" referred to products made of one or more of the essential nutrients, such as vitamins, minerals, and protein. But the 1994 Dietary Supplement Health and Education Act (DSHEA) broadened the definition to include, with some exceptions, any product intended for ingestion as a supplement to the diet. In addition to vitamins, minerals, and proteins, dietary supplements may include herbs, botanicals, and other plant-derived substances, as well as amino acids, concentrates, metabolites, constituents, and extracts of these substances.

It's easy to spot a supplement, because DSHEA requires manufacturers to include the words "dietary supplement" on product labels. Also, a "Supplement Facts" panel (Figure 7-2) is required on the labels of most dietary supplements.

Dietary supplements come in many forms, including tablets, capsules, powders, softgels, gelcaps, and liquids. Though commonly associated with health-food stores, dietary supplements also are sold in grocery, drug, and national discount chain stores, as well as through mail-order catalogs, TV programs, the Internet, and direct sales.

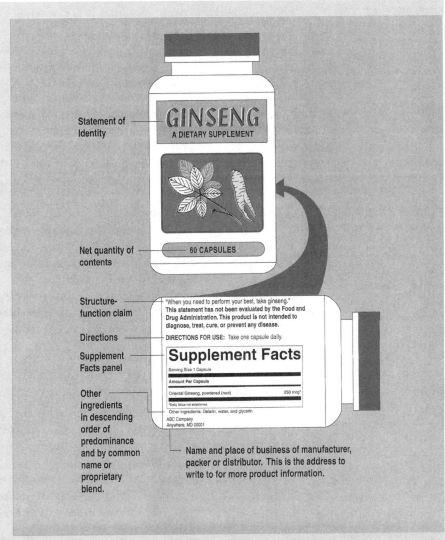

Statement of Identity

Net quantity of contents

Structure-function claim

Directions

Supplement Facts panel

Other ingredients in descending order of predominance and by common name or proprietary blend.

GINSENG
A DIETARY SUPPLEMENT

60 CAPSULES

"When you need to perform your best, take ginseng." This statement has not been evaluated by the Food and Drug Administration. This product is not intended to diagnose, treat, cure, or prevent any disease.

DIRECTIONS FOR USE: Take one capsule daily.

Supplement Facts

Serving Size 1 Capsule

Amount Per Capsule

| Oriental Ginseng, powdered (root) | 250 mcg* |

*Daily Value not established.

Other ingredients: Gelatin, water, and glycerin.

ABC Company
Anywhere, MD 00001

Name and place of business of manufacturer, packer or distributor. This is the address to write to for more product information.

Figure 7-2

Anatomy of the new requirements for dietary supplement labels (effective March 1999)

Source: Food and Drug Administration.

The Food and Drug Administration (FDA) oversees safety and manufacturing and product information, such as claims, in a product's labeling, package inserts, and accompanying literature. The Federal Trade Commission regulates the advertising of dietary supplements.

One thing dietary supplements are not is drugs. A drug, which sometimes can be derived from plants used as traditional medicines, is intended to diagnose, cure, relieve, treat, or prevent disease. Before marketing, a drug must undergo clinical studies to determine its effectiveness, safety, possible interactions with other substances, and appropriate dosages, and the FDA must review these data and authorize the drug's use before it is marketed. The FDA does not authorize or test dietary supplements.

A product sold as a dietary supplement and touted in its labeling as a new treatment or cure for a specific disease or condition would be considered an unauthorized—and thus illegal—drug. Labeling changes consistent with the provisions in DSHEA would be required to maintain the product's status as a dietary supplement.

Another thing dietary supplements are not is replacements for conventional diets. Supplements do not provide all the known—and perhaps unknown—nutritional benefits of conventional food.

As with food, federal law requires manufacturers of dietary supplements to ensure that the products they put on the market are safe. But supplement manufacturers do not have to provide information to the FDA to get a product on the market. FDA review and approval of supplement ingredients and products is not required before marketing.

Under DSHEA, once a dietary supplement is marketed, the FDA has the responsibility for showing that a dietary supplement is unsafe before it can take action to restrict the product's use.

Claims that tout a supplement's healthful benefits have always been a controversial feature of dietary supplements. Manufacturers often rely on them to sell their products. But consumers often wonder whether they can trust the claims.

Under DSHEA and previous food-labeling laws, supplement manufacturers are allowed to use, when appropriate, three types of claims: **nutrient-content claims,** disease claims, and nutrition-support claims, which include "structure-function claims."

Nutrient-content claims describe the level of a nutrient in a food or dietary supplement, and are discussed in Chapter 2. For example, a supplement containing at least 200 milligrams of calcium per serving could carry the claim "high in calcium."

Disease claims show a link between a food or substance and a disease or health-related condition. The FDA authorizes these claims based on a review of the scientific evidence. For example, a claim may show a link between folate in the product and a decreased risk of neural-tube defects in pregnancy if the supplement contains enough folate. Page 62 lists claims currently allowed.

Nutrition support claims can describe a link between a nutrient and the deficiency disease that can result if the nutrient is lacking in the diet. For example, the label of a vitamin C supplement could state that vitamin C prevents scurvy. When these types of claims are used, the label must mention the prevalence of the nutrient-deficiency disease in the United States.

These claims also can refer to the supplement's effect on the body's structure or function, including its overall effect on a person's well-being. These are known as *structure-function claims*. Examples include:

■ Calcium builds strong bones.
■ Antioxidants maintain cell integrity.
■ Fiber maintains bowel regularity.

Manufacturers can use structure-function claims without FDA approval. They base their claims on their review and interpretation of the scientific literature. Like all label claims, they must be true and not misleading. They must also be accompanied by the disclaimer, "This statement has not been evaluated by the Food and Drug Administration. This product is not intended to diagnose, treat, cure, or prevent any disease."

Manufacturers who plan to use a structure-function claim must inform the FDA of the use of the claim no later than 30 days after the product is first marketed. While the manufacturer must be able to substantiate its claim, it does not have to share the substantiation with the FDA or make it publicly available. If the submitted claim promotes the product as a drug instead of a supplement, the FDA can advise the manufacturer to change or delete the claim.

To help protect themselves, consumers should:

■ Look for ingredients in products with the U.S.P. notation, which indicates that the manufacturer followed standards established by the U.S. Pharmacopoeia.
■ Avoid substances that are not known nutrients.
■ Be aware that the term "natural" doesn't guarantee that a product is safe.

■ Consider the name of the manufacturer or distributor. Supplements made by a nationally known food and drug manufacturer have probably been made under tight controls, because these companies already have in place manufacturing standards for their other products.

■ Write to the supplement manufacturer for more information.

Consumers should also be sure to tell their healthcare providers about the supplements they take.

Although most Americans can get needed vitamins and minerals through food, situations do occur when supplements may be needed:

■ Women in their childbearing years and pregnant or lactating women, who may need iron and/or folate.

■ People with known nutrient deficiencies, such as an iron-deficient woman.

■ Elderly people who are eating poorly, have problems chewing, or have other concerns.

■ Drug addicts or alcoholics.

■ People eating less than 1,200 calories a day—such as dieters—who may need supplements because it is hard to get enough nutrients in such low-calorie diets.

■ People on certain medications or with certain diseases.

If you really feel you need additional nutrients, your best bet is to buy a multivitamin-and-mineral supplement that supplies 100 percent of the RDA or AI. It can't hurt, and it may act as a safety net for individuals who eat haphazardly. But keep in mind that more is not always better, and that no supplement can adequately take the place of food and serve as a permanent substitute for improving a poor diet. In other words, use supplements to "supplement" a good diet, not to substitute for a poor diet.

Part Two

Developing and Marketing Healthy Recipes and Menus

Chapter 8
Developing Healthy Menus and Recipes

Nutrition is growing in importance for foodservice and culinary profession-als. In the 1998 Nutritional Menuing survey of 499 operators by *Food Service Director,* 76 percent of operators expected their sales of and profits from nutri-tious entrées, snacks, and beverages to grow. Also, the percentage of opera-tors promoting healthy selections on the menus grew from 84 to 90 percent.

From healthy salads to decadent desserts, taste and presentation are important for all menu items. This chapter will help you to:

■ develop and evaluate healthy menu selections.
■ define seasoning, flavoring, herbs, and spices.
■ suggest ingredients and methods to develop flavor.
■ identify techniques and cooking methods that are healthy.
■ discuss different ways to present foods for maximum eye appeal.
■ give examples of healthy dishes for each section of the menu.
■ identify substitutions for ingredients to make a healthier dish.

Healthy Menus

The healthy menu provides choice: Nutritious dishes are available for guests who desire them. So what is a healthy menu item or meal? Can it be defined? Yes, it can be defined, although in different ways. You may want to define a healthy menu item simply as one that is moderate in the amount of calories, fat, cholesterol, and sodium that it contains; or you may want to be more precise, and use the following guidelines.

A balanced meal will generally have no more than:

■ 30 percent of its total calories from fat, including 10 percent of its total calories from saturated/trans fat
■ 150 milligrams of cholesterol
■ 1,000 milligrams or less of sodium
■ 15 percent or less of its total calories from protein
■ 55 percent of its total calories from carbohydrates (10 percent or less from simple carbohydrates)

You may also use the U.S. Dietary Guidelines for Americans as the basis for calling a menu item or meal healthy.

A healthy menu may simply highlight two or more entrées and one or two appetizers and desserts. To develop some healthy menu items, the first step is to look seriously at your existing menu while engaging in some old-fashioned menu planning. You may go in one of three directions.

1. *Use existing items on your menu.* Certain menu selections, such as fresh vegetable salads or grilled skinless chicken, may already meet your needs.

2. *Modify existing items to make them more nutritious.* For example, sear fish instead of roasting it with a butter sauce. In general, modification centers on ingredients, preparation, and cooking techniques. Modifying an existing item may simply mean offering a half-portion of the protein.

3. *Create new selections.* Many resources are available to obtain healthy recipes: cookbooks, magazines, websites. Or draw on your culinary skills and creativity, and make your own recipes.

Whenever you are involved in menu planning, keep in mind the following considerations.

1. Is the menu item tasty? Taste is the key to customer acceptance and the successful marketing of these items. If the food does not taste delicious and look out of this world, then no matter how nutritious it may be, it is not going to sell.
2. Does the menu item blend with and complement the rest of the menu?
3. Does the menu item meet the food habits and preferences of the guests?
4. Is the food cost appropriate for the price that can be charged?
5. Does each menu item require a reasonable amount of preparation time?
6. Is there a balance of color in the foods themselves and in the garnishes?
7. Is there a balance of textures, such as coarse, smooth, solid, and soft?
8. Is there a balance of shape, with different-sized pieces and shapes of food?
9. Are flavors varied?
10. Are the food combinations acceptable?
11. Are cooking methods varied?
12. Can each menu item be prepared properly by the cooking staff?

To develop healthy menu items that sell and satisfy customers, you need to use the following three steps.

Step 1: Foundations: Flavor

A solid foundation in foods and cooking is necessary in order to develop healthy menus and recipes. You are expected to know basic culinary terminology and techniques and have a working knowledge of ingredients, from almonds to zucchini. A basic culinary skill that needs some further refinement when cooking healthy is that of flavoring. Because you can't rely on more than moderate amounts of fat, salt, or sugar for taste and flavor, you will need to develop excellent flavor-building skills.

Seasonings and **flavorings** are very important in healthy cooking, because they help replace missing ingredients such as fat and salt. Seasonings are used to bring out flavor already present in a dish, whereas flavorings add a new flavor or modify the original one. The difference between them is one of degree.

Herbs and Spices

Herbs and spices are key flavoring ingredients in nutritional menu planning and execution, and are the backbone of most menu items, lending themselves to cultural and regional food styles. Good sound nutritional cooking can be virtually equal to classical cooking in terms of technique, creative seasoning, flavor blending, and presentation. It's helpful when moderating fat, cholesterol, and sugars to enhance recipes with an abundance of seasonings like cinnamon, nutmeg, mace, cloves, vanilla beans, ginger, star anise, juniper, and cardamom. These spices give you a sense of sweet satisfaction as well as bold character to your recipes.

The use of herbs in recipes changes the flavor direction to whichever herb is prominent. For instance, the use of basil, oregano, and thyme in a tomato vinaigrette points the dish to an Italian flavor. Take that same dish and add cilantro and lime juice and you move South of the Border, or add fresh chopped tarragon with shallots and you're in France. There is no end to your creative abilities once you understand the basic format to healthy cooking.

Herbs are the leafy parts of certain plants that grow in temperate climates. **Spices** are the roots, bark, seeds, flowers, buds, and fruits of certain tropical plants. Figure 8-1 shows a number of herbs and spices. Herbs are generally available fresh and dried. Spices are mostly available in the dried form.

Fresh herbs, as opposed to dry, are far superior and more versatile when creating recipes. Herbs commonly available fresh include parsley, basil, dill, chives, tarragon, thyme, and oregano. Fresh herbs are great when you need a crisp clean taste and maximum flavor. They can withstand only about 30 minutes of cooking, so they work best finishing dishes.

Although the use of fresh herbs is not always possible, dry herbs can be substituted with better-than-average results. Dried herbs work well in longer cooking—such as in stocks, stews, and sauces. You can use dried herbs along with fresh herbs toward the end of cooking to get a richer and cleaner flavor.

The real purpose of herbs and spices is not to rescue, remedy, flavor, or season, but to build. Spices and herbs are basically flavor builders. This is their proper use in cooking; they should be cooked with the dish as it is being made so that their flavors blend smoothly, giving character and depth to the dish.

Learning to identify the innumerable different herbs and spices requires a keen sense of taste and smell. Simply looking at them is not enough.

Figure 8-1
Herbs and spices

Courtesy American Spice
Trade Association.

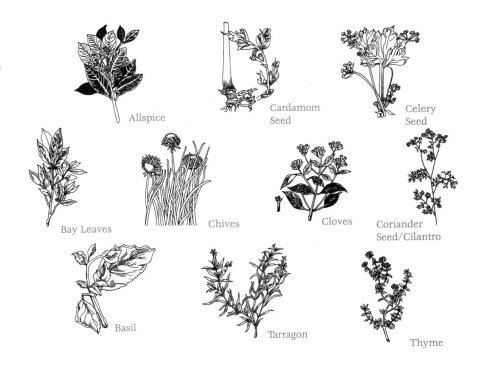

Taste them, smell them, feel them, use them. The key to most of them is aroma, for in their aroma is about 60 percent of their flavor. Their aromatic quality not only adds flavor to the food as it is eaten, but heightens the anticipation of the diner as the food is being cooked and served.

There are many, many herbs and spices. Let's look at those most likely to be found in the kitchen. To help you understand them, we'll sort them into groups, but first let's look at the many forms of pepper.

Pepper comes in three forms: black, white, and green. White and black pepper both come from the oriental pepper plant. Black pepper is the dried unripe berry; white pepper is the kernel of the ripe berry. Green peppercorns are picked before ripeness, and preserved.

Black pepper comes in four forms: whole black peppercorns, crushed, butcher's grind, and table-ground pepper. Whole pepper is used as a flavor builder during cooking, as in making stocks. Crushed black pepper can function as a flavor builder during cooking, as well, or it can be added as flavoring to a finished dish. Many Americans enjoy the flavor contrast of fresh crushed peppercorns straight from the pepper mill on a crisp green salad. The flavor of ground black pepper is characteristic of certain cuisines and certain parts of the country. Cooks catering to these clienteles are likely to add this flavor as they season the food.

As a seasoning, black pepper is used only in dark-colored foods; it spoils the appearance of light-colored foods. *White pepper* is used in light-colored foods because its presence is concealed. White pepper comes in two forms: whole peppercorns and ground white pepper. White peppercorns are used in the same ways as black.

Ground white pepper is good for all-around seasoning. It blends imperceptibly into white dishes both in appearance and in flavor, and it has the strength necessary to season dark dishes. Ground white pepper is chosen by most good cooks as the true seasoning pepper. It is seldom used as a table pepper, as it is expensive.

Green peppercorns are preserved either by packing them in liquid (such as vinegar or brine) or by drying them. They are used in white-tablecloth (luxury) restaurants to complete certain recipes.

Pink peppercorns are not true peppercorns, but they look like peppercorns and have a sweet, slightly peppery taste. They are native to South America, and sometimes mistakenly called Japanese peppers because they are one of the few spices used in Japanese cooking. Possible adverse reactions to pink peppercorns have been reported when added generously to dishes, so use in small amounts.

Red pepper, also called *cayenne,* is completely unrelated to white or black pepper. It comes from dried pepper pods. It is quite hot and easily overused. Added with restraint in soups and sauces, it can lend a spicy hotness. When used without as much restraint, it creates the hot flavor of many foods from Mexico, South America, and India.

Let's look at nine herbs and spices that are used as often for their distinctive flavors as for general flavor enrichment. As a flavor builder, each goes beyond the subtlety of the stock herbs and spices, which gives a definitely different support flavor—even though you can't single it out from the flavor of the dish as a whole. Used in quantities large enough to taste, they become major flavors rather than flavor builders. Several of them can also be used as flavorings.

Basil, oregano, and tarragon are available fresh and also come in the form of crushed dried leaves. They look somewhat alike, but their tastes are very different.

Basil has a warm, sweet flavor that is welcome in many soups, sauces, entrées, relishes, salsa, dressings, and vegetables such as tomatoes, peppers, eggplant, and squash. It blends especially well with tomato, lemons, and oranges. Like many other herbs, it has symbolisms: In India it expresses reverence for the dead; in Italy it is a symbol of love.

Oregano belongs to the same herb family as basil, but it makes a very different contribution to a dish—a strong bittersweet taste and aroma you may have met in spaghetti sauce. It is used in many Italian, Mediterranean, Spanish, South American, and Mexican dishes.

Tarragon has a flavor that is somehow light and strong at the same time. It tastes something like licorice. It is used in poultry and fish dishes, as well as in salads, sauces, and salad dressings.

Rosemary, like bay leaf, is used in dishes where a liquid is involved—soups, stocks, sauces, stews, and braised foods. The leaf of an evergreen shrub of the mint family, it has a pungent, hardy flavor and fragrance. Fresh or dried, it looks and feels like pine needles. It is used mostly with meats, game, poultry, mushrooms, and flavorful ragouts.

Dill and *mustard* have flavors that will be very familiar to you: dill as in pickle and mustard as in hot dog. Fresh or dried dill leaves, often called dill weed, are used in soups, fish dishes, stews, salads, and butters. Whole dill seed is used in some soups and sauerkraut. Dry mustard, a powdered spice made from the seed of the mustard plant, comes in three varieties: white, yellow, and brown. The brown has the sharper and more pungent flavor. All are used to flavor sauces, dips, dressings, and entrées. Prepared mustards are also made from all varieties and serve as excellent flavor enhancers.

Paprika is another powdered spice that comes in two flavors, mild and hot. Both kinds are made from dried pods of the same pepper family as red pepper and cayenne, and they look something like the seasoning peppers, but they do not do the work of seasonings. Hungarian paprika is the hot spicy one; Spanish paprika is mild in flavor and its red color has lots of eye appeal. Hungarian paprika is used to make goulash and other braised meats and poultry. Spanish is used for coloring, blending rubs, and mild seasoning. Paprikas are sensitive to heat and will turn brown if exposed to direct heat.

Still another branch of this same pepperpod family gives us chili peppers, the crushed or dried pods of several kinds of Mexican peppers and Asian dried red chilis. Colors range from red to green and flavors from mild to hot. Chili peppers are used in Mexican, Asian, Thai, Peruvian, Indian, Cuban, and other South American cuisines.

Several spice blends are available. Two of them are standards in any kitchen. *Chili powder* is one, which is a combination of toasted ground dried chili peppers. Chili powder varies from mild to very hot. It is used, of course, in chili, where it functions as a major flavor, and in many Mexican, South American, Cuban, and Southwestern dishes.

Curry powder is a blend of up to 20 spices. In India, where it originated, cooks blend their own curry powders, which may vary considerably. In the

United States curry powder comes premixed in various blends from mild to hot. Curry powders usually include cloves, black and red peppers, cumin, garlic, ginger, cinnamon, coriander, cardamom, fenugreek mustard seeds, turmeric (which provides the characteristic yellow color), and sometimes other spices.

A group of powdered sweet aromatic spices from the tropics are used frequently in baking, in dessert cookery, and occasionally in sauces, vegetables, and entrées. Among these are cinnamon, nutmeg and its counterpart mace, and ginger. *Cinnamon* comes from the dried bark of the cinnamon or cassia tree, nutmeg and mace from the seed of the nutmeg tree, and ginger from the dried root of the ginger plant.

In hot foods, the *nutmeg* flavor goes well with potatoes, dumplings, spinach, quiche, and some soups and entrées. *Mace,* a somewhat paler alternative to nutmeg, has a similar flavor and is used in bratwurst, savory dishes, baked goods, and patés. Ground cinnamon and ginger are used in a variety of cuisines, both sweet and savory. Cinnamon is also available in sticks.

Mint is a sweet herb, with the familiar flavor you meet in toothpaste and chewing gum. Mint is available in many varieties: The most popular are spearmint and peppermint. Others include chocolate, licorice, orange, and pineapple. The flavor of a mint sauce offers a refreshing complement to lamb. Fresh mint makes a good flavoring and garnish for fruits, vegetables, salsas, relishes, salads, dressings, iced tea, desserts, and sorbets.

Many herbs and spices can be combined to produce blends with distinctive flavors. For example, a cattleman's blend with paprika, peppers, chilis, and other dried herbs can be used as a steak seasoning. A seed blend, such as ground cardamom, fennel, anise, star anise, cumin, and coriander, is excellent in soups, marinades, vegetables, and chutneys. Ethnic blends, such as the following, also present many flavoring possibilities.

- Italian: basil, oregano, garlic, onion
- Asian: ginger, five spices, garlic, scallion
- French: tarragon, mustard, chive, chervil, shallot
- South American: chili powder, lime juice, cilantro
- Indian: ground nutmeg, fennel, coriander, cinnamon, fenugreek, curry
- Mediterranean: oregano, thyme, pepper, coriander, onion, garlic

Blends such as these are foundations for starting or finishing a dish, and are wonderful to add depth of flavor. Always check the salt and sugar content of premade blends.

Another way to get maximum flavor out of herbs and spices is to toast them. Table 8-1 lists whole spices that can be toasted. Toast them in a hot

TABLE 8-1 **Examples of Whole Toasted Spices**

- Mustard seed
- Fennel
- Coriander
- Star Anise
- Cardamon
- Caraway
- Cumin
- Chilis
- Juniper
- Allspice

nonstick pan, then grind them and use to season marinades, salad dressings, rubs, soups, stews, ragouts, salsas, and relishes.

Table 8-2 is a reference chart for many herbs and spices.

Juices

Juice can be used as is for added flavor, or can be reduced (boiled or simmered down to a smaller volume) to get a more intense flavor, vibrant color, and syrupy texture. Reduced juices make excellent sauces and flavorings. Use a good-quality juicer or buy quality premade juice.

For example, orange juice can be reduced (simmered) to orange oil, which is excellent in salad dressings, marinades, and sauces. Also, freshly made beet juice can be used to enhance stocks, glazes, and sauces. In a squirt bottle, reduced beet juice can be used lightly on plates for decoration and flavor, especially with salads, appetizers, and entrées. Other juices that are useful in the kitchen are carrot, fennel, celeriac, pomegranate, ginger, celery, asparagus, bell peppers (yellow, red, orange), herbs (watercress, cilantro, parsley, basil, chive), leek, and radish.

Lemon and lime juice are seldom called seasonings, and yet their use as seasonings is not unusual. Many recipes call for small amounts of lemon rind or juice. When used with restraint to spark the flavor of the dish itself and the citrus flavor cannot be perceived, lemon and lime juice are seasonings.

Vinegars and Oils

Various types of vinegars can add flavor to a wide variety of dishes, from salads to sauces. They have a light, tangy taste, and add flavor without fat. Popular vinegars include wine vinegars (made from white wine, red wine, rose wine, rice wine, champagne, or sherry), cider vinegar (made from

TABLE 8-2 Herb and Spice Reference Chart

Product	Market Forms	Description	Uses
Allspice (spice)	Whole, ground	Dried, dark brown berries of an evergreen tree indigenous to the West Indies and Central and South America. Smells of cloves, nutmeg, and cinnamon.	Braised meats, curries, baked goods, puddings, cooked fruit
Anise seed (spice)	Whole, ground	Tiny dried seed from a plant native to Eastern Mediterranean. Licorice (sweet) flavor.	Baked goods, fish, shellfish, soups, sauces
Basil (herb)	Fresh, dried: crushed leaves	Bright green tender leaves of an herb in the mint family. Sweet, slightly peppery flavor.	Tomatoes, eggplant, squash, carrots, peas, soups, stews, poultry, red sauces
Bay leaves (herb)	Whole, ground	Long, dark green, brittle leaves from the bay tree, a small tree from Asia. Pungent, warm flavor when leaves are broken. Always remove bay leaves before serving (due to toughness).	Stocks, sauces, soups, braised dishes, stews, marinades
Caraway seed (spice)	Whole, ground	Crescent-shaped brown seed of a European plant. Slightly peppery flavor.	German and Eastern European cooking (such as sauerkraut and coleslaw), rye bread, pork
Cardamom (spice)	Whole pod, ground seeds	Tiny seeds inside green or white pods that grow on a bush of the ginger family. Sweet and spicy flavor. Very expensive.	Poultry, curries and other Indian cooking, Scandinavian breads and pastries, puddings, fruits
Cayenne, red pepper (spice)	Ground	Finely ground powder from several hot types of dried red chile peppers. Very hot. Use in small amounts.	In small amounts in meats, poultry, seafood, sauces, and egg and cheese dishes
Celery seed (spice)	Whole, ground, ground mixed with salt or pepper	Small, gray-brown seeds produced by celery plant. Distinctive celery flavor.	Dressings, sauces, soups, salads, tomatoes, fish
Chervil (herb)	Fresh, dried: crushed leaves	Fernlike leaves of a plant in the parsley family. Like parsley with slight pepper taste, smells like anise.	Stocks, soups, sauces, salads, egg and cheese dishes, French cooking
Chives (herb)	Fresh, dried	Thin grasslike leaves of a plant in the onion family. Mild onion flavor.	Poultry, seafood, potatoes, salads, soups, egg and cheese dishes

Name	Form	Description	Uses
Cinnamon (spice)	Stick, ground	Aromatic bark of the cinnamon tree, a small evergreen tree of the laurel family. Sweet, warm flavor.	Baked goods, desserts, fruits, lamb, ham, rice, carrots, sweet potatoes, beverages
Cloves (spice)	Whole, ground	Dried flower buds of a tropical evergreen tree. Sweetly pungent and very aromatic flavor.	Stocks, marinades, sauces, braised meats, ham, baked goods, fruits
Coriander seeds (spice)	Whole, ground	Small seeds from the cilantro plant. Mild, slightly sweet and musty flavor.	Pork, pickling, soups, sauces, chutney, casseroles, Indian cooking
Cumin seed (spice)	Whole, ground	Seed of a small plant in the parsley family. Looks like caraway seed but lighter in color. Pungent, strong, earthy flavor.	Used to make curry and chili powders. Cooking of India, Middle East, North Africa, and Mexico; sausage, Munster cheese, sauerkraut
Curry powder (spice blend)	Ground blend	A spice blend of up to twenty spices. Often includes black pepper, cloves, coriander, cumin, ginger, mace, and turmeric. Depending on brand, flavor and hotness can vary tremendously.	Indian cooking, eggs, beans, soups, rice
Dill weed (herb)	Fresh, dried: crushed	Delicate, green leaves of dill plant. Dill pickle flavor.	Salads, dressings, sauces, vegetables, fish, dips
Dill seed (spice)	Whole, ground	Small, brown seed of dill plant. Bitter flavor—much stronger than dill weed.	Pickling, sour dishes, sauerkraut, fish
Epazote (herb)	Fresh	Coarse leaves of a wild plant that grows in the Americas. Strong, exotic flavor.	Mexican and Southwestern cooking
Fennel seed (spice)	Whole, ground	Oval, green-brown seeds of a plant in the parsley family. Mild licorice flavor.	Fish, pork, Italian sausage, tomato sauce, pickles, pastries
Fenugreek seed (spice)	Whole, ground	Small beige seeds of a plant in the pea family. Bittersweet flavor. Smells like curry.	Indian cooking such as curries and chutneys
Ginger (spice)	Fresh whole, dried whole, dried ground (also candied or crystallized)	Dried tan root of the tropical ginger plant. Hot but sweet flavor.	Asian dishes such as curries, baked goods, fruits, beverages
Juniper berries (spice)	Whole	Purple berries of an evergreen bush. Pinelike flavor.	Venison and other game dishes, pork, lamb, marinades

TABLE 8-2 *(continued)*

Product	Market Forms	Description	Uses
Lemon grass (herb)	Fresh stalks, dried: chopped	Tropical and subtropical scented grass. White leaf stalks and lower part are used. Bright, lemon flavor.	Soups, marinades, stir-fries, curries, salads, Southeast Asian cooking
Mace (spice)	Whole blades or ground	Lacy orange covering on nutmeg. Like nutmeg in flavor but less sweet.	Baked products, fruits, pork, poultry, some vegetables
Marjoram (herb)	Fresh, dried: crushed leaves	Leaves from a plant in the mint family. Mild flavor similar to oregano with a hint of mint.	Lamb, poultry, stuffing, sauces, vegetables, soups, stews
Mint (herb)	Fresh, dried: crushed leaves	A family of plants that include many species and flavors such as spearmint, peppermint, and chocolate. Cool, minty flavor.	Lamb, fruits, some vegetables, tea and other vegetables
Mustard seed (spice)	Whole, ground (prepared mustard)	Tiny seeds of various mustard plants, seed may be white or yellow, brown, or black. The darker the seed, the sharper and more pungent the flavor.	Meats, sauces, dreasings, pickling spices, prepared mustard
Nutmeg (spice)		Large brown seed of the fruit from the nutmeg tree. Sweet, warm flavor.	Baked products, puddings, drinks, soups, sauces, many vegetables
Oregano (herb)	Fresh leaves, dried: crushed	Dark green leaves of oregano plant. Pungent, pepperlike flavor.	Italian foods such as tomato sauce and pizza sauce, meats, sauces, Mexican cooking
Paprika (spice)	Ground	Fine powder from mild varieties of red peppers. Two varieties: Spanish and Hungarian. Hungarian is darker in color and much stronger in flavor.	Spanish used mostly as garnish; Hungarian used in braised meats, sauces, gravies, some vegetables
Parsley (herb)	Fresh, dried flakes	Green leaves and stalks of several varieties of parsley plant. Mild, sweet flavor.	Bouquet garni, fines herbes, almost any food
Peppercorns, black and white (spice)	Whole, crushed, ground	Dried, black or white hard berries from same tropical vine that are picked and handled in different ways. Black: pungent earthy flavor. White: similar but more mild than black.	Almost any food

Spice/Herb	Form	Description	Uses
Pink peppercorns (spice)	Whole	Dried or pickled red berries of an evergreen. Not related to black pepper. Bitter flavor, not as spicy as black pepper.	Use in small quantities in meat, poultry, and fish dishes, and in whole pepper mixtures
Poppy seed (spice)	Whole	Tiny cream-colored or deep blue seeds from the poppy plant. Nutty flavor.	On breads and rolls, in salads and noodles, ground poppy seed in pastries
Rosemary (herb)	Fresh, dried leaves	Stiff green leaves that look like pine needles of a shrub. Strong flavor like pine.	Roasted and grilled meats such as lamb, sauces such as tomato, soup
Saffron (herb)	Whole (threads), ground	Dried flower stigmas of a member of the crocus family. Used in very small amounts, has a bitter yet sweet taste, and colors foods yellow. Most expensive spice. Mix with hot liquid before using.	Paella, risotto, bouillabaisse, seafood, poultry, baked products
Sage (herb)	Whole (fresh or dried), rubbed (chopped), ground	Gray-green leaves and blue flowers of a member of the mint family. Strong, musty flavor.	Pork, sausage, stuffing, salads, beans
Savory (herb)	Crushed leaves	Small, narrow leaves of plant in mint family. Summer savory is preferable to winter savory. Bitter flavor.	Meat, poultry, sausage, fish, vegetables, beans
Sesame seeds (spice)	Whole	Creamy oval seeds of tall, tropical sesame plant. Nutty flavor.	On breads and rolls. Ground seeds used to make tahini
Star anise (spice)	Whole, ground	Dried, star-shaped fruit of an evergreen native to China. Dark red. Licoricelike flavor.	Chinese cooking
Tarragon (herb)	Fresh, dried: crushed	Small plant with long narrow leaves and gray flowers. Delicate sweet flavor with hint of licorice.	Bearnaise sauce, vinegars, dressings, poultry, fish, salads
Thyme (herb)	Fresh, dried: crushed or ground	Tiny leaves and purple flowers of a short, bushy plant. Spicy, slightly pungent flavor.	Bouquet garni, meat, poultry, fish, soups, sauces, tomatoes
Turmeric (spice)	Ground	Orange-yellow root of member of ginger family. Musky, peppery flavor. Colors foods yellow.	Curry powder, curries, chutney

apples), and balsamic vinegar. Balsamic vinegar, a dark brown vinegar with a rich sweet-sour flavor, is made from the juice of a very sweet white grape and is aged for at least ten years. Vinegars can also be infused, or flavored, with all sorts of ingredients, such as chili peppers, roasted garlic, or any herbs, vegetables, and fruits. These types of vinegars are called flavored, or infused, vinegars. For example, lemon vinegar works well in salad dressings and cold sauces.

Like vinegar, oils can be infused with ingredients such as ground spices, fresh herbs, juices, and fresh roots. Small amounts of flavored oils can add much flavor to finish sauces, dressings, marinades, relishes, salads, or can be used alone to drizzle over foods ready to be served. Use a neutral oil such as canola, safflower, or corn.

To make a ground spice oil, mix the spice first with water (three tablespoons spice to one tablespoon water) to wake up the flavor. Then mix with about two cups of oil. Let sit for about four to six days at room temperature, shaking several times a day to fuse flavors. Filter the oil and reserve for use.

To make a tender herb oil, blanch the herbs first and then shock in ice water. Drain and dry the herbs, then purée the herbs with oil and strain through a fine filter. Keep in the refrigerator for about one week. Herbs with a harder texture can be chopped and mixed in a food processor with oil. Let sit several days and then strain. Keep in the refrigerator for several weeks.

To make an oil with vegetable and fruit juices, reduce the juice to a syrup, then blend in a food processor with a little oil, Dijon mustard, and a touch of honey. The oil is ready to use.

Stock

Stock, a flavored liquid used in making soups, sauces, stews, sautés, and braised foods, functions as the body of many foods as well as a flavor builder. Body refers to the amount of flavor—its strength or richness. The French call stock *fond de cuisine,* base of cooking, which describes its role exactly.

Stocks are made by simmering water, bones, regular *mirepoix* (onion, carrot, celery, and sometimes tomatoes or leek) or white mirepoix (onion, celery, leek, fennel), herbs, and spices. As they simmer, the flavor-producing substances are extracted from the bones and flavor builders and dissolve in the water. Gelatin is also drawn from the bones, a major source of body. Though it may be imperceptible in a hot stock, gelatin causes the stock to thicken or jell when chilled.

Chicken stock is made using chicken bones. If the mirepoix and bones are caramelized (browned) first before being added to the stock, the result will be a brown chicken stock. Without this browning process, the result is a white chicken stock.

Fish stock is made like a white stock, simply using cleaned fish bones. Fish stock cooks very quickly, in less than an hour at a rolling simmer. Depending upon the types of bones/shells used, you can make clam stock, lobster stock, or other types. Crustacean stock is made usually by caramelizing the shells and mirepoix with tomato and paprika for a rich brown-red color. Fish or chicken stock is added. A blend of both stocks creates most favorable results.

Brown stock is made like white stocks, except that the bones (generally beef or veal) and mirepoix are browned before using. Tomato products and red wine may also be used.

Vegetable stock is made without any animal products. Vegetables are sweated in a touch of oil or stock, then herbs and spices are added. The vegetables are covered with water and simmered for 1-1/2 hours. Wine may be added. Vegetable stock is normally considered a white stock unless tomatoes are added.

In building a good stock, you need to know the nature of the product you are aiming for. Here are the principal measures of stock quality.

- A good stock is fat-free.
- A good stock is clear—translucent and free of solid matter.
- A good stock is pleasant to the senses of smell and taste.
- A good stock is flavorful, but the flavor is neutral. The flavor of the main ingredient, though predominant, is not overpowering. No one flavor builder is identifiable over the flavor of the main ingredient.

And here are some very important guidelines to observe in making stock.

- Use good raw bones—bones that smell pleasant and fresh. Wash chicken and fish bones.
- Remove excess fat from the bones. Fat will produce grease in the stock, spoiling its flavor and appearance.
- Start with cold liquid. Wash chicken and fish bones. They are naturally filled with blood and other impurities. These impurities will dissolve in cold water, and then as the stock is heated, they will become solid and rise to the surface, where you can skim them off. This is especially true with beef and veal stocks. Therefore, a cold-water start will produce a clear stock, whereas starting with hot water will produce a cloudy one.
- Use a tall, narrow pot to minimize evaporation. A certain amount of flavor is lost in evaporation, and the rate of evaporation depends on the surface area of the liquid.
- After bringing the cold water to a boil, reduce the heat to a simmer (about 185 degrees F. or 85 degrees C.). Keep the cooking temperature below the boil. It takes long, slow simmering to extract the flavors you want from the bones and flavor builders. Too high a temperature will increase evaporation and loss of desirable flavors. It will also break down vegetable textures, producing undesirable flavors and a cloudy stock.

- Skim occasionally—that is, remove the impurities that rise to the surface, using a skimmer or ladle.
- Do not salt the stock because it will usually be further cooked—in a sauce or soup, for example. As a stock is cooked further, it reduces, meaning its volume decreases due to evaporation. A stock that tastes lightly salted when prepared will taste much saltier as it is cooked further and reduces in volume.
- If you add kitchen scraps to stock, make sure they are clean and wholesome.
- Cool the stock quickly and store it properly.
- A stock's shelf life is no more than three to five days in the cooler. Stocks can be frozen without loss of quality.
- Degrease the finished stock—that is, remove the fat from the surface. The most effective method is to chill the stock and remove the layer of fat that congeals on top. If you must use the stock immediately, you can skim the hot fat off the top with a ladle.

To reduce the amount of fat in stocks, use only a small amount of oil to sauté the mirepoix, or sweat them in stock, wine, or other liquids.

Stocks are a low-calorie way to support flavor for the recipes they are used in. One cup of stock is about 40 calories at most, or only 5 calories per fluid ounce. Through the use of herbs, spices, and aromatic vegetables, as well as the flavor from caramelized bones, stocks can be made quite flavorful without adding any calories. Also, stocks can be thickened without high-fat roux. Arrowroot or cornstarch does an admirable job of thickening without fat. Puréed vegetables or potatoes can be used to give body, or simply reduce the stock to a glaze.

Glazes are basic preparations in classical cookery and are the forerunners of today's convenience products. They are simply stocks reduced to a thick, gelatinous consistency with addition of flavoring and seasonings. Meat glaze, or *glace de viande* (*vee-end*), is made from brown stock. *Glace de volaille* is made from chicken stock, and *glace de poisson* is made from fish stock. To prepare a glaze, you reduce stock over moderate heat, frequently skimming off the foam and impurities that rise to the top. When the stock has reduced by about half, strain it through cheesecloth into a smaller heavy pan. Place it over low heat and continue to reduce until the glaze will form an even coating on a spoon. Cool, cover, and refrigerate or freeze.

Small amounts of glazes (remember, they are very concentrated!) are used to flavor sauces and other items. They make sauces cleaner and fresher tasting than sauces made with thickeners. They can also be added to soups to improve and intensify flavor. However, they cannot be used to re-create the stock from which they were made; the flavor is not the same after the prolonged cooking at higher temperatures.

Concentrated convenience bases are widely used. The results vary widely, partly because of the bewildering variety of products on the market and partly because they are often misused. Few of them can function as instant stocks. Compare the taste of a convenience base prepared according to instructions with the taste of a stock made from scratch, and you will see why. Many convenience bases have a high salt content and other seasonings and preservatives. This gives them strong and definite tastes that are difficult to work with in building subtle flavors for soups and sauces.

Among the many bases available, take care to choose the highest-quality products, though these are expensive. Look for those that list beef, chicken, or fish extract as the first ingredient (not salt!). Avoid those with lots of chemical additives. If you must use a base as a stock, fortify it using these steps.

1. Dry sauté a white or regular mirepoix (small diced).
2. Add base, water, and sachet with fresh herbs, garlic, shallots, bay leaf, and peppercorns.
3. Simmer for 20 minutes.
4. Strain and reserve for use.

Cooking without fresh stock means using your head. There is no one all-purpose stock substitute. If a recipe calls for stock, you will have to analyze the role the stock is to play in that recipe and choose the type of convenience item accordingly. Use with caution, and taste the product as you go. Remember that salt is the major ingredient in nearly every base, and adjust the amount of salt in your recipe.

Rubs and Marinades

Rubs combine dry ground spices, such as coriander, paprika, chili, and finely cut herbs such as thyme, cilantro, and rosemary. Rubs may be dry or wet. Wet rubs, also called pastes, use liquid ingredients such as mustard or vinegar. Pastes produce a crust on the food. Wet or dry seasoning rubs work particularly well with beef and pork and can range from a mesquite barbecue seasoning rub to a Jamaican jerk rub. To make a rub, mix various seasonings together and spread or pat evenly on the meat, poultry, or fish, just before cooking for delicate items or at least 24 hours before for large cuts of meat. The larger the piece of meat or poultry, the longer the rub can stay on. The rub flavors the exterior of the meat as it cooks.

Marinades, seasoned liquids in which foods are soaked before cooking, are useful for adding flavor as well as for tenderizing meat and poultry. Marinades bring out the biggest flavors naturally so you don't need to drown the food in fat, cream, or sauces. Marinades allow a food to stand on its own with a light dressing, chutney, sauce, or relish. Fish can also be marinated. Although fish is already tender, a short marinating time (about 30 minutes) can develop a unique flavor. A marinade usually contains an acidic ingredient, such as wine, beer, vinegar, citrus juice, or plain yogurt, to break down the tough meat or poultry. The other ingredients add flavor. Without the acidic ingredient, you can marinate fish for a few hours to instill flavor. Oil is often used in marinades to carry flavor, but it isn't essential. Fat-free salad dressings such as Italian work well in marinades. A simple fish marinade is fish stock, lemon rind, white wine, tarragon, thyme, dill, black pepper, shallots, dijon mustard, and a few drops of oil.

To give marinated foods flavor, try citrus zest, diced vegetables, fresh herbs, shallots, garlic, low-sodium soy sauce, mustard, and toasted spices. For example, citrus and pineapple marinades can be flavored with Asian seasonings such as ginger and lemon grass. Tomato juice with allspice, Worcestershire sauce, cracked black pepper, mustard, fresh herbs, and coriander is great for flank steak.

Aromatic Vegetables

Onions and their cousins garlic, scallions, leeks, shallots, and chives are a special category of flavorings that add strong and distinctive flavors and aromas to both cooked and uncooked foods. These bulbous plants of the lily family arrive in the kitchen whole and fresh rather than dried and powdered, though some are available in dried forms. We use them in greater quantity, except for garlic, than the "pinch" or the "few" that is our limit on most herbs and spices.

The onion is the scaly bulb of an herb used since ancient times and grown the world over, the commonest and most versatile flavor builder in the kitchen. Raw onion adds a pungent flavor to salads and cold sauces. Cooked, it has a sweet, mellow, come-on flavor that blends with almost anything. Cooked onions are also served as a vegetable.

Garlic, the bulb of a plant of the same family, is available as cloves (bulblets), chopped, powdered, or in juice form, with the fresh clove having by far the best flavor. Garlic is used as a flavor builder in many preparations, including stocks, stews, sauces, salads, and salad dressings. Roasted garlic has a strong flavor that can successfully replace salt in some recipes. When puréed, it gives a binding, creamy texture to sauces, dressings, beans, grains, and soups, as well as a pleasant flavor.

Chives are another bulbous herb of the onion family, the only one whose leaves rather than the bulb are eaten. Chives are usually used raw, since most of their flavor is lost if they are cooked. They are clipped from the plant and added, usually finely sliced, to many foods.

The leek, a mild-flavored relative of the onion, has a cylindrical bulb. It is the partner of the onion in the mirepoix. Some dark leaves may be used in stock—too many will cause a grey-green color. For most other preparations, the green is cut off because it develops a bitterness. Braised leeks are very popular as a vegetable in France, where they are known as the poor man's asparagus. The leek's triumph is the cold soup known as vichyssoise.

The scallion is a young onion, also known as a green onion or spring onion. It has a mild flavor as onions go. Minced or sliced, it is added to salads, marinades, dressings, salsas, and relishes with all of the white bulb and some green top. It can pinch-hit in cooking for the full-grown onion, and its green top, minced, can substitute for chives in an emergency.

The shallot is a cluster of brown-skinned bulbets similar to garlic. It is somewhere between garlic and onion in both size and flavor, but is milder and more delicate than either. One thinks of shallots with wine cookery and with mushrooms in a marvelous stuffing called duxelles, but they are useful to build flavor in a wide variety of dishes.

Sauce Alternatives: Coulis, Salsa, Relish, Chutney, Compote, and Mojos

Alternatives to classic sauces, many of which are high in fat, take many forms. **Coulis,** a French term, refers to a sauce made of a purée of vegetables or fruits. A vegetable coulis, such as bell pepper coulis, can be served hot or cold to accompany entrées and side dishes. Fruit coulis is usually served cold as a dessert sauce.

A vegetable coulis is often made by cooking the main ingredient, such as tomatoes, with typical flavoring ingredients, such as onions and herbs, in a liquid such as stock. The vegetable and flavorings are then puréed, and the consistency and flavoring of the product are adjusted. A vegetable coulis can also be made by cooking the vegetable with potato or rice (2 ounces per gallon) to give the sauce a silky texture and a smooth consistency once puréed and strained.

The texture of a vegetable or fruit coulis is quite variable, depending on the ingredients and how it is to be used. A typical coulis is about the consistency and texture of a thin tomato sauce. The color and flavor of the main ingredient should stand out.

Salsa and relishes are versatile, colorful, low-fat sauces. Salsas and relishes can be defined as chunky mixtures of vegetables and/or fruits and flavor ingredients. They may feature ordinary ingredients, such as tomatoes, and sometimes not-so-ordinary ingredients, such as papaya or jicama. Salsa is a Latino word, and traditional salsa is made with tomatoes. Relishes are often spicy and made from pickling foods. Served cold, they are excellent as sauces for meat, poultry, and seafood. Since salsas and relishes contain little or no fat, they rely on intense flavor ingredients, such as cilantro, jalapeño peppers, lime juice, lemon juice, garlic, dill, pickling spice, coriander, and mustard.

Chutney, such as tomato-papaya chutney, is made from fruits, vegetables, and herbs, and comes from India originally. Recipe 8-21 explains how to make Papaya and White Raisin Chutney. A compote is a dish of fruit, fresh or dried, cooked in syrup flavored with spices or liqueur. It is often served as dessert or as an accompaniment.

With their many colors, flavors, and textures, each of these items is a thoroughly contemporary addition to sauces.

Alcoholic Beverages

Wines, liqueurs, brandy, cognac, and other spirits are often added as flavorings at the end of cooking. Sherry is a popular American flavoring for sauces. Wine or brandy is often poured over a dish and flamed—set afire—at the time of service. This adds some flavor, but is done more for show.

In such dishes as sauces, wines and spirits may be added during cooking to become part of the total flavor. They are then flavor builders rather than flavorings. The same product can play one role in one dish and a different role in another dish.

Extracts and Oils

Extracts and oils from aromatic plants are used in small quantities primarily in the bake shop—extracts of vanilla, lemon, and almond; oils such as peppermint and wintergreen.

> **■ MINI-SUMMARY**
>
> Table 8-3 lists the powerhouses of flavor described in this section, as well as other possibilities. Flavoring adds a complementary flavor to a dish at the end of its preparation. It creates a blend in which both the original flavor and the added flavor are identifiable, as in the addition of black pepper to a green salad. Most flavorings are products with distinctive tastes, capable of holding their own in a dish.

TABLE 8-3 Powerhouses of Flavor

- Fresh herbs
- Toasted spices
- Herb and spice blends
- Freshly ground pepper
- Citrus juices, citrus juice reductions
- Strong-flavored vinegars and vinaigrettes
- Wines
- Strong-flavored oils such as extra virgin olive oil or walnut oil
- Infused vinegars and oils
- Reduced stock (glazes)
- Rubs and marinades
- Raw, roasted, or sautéed garlic
- Caramelized onions
- Roasted bell peppers
- Chili peppers
- Grilled or oven-roasted vegetables
- Coulis, salsa, relish, chutney, mojos
- Dried foods: tomatoes, cherries, cranberries, raisins
- Fruit and vegetable purées
- Horseradish and Dijon mustard
- Extracts

Step 2: Healthy Cooking Methods and Techniques

Healthy cooking methods and techniques either do not add fat or add moderate amounts of primarily monounsaturated or polyunsaturated fat. Before looking at cooking methods, let's review some techniques that are used often to develop flavor in healthy recipes.

- **Reduction** means to boil or simmer a liquid down to a smaller volume. In reducing, the simmering or boiling action causes some of the liquid to evaporate. The purpose may be to thicken the product or to concentrate the flavor, or both. A soup or sauce is often simmered for one or both reasons. The use of reduction eliminates thickeners, and intensifies and increases flavor so you can use a smaller portion.

- **Searing** means to expose the surfaces of a piece of meat to high heat before cooking at a lower temperature. Searing, sometimes called browning, can be done in a hot pan in a little oil or in a hot oven. Searing is done to give color and to produce a distinctive flavor. Dry searing can be done over high heat in a nonstick pan using vegetable oil spray or a spray of olive oil.

■ **Deglaze** means adding cold liquid to the hot pan used in making sauces and meat dishes. Any browned bits of food sticking to the pan are scraped up and added to the liquid.

■ **Sweat** means to cook slowly in a small amount of fat over low or moderate heat without browning. You can sweat vegetables and other foods without fat. Instead, sweat in stock or wine.

■ **Puréeing** of vegetables or starchy foods is commonly done as a means to thicken soups, stews, sauces, and other foods. While the food processor is useful for this process, portable vertical mixers can make a very smooth purée with a silky texture and smoothness.

Dry-Heat Cooking Methods

Dry-heat cooking methods are acceptable when heat is transferred with little or no fat, and excess fat is allowed to drip away from the food being cooked. Both pan frying and deep frying add varying amounts of fat, calories, and perhaps cholesterol, depending on the source of the fat, and frying is therefore not an acceptable cooking method. Sautéing can be made acceptable by using nonstick pans and little or no vegetable oil.

Roasting Roasting, cooking with heated, dry air, is an excellent method for cooking larger, tender cuts of meat, poultry, and fish that will provide multiple servings. When roasting, always place meat, poultry, or seafood on a rack so the drippings fall to the bottom of the pan, and the meat therefore doesn't cook in its own juices. Also, cooking on a rack allows for air circulation and more even cooking. In addition to meat, poultry, and seafood, vegetables can be oven-roasted to bring out their flavor. The browning that occurs during roasting adds rich flavors to meats, poultry, seafood, and vegetables. For example, potato wedges or slices can be seasoned and roasted. Vegetables and potatoes don't need to be roasted on a rack. Season with herbs, spices, pepper, stock, garlic, onion, and a few sprays of oil.

For an accompaniment to meat or poultry, you may want to simply use the *jus*. To give the jus some additional flavor, add a mirepoix to the roasting pan during the last 30 to 40 minutes. To remove most of the fat from the jus, you can use a fat-separator pitcher or fat-off ladle. If time permits, you can refrigerate the jus and the fat will congeal at the top.

If you prefer to thicken the natural jus to make *jus lié*, first remove the fat from the jus. Because you will be using the roasting pan to make this product, also pat out the fat from the bottom of the pan. Add some of the jus and a little wine to deglaze the pan. Then add some vegetables and cook on a moderately high heat so they brown or caramelize. At this point, add more jus so the vegetables don't burn. Stir the ingredients to release the food from the pan and get its flavor. Continue to add jus, deglaze the pan, and reduce the jus until the color is appropriate. If there is not enough time to reduce the jus down to the proper consistency, you can thicken it with cornstarch.

Another way to thicken the natural jus is to add starchy vegetables such as beans or potatoes to the jus during the last 30 to 45 minutes of cooking. These vegetables can then be puréed to naturally bind the jus.

To develop flavor, rubs and marinades for meat, poultry, and fish are two possibilities that have been discussed. Smoking can also be used to complement the taste of meat, poultry, or fish. Hardwoods or fruitwoods, such as the following, are best for producing quality results:

■ Fruit (apple, cherry, peach) woods work well with light entrées such as poultry or fish.
■ Hickory and sugar maple are more flavorful and work better with beef, pork, sausage, and salmon.
■ Mesquite produces an aromatic smoke that also works well with beef and pork.

To use hardwoods, you need to soak them first for 30 to 40 minutes, then drain. This way the wood does not burn, but smolders.

Smoke-roasting, also called pan-smoking, is not done in the oven, but on top of the stove. It works best with smaller, tender items such as chicken breast or fish fillet, but can be used to add flavor to larger pieces. Place about half an inch of soaked wood chips in the bottom of a roast pan or hotel pan lined with aluminum foil. Next, place the seasoned food on the rack and reserve. Heat the pan over a moderate-high heat until the wood starts to smoke, then lower the heat. Place the rack over the wood and cover the pan. Smoke until the food has the desired smoke flavor, then complete the cooking in the oven if the food is not yet done. Smoking for too long can cause undesirable flavors. Also, be very careful when opening the lid, due to the heat and smoke.

CHEF'S TIPS

■ Trim excess fat before cooking.
■ Roast on a rack and uncovered (otherwise you are steaming the food). Cook at an appropriate temperature to reduce drying out. Basting during cooking will also reduce drying. Use a meat thermometer and allow for carryover cooking when you remove the dish from the oven. Lastly,

don't slice the meat until it has rested sufficiently so you don't lose valuable juices, and slice across the grain to maximize tenderness.

- To develop flavor in meats, poultry, and fish, use rubs and marinades. You can also stuff the food (with vegetables and grains, for example), sear and/or season the food before cooking, or smoke the food over hardwood chips.
- When smoking, add dried pineapple skins, dried grapevines, or dried rosemary sticks to the wood chips for added flavor possibilities.
- Vegetables can be roasted in the oven, which results in wonderful texture and flavor, due largely to caramelization of the natural sugars. Root vegetables, such as celery root or sweet potatoes, are hardier and take longer to oven roast than vegetables such as peppers, radishes, or patty pan squash. Flavorings such as shallots, thyme, and pepper add flavor. Recipe 8-25 features roasted summer vegetables.
- Other accompaniments that add flavor with moderate or no fat are vegetable coulis, chutney, vinaigrettes, salsas, compotes, and mojos.

Broiling and Grilling Broiling, cooking with radiant heat from above, is wonderful for single servings of steak, chicken breast, or fish with a little more fat such as salmon, tuna, or swordfish that can be served immediately. The more well done you want the product, or the thicker it is, the longer the cooking time and the farther from the heat source it should be. Otherwise, the outside of the food will be cooked but the inside will not be done.

Grilling, cooking with radiant heat from below, is also an excellent method for cooking meat, poultry, seafood, and vegetables. Like broiling, grilling browns foods and the resulting caramelization adds flavor.

Once considered a tasty way to prepare hamburgers, steaks, and fish, grilling is now used to prepare a wide variety of dishes from around the world. For example, chicken is grilled to make fajitas (Tex-Mex style of cooking) or to make jerk barbecue (Jamaican style of cooking). Grilling is also an excellent method to bring out the flavors of many vegetables. Grill them with a little oil, vinegar, or lemon juice, and selected seasonings.

Grilling foods properly requires much cooking experience to get the grilling temperature and timing just right. Some general rules to follow:

1. Cook meat and poultry at higher temperatures than seafood and vegetables.
2. For speedy cooking, use boneless chicken that has been pounded flat and meats that are no more than a half-inch thick. Fat should be trimmed off meats.
3. Don't try to grill thin fish fillets, such as striped bass, because they will fall apart. Firm-fleshed, thicker pieces of fish, such as swordfish, salmon,

and tuna, do much better on the grill. Don't turn foods too quickly—they'll stick and tear. Mark foods by turning the food around 90 degrees without turning it over.

To flavor foods that will be broiled or grilled, consider marinades, rubs, herbs, and spices. Lean fish can be marinated, sprinkled with Japanese crumbs, and glazed in a broiler. If grilling, consider placing soaked hardwood directly on the coals to smoke the food.

CHEF'S TIPS

- Keep the grill clean and properly seasoned to prevent sticking. The broiler must also be kept clean and free of fat buildup; otherwise it will smoke.
- Marinating tender or delicately textured foods for broiling or grilling will firm up their texture, so that they are less likely to fall apart during the cooking process.
- Use marinades, rubs, herbs, spices, crumbs, and smoking to add flavor. When grilling, consider using hardwood chips.
- During broiling, butter has traditionally been used to prevent food from drying out from the intense direct heat. In its place, you may spray it with olive oil, or baste it with marinade, stock, wine, or reduced-fat vinaigrette during cooking. To finish, sprinkle a thin layer of Japanese bread crumbs and glaze under the broiler.
- During grilling, the smoke produced when fat drips to coals is a source of cancer-causing chemicals. The risk involved in eating grilled foods are small, and neither the American Cancer Society nor the National Cancer Institute recommends that we avoid grilled foods. To limit the risk, trim meat or poultry of outside fat to reduce the fat that drips on the coals, and also use marinades.
- To retain flavor during cooking, turn foods on the broiler or grill with tongs—not forks, which cause loss of juices.
- If making kabobs, soak the wooden skewers ahead of time so they can endure the cooking heat without excessive drying or burning.
- Prepare these foods to order, and serve immediately.
- Serve with flavorful sauces such as chutney, relishes, or salsa.

Sauté and Dry Sauté Sautéing, to cook food quickly in a small amount of fat over high heat, can be used to cook tender foods that are in either single portions or small pieces. Sautéing can also be used as a step in a recipe to add flavor to foods such as vegetables, by either cooking or reheating. Sautéing adds flavor in large part from caramelization (browning) that occurs during cooking at relatively high temperatures.

When sautéing, use a shallow pan to allow moisture to escape, and allow space between the food items in the pan. Use a well-seasoned or nonstick pan and add about half a teaspoon or two sprays of oil per serving after preheating the pan. Instead of oil, you can use vegetable-oil cooking sprays, which come in a variety of flavors, such as butter, olive, Asian, Italian, or mesquite. A quick two-second spray adds about 1 gram of fat (9 calories) to the product. To use these sprays, spray the preheated pan away from any open flames (the spray is flammable) and then add the food. New pump spray bottles are available that allow you to fill them with the oil of your choice.

For an even lower-in-fat cooking method, use the dry sauté technique. Using this technique, heat a nonstick pan, spray with vegetable-oil cooking spray, then wipe out the excess with a paper towel. Heat the pan again, then add the food.

If browning is not important, you can simmer the ingredient in a small amount of fat-free liquid, such as wine, vermouth, flavored vinegar, or defatted stock, to bring out the flavor. Vegetables naturally high in water content, such as tomatoes or mushrooms, can be cooked with little or no added fluid, at a very high quick heat.

When sautéing or dry sautéing, you can deglaze the pan after cooking with stock, wine, or other low-fat liquid. Then add shallots, garlic, or other seasonings to the sauce.

CHEF'S TIPS

- To add flavor, use marinades, herbs, or spices
- Use high heat or moderately high heat and a well-seasoned or nonstick pan used only for dry sautéing.
- Pound pieces of meat or poultry flat to increase the surface area for cooking and thereby reduce the cooking time.
- To sauté vegetables, simmer them with a liquid such as stock, juice or wine in a nonstick pan, and add seasonings such as cardamom, coriander, or fennel. Cook over a moderately high heat. As the vegetables start to brown and stick to the pan, add more liquid and deglaze the pan. Once the vegetables are ready, add a little bit of butter or flavorful oil, perhaps one teaspoon for four servings, to give the product a rich flavor without much fat, and a shiny texture.
- Another sauté method is to blanch vegetables in boiling water to desired doneness, then shock in ice water. Drain and dry sauté in a hot nonstick pan with stock, wine, fresh herbs, garlic, and shallots (chopped), and finish with fresh black pepper and extra virgin olive oil, butter, or nut oil (one teaspoon for four servings).
- Once prepared, serve sautéed foods immediately. They do not hold well.
- A sautéed entrée goes well with portions of stews, ragouts, pilafs, or risotto.

Stir-frying Stir-frying, cooking small-sized foods over high heat in a small amount of oil, preserves the crisp texture and bright color of vegetables and cooks strips of poultry, meat, or fish quickly. Typically, stir-frying is done in a wok, but a nonstick pan can be used. Steam-jacketed kettles and tilt frying pans can also be used to make quantity stir-fry menu items. Cut up ingredients as appropriate into small pieces, thin strips, or diced portions.

CHEF'S TIPS

- Coat the cooking surface with a thin layer of oil. Peanut oil works well because it has a strong flavor (so you can use just a little) and a high smoking point. You can also use vegetable cooking spray, and wipe away any excess.
- Have your ingredients ready next to you, because this process is fast.
- Partially blanch the vegetables first, so that they will cook completely without excessive browning.
- Preheat the equipment to a high temperature.
- Foods that require the longest cooking times, usually the meat or poultry, should be the first ingredients that you start to cook.
- Stir the food rapidly during cooking and don't overfill the pan.
- Use garlic, scallion, ginger, rice wine vinegar, low-sodium soy sauce, and chicken/vegetable stock for flavor.
- Add a little sesame oil at the end for taste (about 1 teaspoon per four servings).

Moist-Heat Cooking Methods

Moist-heat cooking methods involve water or a water-based liquid as the vehicle of heat transfer, and are often used with secondary cuts of meat and fowl, or legs and thighs of poultry. When moist-heat cooking meat or poultry, the danger is that the fat in the meat or poultry, although leaner, stays in the cooking liquid. This problem can be resolved to a large extent by chilling the cooking liquid so the fat separates and is then removed before the liquid is used. If the liquid needs to be chilled quickly, place in an ice bath for quickest results.

Compared to most dry-heat cooking methods, these methods do not add the flavor that dry-heat cooked foods get from browning, deglazing, or reduction. In order for foods that use moist-heat cooking to be successful, you will need:

- very fresh ingredients
- seasoned cooking liquids using fresh stock, wine, fresh herbs, spices, aromatic vegetables, and other ingredients
- strongly flavored sauces or accompaniments to achieve flavor and balance.

Steaming has been the traditional method of cooking vegetables in many quantity kitchens because it is quick and retains flavor, moisture, and nutrients. It's healthy too, because it requires no fat. The best candidates for steaming include foods with a delicate texture, such as fish, shellfish, chicken breasts, vegetables, and fruits. Be sure to use absolutely fresh ingredients to come out with a quality product.

Fish is great when steamed *en papillote* (in parchment) or in grape, spinach, or cabbage leaves. The covering also helps retain moisture, flavor, and nutrients. Consider marinating fish beforehand to add moisture and flavor.

It is possible to introduce flavor into steamed foods (as well as poached foods) by adding herbs, spices, citrus juices, and other flavorful ingredients to the water. For example, steam halibut over a carrot-lobster broth. Steamed foods also continue to cook after they come out of the steamer, so allow for this in your cooking time.

Poaching, cooking a food submerged in liquid at a temperature of 160 to 180 degrees F. (71 to 82 degrees C.), is used to cook fish, tender pieces of poultry, eggs, and some fruits and vegetables. To add flavor, you can poach the foods in liquids such as chicken stock, fish stock, or wine flavored with fresh herbs, spices, ginger, mirepoix, vegetables such as garlic or shallots, or citrus juices. Serve poached foods with flavorful sauces or accompaniments.

Fish is often poached in a flavored liquid known as court bouillon. The liquid is simmered with vegetables (such as onions, carrots, and celery), seasonings (such as herbs), and an acidic product (such as wine, lemon juice, or vinegar). It may be used to cook fish to be served either hot or cold, but it is not generally used in making sauce. Cook the court bouillon for about 20 minutes, strain out vegetables and herbs, then add the fish to infuse these flavors. Place the fish to be poached on a rack or wrap it up in cheesecloth.

Braising, or stewing (stewing usually refers to smaller pieces of meat or poultry), involves two steps: searing or browning the food (usually meat) in a small amount of oil or its own fat, then adding liquid and simmering until done. Foods to be braised are also often marinated before searing to develop flavor and tenderize the meat. When browning meat for braising, sear it in as little fat as possible without scorching, and then place it in a covered braising pan to simmer in a small amount of liquid. To add flavor, place roasted vegetables, herbs, spices, and other flavoring ingredients in the bottom of the braising pan before adding the liquid.

Following are the steps for braising meat or poultry.

1. Season the meat or poultry. Trim fat.
2. In a small amount of oil in a brazier, sear the meat or poultry to brown it and develop some flavor.

3. Remove the meat and any excess oil from the pan.
4. Put the pan back on the heat. Add vegetables and caramelize (brown) them.
5. Deglaze the pan with wine and stock, being sure to scrape the fond.
6. Add tomato paste, wine, stock, and other aromatic vegetables. Reduce.
7. Return the meat to the brazier and put into the oven, covered, to simmer. Make sure the meat is three-quarters covered with liquid. This allows the meat to cook more evenly and avoids the bottom from getting scorched.
8. When the meat is done (forktender), strain the juices. Skim off fat and reduce juices. Purée the vegetables from the juices and use them (or you can use cornstarch) to thicken the jus to make a light gravy.

Vegetables, beans, and fish can also be braised.

Microwaving is another wonderful method for cooking vegetables, because no fat is necessary and the vegetables' color, flavor, texture, and nutrients are retained. Boiling or simmering vegetables is not nearly as desirable as steaming or microwaving, because nutrients are lost in the cooking water and more time is required.

■ MINI-SUMMARY

Healthy cooking methods include baking, roasting, broiling, grilling, stir-frying, sautéing with little or no oil, boiling, simmering, poaching, microwaving, and braising, when the fat is removed from the cooking liquid. Cooking techniques that are especially useful in preparing healthy items include reduction, searing, deglazing, sweating, and puréeing.

Step 3: Presentation

Here are the main considerations that underlie the art of presentation.

■ **Height** gives a plate interest and importance. A raised surface or high point calls attention to itself. A flat and level surface is monotonous. When actual height is difficult to attain, implied height, or an illusion of height, can often be achieved by causing the eye to focus on a particular point. This can be done in several ways. One is by arranging ingredients in a pattern that guides the eye to that point. Another is to use an eye-catching ingredient or garnish to establish a focal point.

■ **Color** is very, very important. Too many colors tend to confuse the eye and dissipate the attention, so don't overdo it.

■ **Shape** is important too. Vary the shapes on one plate.

■ Match the **layout** of the menu item with the shape of the plate. For example, salads are usually presented on round plates. This means that the lines, forms, and shapes of the salad ingredients must be arranged in a pattern that fits harmoniously into a circle. The pattern may repeat the curve of the plate's edge, or echo its roundness on a smaller scale, or complement it with balance and symmetry. The flow of your food presentation should be tight, meaning having food fairly close together to retain heat. The flow from left to right should curve inward to guide the eye back to the middle of the plate. The pattern begins with the rim of the plate, so never place anything on the rim. It is the frame of your design.

■ The most effective **garnish** is something bright, eye-catching, contrasting in color, pleasing in shape, and simple in design. It should enhance the plate and not be the focus. At times, sauces may act as the garnish.

One of the tricks of presentation with healthy dishes is to make less look like more. When you are serving smaller portions of meat, poultry, or seafood, various techniques can be used to make the protein portion size look larger. By slicing meat or poultry thin, you can fan out the slices on the plate to make an attractive arrangement, and also one that looks plentiful. You can also arrange a piece of meat, poultry, or seafood on a bed of grains, vegetables, and/or fruits, or cover it with sautéed vegetables. In addition, serving larger portions of side dishes with the entrée also helps to make the plate look full. Sauces such as vegetable coulis, salsa, and relishes also help cover the plate, and they provide eye appeal and color.

A common problem that crops up when plating healthy foods is that many dishes lose heat quickly and dry out fast. High-fat sauces help keep a dish hot. When meat is sliced for presentation, it loses more juice and heat, so it dries out quickly. To overcome this problem, chefs often place foods close together on the plate, putting the densest food in the center to keep the other foods warm. When slicing meats for plating, you can slice just part of the meat for appearance and leave the remaining piece whole for the guest to cut.

Keep all garnishes simple. Some dishes, such as angel-food cake with a fruit sauce, are already garnished, although you may top them with a slice or two of fresh fruit.

■ MINI-SUMMARY

Pay special attention to height, color, shapes, unity, and garnishes for maximum plate presentation.

Healthy Recipes and Chef's Tips

You can modify recipes for many reasons, such as to reduce the amount of calories, fat, saturated fat, cholesterol, sodium, or sugar. You may also modify a recipe to get more of a nutrient, such as fiber or vitamin A. Whether modifying a recipe to get more or less, there are four basic ways to go about it.

1. Change/add healthy preparation techniques.
2. Change/add healthy cooking techniques.
3. Change an ingredient by reducing it, eliminating it, or replacing it.
4. Add a new ingredient(s), particularly to build flavor.

If you do decide to modify a recipe, follow these steps.

1. Examine the nutritional analysis of the product and decide how and how much you want to change the product's nutrient profile. For example, in a meat-loaf recipe, you may decide to decrease its fat content to less than 40 percent and increase its complex carbohydrate content to 10 grams per serving.
2. Next, you need to consider flavor. What can you do to the recipe to keep maximum flavor? Should you try to mimic the taste of its original version, or will you have to introduce new flavors? Will you be able to produce a tasty dish?
3. Next, modify the recipe using any of the methods just discussed. When modifying ingredients, think about what functions each ingredient performs in the recipe. Is it there for appearance, flavor, texture? What will happen if less of an ingredient is used, or a new ingredient is substituted? You also have to consider adding flavoring ingredients in many cases.
4. Evaluate your product to see whether it is acceptable. This step often leads to further modification and testing. Be prepared to test the recipe a number of times, and also be prepared for the fact that some modified recipes will never be acceptable.

Of course, if you don't want to go through the trouble of modifying current recipes, you can select and test recipes from healthy cookbooks or other sources, or create your own recipes. When developing new recipes, be sure to choose fresh ingredients and cooking methods and techniques that are low in fat. Also, pay attention to developing flavor, and cook foods to order as much as possible.

Appetizers

Appetizers are a very creative part of the menu. Ingredients might include fresh fruits and vegetables, fresh seafood and poultry, fresh herbs, spices, infused oils, vinegars, and pasta.

Recipes for Roast Chicken and Shredded Mozzarella Tortellinis (Recipe 8-1), Scallop Rolls in Rice Paper (Recipe 8-2), Crab Cakes (Recipe 8-3), Eggplant Rollatini with Spinach and Ricotta (Recipe 8-4), and Mussels Steamed in Saffron and White Wine (Recipe 8-5) are found in this chapter. Additional ideas for appetizers include:

- Ricotta Cheese and Basil Tortellonis with Salsa Cruda and Arugula
- Napoleon of Grilled Vegetables and Wild Mushrooms with Roasted-Pepper Sauce
- Potato and Onion Tart with Forelli Pear Vinaigrette
- Pan-Seared Scallops with Butter Potato-Pancake, Sunsprouts and Blood-Orange Dressing
- Spicy Chicken and Jack Cheese Quesadillas with Tomatilla Salsa
- Maine Crab Cakes with Smoked-Pepper Sauce and Baby Lettuces
- Red Lentil Chili with Baked Whole-Wheat Tortilla Chips

CHEF'S TIPS
- Appetizers can often be sized-down entrées. For example, Maine Crab Cakes with Smoked-Pepper Sauce and Baby Lettuces can be made in larger or smaller portions to appear as entrée or appetizer.
- There are certain ingredients with which you can make a wide variety of appetizers. Consider using wonton skins and rice paper as wrappers, and stuff them with fillings such as white beans and artichokes with roasted garlic, or spiced butternut squash.
- Dried beet chips are a wonderful way to add color to appetizers. To make, slice red beets thin, and dip in simple syrup. Dry on a silk mat. Put in 275°F oven until crisp, about 15–20 minutes.
- Creative sauces and relishes can help sell appetizers. For example, serve smoked shrimp with mango sauce and cucumber black bean relish.

Soups

Soups make up some of the most nutritious meals, from the hearty minestrone to robust butternut squash soup, creamed with nonfat yogurt and spiced with nutmeg. Soups can be balanced into a meal as an appetizer or given first billing as an entrée.

Soups are a wonderful place to spotlight more than just vegetables. Beans, lentils, split peas, and grains such as rice are also healthy ingredients that work well with soups. Also, some of these ingredients are starchy enough to be used to thicken soups, instead of using traditional roux (a thickening agent of fat and flour in a one-to-one ratio by weight), which is high in calories and fat. Examples of starchy foods that work well as thickeners include beans, lentils, rice, other grains, and puréed vegetables such as potatoes and squash. These foods can be used to make soups such as black bean, lentil, split pea, Pasta e Fagioli (Recipe 8-6), and vegetable chili soup.

To make a cream soup such as cream of broccoli without using cream or roux, start by dry sautéing broccoli, onion, herbs, garlic, and shallots in chicken stock and white wine. Deglaze the pan, then add potato and cover with vegetable or chicken stock. Once the potato is done, purée the ingredients to the proper consistency. Garnish the soup with small, steamed broccoli florets.

Rice also is an excellent thickener and lends a creamy texture to the soup. Rice can be used successfully to thicken corn, carrot, and squash soups. Use about six ounces rice to one gallon of stock. See Recipe 8-7, Butternut Squash Bisque, for an example.

CHEF'S TIPS
- Strain soups such as broccoli, celery, and asparagus through a large-holed china cap to remove fibers.
- Also purée bean soups such as black bean or split pea to get a homogenous product. Next, strain to remove skins.
- Rice and potatoes work well as thickeners in many soups.
- Replace ham in bean soups with smoked chilies, or your own smoked turkey, or veal bacon.
- Garnish soups with an ingredient of the soup whenever possible. For example, put pieces of baked tortillas on top of Mexican soup. The use of fresh vegetables, fruits, or herbs as a garnish adds interest, color, and flavor. Also consider garnishing some soups, as appropriate, with a small amount (such as one teaspoon) of cream, nuts, croutons, sour cream, smoked chicken, or avocado.

Salads and Dressings

Components of salads go way beyond simple raw vegetables. Consider using grains, beans, lentils, pasta, fresh fruits and juices, oven-dried vegetables, fresh poultry or seafood, game, herbs and spices such as ginger, Kafir lime leaves, star anise, cardamom, curry, lavender, lemon balm, and fresh cinnamon. Salads are a wonderful place to feature high-fiber, low-fat ingredients.

Recipes for Wild Mushroom Salad (Recipe 8-8) and Baby Mixed Greens with Shaved Fennel and Orange Sections (Recipe 8-9) appear in this chapter. Other possible salad combinations include the following.

- Baby Lentil and Roasted Vegetables
- Yellow and Red Tomato Salad with Fresh Basil, Oregano, and Double-Roasted Garlic
- Haricot Verts with Trio of Roasted Peppers
- Carrot, Celery, Golden Pineapple, and Dried Pear Salad with Sweet-and-Sour Dressing
- Bowtie Pasta with Fresh Tuna and Chives in a Lemon Basil Dressing
- Penne Salad with Tomato and Cilantro
- Yogurt Chicken with Fresh and Dried Fruits
- Wheat Berry Salad with Wild Mushrooms and Rosemary Thyme Dressing
- Organic Baby Lettuces with Marinated Cucumber and Tomato with Classic French Dressing

Dressings are used in much more than salads. They can often be used as an ingredient of entrées, appetizers, relishes, vegetables, and marinades. There are many categories of salad dressings. The best place to start is the basic vinaigrette, because it is simple and you can use ingredients such as herbs and spices to make many variations. The best ingredients to use include good-quality vinegars, first-pressed olive and nut oils, and fresh herbs, because you need the strongest flavor with the least amount of fat. Other good ingredients that add flavor without fat include Dijon mustard, shallots or garlic (which may be roasted for a robust flavor), a touch of honey, reduced vinegars, and lemon or lime juice.

For examples of vinaigrette recipes, see recipes for Basic Herb Vinaigrette (Recipe 8-10), Orange Vinaigrette (Recipe 8-11), and Sherry Wine Vinaigrette (Recipe 8-12). If you look at Basic Herb Vinaigrette, you will notice that instead of a ratio of 3 parts oils to 1 part vinegar, this recipe uses 1 part oil, 1 part vinegar, and about 2 parts thickened chicken or vegetable stock. This results in a satisfactory product whose flavor profile is boosted with fresh herbs and garlic and high-quality olive oil and vinegar.

Other salad dressings often fit into one of one of these categories.

1. *Creamy dressings*—Tofu can be processed to produce a creamy dressing, as in Green Goddess Dressing (Recipe 8-14). Other creamy ingredients are nonfat or low-fat yogurt, nonfat sour cream, or ricotta cheese. Puréed fruits and vegetables are creamy and can be used as emulsifiers in salad dressings.

2. *Puréed dressings* — Examples of puréed dressings include potato vinaigrette, hummus, and smoked-pepper coulis. Some of these dressings, such as hummus, work well as dips.

3. *Reduction dressings* — Examples include orange (see Recipe 8-11), beet, carrot-balsamic, and apple cider. These dressings can be made simply and are powerhouses of flavor.

CHEF'S TIPS

■ As elsewhere in the kitchen, use fresh, high-quality ingredients. Choose ingredients for compatibility of flavors, textures, and colors.

■ Vegetables, both raw and cooked, go well with dressings having an acid taste such as vinegar or lemon.

■ Legumes make wonderful salads. For example, black-eyed peas go well with flageolets and red adzuki beans. To add a little more color and develop the flavor, you might add chopped tomatoes, fresh cilantro (Chinese parsley), haricots vert (green beans), and roasted peppers.

■ The reduction dressings, such as reduced beet juice, can be put into a squirt bottle and used to decorate the plate for a salad, appetizer, or entrée.

■ Plan your presentation carefully in terms of height, color, and composition. Keep it simple. Do not overcolor or overgarnish.

Entrées

Developing healthy entrées will draw upon your entire knowledge of ingredients, cooking techniques and methods, and nutrition. As the popular expression goes: "The sky is the limit!" You have a wide variety of ingredients, cooking techniques, and methods at your disposal to create numerous delicious entrées, ranging from traditional beef to meatless dishes, such as the following.

■ Classic Beef Stew with Horseradish Mashed Potatoes
■ Grilled Pork Chop Adobo with Spicy Apple Chutney (Recipe 8-16)
■ Sautéed Veal Loin with Barley Risotto and Roasted Peppers
■ Pan-Seared Louisiana Spiced Breast of Chicken
■ Grilled Chicken Breast and Quinoa Salad with Cucumber, Tomato, Corn, and Peppers
■ Mediterranean Chicken Breast with Basmati Rice
■ Grilled Chicken/Turkey Burger with Oven-Baked Blue Corn Tortillas and Salsa Roja
■ Stir-Fried Chicken and Garden Vegetables with Brown Rice

- Cedar-Planked Wild Striped Bass with Ratatouille and Red Bean Salad
- Lemon-Scented Tuna with Charred Tomato Salsa
- Basil Angel-Hair Pasta with Fruits of the Sea (Mussels, Scallops, Shrimp, and Calamari)
- Pan-Blackened Florida Grouper with Pepper and Pineapple Relish
- Eggplant Roulade with Purée of White Beans and Artichokes
- Timbales of Quinoa, Brown Rice, Barley, and Roasted Vegetables
- Rice and Wild Mushroom Cakes

Entrée recipes in this chapter include Potato Lasagna (Recipe 8-17), Chicken Sausage (Recipe 8-18), Slates of Salmon (Recipe 8-19), and Braised Lamb (Recipe 8-20).

Chapter 4 reviews the steps in selecting and cooking lean cuts of meat, poultry, and fish. As a review, lean cuts include the following:

- **Beef:** eye of round, inside (top) round, outside (bottom) round, flank steak, top sirloin butt
- **Veal:** any trimmed cut except commercially ground veal patties
- **Pork:** pork tenderloin, pork chop (sirloin), pork chop (top loin), pork chop (loin)
- **Lamb:** shank, sirloin
- **Poultry:** breast (skin removed after cooking)
- **Fish:** all fish and shellfish

It's important to use high-quality products, trimmed of fat as appropriate. Rubs and marinades, as well as cooking methods (such as hardwood grilling), can do much to add flavor and reduce fat.

After reading up to this point, you may have noticed the lack of classical sauces. Rather than foods covered with rich, heavy sauces, the emphasis in this book is on the taste and appearance of the food itself. In sauce making, this has meant a great change in the techniques of thickening. Followers of the new style often use purées and reductions instead of roux to make sauces, as you have seen in vegetable coulis and meat juices that have been thickened with puréed vegetables. Also, stock can be flavored and reduced to make a quality sauce that can be used in many dishes on the menu. This chapter features the following alternatives to traditional sauces.

- Papaya and White Raisin Chutney (Recipe 8-21)
- Papaya Plantain Salsa (Recipe 8-22)
- Red Pepper Coulis (Recipe 8-23)
- Hot and Sour Sauce (Recipe 8-24)

CHEF'S TIPS
- For meat, poultry, and fish entrées, a 3- to 4-ounce cooked portion is adequate.
- Fish is a very versatile and nutritious food. Almost any food, such as rice or beans or pasta, goes with fish. Serve fish on top of a vegetable ragout, or salmon with vegetable curried couscous.
- When choosing legumes for a dish, think color and flavor. Make sure the colors you pick will look good when the dish is complete. Also think of other ingredients you will use for flavor.
- Bigger beans, such as gigante white beans, hold their shape well and lend a hearty flavor to stews, ragouts, and salads.
- When using cheese in an entrée, use a small amount of a strong cheese such as gorgonzola or goat cheese.
- Create new fillings for pasta that don't rely totally on cheese. For example, sweat out puréed butternut squash and potato. Add fresh thyme, roasted shallots, and perhaps a little roasted duck for flavor. Or purée together cooked artichokes, white beans, roasted garlic, and carrots or ratatouille for flavor and color.

Side Dishes

There is no end to the variety of substitutions and side dishes for every entrée. The same dish can take on a new face simply by changing the starch or the vegetable. Besides the traditional side dish of vegetables and potatoes, consider using grains such as wheat berries or barley; try legumes such as black beans or lentils, tofu, or fresh fruits. Also consider techniques such as the following:

- **Puréeing.** For example, purée sweet potatoes, butternut squash, and carrots, flavored with cinnamon, honey, and thyme.
- **Roasting.** For example, roast onions with cinnamon, ginger, vinegar, and sugar. Recipe 8-25 shows how to prepare Roasted Summer Vegetables.
- **Grilling.** For example, grill portabello with polenta, garden tomatoes, and roast elephant garlic.
- **Stir-fry.** Try Hot and Sour Stir-Fry with Seared Tofu and Fresh Vegetables.

Additional examples of side dishes include these:

- Roasted Garlic and Yogurt Red Mashed Potatoes with Dill
- Couscous with Dried Fruit, Cucumber, and Mint
- Seven-Vegetable Stir-Fried Rice
- Ratatouille Strudel with Oven-Dried Tomatoes Wrapped in Phyllo

■ Oven-Baked French Fries with Cajun Rub
■ Swiss-Style Marinated Red Cabbage
■ Sautéed Wild Mushrooms
■ Risotto (a creamy dish originating in Italy that is prepared with rice, butter, and stock) with Spring Onions and Pesto

Grains, such as rice, are versatile and make excellent side-dish ingredients. See Recipe 8-26, which shows how to make Mixed Grain Pilaf.

CHEF'S TIPS
■ When using vegetables, you need to *think*—about what's in season for maximum flavor, about how the dish will look (its colors), and about how the dish will taste. Think flavor and color.
■ Also, think variety. Serving vegetables doesn't mean switching from broccoli to cauliflower and then back to broccoli. There are many, many varieties of vegetables to choose from. Be adventurous.
■ Add grains to vegetable dishes, such as brown rice with stir-fried vegetables.
■ Serve grains and beans. For example, rice and beans is a very popular and versatile dish using grains and legumes. Mix purple rice with white beans, or wild rice with cranberry beans for appearance.
■ Salads can often be used as side dishes.

Desserts

You can make a wide variety of desserts without compromising health. Think of fruits (either fresh or as a key ingredient), quick breads, cobblers, puddings, phyllo strudels, and even some cakes and cookies. The following recipes show how fruit can be used in many forms.

■ Oatmeal-Crusted Peach Pie
■ Apple Strudel with Caramel Sauce
■ Fruit Sorbets
■ Spiced Carrot Cake with Orange Custard Sauce
■ Fresh Fruit Compote with Buttermilk Dumplings
■ Chocolate Raspberry Meringue Torte
■ Banana Ginger Pudding
■ Poached Sickle Pears with Merlot Syrup and Almond Tuille
■ Grilled Peppered Pineapple with Vanilla Ice Cream and Crunchy Cookies
■ Yogurt Cheesecake

Recipe 8-28 highlights a way to use cocoa and bittersweet chocolate to make a chocolate pudding cake that is baked in sugar molds in a bain-marie (hot-water bath). Recipe 8-29 uses soft tofu, egg whites, raspberry purée, and sugar to make ice cream that, with the help of an ice cream machine, can be made right in the kitchen. This raspberry creamed ice goes well with Recipe 8-30, Angel Food Savarin. Recipe 8-31, Heart-Healthy Oatmeal Raisin Cookies, uses mostly applesauce (with a tiny amount of canola oil) to replace the usual fat. This is an excellent cookie for use of a fat substitute, because it is naturally sweet (due to the raisins) and spicy.

CHEF'S TIPS
- To make sorbet without sugar, simply strain and churn puréed fruits. Make sure the fruits are at the peak of ripeness.
- Use angel-food cake as a base to build a dessert. Serve it with fresh fruit sorbet, warm sautéed apples and cranberries, pear and ginger compote, or mango, mint, and papaya salsa.
- Likewise, use phyllo in many ways. For example, stuff it with baked strawberries and nonfat granola, or bake it in a cup and fill it with sautéed apples garnished with dried apple chips to give it some crunch.
- Compote is an additional way to serve fruit. Compote is a dish of fruit, fresh or dried, cooked in syrup flavored with spices or liqueur. When you consider how many fruits you have at your disposal, as well as spices and flavorings, the possibilities are endless.

Breakfasts

Hot and cold cereals have been the foundation of breakfast for generations in many ethnic cultures, from Swiss Birchenmuesli (oats with fresh and dried fruits soaked in milk or cream) to hot English oatmeals to contemporary American granolas. There are numerous variations from these classic foundations. For instance, the liquid used to make hot cereals like oatmeal can be a variety of fruit juices such as apple, pineapple, or orange. They can also be spiced with cinnamon, nutmeg, ginger, allspice, or cloves. For more adventurous customers, you can use jalapeño jack cheese, star anise, lavender, lemon balm, or any fresh herb combination.

The traditional Swiss birchenmuesli is made with rolled oats, heavy cream, sugar, nuts, and dried and fresh fruits. To modify this recipe, more fresh fruits are added to the cereal mixture, skim milk replaces heavy cream, and nonfat yogurt and spices complete the taste needed to make this old-world classic a modern hit.

Pancakes, French toast, and toppings are quite honestly the apple pie of breakfast. It's hard to imagine a breakfast menu without blueberry pancakes or thick crispy French toast with syrup. A typical pancake batter contains whole eggs, oil, and sugar. To use less of these ingredients, you will have to add other ingredients that you can sink your teeth into. For example, you can use wheat germ, rolled oats, stone-ground wheat, or millet to create a hearty texture. You also need to include spices and fruit flavorings. By putting leftover berries into batter or using overripe fruit to make syrups, you utilize your inventory while creating a quality product. You can also fine-cut the fruits and toss them with fresh mint or lemon balm to create a sweet salsa or compote.

Quick breads, muffins, and scones are also staple items for breakfast menus as well as buffets. Fruit juices, concentrates, and purées are a wonderful source of flavor when baking quick breads, muffins, and scones, as well as in hot cereals and pancake batters. Low-fat spreads might include flavored nonfat ricotta cheese or flavored low-fat cream cheeses.

Some menu possibilities for breakfast include the following:

- Glazed Grapefruit and Orange Slices with Maple Vanilla Sauce (Recipe 8-35)
- Crunchy Cinnamon Granola (Recipe 8-33)
- Pineapple Ginger Cream of Wheat
- Jalapeño Jack Cheese Grits
- Spiced Oatmeal with Dried Cranberries
- Stuffed French Toast Layered with Light Cream Cheese and Bananas (Recipe 8-34), topped with strawberry syrup
- Wheat-Berry Pancakes with Fresh Fruits
- German Pancakes with Orange Syrup
- Whole-Wheat Peach Chimichangas
- Banana Ginger Raisin Bread
- Dried-Cherry Scones

Breakfast is a wonderful time for a buffet. It is a time saver for guests to give them wide variety. Consider a buffet with platters of sliced fresh fruits with maple vanilla yogurt, dried-cherry scones, jalapeño jack grits, banana pancakes, classic vegetable frittata, crunchy cinnamon granola, and high-five fruit muffins.

CHEF'S TIPS

- To make an excellent omelet without cholesterol, whip egg whites until they foam. Add a touch of white wine, Dijon mustard, and chives. Whip to a loose meringue. Spray a hot nonstick pan with oil and add your eggs. Cook the way you do a whole egg omelet. When the omelet is close to done, put the pan under the broiler to finish. The omelet will puff up. Stuff the omelet, if desired, with something like grilled, roasted, or sautéed vegetables, then fold over and serve.
- For color and flavor, serve an omelet with spicy vegetable relish poured on top of it, or place the omelet in a grilled blue corn tortilla and serve with salsa roja.

■ When writing breakfast menus, make sure you provide balanced, healthful, and flavorful breakfasts, such as glazed grapefruit and orange slices with maple vanilla sauce, blueberry wheat pancakes with strawberry syrup, and a side dish of chicken hash.

■ Breakfast is probably the best time of day to offer freshly squeezed juices. Make sure they are fresh and that you offer a good variety.

Check-Out Quiz

1. Sweating vegetables can be done in stock.
   a. True   **b.** False

2. Flavorings are used to bring out flavor that is already in a dish.
   a. True   **b.** False

3. Ginger is an example of a fresh herb.
   a. True   **b.** False

4. Basil, oregano, and garlic are often used in Indian dishes.
   a. True   **b.** False

5. Toasted spices can be ground and used in marinades, salad dressings, rubs, and soups.
   a. True   **b.** False

6. Gelatin is what gives stock body.
   a. True   **b.** False

7. A good stock is brimming with flavor.
   a. True   **b.** False

8. The shallot is a green onion.
   a. True   **b.** False

9. Chutney refers to a sauce made of a purée of vegetables or fruits.
   a. True   **b.** False

10. Portable vertical mixers are a great tool for puréeing foods.
   a. True   **b.** False

Activities and Applications

1. Recipe Modification

Use one of the substitutions listed in Table 8-4 to prepare a recipe. Compare the flavor, texture, shape, and color of both products. Ask someone who doesn't know which is the modified version to test the products and determine which one tastes better.

2. Nutrient Content of Modified Recipes

Following is a recipe for Monte Cristo sandwiches that has been modified to make it more nutritious. Calculate and compare the nutrient content before and after modification.

TABLE 8-4 Recipe Substitutions

In Place of	Use
Butter	Margarine
Whole milk	Skim or low-fat milk (Add some nonfat dry milk, if desired, to make it creamy)
1 cup shortening or lard	3/4 cup vegetable oil
1 cup heavy cream	1 cup evaporated skim milk
1 cup light cream	1 cup evaporated skim milk
1 cup sour cream	1 cup nonfat plain yogurt or low-fat cottage cheese blended with 1 to 2 tablespoons buttermilk or lemon until creamy
Cream cheese	Neufchâtel or reduced-calorie cream cheese
1 ounce baking chocolate	3 tablespoons cocoa and 1 tablespoon vegetable oil
1 egg	1/4 cup cholesterol-free egg substitute or 2 egg whites
Whipped cream	Whip 1 cup low-fat ricotta cheese with 3 tablespoons nonfat plain yogurt and 2 tablespoons sugar. Use honey, vanilla, or other extracts to flavor.
Whole-milk mozzarella	Part-skim mozzarella, low moisture
Whole-milk ricotta	Part-skim ricotta
Creamed cottage cheese	Low-fat cottage cheese or pot cheese
Cheddar cheese	Low-fat cheddar cheese
Swiss cheese	Low-fat Swiss cheese, such as Swiss Lorraine
Cream cheese	Neufchâtel or reduced-calorie cream cheese
Ice cream	Ice milk or frozen yogurt
Mayonnaise	Nonfat plain yogurt

Original Monte Cristo Sandwich

Yield: 50 portions

Ham, cooked, boneless	50 1-oz. slices
Turkey, cooked, boneless	50 1-oz. slices
Swiss cheese	50 1-oz. slices
White bread	100 slices
Whole milk	3 cups
Salt	1 teaspoon
Eggs, whole, slightly beaten	1 quart (24 eggs)
Shortening, melted	2 cups

Place one slice each of ham, turkey, and cheese on one slice of bread and top with a second slice. Blend milk, salt, and egg. Dip each side of the sandwich into the egg and milk mixture; drain. Grill each sandwich on a well-greased griddle about 2 minutes on each side, or until golden brown and the cheese is melted.

Modified Monte Cristo Sandwich

Yield: 50 portions

Low-fat or fat-free honey ham	50 1-oz. slices
Turkey, fresh roasted	50 1-oz. slices
Swiss cheese	50 0.5-oz. slices
Rye bread	100 slices
Skim milk	3 cups
Egg whites, lightly beaten	1 quart
Butter-flavored cooking spray	As needed

Place one slice each of ham, turkey, and cheese on one slice of bread and top with a second slice. Blend the milk and egg whites. Dip each side of the sandwich into the egg and milk mixture; drain. Grill each sandwich on a grill sprayed with butter-flavored cooking spray. Cook about 2 minutes on each side, or until golden brown and the cheese is melted.

3. Nutritional Contribution of Ingredients in Recipes

Using nutrient composition information, figure out which ingredient in the recipe for Original Monte Cristo Sandwiches contributes the most fat. Which ingredient contributes the most calories? Which ingredient contributes the most carbohydrates?

4. Flavorings

Pick out five recipes from a traditional cookbook and five recipes from a healthy cookbook. Determine the sources of flavor for each recipe and record them. Compare the ingredients used to flavor each set of recipes. Does either set of recipes tend to use more fat or more herbs and spices?

5. Computerized Nutrient Analysis

Do a nutrient analysis of a recipe from a cookbook that has all the recipes analyzed. Compare your results to the book's nutrient analysis. Are they close? Why or why not?

6. Menu Planning Exercise

Using a menu from a restaurant or foodservice, recommend two healthy entrées and two healthy desserts that would fit well on this menu. Be ready to explain why you selected these menu items and how they fit with the rest of the menu and the clientele.

Nutrition Web Explorer

Cooking Light magazine http://www.cookinglight.com
On the home page of this popular magazine on healthy cooking and fitness, click on "Food." Print out at least two recipes, and read how a recipe was modified under "Reader Requests."

Cooking.com http://www.cooking.com
This popular site features cooking equipment, recipes, and many other resources for cooks. On its home page, find out which equipment is being featured for light cooking.

Chefwriter.com http://www.chefwriter.com
At this website, click on "Ask the CMC (Certified Master Chef)," and ask a question about healthy cooking or ingredients. Bring the CMC's response to class.

Recipes

8-1 Roast Chicken and Shredded Mozzarella Tortellinis

8-2 Scallop Rolls in Rice Paper with Vegetable Spaghetti

8-3 Crab Cakes

8-4 Eggplant Rollatini with Spinach and Ricotta

8-5 Mussels Steamed in Saffron and White Wine

8-6 Pasta e Fagioli

8-7 Butternut Squash Bisque

8-8 Wild Mushroom Salad

8-9 Baby Mixed Greens with Shaved Fennel and Orange Sections

8-10 Basic Herb Vinaigrette

8-11 Orange Vinaigrette

8-12 Sherry Wine Vinaigrette

8-13 Ginger Lime Dressing

8-14 Green Goddess Dressing

8-15 Capistrano Spice Rub

8-16 Grilled Pork Chop Adobo with Spicy Apple Chutney

8-17 Potato Lasagne

8-18 Chicken Sausage

8-19 Slates of Salmon

8-20 Braised Lamb

8-21 Papaya and White Raisin Chutney

8-22 Papaya-Plantain Salsa

8-23 Red Pepper Coulis

8-24 Hot-and-Sour Sauce

8-25 Roasted Summer Vegetables

8-26 Mixed-Grain Pilaf

8-27 Sautéed Cabbage

8-28 Warm Chocolate Pudding Cake with Almond Cookie and Raspberry Sauce

8-29 Raspberry Creamed Ice

8-30 Angel Food Savarin

8-31 Fresh Berry Phyllo Cones

8-32 Heart-Healthy Oatmeal Raisin Cookies

8-33 Crunchy Cinnamon Granola

8-34 Stuffed French Toast Layered with Light Cream Cheese and Bananas

8-35 Glazed Grapefruit and Orange Slices with Maple Vanilla Sauce

Appetizers

Recipe 8-1 *Roast Chicken and Shredded Mozzarella Tortellinis*

Category: Appetizer   Yield: 6 portions

INGREDIENTS

4 oz. chicken breast
1/2 teaspoon canola oil
Pinch fresh oregano, chopped
Pinch fresh basil, chopped
Pinch paprika
1/2 cup reduced-fat shredded mozzarella
1/2 cup skim-milk ricotta

1/4 teaspoon fresh garlic, chopped
2 pinches freshly ground black pepper
1 teaspoon fresh basil, chopped
1/2 teaspoon fresh parsley, chopped
18 wonton skins
1 cup water
18 arugula leaves
1 tomato, chopped

STEPS

1. Preheat oven to 350°F.
2. Coat chicken breast with oil, herbs, and paprika. Roast about 5 minutes, or hardwood grill for one minute. Let cool and fine julienne. Reserve.
3. In bowl, mix cheeses, garlic, pepper, and herbs. Add chicken and mix well.
4. Lay out wonton skins and paint with water.
5. Place 1 full teaspoon of chicken mixture per skin, or more, to use up all the filling.
6. Fold skin to make a triangle, then connect end in opposite direction to make tortellini shape.
7. Poach tortellinis for 3 minutes in boiling water or steam for 2 minutes.
8. While pasta is cooking, chop tomato and toss with black pepper. Let juice accumulate. Paint leaves of arugula lightly with juice of tomato for flavor.
9. Arrange each plate with 3 arugula leaves facing out like spokes of a wheel, with a tortellini between each 2 leaves. Place red pepper coulis (see Recipe 8-23) and a spoonful of chopped tomato in the middle.

NUTRITIONAL ANALYSIS:

Calories:	Protein (gm):	Fat (gm):	Carbo (gm):	Sodium (mg):	Chol (mg):
150	12	4	15	221	24

Recipe 8-2 *Scallop Rolls in Rice Paper with Vegetable Spaghetti*

Category: Appetizer   Yield: 15 portions

INGREDIENTS

1 large carrot
1 large daikon
2 large zucchini
4 tablespoons ginger lime dressing
1 pound tofu
3 cups bean sprouts
2 whole red peppers, julienne
2 cups spinach leaves

1/2 cup snow peas, julienne
1/2 small head of white cabbage,
   shredded
1/4 cup ginger lime dressing
22 ounces sea scallops (30 scallops),
   sliced in half
15 sheets rice paper
2 tablespoons cilantro, whole leaves
2 ounces hot-and-sour sauce

STEPS

1. Make carrot and daikon spaghetti with oriental spaghetti machine.
2. Make zucchini noodles with mandoline.
3. Toss spaghetti and noodles with 2 tablespoons ginger lime dressing.
4. Grill tofu and marinate with 2 tablespoons ginger lime dressing. Dice and reserve for service.
5. Marinate bean sprouts, red pepper, spinach leaves, snow peas, and white cabbage in 1/4 cup ginger lime dressing while you do the next step.
6. Grill scallops on one side for color.
7. Soak sheets of rice paper in warm water one at a time for next step.
8. Layer scallops, vegetables, and cilantro leaves on moistened rice paper and fold like an eggroll. Steam about 3 minutes.
9. In a soup bowl, rest the rice-paper roll against the marinated vegetable noodles. Sprinkle tofu around the vegetable noodles. Pour hot-and-sour sauce over mixture.

NUTRITIONAL ANALYSIS:

Calories:	Protein (gm):	Fat (gm):	Carbo (gm):	Sodium (mg):	Chol (mg):
107	10	3	11	237	13

Recipe 8-3 *Crab Cakes*

Category: Appetizer   Yield: 10 servings

INGREDIENTS

1-1/4 pounds jumbo lump crabmeat
1-1/2 teaspoons chives, chopped fine
1-1/2 teaspoons parsley, chopped fine
1 tablespoon old bay seasoning
3/4 teaspoon Dijon mustard
3/4 teaspoon thyme, chopped
5 ounces potatoes, cooked and mashed

1-1/2 egg whites
3/4 teaspoon lemon juice
1-1/2 teaspoon white wine
1/2 egg white
Japanese bread crumbs, as needed
3/4 teaspoon chives, chopped fine
Vegetable-oil cooking spray

STEPS

1. Pick crabmeat to remove bits of shell.
2. Add in chives, parsley, old bay seasoning, mustard, thyme, and riced potatoes. Mix gently.
3. In a stainless-steel bowl, add 1-1/2 egg whites, lemon juice, and white wine. Whip to form stiff peaks.
4. Fold whipped egg whites into crab mixture and mold into 3-ounce crab cakes.
5. In a metal pan place 1/2 egg white. Dip the crab cakes in the egg white, then in the crumbs mixed with chives.
6. Sauté the crab cakes to a crisp golden brown in a nonstick pan sprayed with vegetable spray. Serve with salsa or mojo of your choice.

NUTRITIONAL ANALYSIS:

Calories:	Protein (gm):	Fat (gm):	Carbo (gm):	Sodium (mg):	Chol (mg):
97	12	2	7	266	24

Recipe 8-4 *Eggplant Rollatini with Spinach and Ricotta*

Category: Appetizer **Yield: 20 servings**

INGREDIENTS

3 lbs. fresh spinach

1-1/2 pounds eggplant, peeled and cut lengthwise

1/4 cup balsamic vinegar

olive-oil spray

vegetable-oil cooking spray

4 ounces onions, finely chopped

2 teaspoons garlic, chopped

1 teaspoon cracked black pepper

1 teaspoon garlic herb seasoning

1/4 cup Italian parsley, chopped

2 tablespoons fresh basil, chopped

16 ounces skim milk ricotta

2 ounces feta cheese, crumbled

2 tablespoons grated Parmesan cheese

3/4 cup whole-wheat bread crumbs

1 egg

1 egg white, slightly beaten

1 quart tomato sauce

1 teaspoon chives, chopped

STEPS

1. Steam spinach. Drain well and rough chop. Reserve.
2. Paint eggplant with balsamic vinegar on both sides, and spray with olive oil. Place on sheet pan sprayed with vegetable-oil cooking spray.
3. Bake in 400°F oven for 10 minutes. Remove and flip over. Finish baking until tender. Reserve.
4. Spray nonstick skillet with vegetable-oil cooking spray. Quickly sauté onions and garlic. Add drained spinach. Remove from heat and let cool. Add in herb seasoning herbs, cheeses, and bread crumbs. Mix in egg and beaten egg white. Chill mixture.

5. Place two heaping tablespoons of spinach mixture on each eggplant slice. Roll up and place open end down in casserole dish. Cover with tomato sauce. Bake in 350° oven for 30 minutes.

6. To serve, place 1 rollatini on plate. Top with tomato sauce and sprinkle with chives.

NUTRITIONAL ANALYSIS:

Calories:	Protein (gm):	Fat (gm):	Carbo (gm):	Sodium (mg):	Chol (mg):
121	7	3	18	154	21

Recipe 8-5 *Mussels Steamed in Saffron and White Wine*

Category: Appetizer   Yield: 10 servings

INGREDIENTS

Broth
1 tablespoon garlic, chopped
1 teaspoon olive oil
7 ounces white wine
1 gram saffron
1 ounce fresh lemon juice
2 ounces chicken stock
1/2 teaspoon pepper, fresh ground
1 teaspoon fresh thyme, chopped
Mussels
80 fresh New Zealand mussels

Garlic Rouillé
2 teaspoons roasted garlic
1 cup Yukon Gold potatoes, cooked and riced
1 tablespoon chicken stock
1/2 teaspoon fresh lemon juice
1 teaspoon red chili paste
1 teaspoon chives
10 slices of whole-wheat bread, cut into 2-1/2-inch rounds
1 cup carrots, blanched, julienne
1 cup celery, blanched, julienne
2 tablespoons chives

STEPS

1. For the broth, sauté garlic in olive oil. Add wine and saffron. Let reduce for 2 minutes.

2. Add lemon juice, stock, black pepper, and fresh thyme. Cook over low heat for 1 minute. Cool and store for service.

3. Scrub mussels and remove beard.

4. Make garlic rouillé by puréeing garlic and mixing with potatoes, chicken stock, lemon juice, and red chili paste. Fold in chives and spread on whole-wheat bread, and toast.

5. At service, steam carrots, celery, and mussels in saffron broth. Place mussels in a large bowl. Finish with chives and broth. Garnish with toasted whole-wheat crouton spread with garlic rouillé.

NUTRITIONAL ANALYSIS:

Calories:	Protein (gm):	Fat (gm):	Carbo (gm):	Sodium (mg):	Chol (mg):
139	5	2	22	448	3

Soups

Recipe 8-6   *Pasta e Fagioli*

Category: Soup   Yield: 16 servings (approximately 1 cup)

INGREDIENTS

1 large yellow onion, chopped
6 cloves garlic, chopped
2 tablespoons olive oil, extra virgin
2 pounds pinto, kidney, and black beans,
   dried and soaked overnight
1/2 cup tomato paste
2 quarts vegetable or chicken stock,
   defatted

1-1/2 pounds plum tomatoes, peeled,
   seeded
2 bay leaves
1 tablespoon red pepper flakes
1 tablespoon fresh cracked black pepper
2 tablespoons fresh oregano, chopped
2 tablespoon fresh thyme, chopped
1 tablespoon fresh rosemary, chopped

STEPS

1. In a large pot, sauté chopped onion and garlic in olive oil.
2. Drain the beans and add to the pot. Add in tomato paste to blend. Sauté.
3. Add 2 quarts of stock and bring to a boil. Reduce to simmer and stir occasionally.
4. Add chopped plum tomatoes, bay leaves, red pepper flakes, and black pepper.
5. Simmer for 1 to 1-1/2 hours until beans are soft. Add herbs during last 30 minutes.
6. Remove one cup of beans and purée. Return to pot. Purée more beans to achieve desired thickness. Remove bay leaves.
7. Adjust flavor, if necessary, by adding more herbs or balsamic vinegar.

NUTRITIONAL ANALYSIS:

Calories:	Protein (gm):	Fat (gm):	Carbo (gm):	Sodium (mg):	Chol (mg):
242	13	3	42	281	0

Recipe 8-7   *Butternut Squash Bisque*

Category: Soup   Yield: 24 portions

INGREDIENTS

1 ounce sweet butter
2 teaspoons shallots, chopped
1 cup celery, chopped
1 cup onion, chopped

6 ounces rice
2 quarts chicken stock
1/2 teaspoon cinnamon
1/2 teaspoon nutmeg
1 bay leaf

2 teaspoons garlic, chopped

1 tablespoon fresh thyme

4 pounds butternut squash, cleaned and diced

Cinnamon, as needed for garnish

Nonfat plain yogurt, as needed for garnish

Chives, as needed for garnish

STEPS

1. In soup pot, melt butter. Sauté shallots, celery, onion, garlic, and thyme.
2. Add butternut squash, rice, chicken stock, cinnamon, nutmeg, and bay leaf. Cook until rice is tender.
3. Puree with blender and strain.
4. Garnish with cinnamon, nonfat plain yogurt, and chives.

NUTRITIONAL ANALYSIS:

Calories:	Protein (gm):	Fat (gm):	Carbo (gm):	Sodium (mg):	Chol (mg):
101	2	1	15	207	3

Salads and Dressings

Recipe 8-8 *Wild Mushroom Salad*

Category: Salad or Appetizer   Yield: 1 portion

INGREDIENTS

Dressing:

1 tablespoon sherry vinegar

3 twists black pepper

1 teaspoon Dijon mustard

1 teaspoon white wine

1/2 teaspoon shallots

1/4 teaspoon fresh rosemary, chopped

1 teaspoon apple juice

1/2 teaspoon fresh thyme, chopped

Salad:

1 cup of greens: either red oak leaf, frisée lettuce, mustard greens, mesclun lettuce mix

1/2 cup of mushrooms: shiitake mushrooms, oyster mushrooms, domestic mushrooms (can be any mix of mushrooms)

1 tablespoon diced peppers

1 teaspoon fresh chives, chopped

STEPS

1. Incorporate all dressing ingredients together in a bowl.
2. Clean and wash lettuce, and arrange on plate.
3. Toss mushrooms and peppers, in bowl with dressing. Sear on flat top or hot sauté pan.
4. Place warm (or chilled, if preferred) mushrooms in center.
5. Top with fresh chives.

NUTRITIONAL ANALYSIS:

Calories:	Protein (gm):	Fat (gm):	Carbo (gm):	Sodium (mg):	Chol (mg):
29	1	1	5	34	0

Recipe 8-9 *Baby Mixed Greens with Shaved Fennel and Orange Sections*

Category: Salad or Appetizer   **Yield: 10 servings**

INGREDIENTS

14 cups mixed greens such as baby Bibb, baby romaine, frisée, lolla rosa, tatsoi

3 heads fresh fennel

5 fresh oranges

20 ounces sherry-wine vinaigrette

STEPS

1. Prepare salad greens and reserve for service.
2. Cut fennel in half and shave paper-thin on a meat slicer. Place in ice water to crisp. Drain and reserve for service.
3. Dry some fennel slices in a low oven to use as garnish (about 30 minutes).
4. Section oranges.
5. In mixing bowl, toss fennel lightly in sherry wine vinaigrette.
6. Toss mixed greens with dressing. Use only 2 ounces of dressing per person.
7. Coat orange sections lightly with dressing.
8. Arrange orange sections on outside of plate. Make a tower of lettuce and fennel. In the center of the plate top with fennel slices that have been dried in the oven. Serve immediately.

NUTRITIONAL ANALYSIS:

Calories:	Protein (gm):	Fat (gm):	Carbo (gm):	Sodium (mg):	Chol (mg):
63	3	0.5	13	35	0

Recipe 8-10 *Basic Herb Vinaigrette*

Category: Dressing   **Yield: Approximately 1 gallon, 2-ounce portion size**

INGREDIENTS

9 cups chicken stock or vegetable stock

4 cups balsamic vinegar

1/2 ounce cornstarch slurry (cornstarch with cold water)

2 tablespoons fresh chopped garlic

2 teaspoons black pepper, coarsely ground

2 tablespoons fresh thyme, chopped

2 ounces fresh parsley, chopped

2 tablespoons, fresh oregano, chopped

1-1/2 ounces fresh basil, chopped

2 ounces chives, chopped

4 cups extra virgin olive oil

STEPS

1. Use stock or a quality low-sodium vegetable base. Heat to a rolling boil.
2. Thicken with cornstarch slurry, to a nappe (to coat) consistency.

3. Cool, and add garlic, pepper, herbs, and vinegar.
4. Whisk olive oil into the mixture to emulsify.
5. Cool and store.

You can substitute different flavored vinegars or virgin nut oil to create alternate dressing options.

NUTRITIONAL ANALYSIS:

Calories:	Protein (gm):	Fat (gm):	Carbo (gm):	Sodium (mg):	Chol (mg):
149	1	14	5	39	4

Recipe 8-11 *Orange Vinaigrette*

Category: Dressing   **Yield: 10 2-ounce servings**

INGREDIENTS

1 pint orange juice
1 tablespoon Dijon mustard
1 tablespoon honey
1 tablespoon shallots, finely diced
1/2 teaspoon coriander, ground

1/2 teaspoon black pepper, ground
2 ounces white-wine vinegar
2 ounces extra virgin olive oil
1 teaspoon thyme, fresh, chopped
1 teaspoon chives, fresh, chopped
1 teaspoon basil, fresh, chopped

STEPS

1. In a saucepan, reduce orange juice to 1/2 cup, or use 1/2 cup of concentrate.
2. Place cooled orange syrup in food processor. Add mustard, honey, shallots, coriander, black pepper, and white-wine vinegar.
3. Start food processor and slowly add oil to emulsify.
4. Add fresh herbs last, but do not purée.
5. Reserve in refrigerator until needed.

NUTRITIONAL ANALYSIS:

Calories:	Protein (gm):	Fat (gm):	Carbo (gm):	Sodium (mg):	Chol (mg):
68	0.4	6	7	10	0

Recipe 8-12 *Sherry Wine Vinaigrette*

Category: Dressing   **Yield: 10 1 1/2-ounce servings**

INGREDIENTS

1/2 cup sherry vinegar
10 twists fresh ground black pepper
3-1/3 tablespoons Dijon mustard
3-1/3 tablespoons white wine
1-2/3 tablespoons shallots

2-1/2 teaspoons rosemary, chopped
2 ounces apple juice
1-2/3 tablespoons fresh thyme, chopped
3-1/3 tablespoons chives, finely chopped
1/2 cup peppers, tricolored, diced

STEPS

1. Mix all ingredients together in a bowl. Hold for service. Lasts 1 week in refrigerator.

NUTRITIONAL ANALYSIS:

Calories:	Protein (gm):	Fat (gm):	Carbo (gm):	Sodium (mg):	Chol (mg):
21	0.5	0.5	3	30	0

Recipe 8-13   *Ginger Lime Dressing*

Category: Dressing   Yield: 8 2-ounce servings

INGREDIENTS

1 cup soy sauce, low sodium
1/2 cup lime juice
Grated rind from 2 limes
1/4 cup scallions, finely sliced
1/2 cup water
1 teaspoon garlic, chopped
1 teaspoon ginger, chopped

STEPS

1. Mix all ingredients together in a bowl. Let sit overnight. Lasts 2 weeks in refrigerator.

NUTRITIONAL ANALYSIS:

Calories:	Protein (gm):	Fat (gm):	Carbo (gm):	Sodium (mg):	Chol (mg):
26	2	0	5	800	0

Recipe 8-14   *Green Goddess Dressing*

Category: Dressing   Yield: 6 portions

INGREDIENTS

4 ounces tofu (firm), well drained
1/2 cup cider vinegar
2 stalks celery
1/2 cup spinach leaves, washed and dried
1/3 cup fresh parsley (no stems)
1 tablespoon lemon juice
2 scallions
Fresh tarragon leaves
Fresh ground pepper

STEPS

1. Blend together in food processor until smooth. Use just enough tarragon and pepper to taste. This dressing will last 2 days in the refrigerator.

NUTRITIONAL ANALYSIS:

Calories:	Protein (gm):	Fat (gm):	Carbo (gm):	Sodium (mg):	Chol (mg):
23	2	1	3	19	0

Recipe 8-15 *Capistrano Spice Rub*

Category: Rub   Yield: 5 cups

INGREDIENTS

1 cup dried oregano
1 cup dried basil
1/2 cup dried thyme

1 cup fresh rosemary, chopped
1/2 cup black pepper, butcher's ground
1/2 cup garlic powder

STEPS

1. Blend all ingredients together. Store in an airtight container to preserve freshness.
2. Coat meat, fish, or poultry generously with spice rub prior to cooking. Can also be used to spice up soups, dressings, and vegetable dishes.

Entrées

Recipe 8-16 *Grilled Pork Chop Adobo with Spicy Apple Chutney*

Category: Entrée   Yield: 20 portions

INGREDIENTS

Adobo Spice Rub:
3 fresh green chiles (pablano)
2 fresh jalapeño peppers
10 garlic cloves
3 tablespoons fresh oregano
2 tablespoons ground cumin
2 tablespoons freshly ground black pepper
1 teaspoon ground cinnamon
1/2 pound tomatillos, husks removed
1 cup red wine vinegar
Pork:
5 pounds pork loin, center cut, boneless
Chutney:
1 chipotle pepper

1 onion, diced
1/2 tablespoon garlic, chopped
8 apples, cored and diced
1/2 cup raisins
1 ounce lemon juice
zest from 1 lemon
1 tablespoon ground cardamon
3 tablespoons sugar
1 cup cider vinegar
1/2 tablespoon ground fennel seed
1/4 teaspoon mace
1/2 bunch fresh cilantro, chopped

STEPS

1. Combine all the adobo spice rub ingredients in a food processor and blend until a paste forms.
2. Generously rub the pork loin with the paste and let marinate overnight in refrigerator.

3. Rehydrate the chipotle pepper in hot water until softened. Cut in half, remove seeds, and chop.
4. In a sauté pot, combine all chutney ingredients except cilantro.
5. Let simmer for 20 minutes until slightly thickened. Finish with cilantro. Cool and serve at room temperature. (The chutney can be made in advance and refrigerated.)
6. Sear pork loin until nicely caramelized. Set up smoking station with wood chips soaked in water. Smoke with medium smoke for about 15 minutes.
7. Remove and finish in a slow-roasting oven at 300°F until at 155°F.
8. Let rest 15 minutes before slicing. Serve with apple chutney.

NUTRITIONAL ANALYSIS:

Calories:	Protein (gm):	Fat (gm):	Carbo (gm):	Sodium (mg):	Chol (mg):
144	17	3	12	34	48

Recipe 8-17 *Potato Lasagne*

Category: Entrée (or Appetizer)   Yield: 6 servings

INGREDIENTS

16 ounces skim-milk ricotta
1/4 teaspoon fresh garlic, chopped
1/2 teaspoon fresh basil, chopped
1 teaspoon fresh parsley, chopped
1/4 teaspoon oregano leaves, dried

1/8 teaspoon black pepper, freshly ground
3 large Idaho potatoes, peeled
3 to 4 cups tomato sauce
8 ounces skim-milk mozzarella
Fresh basil leaves, for garnish

STEPS

1. Combine ricotta cheese, garlic, basil, parsley, oregano, and black pepper.
2. Coat baking dish with nonstick vegetable-oil spray.
3. Cut potatoes lengthwise into 1/8-inch slices.
4. Heat oven to 350°F.
5. To assemble lasagne, spread about 1/3 of the potatoes evenly over bottom of baking dish. Top with 1/2 of the ricotta cheese mixture, 1/3 of the tomato sauce, and 1/2 of the mozzarella cheese. Repeat the layers. Place remaining potatoes evenly over top, and spoon on remaining tomato sauce.
6. Cover with foil. Bake 1 hour or until potatoes are tender.
7. Uncover and allow to cool 10 minutes before serving.
8. Cut lasagne into 6 servings. Garnish each serving with basil leaves.

This recipe can be made a day in advance.

NUTRITIONAL ANALYSIS:

Calories:	Protein (gm):	Fat (gm):	Carbo (gm):	Sodium (mg):	Chol (mg):
321	21	12	34	460	45

Recipe 8-18 *Chicken Sausage*

Category: Entrée   Yield: 5 servings

INGREDIENTS

2-1/2 ounces wheat bread, diced
2 egg whites
2 ounces nonfat plain yogurt
1 pound chicken breast, cleaned and
  diced

Freshly ground black pepper to taste
1 teaspoon fresh oregano, chopped
1/2 teaspoon fresh thyme, chopped
1 teaspoon fresh basil, chopped
2 teaspoons roasted garlic, sliced
2 tablespoons dried tomatoes, diced

STEPS

1. In a bowl, soak wheat bread with egg whites and nonfat yogurt to make *panada*.
2. In a food processor, pulse chicken meat to purée. Add panada, pepper, and herbs. Pulse to a smooth consistency, adding about 2 tablespoons of crushed ice to keep temperature down.
3. Fold in roasted garlic and tomatoes.
4. Pipe into sausage casing or plastic wrap. Tie and poach in 160°–170°F water until an internal temperature of 160°F is reached. Shock in ice water. Cool and unwrap for service.
5. Grill to reheat and finish cooking.
6. Serve with greens, tomato salad, roasted onions, German-style potato salad, or grilled vegetables.

NUTRITIONAL ANALYSIS:

Calories:	Protein (gm):	Fat (gm):	Carbo (gm):	Sodium (mg):	Chol (mg):
208	32	4	9	372	77

Recipe 8-19 *Slates of Salmon*

Category: Entrée   Yield: 1 serving

INGREDIENTS

4 ounces salmon steak—cut very thin on
  bias
1 teaspoon olive oil
Pinch black pepper
1 cup arugula

1/2 cup endive
1 tablespoon plum tomatoes, seeded and
  diced
1 tablespoon cucumbers, peeled, seeded,
  and diced
2 ounces Green Goddess Dressing (see
  Recipe 8-14)

STEPS

1. Paint the salmon with olive oil and black pepper. Grill to desired temperature.

2. Toss greens with Green Goddess Dressing and place in middle of plate. Place salmon against greens and sprinkle with tomato and cucumbers.

NUTRITIONAL ANALYSIS:

Calories:	Protein (gm):	Fat (gm):	Carbo (gm):	Sodium (mg):	Chol (mg):
187	24	9	3	89	59

Recipe 8-20 *Braised Lamb*

Category: Entrée   Yield: 12 portions

INGREDIENTS

12 lamb hind shanks
1 tablespoon olive oil
4 tablespoons vegetable seasoning
5 cloves garlic, sliced
2 onions, diced
4 carrots, sliced
6 celery stalks, diced
3 bay leaves
2 whole thyme sprigs

2 whole rosemary sprigs
8 ounces red wine
8 ounces fresh tomatoes
1/2 cup tomato paste
1 gallon lamb stock, defatted
3 tablespoons Worcestershire sauce
1 tablespoon cracked black pepper
36 large shiitake mushroom caps
24 ounces cannellini beans, cooked
3 tablespoons garlic herb seasoning

STEPS

1. Trim all fat from lamb shanks.
2. Place olive oil into large heated braising pan. Season shanks with vegetable seasoning. Sear in the large pot until brown. Remove shanks.
3. Quickly sauté garlic, onions, carrots, celery, bay leaves, thyme, and rosemary in same pot. Deglaze with red wine, fresh tomatoes, and tomato paste, and reduce.
4. Return shanks to pot with defatted lamb stock, Worcestershire sauce, and pepper. Bring to boil, reduce heat, and simmer gently for 90 minutes. Take out shanks. Strain and put sauce through a food mill. Reduce to proper consistency and skim. Pour over shanks. Cool.
5. For each order, cook 3 shiitake caps with 2 ounces cannellini beans in 3 ounces of lamb stock and 1 teaspoon garlic herb seasoning. Cook lightly to reduce. Heat each shank in sauce slowly.
6. Place 1 piece of shank on plate with vegetables and sauce. Top with shiitakes and beans.

This dish can be served with a variety of vegetables, beans, and grains.

NUTRITIONAL ANALYSIS:

Calories:	Protein (gm):	Fat (gm):	Carbo (gm):	Sodium (mg):	Chol (mg):
391	40	10	33	471	107

Relishes, Salsas, Coulis, Chutneys, and Sauces

Recipe 8-21 *Papaya and White Raisin Chutney*

Category: Relish, Salsa, Coulis, and Chutney   Yield: 16 servings

INGREDIENTS

6 pounds very ripe papaya, diced

2 cups onion, diced

1 tablespoon garlic, chopped

1/2 cup brown sugar

1/2 cup white sugar

1 cup raisins, seedless

1 cup white wine vinegar

1 teaspoon cardamom

1 teaspoon cinnamon

2 bay leaves

2 teaspoons fresh thyme, chopped

STEPS

1. In a small saucepot, mix all ingredients except thyme.
2. Reduce to a thick paste. Add thyme while chutney is still hot.
3. Let cool; remove bay leaves. Can be stored up to 2 weeks.

NUTRITIONAL ANALYSIS:

Calories:	Protein (gm):	Fat (gm):	Carbo (gm):	Sodium (mg):	Chol (mg):
126	1	0	31	8	0

Recipe 8-22 *Papaya–Plantain Salsa*

Category: Relish, Salsa, Coulis, and Chutney   Yield: 10 servings

INGREDIENTS

1 plantain, finely diced

1 teaspoon extra virgin olive oil

1 papaya, finely diced

1/2 peppers, red and orange, finely diced

1/2 red onion, finely diced

2 teaspoons cilantro, chopped

2 teaspoons chives, finely sliced

1 cup white wine vinegar

2 teaspoons honey

Lime juice from 2 limes

5 teaspoons extra virgin olive oil

STEPS

1. In a nonstick sauté pan, toast the diced plantain in 1 teaspoon of extra virgin olive oil until crisp outside and tender inside.
2. In a stainless steel mixing bowl, place half the papaya, peppers, and red onion. Add the toasted plantain, cilantro, and chives. Reserve.
3. In a food processor, place the remaining papaya, vinegar, honey, and lime juice. Purée until smooth, adding 5 teaspoons olive oil.
4. Add more vinegar if too thick.
5. Strain through a fine sieve into the reserved plantain-papaya mixture.
6. Reserve in refrigerator for use. Lasts about 5 days.

NUTRITIONAL ANALYSIS:

Calories:	Protein (gm):	Fat (gm):	Carbo (gm):	Sodium (mg):	Chol (mg):
75	0.6	3	11	4	0

Recipe 8-23 *Red Pepper Coulis*

Category: Relish, Salsa, Coulis, and Chutneys   Yield: 36 ounces, or 18 2-ounce servings

INGREDIENTS

4 pounds red peppers (or substitute other vegetables)
1 ounce minced shallots
1 tablespoon minced garlic
1/2 teaspoon minced jalapeño pepper
2 ounces olive oil

2 ounces tomato paste
2 90-count potatoes, peeled and diced
18 ounces chicken or vegetable stock
2 tablespoons fresh basil, chopped
2 teaspoons fresh thyme, chopped
2 teaspoons fresh oregano, chopped
1-3/4 ounces balsamic vinegar

STEPS

1. Cut peppers in half and remove seeds. Place on oiled sheet pans and roast in hot oven, or grill the peppers. Weigh out 3-1/4 pounds of grilled red peppers.
2. Sauté shallots, garlic, and peppers in oil.
3. Add tomato paste and sauté. Do not brown.
4. Add red peppers, potatoes, and stock. Simmer until peppers and potatoes are tender.
5. Add basil, thyme, and oregano. Cook 5 more minutes.
6. Purée in a blender.
7. Finish with vinegar, and strain through large-hole china cap.

NUTRITIONAL ANALYSIS:

Calories:	Protein (gm):	Fat (gm):	Carbo (gm):	Sodium (mg):	Chol (mg):
67	2	3	8	26	0

Recipe 8-24 *Hot-and-Sour Sauce*

Category: Sauce   Yield: 16 2-ounce portions

INGREDIENTS

24 ounces chicken stock, defatted
2 ounces sesame oil
4 ounces white wine
3 ounces rice wine vinegar
1-1/2 tablespoons cornstarch
1 tablespoons lime juice, fresh squeezed

1/3 teaspoon lime rind
2 ounces scallions, chopped
2 teaspoons ginger, chopped
1/2 teaspoon jalapeño pepper
1/4 teaspoon cumin
1 teaspoon cilantro

STEPS

1. Heat chicken stock and wine. Thicken with cornstarch slurry made with vinegar.
2. Take off heat and cool. Add remaining ingredients except oil. Whip in oil at the end. Refrigerate for later service.

NUTRITIONAL ANALYSIS:

Calories:	Protein (gm):	Fat (gm):	Carbo (gm):	Sodium (mg):	Chol (mg):
54	1	4	3	101	5

Side Dishes

Recipe 8-25 *Roasted Summer Vegetables*

Category: Side Dish   Yield: 10 servings

INGREDIENTS

2 pounds carrots, peeled and cut oblique

1 pound red bell peppers

1 tablespoon olive oil

2 tablespoons shallot, chopped

4 cloves garlic, chopped

1 pound corn, cleaned

1-1/2 pound assorted radishes, sliced 1/4"

1/2 pound patty-pan squash

1/2 tablespoon thyme

2 bay leaves

black pepper to taste

4 ounces chicken stock

1 tablespoon basil leaves, chopped

1/2 tablespoon rub of your choice

STEPS

1. Blanch carrots for 2 minutes.
2. Cut peppers in half and remove seeds. Place on oiled sheet pans and roast in hot oven or grill the peppers. Remove skins.
3. Mix all ingredients except basil and rub together and place in hot roasting pan in a 350°F oven. Let roast, stirring occasionally, for 30 minutes. Add more stock if needed to prevent burning. If vegetables are getting too brown, cover with foil.
4. Add basil and rub at end for more flavoring.

NUTRITIONAL ANALYSIS:

Calories:	Protein (gm):	Fat (gm):	Carbo (gm):	Sodium (mg):	Chol (mg):
123	3	2	26	52	0

Recipe 8-26 *Mixed-Grain Pilaf*

Category: Side Dish   Yield: 10 servings

INGREDIENTS

1 cup wheat berries

2 cups water

2 thyme stems

1 bay leaf

1 cup quinoa

1 cup hot water

1 bay leaf

1/2 ounce chicken stock, defatted

1 ounce onions, diced

3 ounces brown rice

1 pint chicken stock, defatted

2 thyme stems

1 bay leaf

1 head roasted garlic

2 teaspoons olive oil

2 ounces onions, chopped

2 tablespoons chopped basil, oregano, and chives

Chicken stock, as needed

STEPS

1. In a sauce pot of boiling water, add wheat berries with thyme stems and bay leaf and cook for about 1 hour. Drain and cool.
2. Soak quinoa in cold water to take out any impurities. Steam quinoa in water with bay leaf. Cover and let steam for about 20 minutes.
3. Heat 1/2 ounce of chicken stock in a stockpot. Add onions and sweat until they are translucent.
4. Add rice, 1 pint stock, thyme, and bay leaf. Simmer and cover pilaf style in a 350°F oven for about 20 minutes, or until done.
5. Purée roasted garlic.
6. Heat oil in stockpot. Sweat onions until translucent. Add puréed garlic, wheat berries, quinoa, brown rice, and herbs.
7. Let simmer for 5 minutes. Adjust consistency with stock if needed.

NUTRITIONAL ANALYSIS:

Calories:	Protein (gm):	Fat (gm):	Carbo (gm):	Sodium (mg):	Chol (mg):
211	7	3	39	31	0

Recipe 8-27 *Sautéed Cabbage*

Category: Side Dish   **Yield: 8 portions**

INGREDIENTS

32 ounces shredded cabbage

1 quart chicken stock

1 quart fish stock

2 tablespoons ground caraway seeds

1 tablespoon juniper berries

1 tablespoon peppercorns

2 sprigs fresh thyme

3 each bay leaves

2 ounces dijon mustard

3 ounces sugar

1/2 cup white wine vinegar

4 turns fresh black pepper, milled

STEPS

1. In saucepan, mix cabbage, stocks, and ground caraway seeds.
2. Place juniper berries, peppercorns, thyme, and bay leaves in a sachet bag. Place in saucepan with cabbage.
3. Mix mustard, sugar, vinegar, and pepper. Add to cabbage. Simmer over low heat until tender, about 15 to 20 minutes.

NUTRITIONAL ANALYSIS:

Calories:	Protein (gm):	Fat (gm):	Carbo (gm):	Sodium (mg):	Chol (mg):
79	99	2	7	271	25

Desserts

Recipe 8-28 *Warm Chocolate Pudding Cake with Almond Cookie and Raspberry Sauce*

Category: Dessert   **Yield: 10 portions**

INGREDIENTS

1 tablespoon orange zest
3 cups skim milk
3-1/2 ounces sugar
1 ounce cocoa powder
3 ounces cornmeal
10 egg whites
3-3/4 ounces sugar
3 ounces bittersweet chocolate
1 pint raspberries
1 ounce Kirschwasser
4 ounces white wine

2-3/4 ounces honey
1-1/2 ounces sugar
1 ounce almond paste
1/2 ounce bread flour
1/4 teaspoon cinnamon
1 egg white
pinch salt
2 teaspoons cream
2 teaspoons skim milk
Confectioners' sugar, about 1 tablespoon
   for dusting cookies

STEPS

1. Steep orange zest in milk and bring to a boil. Simmer.
2. Combine sugar, cocoa, and cornmeal. Pour in steady stream into simmering milk. Stir until thick and cornmeal is cooked. Allow to cool.
3. Whip egg whites and sugar (3-3/4 ounces). Fold into base. Then fold in bittersweet chocolate.
4. Pour mixture into sugared molds and bake in bain marie about 20 to 24 minutes at 400°F. Keep warm.
5. Purée 3/4 of raspberries (reserve others for garnish) in food processor with the Kirschwasser, wine, and honey. Strain and reserve.
6. Cream sugar, almond paste, flour, and cinnamon. Add remaining ingredients, except confectioners' sugar, and allow to rest. Spread paste in 10 three-inch circles on silicon. Bake in 350°F oven until edges are brown. Curve the circles on a rolling pin while still hot to make a decorative tuille garnish.
7. Cool and dust with confectioners' sugar.
8. Pool sauce on plate. Unmold cake on sauce. Place cookie on top of cake and garnish with raspberries.

NUTRITIONAL ANALYSIS:

Calories:	Protein (gm):	Fat (gm):	Carbo (gm):	Sodium (mg):	Chol (mg):
235	9	2	45	106	3

Recipe 8-29 *Raspberry Creamed Ice*

Category: Dessert   Yield: 1 quart or 4 1-cup servings

INGREDIENTS

1 pound soft tofu

1 pint raspberry purée (or substitute any
    other fruit)

6 ounces egg whites

1 ounce turbinado sugar

STEPS

1. Cream tofu in processor until smooth.
2. Add raspberry purée slowly to achieve creamy texture.
3. Whip egg whites until they form soft peaks. Add sugar until the meringue forms peaks.
4. Churn in an ice-cream machine until the mixture reaches the desired consistency.
5. Freeze in an airtight container.

NUTRITIONAL ANALYSIS:

Calories:	Protein (gm):	Fat (gm):	Carbo (gm):	Sodium (mg):	Chol (mg):
148	12	5	17	79	0

Recipe 8-30 *Angel Food Savarin*

Category: Dessert   Yield: 8 4-inch savarin rings

INGREDIENTS

2 ounces cake flour

6 ounces 10X sugar

6 ounces egg whites

1 teaspoon cream of tartar

1/2 teaspoon lemon rind

STEPS

1. Sift flour with 3 ounces of sugar. Repeat, and set aside.
2. Whip egg whites and cream of tartar to soft peaks, then gradually add remaining 3 ounces of sugar. Whip until stiff and glossy.
3. Gently fold in sifted ingredients and lemon rind.
4. Pipe into savarin molds that have been sprayed with cold water.
5. Bake in 350°F oven until light golden brown on top.
6. Cool completely, then remove from mold.

NUTRITIONAL ANALYSIS:

Calories:	Protein (gm):	Fat (gm):	Carbo (gm):	Sodium (mg):	Chol (mg):
119	3	0	27	35	0

Recipe 8-31 *Fresh Berry Phyllo Cones*

Category: Dessert   Yield: 8 portions

INGREDIENTS

Phyllo
Vegetable-oil cooking spray
8 phyllo sheets
8 teaspoons melted butter
Bavarian Mix
7 ounces maple syrup
9 ounces skim-milk ricotta cheese
13 ounces nonfat plain yogurt

1 teaspoon vanilla extract
2 teaspoons gelatin
4 teaspoons water
Fruit and Garnish
4 cups fresh berries (strawberries,
    blueberries, raspberries)
Powdered sugar, for dusting
8 each mint leaves

STEPS

1. Make 2-1/2-inch circle cones out of aluminum foil. (Wrap foil around a small juice or soup can.)
2. Spray foil lightly with vegetable-oil spray.
3. Cut phyllo into 2-1/2-inch widths. Paint with melted butter between layers. Use 1 sheet per cone.
4. Wrap phyllo around foil and take the cone off the can.
5. Place cone on baking sheet. Repeat process 8 times. These can be made one day in advance.
6. Bake phyllo cones at 375°F for about 10 minutes. Let cool. Take off foil and reserve it for future use.
7. Mix maple syrup, ricotta cheese, yogurt, and vanilla in blender. Whip until smooth.
8. Soften gelatin in water. Warm to dissolve.
9. Add a little of the dessert base to the dissolved gelatin to temper the mixture. Mix the rest of the dessert base into tempered mixture. Fold in 2 cups of fresh berries. Reserve other fruit for garnish.
10. Place a cone on a dessert dish. Scoop 3 ounces of the Bavarian mixture into the cone.
11. Top with fresh berries. Dust with powdered sugar. Garnish with mint leaf.

NUTRITIONAL ANALYSIS:

Calories:	Protein (gm):	Fat (gm):	Carbo (gm):	Sodium (mg):	Chol (mg):
262	9	8	39	212	22

Recipe 8-32 *Heart-Healthy Oatmeal Raisin Cookies*

Category: Dessert   Yield: 40 1-ounce cookies

INGREDIENTS

1 pound brown sugar
14 ounces applesauce
1/2 ounce canola oil
1 ounce egg whites

12 ounces all-purpose flour
7 ounces oatmeal
1 tablespoon baking soda
2-1/2 cups raisins
1 teaspoon cinnamon

STEPS

1. Mix together the sugar, applesauce, oil, and egg whites until smooth.
2. Combine the flour, oatmeal, baking soda, raisins, and cinnamon and add to the applesauce mixture. Mix until well blended.
3. Spoon onto a parchment-lined sheet pan and bake at 325°F until golden brown.

NUTRITIONAL ANALYSIS:

Calories:	Protein (gm):	Fat (gm):	Carbo (gm):	Sodium (mg):	Chol (mg):
130	2	1	30	142	0

Breakfasts

Recipe 8-33 *Crunchy Cinnamon Granola*

Category: Breakfast   Yield: About 6 cups, or 12 half-cup servings

INGREDIENTS

2 cups rolled oats
1 cup wheat flakes
1 cup wheat germ
1/4 cup unsalted sunflower seeds
1 tablespoon cinnamon

3/4 cup apple juice
1/2 cup prune juice
1 cup assorted dried fruit (such as raisins, apricots, dates, figs, pears)
1 tablespoon honey
1/2 cup skim milk

STEPS

1. Mix all dry ingredients together in a large mixing bowl.
2. In a saucepan, place juices over a medium heat and reduce by one-third.
3. Add dried fruits to hot liquid and cook slowly for one minute.
4. Pour fruits and juices over dry mixture. Add honey and skim milk, and toss to moisten oats.
5. Pour onto a vegetable-oil-sprayed cookie sheet with sides. Place in 325°F oven to toast, and stir to keep oats from burning and to brown the granola evenly. This takes about 30 minutes.

6. Let cool and store in airtight container. You can grind in the food processor after cooling to get a fine-textured cereal. Serve at breakfast with sliced fresh fruit and nonfat yogurt.

NUTRITIONAL ANALYSIS:

Calories:	Protein (gm):	Fat (gm):	Carbo (gm):	Sodium (mg):	Chol (mg):
169	6	3	32	31	0

Recipe 8-34 *Stuffed French Toast Layered with Light Cream Cheese and Bananas*

Category: Breakfast   **Yield: 4 servings**

INGREDIENTS

8 slices whole-wheat bread
2 ounces light cream cheese
4 small bananas
Batter:
1-1/2 cups egg substitute
3/4 cup skim milk

1 teaspoon vanilla
1/2 teaspoon cinnamon
Syrup:
1 cup fresh strawberries, cleaned and cut
1/4 cup strawberry all-fruit jam
1 tablespoon lemon juice

STEPS

1. Lay out bread and spread all 8 slices with cream cheese evenly.
2. Slice bananas paper-thin and layer on 4 slices of bread, overlapping slightly, and top each slice with other side of bread; press down lightly.
3. Whip together batter ingredients to a smooth consistency.
4. Heat a nonstick pan and spray with vegetable-oil spray.
5. Dip the banana sandwiches carefully in batter on both sides to absorb batter. Place in nonstick pan and brown nicely on both sides, about 1-1/2 to 2 minutes per side. Do not let the pan get too hot, or the toast will brown without cooking in the middle. Transfer to an oven-safe dish.
6. Warm slightly in the oven to crisp before serving.
7. While the French toast is in the oven, blenderize strawberries, jam, and lemon juice until smooth. Serve as syrup on the side.

NUTRITIONAL ANALYSIS:

Calories:	Protein (gm):	Fat (gm):	Carbo (gm):	Sodium (mg):	Chol (mg):
526	23	11	86	552	10

Recipe 8-35 *Glazed Grapefruit and Orange Slices with Maple Vanilla Sauce*

Category: Breakfast   **Yield: 4 portions**

INGREDIENTS

2 grapefruit, peeled and sliced
3 oranges, peeled and sliced
1 ounce maple syrup

2 ounces apple juice
1 ounce water
1 tablespoon granola
4 sprigs mint

STEPS

1. Alternate orange slices and grapefruit slices on individual plates like a wheel, until you have 4 portions.
2. Heat maple syrup, apple juice, and water. Reduce glaze by one half. Pour lightly over fruits. Top with granola and very thin strips of mint.

NUTRITIONAL ANALYSIS:

Calories:	Protein (gm):	Fat (gm):	Carbo (gm):	Sodium (mg):	Chol (mg):
124	2	1	30	2	0

Chapter 9

Marketing Healthy Menu Options

Marketing nutrition has become increasingly common as interest has increased. According to a National Restaurant Association report, more than half of consumers 35 and older look for lower-fat menu options when eating out. Also, two out of every five 18- to 34-year-olds are doing the same. Restaurateurs report that their customers are increasingly requesting meatless meals. Add to this discussion the number of customers, particularly seniors over 65 years of age, who make requests such as no salt or no butter.

Restaurants offer much more to nutrition-conscious customers than the old-fashioned diet plate consisting of cottage cheese with fruit on top of a lettuce leaf. Menus now carry items ranging from low-fat, low-calorie tostados to full-course meals that do it all for under 600 calories. Restaurants market their nutritionally modified dishes on menus and table tents, by radio advertising, and in other ways. In 1999, 91 percent of surveyed noncommercial foodservices promoted nutritious items on menus, menu boards, and newsletters and by other methods.

Marketing means finding out what your customers need and want, and then developing, promoting, and selling the products and services they desire. Keeping in touch with your customers is crucial, and can be done without much fuss, by talking with customers, finding out from waitstaff what customers have been requesting, doing a survey, and so on.

This chapter will help you to:

■ Describe two methods a foodservice operator can use to gauge customers' needs and wants.
■ Give three examples of how to draw attention to healthy menu options.
■ Discuss effective ways to communicate and promote a nutrition program to customers.
■ Explain the importance and extent of staff training needed to successfully implement healthy menu options.
■ Describe two methods used to evaluate healthy menu options.
■ Discuss how nutrition labeling laws affect restaurant menus.

Gauging Customers' Needs and Wants

Most foodservice operators who have successfully implemented healthy menu options have done so through reviewing eating trends, examining what other operators are doing, and keeping abreast of their customers' requests for healthy foods. To determine customer wants, foodservice operators could interview the waitstaff about customer requests, for example, for light foods such as broiled meat, poultry, or fish; dishes prepared without salt; sauces and gravies removed or served on the side; butter substitutes; reduced-calorie salad dressing; or skim or low-fat milk.

Another way to gauge customer needs is to do a survey, as shown in Figure 9-1. At the same time, answers to the following questions need to be considered.

1. What are the majority of requests made during a particular meal?
2. Which items are most frequently requested?
3. How much time does your cooking staff and waitstaff have available to meet these special requests?
4. Which requests are easy to meet? Which are very time-consuming?

Answers to these questions can help you decide which types of healthy menu items to offer.

If market research demonstrates a sizable need for healthy entrées and the like, and there is enough time and staff to commit to this project, then now may be the time to do more than meet customers' special requests.

Figure 9-1

Customer survey

1. How often do you visit this restaurant?
   First visit
   Once or twice a year
   Once every three months
   Once every two months
   Once a month
   Two or three times a month
   Once a week
   More than once a week
2. Today I came for:
   Breakfast
   Lunch
   Dinner
   Snack
3. Are you here during your work day?
   Yes   No
4. Are you here for social reasons?
   Yes   No
5. Have you ever been to a restaurant that offers light and nutritious menu choices?
   Yes   No
6. Would you order light and nutritious foods if they were offered here?
   Yes, frequently
   Yes, sometimes
   No
   Not sure
7. How likely would you be to try the following nutritious menu choices?

Menu Choice	Very Likely	Likely	Unlikely
A. Broiled fish without butter			
B. Reduced-calorie salad dressing			
C. Vegetables with no added salt			

■ MINI-SUMMARY

Customer interest in nutritious menu items can be gauged through waitstaff feedback and customer surveys/feedback.

Developing and Implementing Healthy Menu Options

Various personnel are normally involved in the development and implementation phase: foodservice operators, directors, and managers; chefs and cooking staff; and nutrition experts such as registered dietitians. Chefs and cooking staff are valuable resources in modifying recipes or creating new ones, and may be given much of this responsibility. Nutrition experts are needed to provide accurate nutrient analysis data as well as suggestions for modifying dishes. In larger companies, personnel responsible for training, advertising and publicity, marketing, menu planning, and recipe development may also be involved.

Chapter 8 covered the basics of developing healthy menu items and modifying recipes. Once you know what you want to offer, you need to think about how to inform your clientele of these options. Here are some suggestions.

1. Highlight nutritious menu selections with symbols or words such as "light." For example, put a picture of wheat next to nutrition selections that meet specific nutrition goals, usually described at the bottom of the menu (Figure 9-2).
2. Include a special, separate section on the regular menu. With this format, customers are certain to see the nutritious options, and see them as being integrated into the foodservice concept. A heading for this section might be "Fit Fare."
3. Add a clip-on to the regular menu and/or a blackboard or lightboard. This method requires no alterations to the menu and is particularly useful in that it is flexible and inexpensive. Healthy selections can be changed without involving much time or money. In some operations, treating the nutrition selections like daily specials has increased their selling power.
4. Use the waitstaff to offer and describe nutritious menu options. In some instances, healthier preparation methods can be suggested for regular menu items.

ST. ANDREW'S DINNER

WOOD-FIRED PIZZAS
Small   4.50   Large   6.75

Pizza Margherita - Roasted Tomatoes and Mozzarella

Special Pizza of the Day

STARTERS

Wild Mushroom Risotto with White Truffle Oil   5.00

Oven-roasted Clams in "Casino" style Broth   5.50

Smoked Shrimp with Cucumber Salad and Curry Vinaigrette   6.00

Grilled Moroccan-style Quail with Whole Grain Salad   7.50

Thai-style Mussels and Prawns in Spicy Coconut Broth   5.00

SOUPS AND SALADS

Warm Carrot Soup with Ginger Cream   3.00

Mediterranean Vegetable Soup   3.00

Shrimp and Chicken Gumbo   3.50

Soup Sampler - A Taste of Each Soup   4.00

Organic Mesclun Greens with Choice of Creamy Blue Cheese,
Lemon-Thyme Vinaigrette or Balsamic Vinaigrette   3.50

Roasted Beet and Fresh Citrus Salad with Blue Cheese Crumbles   4.00

Figure 9-2

St. Andrew's Cafe menu.

Courtesy: The Culinary Institute of America, Hyde Park, NY.

MAIN COURSES

Potato Gnocchi with Calamata Olives and Winter Vegetables 8.00

Chicken Pot-au-Feu 12.00

Grilled Venison and Green Peppercorn Sauce 15.00

Seared Tuna Steak with Golden Pepper-Potato Puree, Scallions
and Sundried Tomatoes 15.00

Wood-fired Striped Bass Fillets with Roasted Sweet Onions and Fennel 13.50

Bouillabaise with Shrimp and Mussels, Local Potatoes and Braised Fennel 12.00

Grilled Salmon with Lemon and Herb Fettuccini 15.00

Indicates Vegetarian Selections

Michael Garnero, Chef-Instructor
Carleen von Eikh, Maître d'Hôtel Instructor

The St. Andrew's Cafe is a non-smoking restaurant. Thank you for your cooperation.

VISIT OUR OTHER RESTAURANTS:
the American Bounty Restaurant, the Caterina de Medici Restaurant & the Escoffier Restaurant

A 12% service charge has been added to your check. It is used to fund student scholarships, purchase graduation jackets and support student activities. Tips in excess of 12 % will go directly to the students serving you. The service charge amount reflected on your check is not mandatory. If you are not satisfied with your dining experience, the charge may be adjusted. If you have any questions, please see the Maître d'Hôtel Instructor. Thank you.

Brown Rice Pudding with Dried Cherries   4.00

"Crème Brûlée"   4.00

Warm Apple Strudel with Ricotta Glace
and Caramel Sauce   4.00

Poached Pears and Apple Sorbet   4.00

Andrew Quady Muscat Sampler
Electra Orange and Elysium Black   4.25

Harvey's Bristol Cream Sherry   3.25 *glass*

Fonseca Tawny Porto   3.00 *glass*

Coffee or Tea   1.50
Espresso   2.00
Cappuccino   2.50
Cappuccino of the Day made with Torani Syrups   3.50

For additional dessert beverages, please refer to the Wine & Beverage List

Figure 9-2 *(continued)*

No matter which method you use to include healthy selections on the menu, you must consider how thorough a description is appropriate. In general, customers do not want calorie counts, fat, cholesterol, or sodium content on the menu, but prefer simply a good description of the ingredients, portion size, and preparation method. Menu items are more effectively promoted by giving customers this information and by emphasizing quality and variety rather than nutrition.

Marketing healthy menu items can be done in a positive manner. When you tout foods as "heart healthy," you may be approaching some customers in a negative manner. When you market freshly squeezed fruit juices, you are approaching customers in a positive manner.

Promotion

Three methods of promoting a nutrition program are advertising, sales promotion, and publicity. Advertising can be done through magazines and newspapers, radio and television, outdoor displays (posters and signs), indoor table tents and posters, direct mail, and novelties (such as matchboxes). Direct mail works well when targeted to current customers.

Advertising messages should say something desirable, beneficial, distinctive, and believable about the nutritious dining program. For example, the new menu selections could be advertised as healthy and using only the freshest, most exotic ingredients. Because foodservice operators need to get the best advertising for the money, hiring a reputable advertising company may be the best option.

Sales promotions can include coupons, point-of-purchase displays (such as a blackboard at the dining room entrance listing the nutrition selections), or contests (such as having customers guess the number of calories in a nutritious dining entrée to win a free meal).

Publicity involves obtaining free editorial space or time in various media. Many foodservice operators do their own publicity. However, if you wish to obtain the advice of outside publicity consultants, O'Dwyers Directory includes most public relations firms. Here are some ideas for publicizing your nutrition program.

Publicity—Obtaining free space or time in various media to get public notice of a program, book, and so on.

1. Send a **press release** about your healthy dining options to the appropriate contact person by name, not title, as indicated in the following list:

   Television and radio news: Assignment editor or specialty reporters appropriate to your story, such as health and food editors

   Television and radio talk shows: Producer

Newspapers: Section editors (food, health and science, lifestyle), or city desk editor for special events

Magazines and trade publications: Managing editor, articles editor, or specialty editors appropriate to your story

Local publications and newsletters: Corporate employee or customer newsletters, or supermarket, utility company, bank, school, or church publications

Follow up each press release with a phone call. Editors are always looking for article ideas and just may pick up on your story.

2. Offer to write a column on nutritious meal preparation for a local newspaper.

3. Offer cooking demonstrations or on-site classes, or volunteer to conduct classes for health associations, retail stores, or supermarkets.

4. Invite local media and community leaders for the opening day of your new program and let them taste some nutritious menu selections.

5. Contact the foodservice director of a medical center or the public relations director of a health maintenance organization and offer to cosponsor a health or nutrition event such as a bike race or health fair. Check for local health and sporting events in which you can participate.

6. Contact your local American Heart Association and ask if it has a "Dining Out Guide" in which you may feature your restaurant.

7. Develop a newsletter for your operation and use it to publicize the new program (include some of your nutritious recipes). Newsletters help to build loyal customers.

There are many sources for promotional materials, such as table tents, posters, buttons, menu clip-ons, point-of-sale materials, and artwork (Figure 9-3). Food manufacturers, foodservice distributors, and food marketing boards and associations are excellent sources of promotional materials.

Staff Training

Staff training centers on the waitstaff and the cooking staff. Before training begins, involve the waitstaff as much as possible in the development of your nutrition program so that they feel a part of it and take some ownership. They can be a valued resource in designing the program, because they make daily contact with the customers both in selling and serving as well as in listening to requests, compliments, and complaints. During training, the waitstaff needs to understand:

Figure 9-3

Artwork from International
Apple Institute

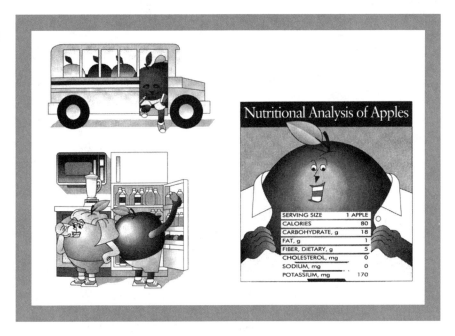

- the scope and rationale for the nutrition program.
- grand-opening details.
- the ingredients, preparation, and service for each menu item.
- some basic food and nutrition concepts so they can help guests with special dietary concerns, such as food allergies.
- how to handle special customer requests, such as orders for half portions.
- merchandising and promotional details.

Table 9-1 gives specific learning objectives for the service staff.

A poorly trained waitstaff will confuse the customer and, quite frankly, doom the program instead of knowledgeably promoting it. Conversely, a properly trained waitstaff can function as excellent sales agents and solicit feedback, including customer recommendations.

The cooking staff also needs training. Their training needs center on

- the scope and rationale for the nutrition program.
- grand-opening details.
- the ingredients, preparation, portion size, and plating of each new menu item.
- some basic food and nutrition concepts so they can help guests with special dietary concerns, such as food allergies.
- how to respond to special dietary requests.

TABLE 9-1 Learning Objectives for Service Staff

1. Servers must be able to respond to consumer health concerns by providing menu suggestions that meet their dietary needs. They should be able to make menu suggestions for the following dietary restrictions:
   Low calorie
   Low sodium
   Low cholesterol and low fat
   Low sodium, low cholesterol, and low fat
   High fiber
2. The waitstaff should be able to describe healthful dining options in straightforward, appropriate language to patrons. Servers must be able to provide information on ingredients, methods of preparation, portion sizes, and how the menu items are served.
   Ingredients: The waitstaff should be knowledgeable about details regarding ingredient usage: addition of fat or the type of fat used in cooking, the use of salt or high-sodium seasonings, cuts of meats used, the type of liquids used to prepare menu items, fats and thickening agents used in sauces, and the use of sugar or sugar substitutes.
   Cooking methods: Patrons commonly need to know not only what the composition of a menu item is but how it was prepared. Was the food fried in vegetable oil or animal shortening? Was the food prepared by pan frying, broiling, baking, poaching, or sautéing? Can the item be broiled without added butter? Is fat removed from meat juices or stocks before using them for sauces or soups?
   Presentation and portions: Patrons frequently want to know how the item will be served when making their menu selections. The waitstaff should be prepared to answer the following questions: What is the portion size of the item? What accompaniments are served with the item? Are special food items available to accompany the item? For example, is light syrup available for the light pancakes? Is a fruit spread available instead of jam for the whole-wheat breads? Can toast be served dry instead of buttered? Can salad dressing be served on the side?
3. The waitstaff should be able to explain the nutritional basis for menu items designated as light or healthful in terms of caloric, fat, cholesterol, and/or sodium content. They may need to answer questions about the program rationale. They should know the nutritional guidelines (US Dietary Guidelines, American Heart Association guidelines, and/or National Cancer Institute recommendations) that provide the basis for the program.
4. Servers should be able to respond to patron inquiries about the availability of special foods or beverages. Does the restaurant serve brewed decaffeinated coffee? Are diet salad dressings available for the light salad entrées? Is margarine available instead of butter? Are herb seasonings available instead of salt for adjusting seasonings at the table? Does the restaurant serve skim milk?
5. The waitstaff should be able to respond to questions concerning substitutions of meal accompaniments. Can an entrée be served with two vegetables instead of a vegetable and a starch? Can a salad be substituted for the starch or vegetable served with the entrée? Can a fruit appetizer be served for dessert?
6. The waitstaff need to know what special requests the foodservice operation can accommodate. For example, can margarine or vegetable oil be used instead of butter in preparing foods? Can entrées be broiled instead of fried? Are smaller portion sizes available? Can sauces and salad dressing be served on the side?
7. The waitstaff should be able to recommend other foods and beverages that complement menu choices, including appetizers, soups, salads, desserts, and beverages that are light or meet the dietary restrictions of the patron.
8. The waitstaff should be knowledgeable about what is served with light menu items and what the correct portion sizes are for these items. Servers will act as the final quality-control agents prior to the serving of the foods. If light menu items are similar to traditional offerings, the waitstaff should be able to distinguish between the two items.

TABLE 9-1 *(continued)*

9. Staff members should respond politely and accurately to guest questions about healthful dining options. It should be emphasized that the proper response to a patron's inquiry is "I can find out for you," not "I don't know." When uncertain of the answer, the staff member should ask the kitchen manager, manager, or chef.

Source: Ganem, Beth Carlson. 1990. *Nutritional Menu Concepts for the Hospitality Industry.* New York: Van Nostrand Reinhold. Reprinted with permission.

Table 9-2 outlines seven ways that employees learn best.

The cooking staff need to understand the prime importance of using only the freshest ingredients, using standardized recipes, measuring and weighing accurately, and attractive presentation.

Training the cooking staff to prepare healthy dishes correctly can be challenging. As managers have found during nutrient-content analysis, cooks do not always prepare recipes exactly as called for. Perhaps a key ingredient was unavailable, time was tight, or the cook forgot a step. Healthy menu items often are more labor-intensive, and more training and coaxing are needed. In any case, cooking staff need training not only in making new menu items but also in understanding the importance of following recipes and serving the correct portion size.

In 1995, the Center for Science in the Public Interest analyzed seventeen healthy entrées from seven of the largest restaurant chains. Although the healthy items had less fat, fewer calories, and more vegetables and fruit than other menu items, several promised less fat than was actually found in the entrée. For example, a fajita from the menu of a Mexican restaurant was supposed to have only 17 grams of fat, but actually had 30 grams of fat. Problems such as this point to the importance of training and retraining cooking staff.

Program Evaluation

The healthy menu program should be evaluated much like any other program. Key questions for evaluation include:

1. How did the program do operationally? Did the cooks prepare and plate foods correctly? Did the waitstaff promote the program and answer questions well?
2. Did the food look good and taste good?

TABLE 9-2 How Employees Learn Best

1. When employees participate in their own training, they tend to identify with and retain the concepts being taught. To get employees involved, choose appropriate training methods.
2. Employees learn best when training material is practical, relevant, useful, and geared to an appropriate level. Learning is facilitated, too, when the material is well organized and presented in small, easy-to-grasp steps. Adult learners are selective about what they will spend time learning, and learning must be especially pertinent and rewarding for them. Adults also need to be able to master new skills at their own pace.
3. Employees learn best in an informal, quiet, and comfortable setting. Your effort in selecting and maintaining an appropriate training environment shows employees that you think their training is important. When employees are stuffed into a crowded office or a noisy part of the kitchen, or when the trainer is interrupted by phone calls, they may rightly feel that their training isn't really important. Employees like to feel special; so, when possible, find a quiet setting. Of course, much training, such as on-the-job training, necessarily takes place in the work environment.
4. Employees learn best when they are being paid for time spent in training.
5. Employees learn best with a good trainer. Although you may not ever find a person with all these qualities, you can use this list to evaluate potential trainers:

   A Successful Trainer
   - Is knowledgeable
   - Displays enthusiasm
   - Has a sense of humor
   - Communicates clearly, concisely, straightforwardly
   - Is sincere, caring, respectful, responsive to employees
   - Encourages employee performance; is patient
   - Sets an appropriate role model
   - Is well organized
   - Maintains control, frequent eye contact with employees
   - Listens well
   - Is friendly and outgoing
   - Keeps calm; is easygoing
   - Tries to involve all employees
   - Facilitates the learning process
   - Positively reinforces employees

6. Employees learn best when they receive awards or incentives. For example, when an employee has completed training for the position of cook, you can send a letter of recognition to the employee, which can also be put into the personnel file. The largest franchisee of Arby's awards employees a progression of bronze, silver, and gold name tags, as well as pay increases, as they learn each area of the restaurant. When the employee has learned all areas, he or she is promoted to the position of crew leader.
7. Employees learn best when they are coached on their performance on the job.

Source: Drummond, Karen. 1992. *Retaining Your Foodservice Employees.* © 1992 by John Wiley & Sons, Inc. Adapted by permission of John Wiley & Sons, Inc.

3. How well did each of the healthy menu options sell? How much did each item contribute to profits? How did the overall program affect profitability?

4. Did the program increase customer satisfaction? What was the overall feedback from customers? Did the program create repeat customers?

Proper program evaluation requires much time observing and talking with staff and customers, as well as going over written records, such as sales records.

Once a program has been evaluated, certain changes to fine-tune the program might be necessary. Here are some suggestions:

■ Develop ongoing promotions to maintain customer interest.
■ Add, modify, or delete certain menu items.
■ Change pricing.
■ Improve the appearance of healthy items.
■ Listen to customers more to get future menu and merchandising ideas.

> ■ **MINI-SUMMARY**
>
> Evaluation is needed to determine program effectiveness from customer, employee, and management viewpoints.

Restaurants and Nutrition Labeling Laws

Foods prepared and served in restaurants or other foodservice operations are exempt from mandatory nutrition labeling found on packaged foods. However, restaurants are not exempt from Food and Drug Administration (FDA) rules concerning nutrient claims and health claims (discussed in Chapter 2) when used on menus, table tents, posters, or signs. In addition, restaurants must have nutrition information available upon request for any menu item using nutrient or health claims.

Nutrient content claims such as "good source of calcium" or "fat free" can appear only if they follow legal definitions (see Table 2-10 in Chapter 2). For example, a food that is a good source of calcium must provide 10 to 19 percent of the Daily Value for calcium in one serving. Nutrient content claims are based on what the FDA has defined as a standardized serving size, called a Reference Amount. Standardized serving sizes or Reference Amounts are frequently measured in grams, milliliters, or cups in order to be very accurate. For example, the Reference Amount for cookies is 30 grams.

In addition to nutrient content claims based on a single serving or Reference Amount, there are content claims for main dishes and meals. A main dish must weigh at least six ounces, be represented on the menu as a main dish, and contain no less than 40 grams each of at least three different foods from at least two food groups. Meals are defined as weighing at least 10 ounces and containing no less than 40 grams each of at least three different foods from at least two food groups. In general, for main dishes and meal products the nutrient content claim is based on the nutrient amount per 100 grams of the food. For example, a "lowfat" food must contain 3 grams of fat or less. Main dishes and meals must therefore contain 3 grams of fat or less per 100 grams, and not more than 30 percent of calories from fat.

Claims that promote a health benefit must meet certain criteria, as described in Chapter 2 and Table 2-11. In addition, any food being used in a health claim may not contain more than 20 percent of the Daily Value for fat, saturated fat, cholesterol, or sodium. Table 9-3 shows the maximum amount of these nutrients that are permitted in a single food serving, main dish, or meal for foods with health claims.

When providing nutrition information for a nutrient or health claim, restaurants do not have to provide the standard nutrition information profile and more exacting nutrient content values required in the Nutrition Facts panel of packaged foods. Instead, restaurants can present the information in any format desired, and they have to provide only information about the nutrient or nutrients that the claim is referring to. Restaurants also are not required to do chemical analyses to determine the nutrient values of their foods. They can use nutrient analysis software, books with nutrient composition information, or cookbooks with reliable nutrient analysis data.

Restaurants may use symbols on the menu to highlight the nutritional content of specific menu items. When doing so, they are required to explain

TABLE 9-3 Maximum Allowable Amount of Fat, Saturated Fat, Cholesterol, and Sodium for Foods with Health Claims

	Total Fat	Saturated Fat	Cholesterol	Sodium
Single Serving*	13 grams	4 grams	60 milligrams	480 milligrams
Main Dish	19.5	6	90	720
Meal	26	8	120	960

* Per Reference Amount and per 50 g when Reference Amount is 30 grams or less *or* 2 tablespoons or less.

the criteria used for the symbols. Restaurants may also highlight dishes that meet criteria set down by recognized organizations, such as the American Heart Association or a medical center. In these cases, the menu must explain that the items meet the dietary guidelines of that organization. Lastly, menus can use references or symbols to show that a food or meal is based on the Dietary Guidelines for Americans.

Check-Out Quiz

1. Marketing includes:
   a. finding out what consumers want and need.
   b. developing a product consumers want and need.
   c. promoting the product.
   d. all of the above.
2. When hearing descriptions of healthy menu entrées, most customers want:
   a. complete nutrient information.
   b. fat, saturated fat, and cholesterol information.
   c. good descriptions of the ingredients, portion size, and method of preparation.
   d. none of the above.
3. When you tout foods as "heart healthy," you are approaching customers in a negative manner.
   a. True   b. False
4. An example of publicity is:
   a. radio advertising.
   b. a press release.
   c. recipes from a food association.
   d. a point-of-purchase display.
5. When you evaluate the success or failure of healthy menu options, you need to get feedback from:
   a. customers.   b. staff.   c. managers.   d. all of the above.

Activities and Applications

1. Restaurant Menu Check
Study the menus from five different foodservices, including a quick-service business, and identify any menu items that are healthy. How do menu items appear on the menu? Would customers know they are healthy? What information is included?

2. Restaurant Promotion

Check restaurant advertisements in the newspaper, on radio, or in other media and watch for any advertising of healthy and nutritious foods. What do the advertisements state? To which market segments are these advertisements targeted?

3. Restaurant Visit

Visit a local restaurant/foodservice that offers healthy menu options. Find out how much it sells of its healthy menu options, and the profile of the typical customer buying these items. Also find out how these products are marketed and evaluated.

Nutrition Web Explorer

A restaurant chain such as   www.Fridays.com or www.chilis.com.
Go to the website of a restaurant chain, such as TGI Fridays or Chili's, and print out their menu items, which come with descriptions. Read the descriptions. Circle the menu items that appear directed to nutritionally conscious customers. For two of the recipes in Chapter 8, write menu descriptions that make the foods sound appealing and also describe their nutrient contribution.

Food marketing boards and associations
American Egg Board   www.aeb.org
Mann Packing   www.broccoli.com
Grains Nutrition Information Center   www.wheatfoods.org
Milk   www.whymilk.com
National Pork Producers Council   www.nppc.org
National Turkey Federation   www.turkeyfed.org
Produce Marketing Association   www.pma.com

Pick one of these food marketing board/association websites to visit. Write a brief report about the website, including the name of the board/association, website address, and list of items available that a foodservice operator could use (such as recipes), and attach a sample material.

Chapter 10

Light Beverages and Foods for the Beverage Operation

Heavy foods and strong alcoholic beverages were a way of life in most beverage operations until about 1970. At that time, the sale of spirits began to decline, whereas sales of "lighter" drinks, such as beer and wine, began to increase. Next, beer went light—a well-received change—and wine by the glass was promoted. Mocktails—nonalcoholic beverages—began to appear as well, as part of the movement against alcohol abuse. In addition, the drinking age was raised to 21 in many states, warning labels appeared on liquor bottles, and successful lawsuits were waged against beverage operations by victims of drunk drivers—all part of a movement aimed not at the use but at the abuse of alcohol. There is a big difference between alcohol abuse and moderate use. In fact, the 1995 edition of the Dietary Guidelines for Americans, for the first time, indicated that moderate drinking is associated with a lower risk for coronary heart disease in some individuals.

In an effort to curb drunk-driving accidents, beverage operations have changed their emphasis from all-you-can-drink specials (and the like) to foods, entertainment, music, contests, games, and theme parties, in order to overcome an association with heavy drinking. Themes might be international, such as Mexican, with tacos and burritos and staff dressed in Mexican outfits. A Workout Night theme might involve asking customers to come dressed in their workout clothes.

This chapter first looks at lower-calorie and lower-alcohol beverages and nutritious food choices, then goes on to discuss marketing these selections. This chapter will help you to:

- list lower-calorie and lower-alcohol drink options.
- state the number of calories in one gram of alcohol.
- define various types of bottled water.
- identify healthy snacks/appetizers.
- describe the steps a foodservice operator takes in order to market light beverages and foods.

Lower-Calorie and Lower-Alcohol Drink Options

Bar customers who want a lower-calorie beverage or a drink with little or no alcohol have a number of choices: fruit juices and fruit-juice-based beverages, creamy drinks, **alcohol-free wines, light beers** and **nonalco-**

holic malt beverages, low-alcohol refreshers, mocktails, bottled waters, and other beverages. As Table 10-1 indicates, when the alcohol content of a drink increases, so do calories. Alcohol contains 7 calories per gram, compared to 4 calories per gram of carbohydrate or protein.

Fruit Juices and Fruit-Juice-Based Beverages

Fruit juices serve as a base for many different nonalcoholic drinks, such as fruit juice served with seltzer (water with bubbles or carbonation). Fruit may serve as the basis for blended drinks, such as strawberry coolers made with fresh strawberries, sugar, sparkling water, and ice cubes.

Nonalcoholic sparkling cider, a soft, fruity, and light drink, is made in France and in the United States. Most sparkling ciders contain no alcohol, but it is available with 5 to 12 percent alcohol. Serve it in stemmed glasses with a fresh fruit garnish to merchandise its fresh, fruity taste.

Creamy Drinks

Creamy drinks are made with milk, yogurt, cream, ice cream, or other dairy products. A good example is a category of creamy drinks called **fruit smoothies,** frozen blends of fruit, milk, and/or fresh or frozen yogurt. When regular dairy products such as cream and ice cream are used, these drinks are not low in calories; however, low-fat or nonfat yogurt, ice milk, and skim or 1 percent milk can be used to make drinks with fewer calories and less fat.

Alcohol-free Wines

Alcohol-free wine has had its alcohol removed after fermentation. It contains up to 0.5 percent alcohol and is available in red, pink, and white varieties, either still or sparkling. Varietal wines with one predominant grape variety, such as gamay or Riesling, are available as well.

Light Beers and Nonalcoholic Malt Beverages

Several varieties of beer with varying alcohol and calorie contents are available, as seen in Table 10-1. Different brands of light beer can vary tremendously in calorie content, but most have one-third to one-half the alcohol and calories of regular beers. Nonalcoholic brews, produced by removing

Alcohol-free wine—Wines whose alcohol is removed after fermentation; they must be 99.5 percent alcohol free.

Light beer—Beer with one-third to one-half less alcohol and calories than regular beers.

Nonalcoholic malt beverages—A beerlike product with only 0.5 percent or less alcohol.

Low-alcohol refreshers—A category of drinks made with fruit juices, carbonated water, and sugar, with an alcohol base of wine, distilled spirits, or malt; the group includes wine coolers.

Mocktails—Drinks made to resemble mixed alcoholic drinks but containing no alcohol.

Fruit smoothies—Frozen blends of fruit, milk, and/or fresh or frozen yogurt.

TABLE 10-1 Calories in Alcoholic and Nonalcoholic Beverages

Beverage	Amount (fluid ounces)	Calories	Alcohol
Mixed Drinks			
Bloody Mary	5 fl. oz.	116	11.7%
Bourbon and soda	4 fl. oz.	105	16.1%
Daiquiri	2 fl. oz.	111	28.3%
Gin and tonic	7.5 fl. oz.	171	8.8%
Manhattan	2 fl. oz.	128	36.9%
Martini	2.5 fl. oz.	156	38.4%
Piña colada	4.5 fl. oz.	262	12.3%
Screwdriver	7 fl. oz.	174	8.2%
Tequila sunrise	5.5 fl. oz.	189	13.5%
Tom Collins	7.5 fl. oz.	121	9.0%
Whiskey sour	3 fl. oz.	123	20.6%
Distilled Liquors			
Gin, 90 proof	1.5 fl. oz.	110	45%
Rum, 80 proof	1.5 fl. oz.	97	40%
Vodka, 80 proof	1.5 fl. oz.	97	40%
Whiskey, 86 proof	1.5 fl. oz.	105	43%
Wine			
Table wine, red	3.5 fl. oz.	74	11.5%
Table wine, rosé	3.5 fl. oz.	73	11.5%
Table wine, white	3.5 fl. oz.	70	11.5%
Beer			
Beer	12 fl. oz.	146	4.5%
Light beer	12 fl. oz.	100	2.2–4.4%
Nonalcoholic malt beverage	12 fl. oz.	50	Less than 0.5%
Nonalcoholic Beverages			
Cola	12 fl. oz.	151	
Ginger ale	12 fl. oz.	124	
Lemon-lime soda	12 fl. oz.	149	
Orange soda	12 fl. oz.	177	
Root beer	12 fl. oz.	152	
Tonic water	12 fl. oz.	125	
Club soda, seltzer, mineral, or sparkling waters	12 fl. oz.	0–10	
Cranberry-apple juice drink	6 fl. oz.	123	
Cranberry juice cocktail	6 fl. oz.	108	
Grape juice drink	6 fl. oz.	94	
Lemonade	8 fl. oz.	100	
Tomato vegetable juice cocktail	8 fl. oz.	44	

Source: United States Department of Agriculture Handbook #8–14.

the alcohol after brewing or by stopping the fermentation process before alcohol forms, contain one-half of 1 percent or less of alcohol. In either case, these products have about half the calories of light beer and one-third the calories of regular beer.

Low-Alcohol Refreshers

Wine coolers—Mixes of wine, fruit flavor, and carbonated water or plain water.

Wine spritzers—Drinks made with wine and club soda.

This group started out as wine coolers, but has grown to include other drinks. **Wine coolers** are a mix of wine, fruit flavor, and carbonated or plain water. They usually contain fructose or high-fructose corn syrup for sweetening, and a few also have preservatives added. They generally have half the alcohol content of wine. Many flavors are on the market, and some brands are available in kegs. Wine coolers can also be made on site and served on the rocks or over shaved ice. **Wine spritzers** are made with wine and club soda and are mixed at the bar. Other low-alcohol refreshers are made with fruit juices, carbonated water, sugar, and distilled spirits or malt instead of wine.

Mocktails

Mocktails are much more than just a Virgin Mary, a bloody Mary without the vodka, or simply tomato juice! Mocktails attempt to imitate the real thing. For instance, a mockarita, an imitation margarita, is blenderized with lemonade, lime juice, and ice cubes and served in a margarita glass with a salted rim. Hot mocktails, such as mock Irish coffee, can be made as well. Some bartenders substitute lower-proof products in cocktails, such as Chablis for tequila to make a white wine margarita; such drinks are not true mocktails, but this substitution does lower both alcohol and calories. Nonalcoholic "liqueurs," such as crème de menthe and crème de cacao, are also available for use in mocktails or lower-alcohol cocktails.

Other Choices

Other lower-calorie and/or no-alcohol alternatives include bottled waters (see the "Bottled Waters" section in this chapter), vegetable juices, diet soft drinks, iced tea flavored with lemon or other fruit juice, and some specialty coffees such as espresso or cafe latte (see Table 10-2). Other flavorings for lower-alcohol drinks include vanilla and rum extracts or nonalcoholic fruit syrups such as cassis.

TABLE 10-2 Espresso Drinks	
Name	Description
Espresso	The word espresso refers to the unique process used to brew this coffee. Hot water is forced under high pressure through finely ground coffee that has been compacted. The coffee is made from beans that have been roasted until very dark. The serving size is about 1-1/2 fluid ounces (45 mL).
Cappuccino	1-1/2 fluid ounces (45 mL) of espresso, topped by frothed milk that is separated into hot milk and foam, in an approximate ratio of one part espresso, one part milk, and one part foam. Usually served in a glass mug.
Latte	A single or double serving of espresso, topped by frothed milk, in a ratio of one part espresso to three parts milk. Usually served in a wide-mouthed glass.
Mocha	A single serving of espresso mixed with about 5 fluid ounces of steamed milk and chocolate syrup to taste. Often topped with whipped cream.

■ MINI-SUMMARY

Drinks with fewer calories and/or less alcohol include fruit juices and fruit-juice-based beverages, creamy drinks, alcohol-free wines, light beers and nonalcoholic malt beverages, low-alcohol refreshers, mocktails, bottled waters, vegetable juices, diet soft drinks, iced tea, and specialty coffees.

Bottled Waters

People buy **bottled water** for what it does not have—calories, sugar, caffeine, additives, preservatives, and, in most cases, not too much sodium. Bottled water cannot contain sweeteners or chemical additives (other than flavors, extracts, or essences) and must be calorie-free and sugar-free. Flavors, extracts, and essences—derived from spice or fruit—can be added to bottled water, but these additions must comprise less than 1 percent by weight of the final product. Beverages containing more than the 1 percent-by-weight flavor limit are classified as soft drinks, not bottled water. In addition, bottled water may be sodium-free or contain "very low" amounts of sodium. Some bottled waters contain natural or added carbonation.

The Food and Drug Administration (FDA) has published standard definitions for different types of bottled water to promote honesty and fair dealing in the marketplace. Bottled water, like all other foods regulated by the FDA, must be processed, packaged, shipped, and stored in a safe and sanitary manner and be truthfully and accurately labeled. Bottled water products must also meet specific FDA quality standards for contaminants, set in response to requirements that the Environmental Protection Agency (EPA) has set for tap water.

To help resolve possible confusion, the FDA has set the following definitions:

Bottled water—Water that is bottled in a sanitary container and sold.

Spring water—Water collected from a spring as it flows to the surface.

Mineral water—Water that comes from a protected underground source and contains at least 250 parts per million in total dissolved solids.

Sparkling water—Any water with carbonation.

Seltzer—Filtered, artificially carbonated tap water.

Club soda—Filtered, carbonated tap water to which mineral salts are added.

Tonic water—A carbonated water containing lemon, lime, sweeteners, and quinine.

■ **Spring water** is water collected as it flows naturally to the surface, or when pumped through a hole from the spring source. The FDA regulation allows labeling to describe how the water came to the surface; for example, "naturally flowed to the surface, not extracted."

■ **Mineral water** must come from a protected underground source and contain at least 250 parts per million in total dissolved solids. No minerals can be added.

■ Water bottled from municipal water supplies must be clearly labeled as such, unless it is processed sufficiently to be labeled as "distilled" or "purified" water.

■ **Sparkling water** is any carbonated water.

Different mineral and carbonation levels of bottled waters make them appeal to different customers and eating situations. For instance, a heavily carbonated (also called sparkling) water such as Perrier is excellent as an aperitif, yet some customers may prefer the lighter sparkle of San Pellegrino. Still waters such as Evian are generally more popular and appropriate to have on the table during the meal.

FDA regulations for bottled waters do not cover any soft drinks or similar beverages that do not highlight a water ingredient. Some popular beverages that are not considered bottled waters include:

■ **Seltzer** is filtered, artificially carbonated tap water that generally has no added mineral salts. It is available with assorted flavor essences, including vanilla cream, black cherry, orange, raspberry, lemon lime, root beer, and cola. If seltzer contains sweeteners (and therefore calories), it must be called a "flavored soda."

■ **Club soda,** sometimes called "soda water" or "plain soda," is filtered, carbonated tap water to which mineral salts are added to give it a unique taste. Most average 30 to 70 milligrams of sodium per 8 ounces.

■ **Tonic water** is not really water or low in calories. It contains 84 calories per 8 ounces. Diet tonic water that uses sugar substitutes is available.

Healthy Snacks and Appetizers

Nutritious food choices for beverage operations need not be dull, monotonous, or unappetizing. Many snack foods and appetizers can easily fit in, such as a variety of dips served with crudités (raw, cut vegetables), fresh fruit, and/or baked tortilla chips. Other choices may include some from this list.

Afternoon Snacks
Fresh Fruit Kabobs with Yogurt Dipping Sauce
Vegetable Tortilla Rolls with Dill Dressing
Whole Fresh Fruits with Sliced Banana Breads and Reduced-Fat Cream Cheese
Air-Popped Popcorn Sprinkled with Parmesan Cheese
Macédoine of Fresh Fruits Marinated with Mint and Sliced Angel-Food Cake
Baked Assorted Tortilla Chips with Tofu Dips
Cinnamon Oatmeal Cookies, Date Granola Bars
Turkey and Stone-Ground Wheat Bread Sandwiches with Dijon Mustard Sauce

Canapés
Baked Pita Chips with Caponata
Stuffed Celery with Herbed Cream Cheese
Tomato, Spinach, and Turkey Rolls with White-Bean Spread
Oat-Bran Mini Muffins with Whole-Grain Honey Mustard Spread and Fresh Turkey
Oven-Dried Vegetable Chips with Hummus

Hot Hors D'Oeuvres
Pan-Seared Vegetable Spring Rolls
Spinach and Feta (Cheese) Wrapped in Phyllo (Dough)
Teriyaki Steak Skewers
Hummus Bruschetta with Roasted Peppers
Grilled Chicken Sausages with Wholegrain Breads
Jalapeño Pancakes with Corn Relish
Grilled Pork Quesadillas with Onions, Peppers, and Mushrooms
Mini Meatballs (Asian or Swedish)

Late-Night Snacks
Meatless Chili with Baked Corn Tortillas
Carrot and Celery Sticks with Garlic Yogurt Sauce

Corn-Flake-Crusted Chicken Tenders with Pineapple Chili Dipping Sauce

Assorted Oat-Bran Fruit Bars with Dried-Fruit Leather

Baked Stuffed Potato Skins with Broccoli, Cheddar Cheese, and Tomato Vinaigrette

Marketing Light Beverages and Foods

The first step in marketing light beverages and foods involves analyzing the surrounding market and clientele needs and wants; projecting sales volume, pricing, and profit structures; and seeing how such options would fit into the operation in terms of ordering, inventory, and menus. A beverage operator might start by choosing five items, perhaps a wine cooler, a de-alcoholized wine and beer, a mineral water, and a juice-based drink. In a more developed market, the choices might be enlarged. Using cookbooks or other resources, you can select, implement, and evaluate lower-calorie foods that fit well into your operation. The production staff must be involved in these decisions. For both drinks and food, variety and quality are of the utmost importance.

Presentation of lower-calorie and/or lower-alcohol drinks is just as important as it is for regular drinks. For instance, serve sparkling cider in stemmed glasses or champagne flutes, and garnish with an apple ring and a sprig of mint.

Bartenders and servers need to be trained about the new offerings as well as tasting them. A well-trained staff can suggest options appropriately, but they must learn to present lower-calorie and/or lower-alcohol drinks as positively as regular drinks. If servers lose tips by selling these new items, a bonus or incentive may be needed in order for the program to be successful.

The proposed program must be merchandised with flair during good selling times to enhance its success. During these times, menu clip-ons, blackboards, lightboards, table tents, posters, or other techniques let customers know what is available. A special dinner menu may be used, and promotions such as water tastings and "nonalcoholic beer of the month" offerings can be effective.

Once the program is launched, getting customer feedback is very important, as well as analyzing sales and profits. After enough time, the success of the program needs to be evaluated, just like any other program.

■ **MINI-SUMMARY**

As with other programs, light beverages and foods must first be selected and developed to meet clientele needs and wants and, hopefully, to make money. Staff must be trained, and merchandising and promotion techniques selected, before the program is launched. When the program has been in place long enough, sales and profits can be evaluated.

Check-Out Quiz

1. Which is lower in calories?
   a. 12 fluid ounces of light beer
   b. 7 fluid ounces of gin and tonic
   c. 6 fluid ounces of white wine
   d. 12 fluid ounces of tonic water
2. Drinks that taste like mixed drinks but contain no alcohol are called:
   a. coolers
   b. de-alcoholized spirits
   c. mocktails
   d. smoothies
3. The emphasis in beverage operations has changed to:
   a. food
   b. theme events
   c. entertainment
   d. all of the above
4. Which of the following foods could not be served as a nutritious snack?
   a. raw vegetables with yogurt dip
   b. hot, soft pretzels
   c. vegetarian pizza on pita bread
   d. deep-fried mozzarella sticks
5. Ordinary water that has been filtered and artificially carbonated is:
   a. club soda
   b. mineral water
   c. seltzer
   d. spring water

Activities and Applications

1. Supermarket Sleuth

Go to a supermarket and examine the selection of bottled waters and de-alcoholized beers and wines. Read the labels carefully.

2. Taste Testing: De-alcoholized Beer and Wine

Have a taste-testing of de-alcoholized beers and wines. Examine the de-alcoholized beers in terms of how close they taste to real beer, calories per bottle, appearance, foam, and flavor. Examine the de-alcoholized wines in terms of how close they taste to real wine, calories per serving, sweetness or dryness, and appearance.

3. Taste Testing: Bottled Waters

Have a taste-testing of bottled waters, including at least two from each of these categories: mineral water, spring water, seltzer, and club soda.

4. Develop Light Beverage and Food Options

You are in charge of developing five food and five beverage selections to be featured in a special Friday-night promotion using a workout theme. Customers will be asked to come in their workout clothes, and more healthful food and beverage selections will be featured.

Nutrition Web Explorer

International Bottled Water   www.bottledwater.org

When you visit this website, click on "Bottled Water Facts" and then find out how to store bottled water in "Frequently Asked Questions."

Part Three

Nutrition's Relationship to Health and Lifespan

Chapter 11
Nutrition and Health

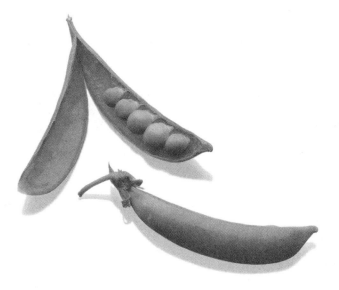

The two leading causes of death in the United States are cardiovascular disease (including coronary heart disease, strokes, and high blood pressure) and cancer. Since 1900, cardiovascular disease has been the number-one killer in the U.S. every year except one (1918). In 1997, there were almost one million deaths from cardiovascular disease, which accounted for 41 percent of all deaths, and about half a million deaths from cancer. In addition to cardiovascular disease and cancer, more and more people are being diagnosed with diabetes.

These diseases all have two things in common: they have great financial and emotional costs, and their prevention and treatment have a dietary component. This chapter will look at coronary heart disease, stroke, high blood pressure, cancer, and diabetes, as well as the healthfulness of the vegetarian diet. This chapter will help you to:

■ list three risk factors for cardiovascular disease
■ list three common forms of cardiovascular disease
■ explain how diet can play a role in the prevention and treatment of cardiovascular disease and cancer
■ distinguish between the Step I and Step II diets and explain when they are used
■ list five menu-planning guidelines to lower cardiovascular risk
■ list five lifestyle modifications for hypertension control
■ define cancer
■ list five menu-planning guidelines to lower cancer risk
■ distinguish between Type 1 and Type 2 diabetes mellitus
■ discuss three principles of planning meals for people with diabetes
■ name the major types of vegetarian eating styles
■ state the health benefits of the vegetarian diet
■ list five menu-planning guidelines for vegetarians

Nutrition and Cardiovascular Disease

Cardiovascular disease (CVD) is a general term for diseases of the heart and blood vessels, as seen in the following:

■ coronary artery disease
■ stroke
■ high blood pressure
■ rheumatic heart disease
■ congenital heart defects

This section will discuss the first three diseases.

Smoking, high blood pressure, and high blood cholesterol are three major **risk factors** for cardiovascular disease. A risk factor is a habit, trait, or condition associated with an increased chance of developing a disease. Preventing or controlling risk factors generally reduces the probability of illness. These three risk factors are modifiable to some extent. Other risk factors are obesity, age (CVD increases with age), a family history of premature CVD (heart attack in a father before age 55, or before 65 in a mother), and diabetes.

The two medical conditions that lead to most cardiovascular disease are atherosclerosis and high blood pressure. Let's take a look at atherosclerosis first. **Atherosclerosis,** a condition characterized by plaque buildup along the artery walls, is the most common form of artery disease. (*Arteriosclerosis* is a general medical term that includes all diseases of the arteries involving hardening and blocking of the blood vessels.) Atherosclerosis affects primarily the larger arteries of the body. In this condition, arterial linings become thickened and irregular with deposits called **plaque.** Plaque contains cholesterol, fat, fibrous scar tissue, calcium, and other biological debris. Why plaque deposits are formed and what role fat and cholesterol play in its formation are questions with only partial answers.

Atherosclerosis develops by a process that is totally silent. At birth the blood vessels are clear and smooth. As time goes on, plaque builds up, resulting in narrower passages and less elasticity in the vessel wall, both of which contribute to high blood pressure (Figure 11-1). What's even more dangerous is when the plaque closes off blood flow completely, or the rough

Plaque—Deposits on arterial walls that contain cholesterol, fat, fibrous scar tissue, and other biological debris.

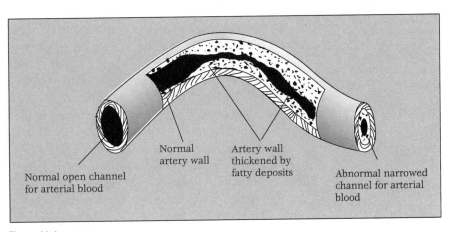

Figure 11-1

Cross-sectional representation of a coronary artery partially closed with plaque

plaque surface provides a place for a blood clot to form and block blood passage in a partially closed artery. If the artery takes blood to the brain, then a stroke occurs. If the closed artery is in the heart, then a heart attack occurs, the next topic.

Coronary Heart Disease

The heart is like a pump, squeezing and forcing blood throughout the body. Like all muscles in the body, the heart must have oxygen and nutrients in order to do its work. The heart cannot use oxygen and nutrients directly from the blood within the chambers of the heart. Instead, nutrients and oxygen are furnished by three main blood vessels outside the heart, referred to as coronary arteries. **Coronary heart disease (CHD)** is a broad term used to describe damage to or malfunction of the heart caused by narrowing or blockage of the coronary arteries.

More than two-thirds of a coronary artery may be filled with fatty deposits without causing symptoms. Symptoms may manifest themselves as chest pain, as in **angina,** or as a heart attack. Angina refers to the symptoms of pressing, intense pain in the area of the heart when the heart muscle is not getting enough blood. Sometimes stress or exertion can cause angina.

Most heart attacks are caused by a clot in a coronary artery at the site of narrowing and hardening that stops the flow of blood. Clots normally form and dissolve in response to injuries in the blood vessels, but in atherosclerosis blood clots appear to form in response to plaque when they are not needed. If an area of the heart is supplied by more than one vessel, the heart muscle may live for a period of time even if one vessel becomes blocked. The extent of heart muscle damage after a heart attack depends on which vessel is blocked, whether it is big or small, and on the remaining blood supply to that area. When the heart muscle does not get adequate oxygen and nutrients, it may die—this is what a heart attack actually is. A further area of heart muscle may be deprived of blood flow and oxygen to a lesser degree, causing a temporary injury called **myocardial ischemia.** This dead or injured heart muscle causes the heart to lose some of its effectiveness as a pump, because reduced muscle contraction means reduced blood flow.

Coronary heart disease is the number-one killer of women. Every year, just as many women as men die from coronary heart disease. Whereas most men's heart attacks are experienced at 40 years of age and older, women do not usually experience heart attacks until after menopause.

Coronary heart disease—Damage to or malfunction of the heart caused by narrowing or blackage of the coronary arteries.

Angina—Symptoms of pressing, intense pain in the heart area caused by insufficient flow blood to the heart muscle.

Myocardial ischemia—A temporary injury to heart cells caused by a lack of blood flow and oxygen.

■ MINI-SUMMARY

Cardiovascular disease includes diseases of the heart such as coronary artery disease. Major risk factors include smoking, high blood pressure, and high blood cholesterol. Atherosclerosis is characterized by plaque buildup on arterial walls, which can eventually close off blood circulation.

Nutrition and Coronary Heart Disease

Much research over the past 40 years has shown that a diet high in saturated fat, trans-fatty acids, and cholesterol contributes to high blood cholesterol. Cholesterol is carried in the blood in the form of substances called *lipoproteins*. CHD risk can be assessed by measuring total blood cholesterol as well as the proportions of the various types of lipoproteins. Total cholesterol refers to the overall level of cholesterol in the blood. High-density lipoprotein (HDL) is often referred to as "good" cholesterol, beause high levels of HDL are associated with lowered CHD risk. High levels of low-density lipoprotein (LDL)—often referred to as "bad" cholesterol—increase CHD risk.

The National Cholesterol Education Program (NCEP) recommends that all adults age 20 and older have their total cholesterol and HDL measured at least once every five years. For people without CHD, a total blood cholesterol level of less than 200 mg/dL is considered desirable; from 200 to 239 is borderline high; and 240 or more is high. An HDL level of less than 35 mg/dL is defined as low, and is considered a CHD risk factor. Lipoprotein analysis, which measures LDL as well as HDL, is recommended for people with CHD or for those at very high risk of developing CHD.

Dietary therapy is the mainstay of treatment of high blood cholesterol at every age. Unless a young adult (a man under age 35 or a premenopausal woman) is at very high risk of CHD, with a total cholesterol level of more than 300 mg/dL, the NCEP recommends that drug therapy be delayed and dietary modification and lifestyle change be attempted first.

Dietary therapy is prescribed in two steps, called the Step I and Step II Diets. These are designed to help reduce saturated fat and cholesterol intake, and to help achieve a desirable weight by eliminating excess calories. The Step I Diet is usually the starting point of dietary therapy. In the Step I Diet, no more than 8 to 10 percent of calories are in the form of saturated fat, 30 percent or less of calories come from total fat, and less than 300 milligrams of cholesterol are allowed each day.

If the patient is already following the Step I Diet and cholesterol continues to be above desirable levels, then the Step II Diet should be instituted, according to the NCEP report. People with high cholesterol who have CHD or other atherosclerotic disease should begin this diet immediately, with physician guidance. The Step II Diet calls for reducing daily saturated fat intake to less than 7 percent of calories, and cholesterol to less than 200 mg. Drug treatment is considered appropriate for adults who have a very high LDL level, especially if they also have other CHD risk factors. The goals of drug therapy are the same as those of dietary therapy: to lower LDL cholesterol to below 160 mg/dL, or 130 mg/dL if two other risk factors are present.

> **■ MINI-SUMMARY**
>
> A diet high in saturated fat and trans-fatty acids contributes to high blood cholesterol. The National Cholesterol Education Program recommends that all adults age 20 and older have their total cholesterol and HDL measured at least once every five years. For people without CHD, a total blood cholesterol level of less than 200 mg/dL is considered desirable; from 200 to 239 is borderline high; and 240 or more is high. An HDL level of less than 35 mg/dL is defined as low and is considered a CHD risk factor. An LDL level of 160 mg/dL is desirable. On the Step I Diet, no more than 8 to 10 percent of calories from saturated fat, 30 percent or less of calories from total fat, and less than 300 mg of cholesterol are allowed. The Step II Diet calls for reducing saturated fat to less than 7 percent of calories, and cholesterol to less than 200 mg.

Stroke

Stroke—Damage to brain cells resulting from an interruption of blood flow to the brain.

Cerebral hemorrhage—A stroke due to a ruptured brain artery.

A **stroke** is damage to brain cells resulting from an interruption of the blood flow to the brain. The brain must have a continual supply of blood rich in oxygen and nutrients for energy. Although the brain constitutes only 2 percent of the body's weight, it uses about 25 percent of the oxygen and almost 75 percent of the glucose circulating in the blood. Unlike other organs, the brain cannot store energy. If deprived of blood for more than a few minutes, brain cells die from energy loss and from certain chemical interactions that are set in motion. The functions these cells control—speech, muscle movement, comprehension—die with them. Dead brain cells cannot be revived.

The majority of strokes are caused by blockages in the arteries that supply blood to the brain. These blockages may be caused by a clot that forms on the inner lining of a brain or neck artery already partly clogged by plaque. The most serious kinds of stroke occur not from blockage but from hemorrhage, when a spot in a brain artery weakened by disease—

usually atherosclerosis or high blood pressure—ruptures or begins to leak blood. If an artery inside the brain ruptures, it is called a **cerebral hemorrhage.** Hemorrhagic strokes account for less than 20 percent of all types of strokes but are far more lethal, with a death rate of over 50 percent. Strokes caused by clots or hemorrhage usually strike suddenly, with little or no warning, and do all their damage in a matter of seconds or minutes.

Because blood clots play a major role in causing strokes, drugs that inhibit blood coagulation may prevent clot formation. Physicians have several drugs at their disposal—including aspirin—to treat those at risk. Aspirin works by preventing blood platelets from sticking together. Controlling blood pressure is also important.

Most people who have had mild strokes, and about half of those who have had moderate or severe paralysis on one side, recover enough to walk out of the hospital under their own steam or with some mechanical aid and resume their lives, though with certain limitations. Others are not so lucky.

Five treatable risk factors associated with stroke include high blood pressure, cigarette smoking, heart disease, history of stroke, and diabetes. High blood pressure, our next topic, is by far the most important risk factor.

> ■ **MINI-SUMMARY**
>
> The majority of strokes are caused by blockages in the arteries that supply blood to the brain. Another type of stroke is called a hemorrhagic stroke, or cerebral hemorrhage.

High Blood Pressure

As many as 50 million Americans have high blood pressure (also called **hypertension**) or are taking antihypertensive medications. Because high blood pressure usually doesn't give early warning signs, it is known as the "silent killer." High blood pressure is one of the major risk factors for coronary heart disease and stroke. All stages of hypertension are associated with increased risk of nonfatal and fatal cerebrovascular disease and renal disease.

Hypertension—High blood pressure.

Arterial blood pressure—The pressure of blood within arteries as it is pumped through the body by the heart.

Arterial blood pressure is the pressure of blood within arteries as it's pumped through the body by the heart. Whether your blood pressure is high, low, or normal depends mainly on several factors: the output from your heart, the resistance to blood flow by your blood vessels, the volume of your blood, and blood distribution to the various organs.

Everyone experiences hourly and even moment-by-moment blood pressure changes. For example, your blood pressure will temporarily rise with strong emotions such as anger and frustration, with water retention caused by too much salty food that day, and with heavy exertion, which makes your heart beat harder and faster, increasing its output by pushing more blood into your arteries. These transient elevations in blood pressure usually don't indicate disease or abnormality.

Blood pressure is represented as a fraction, as in 120/80. The top number, 120, is called the **systolic pressure**—the pressure of blood within arteries when the heart is pumping. The bottom number, 80, is called the **diastolic pressure**—the pressure in the arteries when the heart is resting between beats. Both blood pressure numbers are measured in millimeters of mercury, abbreviated mm Hg.

Normal blood pressure varies from person to person. High blood pressure occurs when the blood pressure stays too high, and is defined as systolic pressure greater than 140 mm Hg and/or a diastolic pressure greater than or equal to 90 mm Hg. Table 11-1 classifies blood pressure readings for adults. Even though hypertension starts with a systolic reading of 140 or higher, and/or a diastolic reading of 90 or higher, optimal blood pressure is about 120/80. People with high normal pressure account for more than one-third of preventable deaths related to blood pressure.

When persistently elevated blood pressure is due to a medical problem, such as hormonal abnormality or an inherited narrowing of the aorta (the

Systolic pressure—The pressure of blood within arteries when the heart is pumping—the top blood-pressure number.

Diastolic pressure—The pressure in the arteries when the heart is resting between beats—the bottom number in blood pressure.

TABLE 11-1   Classification of Blood Pressure for Adults Age 18 and Older*

Category	Systolic mm Hg	Diastolic mm Hg
Normal	<130	<85
High normal	130–139	85–89
Hypertension**		
STAGE 1 (Mild)	140–159	90–99
STAGE 2 (Moderate)	160–179	100–109
STAGE 3 (Severe)	180–209	110–119

* Not taking antihypertensive drugs and not acutely ill. When systolic and diastolic pressure fall into different categories, the higher category should be selected to classify the individual's blood pressure status.

** Based on the average of two or more readings taken at each of two or more visits following an initial screening.

Source: National Institutes of Health and National Heart, Lung, and Blood Institute.

largest artery leading from the heart), it is called **secondary hypertension.** That is, the high blood pressure arises secondary to another condition. Only 5 percent of individuals with hypertension have secondary hypertension. The remaining 95 percent have what is called **primary,** or **essential, hypertension.** The cause of essential hypertension is unknown.

The prevalence of high blood pressure increases with age, is greater for blacks than for whites, and in both races is greater in less educated than in more educated people. It is especially prevalent and devastating in lower socioeconomic groups. In young adulthood and early middle age, high blood pressure prevalence is greater for men than for women; thereafter, the reverse is true.

The following lifestyle modifications offer some hope for prevention of hypertension and are effective in lowering the blood pressure of many people who follow them (Table 11-2):

- weight reduction
- increased physical activity
- adequate dietary intake of potassium, calcium, magnesium, and vitamin C
- moderation of alcohol intake
- moderation of dietary sodium (for some individuals)

These lifestyle modifications can also reduce other risk factors for premature cardiovascular disease. Their ability to reduce death and disease in those with elevated blood pressure has not been conclusively documented. However, because of their ability to improve the cardiovascular risk profile, they offer many benefits at little cost and with minimal risk.

TABLE 11-2 Lifestyle Modifications for Hypertension Control and/or Overall Cardiovascular Risk

- Lose weight if overweight.
- Limit alcohol intake to no more than 1 ounce of ethanol per day (24 ounces of beer, 8 ounces of wine, or 2 ounces of 100-proof whiskey).
- Exercise (aerobic) regularly.
- Reduce sodium intake to less than 2.3 grams per day.
- Maintain adequate dietary potassium, calcium, magnesium, and vitamin C intake.
- Stop smoking, and reduce dietary saturated fat and cholesterol intake for overall cardiovascular health. Reducing fat intake also helps reduce caloric intake—important for control of weight and Type II diabetes.

Source: National Institutes of Health and National Heart, Lung, and Blood Institutes.

Excess body weight is correlated closely with increased blood pressure. Weight reduction reduces blood pressure in a large proportion of hypertensive individuals who are more than 10 percent above ideal weight.

Excessive alcohol intake can raise blood pressure and cause resistance to antihypertensive therapy. Hypertensive patients who drink alcohol-containing beverages should be counseled to limit their daily intake to 1 ounce of ethanol (2 ounces of 100-proof whiskey), 8 ounces of wine, or 24 ounces of beer.

Regular aerobic physical activity, adequate to achieve at least a moderate level of physical fitness, may be beneficial for both prevention and treatment of hypertension. Sedentary and unfit individuals with normal blood pressure have a 20 to 50 percent increased risk of developing hypertension during follow-up when compared with their more active and fit peers.

Individuals vary in their blood pressure response to changes in dietary sodium chloride. Blacks, older people, and patients with hypertension are more sensitive to changes in dietary sodium chloride. Because the average American consumption of sodium is in excess of 3 grams of sodium per day, a level of less than 2.4 grams of sodium per day is recommended.

Calcium, potassium, magnesium, and vitamin C also need to be in the diet in adequate supply to normalize blood pressure. In fact, diet called the DASH diet (Table 11-3, Dietary Approaches to Stop Hypertension) recommends 8 to 10 servings of fruits and vegetables (great sources of potassium, vitamin C, and magnesium) and 2–3 servings of dairy products daily.

When lifestyle modifications do not succeed in lowering blood pressure enough, drugs are the next step. Reducing blood pressure with drugs clearly decreases the incidence of cardiovascular death and disease.

■ MINI-SUMMARY

High blood pressure occurs when the blood pressure stays too high and is defined as systolic pressure greater than 140 mm Hg, and/or a diastolic pressure greater than or equal to 90 mm Hg. Table 11-2 lists the components of lifestyle modifications that can lower blood pressure. Table 11-3 shows the DASH diet, which has been shown to reduce blood pressure. If they do not bring blood pressure down enough, drug treatment is the next choice.

Menu Planning for Cardiovascular Diseases

Menu planning for cardiovascular diseases revolves around offering dishes rich in complex carbohydrates and fiber and using small amounts of fat, saturated fat, cholesterol, and sodium. Too little folate, omega-3 fatty acids, and antioxidants such as vitamin E may very well leave the heart unprotected.

TABLE 11-3 Dietary Approaches to Stop Hypertension (DASH) Diet		
This meal plan is based on 2,000 calories a day. Depending on your calorie needs, your number of daily servings may vary from those listed. Consult your doctor or a dietitian to determine your calorie needs.		
Food Group	Daily Servings	Serving Size
Grains and grain products	7 to 8	1 slice bread 2 to 1-1/4 cup dry cereal 2 cup cooked rice, pasta, or cereal
Vegetables	4 to 5	1 cup raw leafy vegetables 2 cup cooked vegetable 6 oz vegetable juice
Fruits	4 to 5	6 oz fruit juice 1 medium fruit 1/4 cup dried fruit 2 cup fresh, frozen, or canned fruit
Low-fat or nonfat dairy foods	2 to 3	8 oz milk 1 cup yogurt 1-1/2 oz cheese
Meats, poultry, fish	2 or fewer	3 oz cooked lean meat, poultry (skinless white meat), or fish
Nuts, seeds, and dry beans	4 to 5 per week	1/3 cup nuts 2 Tbsp seeds 1/2 cup legumes
Fats and oils	2 to 3	1 tsp soft margarine or butter 1 tsp regular mayonnaise *or* 1 Tbsp low-fat mayonnaise 1 Tbsp salad dressing *or* 2 Tbsp "light" salad dressing 1 tsp oil (olive, corn, canola, safflower, or other)
Sweets	5 per week	1 Tbsp maple syrup, sugar or jelly 1/2 cup sherbet 3 pieces of hard candy

Source: National Heart, Lung, and Blood Institute.

General Recommendations

■ Decrease or replace salt in recipes by using vegetables, herbs, spices, and flavorings.

■ Offer salt-free seasoning blends and lemon wedges.

Breakfast

■ Offer fresh and canned fruits and juices.

■ Almost all cold and hot cereals are great choices. Granola cereals tend to be high in fat unless labeled as reduced fat.

■ Most breads are low in fat except for croissants, brioche, cheese breads, and many biscuits. Bagels, low-fat muffins, and baguettes are good choices.

■ Have reduced-fat margarine and light cream cheese available to spread on bagels or toast.

■ Serve chicken filets, poultry sausages, low-fat ham slices, or fish as leaner sources of protein than the traditional bacon and pork sausage.

■ Offer egg substitutes for scrambled eggs and other egg-based items. Egg substitutes taste better when herbs, flavorings, and/or vegetables are cooked with them. Instead of egg substitutes, you can offer to make scrambled eggs and omelets by mixing one whole egg to two egg whites.

■ Serve an omelet with blanched vegetables such as chopped broccoli or spinach and low-fat cheese instead of regular cheese.

■ As spreads or toppings on pancakes and waffles, offer sauces combining low-fat or nonfat yogurt with a fruit puree.

■ Serve a breakfast buffet with loads of fruits, low-fat dairy products, and cereals.

Appetizers and Soups

■ Offer juices and fresh sliced fruits. Fresh sliced fruits can be served with a yogurt dressing flavored with fruit juices.

■ Offer raw vegetables with dips using low-fat yogurt, low-fat cottage cheese, or ricotta cheese as the base, rather than dips using sour cream, cheeses, cream, or cream cheese. Try hummus, a chickpea-based dip, or salsa made from tomatoes, onions, hot peppers, garlic, and herbs.

■ Offer grilled chicken, broiled Buffalo-style chicken wings, or steamed seafood such as shrimp.

■ Dish up baked (rather than fried) potato skins and baked corn tortillas for tortilla chips. Sprinkle with grated cheese, or garlic, onion, or chili powder.

■ Feature soups that use stock as the base and vegetables and grains as the ingredients. Dried beans, peas, and lentils make great soups when cooked and puréed, without using cream or high-fat thickeners such as roux.

Salads

■ Offer salads with lots of vegetables and fruits.

■ Use only small amounts, if any, of bacon, meat, cheese, eggs, or croutons. Choose cooked beans and peas or low-fat cheeses.

■ Offer reduced-calorie or nonfat salad dressings. Place on the side when desired.

■ Make tuna fish salad and similar salads with low-fat mayonnaise.

■ Use cooked salad dressing that contains little fat for Waldorf and other salads. It has a tarter flavor than mayonnaise.

Breads

■ Most breads are low in fat. Breads with more fat include biscuits, cheese breads, croissants, popovers, brioche, corn bread, and many commercial crackers (although low-fat varieties are available or can be made with less fat).

Entrées

■ Serve combination dishes with small amounts of meat, poultry, or seafood with whole grains such as rice, legumes, vegetables, and/or fruits.

■ Offer moderate portions of broiled, baked, stir-fried, or poached seafood, white-meat poultry without skin, and lean cuts of meat (see Chapter 4).

■ Offer fresh meat, poultry, or seafood instead of canned, cured, smoked, or salty items (such as ham, corned beef, smoked turkey, dried cod, and most luncheon meats). Cheeses are high in sodium, so use only small amounts in sandwiches.

■ Feature freshly made entrées instead of processed or prepared foods.

■ Offer a hamburger, meat loaf, or other ground-beef dishes made with low-fat ground beef.

■ Feature one or more meatless entrées, such as vegetarian burgers. Vegetarian burgers either try to imitate beef burgers (usually through the use of soy products) or are real veggie burgers (made of vegetables, especially mushrooms).

■ For sauced entrées (or side dishes), feature sauces thickened with flour, cornstarch, or vegetable purées. Salsas, chutneys, relishes, and coulis also work well.

■ Offer sandwiches made with roasted turkey, chicken, water-pack tuna fish salad, lean roast beef made from the round, or a spicy bean or lentil spread.

■ For sandwich spreads, use reduced-calorie mayonnaise, French or Russian-style salad dressing, mustard, ketchup, barbecue sauce, or salsa.

■ Feature lots of different vegetables in sandwiches.
■ Instead of high-sodium accompaniments to sandwiches such as pickles, olives, and potato chips, serve fresh vegetables, cole slaw made with reduced-calorie mayonnaise, or another healthful salad.

Side Dishes

■ Most side dishes of vegetables, grains, and pasta are good choices as long as little fat is added during preparation.
■ Serve grilled potato halves instead of French fries, as well as other grilled vegetables.

Desserts

■ Offer fruit-based desserts such as apple cobbler.
■ Spotlight sorbets, sherbets, frozen yogurt, and ice milk. All contain less fat than ice cream does.
■ Feature desserts made from fat-free egg whites, such as angel food cake and meringues. Serve with a fruit sauce.
■ Offer puddings made with skim milk.
■ Serve low-fat cookies such as ladyfingers, biscotti, gingerbread, and fruit bars.

Beverages

■ Offer 1% or skim milk.

> ■ **MINI-SUMMARY**
>
> Menu planning for cardiovascular diseases revolves around offering dishes rich in complex carbohydrates and fiber and using small amounts of fat, saturated fat, cholesterol, and sodium.

Nutrition and Cancer

Following heart disease, **cancer** is the second leading cause of death in the United States. The good news about cancer is that people being treated for cancer are living longer. The bad news is that more of it is being diagnosed; however, there is no evidence of a cancer epidemic. The most prevalent form of cancer is skin cancer, which is quite curable when treated early.

Cancer—A group of diseases characterized by unrestrained cell division and growth that can disrupt the normal functioning of an organ and also spread beyond the tissue in which it started.
Carcinogen—Cancer-causing substance.
Promoters—Substances such as fat that advance the development of mutated cells into a tumor.
Metastasis—The condition when a cancer spreads beyond the tissue in which it started.

More than 90 percent of skin cancers are completely cured. The most frequent malignant cancer found in women is breast cancer.

Cancer is a group of diseases characterized by unrestrained cell division and growth that can disrupt the normal functioning of an organ and spread beyond the tissue in which it started. Figure 11-2 is a diagram of this process. Cancer is basically a two-step process. First, a **carcinogen,** such as an X ray, initiates the sequence by altering the genetic material of a cell, the deoxyribonucleic acid (DNA), and causing a mutation. Such cells are generally repaired or replaced. When repair or replacement does not occur, however, **promoters** such as alcohol can advance the development of the mutated cell into a tumor. Promoters do not initiate cancer but enhance its development once initiation has occurred. The tumor may disrupt normal body functions and leave the tissue for other sites, a process called **metastasis.**

Cancer develops as a result of interactions between environmental factors (such as diet, smoking, alcohol, and radiation) and genetic factors. Research suggests that diet plays a role in the cause of certain cancers. The American Cancer Society estimates that about 33 percent of all cancer deaths are associated with the typical American diet, which is high in fat, red meat, and calories, and low in fiber, fruits, and vegetables. The most common cancers are cancer of the lungs, breast, colon and rectum, prostate, bladder, and skin. Of these cancers, all but one (skin cancer) is associated with diet.

Figure 11-2
Process of cancer.

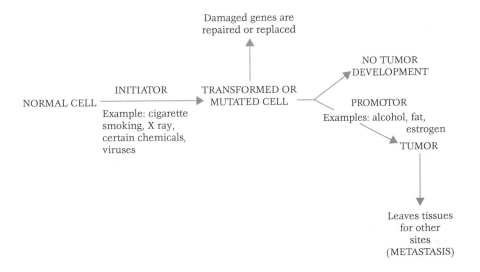

The American Cancer Society has four guidelines to reduce cancer/risk.

1. Choose most of the foods you eat from plant sources.
   ■ Eat five or more servings of fruits and vegetables each day.
   ■ Eat other foods from plant sources, such as breads, cereals, grain products, rice, pasta, or beans several times each day.
2. Limit your intake of high-fat foods, particularly from animal sources.
   ■ Choose foods low in fat.
   ■ Limit consumption of meats, especially high-fat meats.
3. Be physically active: achieve and maintain a healthy weight.
   ■ Be at least moderately active for 30 minutes or more on most days of the week.
   ■ Stay within your healthy weight range.
4. Limit consumption of alcoholic beverages, if you drink at all.

Fruits and vegetables help reduce the risk of cancer because they are rich sources of carotenoids, vitamin C, and fiber. Beta-carotene, the most abundant carotenoid, may inhibit the initiation and promotion of cancers because it is a powerful antioxidant. Antioxidants combine with oxygen, so oxygen is not available to oxidize, or destroy, important substances. Antioxidants prevent the oxidation of unsaturated fatty acids in the cell membrane, DNA (the genetic code), and other cell parts that substances called free radicals try to destroy. In the absence of antioxidants, free radicals may destroy cells (possibly accelerating the aging process) and alter DNA (possibly increasing the risk for cancerous cells to develop). Free radicals are produced in the body through normal metabolism or as a result of exposure to cigarette smoke, radiation, ultraviolet light, air pollutants, some pesticides, or alcohol.

Vitamin C also acts as an antioxidant by preventing the oxidation of certain chemicals, such as nitrosamines, to active carcinogens. Nitrates and nitrites are used in the curing of cold cuts, frankfurters, bacon, and other cured meats. Nitrates can be converted to nitrites by bacteria in the mouth or gastrointestinal tract. Nitrites can then be converted into nitrosamines in the mouth, stomach, and colon. Vitamin C inhibits nitrosamine formation in the gastrointestinal tract.

Phytochemicals—Minute substances in plants that may reduce risk of cancer and heart disease when eaten often.

Cruciferous vegetables—Members of the cabbage family that contain phytochemicals that might help prevent cancer.

Some vegetables, as well as other plant foods (such as fruits, grains, herbs, and spices), contain **phytochemicals,** minute plant compounds that fight cancer formation. For instance, broccoli contains the chemical sulforaphane, which seems to initiate increased production of cancer-fighting enzymes in the cells. Isoflavonoids, found mostly in soy foods, are known as plant estrogens or phytoestrogens because they are similar to estrogen and interfere with its actions (estrogen seems to promote breast tumors). Members of the cabbage family (cabbage, broccoli, cauliflower, mustard greens, kale), also called **cruciferous vegetables,** contain phytochemicals such as indoles and dithiolthiones. They activate enzymes that destroy carcinogens.

Some consumers are concerned about eating more fruits and vegetables that may contain carcinogenic pesticides. The National Academy of Sciences, along with other organizations, feels that the health benefits of eating fresh fruits and vegetables far outweigh any risk associated with pesticide residues. The federal government strictly regulates the kinds and amounts of pesticides used on field crops. The tiny amounts of pesticide residues found on produce are set hundreds of times lower than the amounts that would actually pose any health threat.

■ **MINI-SUMMARY**

Cancer begins as depicted in Figure 11-2. The components of American diet that raise our cancer risk include too much fat, too much red meat, too little fiber, and too few fruits and vegetables. Fiber, fruits, and vegetables lower cancer risk. Beta-carotene and vitamin C in fruits and vegetables act as antioxidants. Some fruits and vegetables also contain phytochemicals thought to fight the formation of cancer.

Menu Planning to Lower Cancer Risk

In the long run, prevention of cancer, and therefore eating a healthy diet, will play the major role in its control. Use the following guidelines to plan menus to lower cancer risk.

1. Offer lower-fat menu items. See Chapter 4 for tips on lowering fat and saturated fat. Also offer more plant-based menu items.
2. Avoid salt-cured, smoked, and nitrite-cured foods. These foods, which are also high in fat, include anchovies, bacon, corned beef, dried chipped beef, herring, pastrami, processed lunch meats such as bologna and hot dogs, sausage such as salami and pepperoni, and smoked meats and cheeses. Conventionally smoked meats and fish contain tars that are thought to be carcinogenic due to the smoking process. Nitrites are known carcinogens.
3. Offer high-fiber foods. For example:
   ■ Use beans and peas as the basis for entrées, and add them to soups, stews, casseroles, and salads. Nuts and seeds are high in fiber but also contain a significant amount of fat and calories, so use them sparingly.
   ■ Serve wholegrain breads, rolls, crackers, cereals, and muffins. Bran or wheat germ can be added to some baked goods to increase the fiber content.
   ■ High-fiber grains such as brown rice and bulgur (cracked wheat) can be used as side dishes instead of white rice.

■ Leave skins on potatoes, fruits, and vegetables as much as possible.

■ Offer salads using lots of fresh fruits and vegetables. Omit shredded cheese, chopped eggs, and bacon bits, all of which contribute fat.

4. Include lots of vegetables, especially cruciferous vegetables. Cruciferous vegetables contain substances that are natural anti-carcinogens. These vegetables include broccoli, brussels sprouts, cabbage, cauliflower, bok choy, kale, collards, kohlrabi, mustard, rutabagas, spinach, and watercress.

5. Offer foods that are good sources of beta-carotene and vitamins C and E. Excellent sources of beta-carotene include dark green, yellow, and orange vegetables and fruits such as broccoli, cantaloupe, carrots, spinach, squash, and sweet potatoes. Good sources include apricots, beet greens, brussels sprouts, cabbage, nectarines, peaches, tomatoes, and watermelon. Excellent sources of vitamin C include citrus fruits and juices, any other juices with vitamin C added, strawberries, tomatoes, and broccoli. Good vitamin C sources include berries, brussels sprouts, cabbage, melons, cauliflower, and potatoes. Vitamin E is found in vegetable oils and margarines, wholegrain cereals, wheat germ, soybeans, leafy greens, and spinach.

6. Offer alternatives to alcoholic drinks. Heavy drinkers are more likely to develop cancer in the gastrointestinal tract, such as cancer of the esophagus and stomach. Chapter 10 presents nonalcoholic alternatives.

■ **MINI-SUMMARY**

The keys to menu planning are lower-fat foods; foods that have not been salt-cured, nitrite-cured, or smoked; high-fiber foods; lots of fruits and vegetables including cruciferous vegetables; and nonalcoholic drinks for heavy drinkers.

Nutrition and Diabetes Mellitus

Diabetes mellitus—A disorder of carbohydrate metabolism characterized by high blood sugar levels and inadequate or ineffective insulin.

Diabetes mellitus gets its name from the ancient Greek word for siphon (tube), because early physicians noted that diabetics tend to be unusually thirsty and to urinate a lot, as if a tube quickly drained out everything they drank. *Mellitus* is from the Latin version of the ancient Greek word for honey, used because doctors in centuries past diagnosed the disease by the sweet taste of the patient's urine.

Diabetes is a disease in which there is insufficient or ineffective **insulin,** a hormone that helps regulate blood sugar level. When the blood sugar is

Hyperglycemia—High levels of blood sugar.

Type 1 diabetes—A form of diabetes seen mostly in children and adolescents. These patients make no insulin and therefore require frequent injections of insulin to maintain a normal level of blood glucose.

Type 2 diabetes—A form of diabetes seen most often in overweight adults. These patients make insulin but his or her tissues aren't sensitive enough to the hormone and so use it inefficiently.

above normal, such as after eating a meal, the pancreas releases insulin. The insulin facilitates the entry of glucose into body cells to be used for energy. If there is no insulin or if the insulin is not working, sugar cannot enter the cells. Thus high blood sugar levels (called **hyperglycemia**) result and sugar spills into the urine.

About 13 million Americans are diabetic—about 6 percent of the population. The life expectancy for a diabetic is only two-thirds that of the non-diabetic, and diabetes is the fourth leading cause of death. Risk factors for the most common type of diabetes include advanced age, family history, obesity, high waist-to-hip ratio, and a high-fat/low-carbohydrate diet.

Diabetics are more vulnerable to many kinds of infections and to deterioration of the kidneys, heart, blood vessels, nerves, and vision. The National Institutes of Health estimates that more than 250,000 Americans per year die from the complications of this illness, largely because it doubles their chances of having heart attacks and strokes. In addition, diabetes is the nation's leading cause of kidney failure and adult blindness. Because of the damage diabetes can do to the blood vessels and nerves of the lower limbs, only accidents necessitate more amputations of the toes, feet, and legs.

There are two types of diabetes.

1. **Type 1 (insulin-dependent diabetes mellitus).** This form of diabetes is seen mostly in children and adolescents. These patients produce no insulin at all and therefore require frequent insulin injections to maintain a normal level of blood glucose. Fewer than 10 percent of Americans who have diabetes have Type 1.

2. **Type 2 (noninsulin-dependent diabetes mellitus).** This form of diabetes is a separate disease from Type 1. Patients are usually older and obesity is common; in fact, 80 percent of patients weigh 20 percent or more than they should. In Type 2, the person's beta cells do, in fact, produce insulin and may make too much. The problem here is that the patient's tissues aren't sensitive enough to the hormone and so use it inefficiently. Some of these patients require insulin, but many do not. Treatment is with diet, weight reduction when appropriate, exercise, and if necessary, oral hypoglycemic agents (medications taken by mouth that can stimulate the release of insulin and improve the body's sensitivity to insulin). Most diabetics fall into this category.

Although both types of diabetes are popularly called "sugar diabetes," they are not caused by eating too many sweets. High sugar levels in the blood and urine are a result of these illnesses; their exact causes are unknown. What is clear is that Type 2 runs in families far more often than

Type 1 does, and—unlike Type 1, which cannot be prevented—it can frequently be avoided by staying in shape.

Symptoms of Type 1 diabetes typically appear abruptly and include excessive, frequent urination, insatiable hunger, and unquenchable thirst. Unexplained weight loss is also common, as are blurred vision (or other vision changes), nausea and vomiting, weakness, drowsiness, and extreme fatigue.

The most immediate and life-threatening aspect of Type 1 diabetes is the formation of poisonous acids called *ketone bodies*. They occur as an end product of burning fat for energy, because glucose does not get into the cells to be burned for energy. Like glucose, ketone bodies accumulate in the blood and spill into the urine.

Symptoms of Type 2 diabetes may include any or all of those of Type 1 except that the problem with ketone bodies is rare. Symptoms are often overlooked because they tend to come on gradually and be less pronounced. Other symptoms are tingling or numbness in the lower legs, feet, or hands; skin or genital itching; and gum, skin, or bladder infections that recur and are slow to clear up. Again, many people fail to connect these symptoms with possible diabetes.

Measuring glucose levels in samples of the patient's blood is key to diagnosing both types of diabetes. This is first done early in the morning on an empty stomach. For more information, the blood test is repeated, usually on another day, before the patient has drunk a liquid containing a known amount of glucose and at intervals thereafter.

Treatment for either type seeks to do what the human body normally does naturally: maintain a proper balance between glucose and insulin. The guiding principle is that food makes the blood glucose level rise, whereas insulin and exercise make it fall. The trick is to juggle the three factors to avoid both hyperglycemia (a blood glucose level that is too high) and hypoglycemia (one that is too low). Either condition causes problems for the patient.

The cornerstone of treatment is diet, exercise, and medication. The exact nature of the diabetic diet has changed over the years from starvation diets to more liberal diets. The current diabetic diet is based on the following principles:

■ There is no one diet suitable for every diabetic. The diet needs to be individualized based on each person's type of diabetes, food preferences, culture, age, lifestyle, medication, other health concerns, education, nutrition status, medical treatment goals, and other factors.

■ The goals for meal planning are to maintain the best glucose control possible, keep blood levels of fat and cholesterol in normal ranges, maintain or get body weight within a desirable range, and meet all nutrient needs.

■ Sugar need not be avoided, because sucrose and other sugars do not impair blood sugar control any more than starchy foods do. Sugars can be incorporated into the diabetic diet as part of the total carbohydrate allowance. The total amount of carbohydrate consumed, rather than the source of carbohydrate, should be the priority. Because sugars appear in foods that usually contain a lot of calories and fat, sweets can be eaten occasionally.

■ Instead of setting rigid percentages of protein, fat, and carbohydrates, the new guidelines recommend that protein make up 10 to 20 percent of calories consumed. Saturated fat and polyunsaturated fat should be maintained at less than 10 percent each. The remaining 60 to 70 percent of calories should come from carbohydrates and monounsaturated fats. Caloric distribution depends on the individual's nutritional assessment and treatment goals.

■ **MINI-SUMMARY**

There are two classifications of diabetes: Type 1 (insulin-dependent), seen mostly in children, and Type 2 (non-insulin-dependent), seen mostly in overweight older adults. The life expectancy for a diabetic is only two-thirds that of the non-diabetic, and diabetes is the fourth leading cause of death. Diabetics are more vulnerable to many kinds of infections and to deterioration of the kidneys, heart, blood vessels, nerves, and vision. The cornerstone of treatment is diet, exercise, and medication. Diets are individualized for each person. Sugar need not be avoided but can be incorporated into the diabetic diet as part of the total carbohydyrate allowance.

Vegetarian Eating

The number of vegetarians in the United States has been increasing. Vegetarians do not eat food that requires the death of, or injury to, an animal. Instead of eating the meat entrées that have traditionally been the major source of protein in the American diet, they dine on main dishes emphasizing legumes (dried beans and peas), grains, and vegetables. Vegetarian entrées, such as red beans and rice, can supply adequate protein with less fat and cholesterol, and more fiber than their meat counterparts (see Table 11-4).

TABLE 11-4 Fat, Saturated Fat, Protein, and Cholesterol in Animal and Plant Foods

Animal Foods	Fat (grams)	Saturated Fat (grams)	Protein (grams)	Cholesterol (milligrams)
Beef, ground broiled, 3 oz.	16	6	21	74
Chicken breast, roasted, 3 oz.	3	1	27	73
Cod, baked, 3 oz.	1	0	19	47
Milk, 2%, 8 fl. oz.	5	3	8	18
Cheese, American, 1 oz.	9	6	6	27
Egg, 1	6	2	6	274

Plant Foods	Fat (grams)	Saturated Fat (grams)	Protein (grams)	Cholesterol (milligrams)
Lentils, cooked, 1/2 cup	0	0	9	0
Brown rice, cooked, 1/2 cup	1	0	2	0
Spaghetti, whole wheat, 1 cup	1	0	7	0
Whole wheat bread, 2 slices	2	0	5	0
Broccoli, chopped 1 cup	0	0	6	0
Apple, 1 medium	0	0	0	0

Sources: United States Department of Agriculture Handbooks. Numbers 8-1, 8-5, 8-13, 8-15, 8-20.

Lacto-ovovegetarians— Vegetarians who do not eat meat, poultry, or fish but do consume animal products in the form of eggs, milk, and milk products.

Lactovegetarians— Vegetarians who do not eat meat, poultry, or fish but do consume animal products in the form of milk and milk products.

Vegans— Individuals eating a type of vegetarian diet in which no eggs or dairy products are eaten; their diet relies exclusively on plant foods.

Whereas vegetarians do not eat meat, poultry, or fish, the largest group of vegetarians, referred to as **lacto-ovovegetarians,** do consume animal products in the form of eggs (ovo) and milk and milk products (lacto).

Another group of vegetarians, **lactovegetarians,** consume milk and milk products but forgo eggs. Most vegetarians are either lacto-ovovegetarians or lactovegetarians. **Vegans,** a third group of vegetarians, do not eat eggs or dairy products and therefore rely exclusively on plant foods to meet protein and other nutrient needs. Vegans are a small group, and it is estimated that only 4 percent of vegetarians are vegans.

In addition, some vegetarians (**pescovegetarians**) eat seafood. Also, some vegetarian diets restrict certain foods and beverages, such as highly processed foods containing certain additives and preservatives, foods that contain pesticides and/or have not been grown organically, or caffeinated or alcoholic beverages.

The number-one reason people give for being vegetarian is health. Being a vegetarian has health benefits. Vegetarians tend to be leaner and to keep their body weight and blood lipid levels closer to desirable levels than nonvegetarians. Vegetarians tend to have a lower incidence of the following diseases:

Pescovegetarians—

Vegetarians who will eat
fish.

1. hypertension
2. coronary artery disease
3. colon and lung cancer
4. non-insulin-dependent (Type 2) diabetes mellitus
5. diverticular disease of the colon

A comprehensive study of rural Chinese suggested that eating much less animal protein and fat (and more complex carbohydrates) results in reductions of blood cholesterol levels and the chronic degenerative diseases (such as heart disease and cancer) associated with high blood cholesterol levels.

Being vegetarian does not necessarily mean that you automatically get these benefits. There are vegetarians who eat well-balanced and varied diets, and then there are vegetarians who eat eggs for breakfast, peanut butter for lunch, and pizza for supper. In other words, it is possible to be a vegetarian and still eat too much fat, saturated fat, and cholesterol. It's probably the exception, rather than the rule, but it is still possible. Being vegetarian does not guarantee that your diet will meet current dietary recommendations. Some other reasons for becoming vegetarian include the following:

1. *Ecology.* For ecological reasons, vegetarians choose plant protein because livestock and poultry require much land, energy, water, and plant food (such as soybeans), which they consider wasteful. According to the North American Vegetarian Society, grains and soybeans that are fed to U.S. livestock could feed 1.3 billion people. Livestock product also wastes loads of water—it takes 2,500 gallons of water to produce one pound of meat, but only 25 gallons to produce one pound of wheat.

2. *Economics.* A vegetarian diet is more economical—in other words, less expensive. This can be easily demonstrated by the fact that in a typical foodservice operation, the largest component of food purchases is for meats, poultry, and fish.

3. *Ethics.* Vegetarians do not eat meat for ethical reasons; they believe that animals should not suffer or be killed unnecessarily. They feel that animals suffer real pain in crowded feed lots and cages, and that both their transportation to market and their slaughter are traumatic.

4. *Religious beliefs.* Some vegetarians, such as the Seventh Day Adventists, practice vegetarianism as a part of their religion, which also encourages exercise and forbids smoking and drinking alcohol.

Nutritional Adequacy of Vegetarian Diets

In 1995, the Dietary Guidelines addressed the topic of vegetarian diets, with assurances that they can be nutritionally adequate when varied and adequate in calories (except for vegan diets, which need supplementation with vitamin B_{12}). Most vegetarians get enough protein, and their diets are typically lower in fat, saturated fat, and cholesterol.

As discussed in Chapter 5, most plant proteins are considered incomplete proteins, but this doesn't mean they are low in quality. When plant proteins are eaten with other foods, the food combinations usually result in complete protein. For example, when peanut butter and whole-wheat bread are eaten over the course of a day, the limiting amino acid in each of these foods is supplied by the other food. Such combinations are called complementary proteins. Eating complementary proteins at different meals during the day generally assures a balance of dietary amino acids. Some vegetable proteins, such as the grains amaranth and quinoa and soybeans, are complete proteins.

Let's take a look at some of the nutrients that need some special attention.

1. *Vitamin B_{12}.* Vitamin B_{12} is found only in animal foods. Lacto-ovovegetarians usually get enough of this vitamin, unless they limit their intake of dairy products and eggs. Vegans definitely need either a supplement or vitamin B_{12}–fortified foods, such as most ready-to-eat cereals, most meat analogs, some soy beverages, and some brands of nutritional yeasts.

2. *Vitamin D.* Milk is fortified with vitamin D, and vitamin D can be made in the skin with sunlight. Generally, only vegans without enough exposure to sunlight need a supplementary source of vitamin D. Some ready-to-eat breakfast cereals and some soy beverages are fortified with vitamin D.

3. *Calcium.* Lactovegetarians and lacto-ovovegetarians generally don't have a problem here, but vegans sometimes do if they don't eat enough other calcium-rich foods. Good choices include calcium-fortified soy milk or orange juice and tofu made with calcium. Some green leafy vegetables (such as spinach, beet greens, Swiss chard, sorrel, and parsley) are rich in calcium, but they also contain a binder (called oxalic acid) that prevents some of the calcium from being absorbed. Dried beans and peas are moderate sources of calcium. Without calcium-fortified drinks or calcium supplements, it can be difficult to take in enough calcium.

4. *Iron.* Interestingly enough, vegetarians do not experience any more problems with iron-deficiency anemia than their meat-eating counter-

parts—don't forget, meat is rich in iron. Iron is widely distributed in plant foods, and its absorption is greatly enhanced by vitamin C-containing fruits and vegetables. Vegetarians get iron from eating dried beans and peas, green leafy vegetables, dried fruits, many nuts and seeds, and enriched and whole-grain products.

5. *Zinc.* Zinc is found in many plant foods, such as whole grains, legumes, and nuts and seeds (especially peanut butter). Its absorption into the body is reduced by plant substances, such as phytate. Children may need zinc supplements.

Infants, children, and adolescents can follow vegetarian diets, even vegan diets. For growing youngsters, however, vegetarian diets need to be well planned, varied, and adequate in calories. In the case of a vegan diet, special attention should be focused on getting enough calories, vitamin B_{12}, vitamin D, calcium, iron, zinc, and linolenic acid. A reliable source of vitamin B_{12} is particularly essential. Soy milk that has been fortified with calcium and vitamin D will help meet the needs for those vitamins. Meat analogs, soy products, legumes, and nut butters are valuable sources of protein, and some provide iron, zinc, and/or linolenic acid. If inadequate amounts of these nutrients are taken in from food, supplements are always available.

> ■ **MINI-SUMMARY**
>
> Reasons for becoming vegetarian may be related to health benefits, economics, ecology, ethics, or religious beliefs. Vegetarian diets can be nutritionally adequate when varied and adequate in calories, except for vegan diets, which need supplementation with vitamin B_{12}. Nutrients of special interest to vegetarians are vitamin B_{12}, vitamin D, calcium, iron, zinc, and linolenic acid.

Vegetarian Food Pyramid The Food Guide Pyramid in Chapter 2, when modified as described here, works well for vegetarians.

1. *Breads, cereals, rice, and pasta* (6–11 servings per day). This group stays the same. Whole-grain products are recommended. Vitamin B_{12}–fortified breakfast cereals are important for vegans. Some vegans with large calorie needs may eat more than 11 servings.
2. *Fruits* (2 or more servings). This group also stays the same, except that no limit is placed on consumption.
3. *Vegetables* (3 or more servings). As with fruits, no limit is placed on the number of servings.

4. *Legumes, Nuts, Seeds, Eggs, and other Meat Substitutes* (2–3 servings). This group obviously omits any meats, poultry, or fish and instead concentrates on substitutes such as cooked dry beans, peas, or lentils; tofu and other soybean products; nuts and seeds, and butters made from them; meat analogs; and eggs for lacto-ovovegetarians. One serving is 1/2 cup of cooked beans, peas, or lentils; 4 ounces of tofu; 1/4 cup of shelled nuts; 1/8 cup of seeds; 2 tablespoons of peanut butter; or 1 egg. One serving of legumes should be served daily.

5. *Milk, cheese, and yogurt* (2 to 3 servings for adults; 3 servings for pregnant and lactating women, teenagers, and young adults up to age 24). This group is enlarged to include soy milk fortified with calcium and vitamin D (and vitamin B_{12} for vegans), and soy cheese fortified with calcium and vitamin D. One serving is 1 cup of fortified soy milk or 1-1/2 ounces of soy cheese. One cup of cooked broccoli or greens can be substituted for 1 cup of milk. If a vegan does not drink soy milk or eat soy cheese, a carefully planned diet with sources of the missed nutrients, or supplements, is necessary.

Menu-Planning Guidelines

Variety is a key word to remember when planning vegetarian menu items.

1. *Use a variety of plant protein sources at each meal: legumes, grain products (preferably whole-grain), nuts and seeds, and/or vegetables.* Vegetarian entrées commonly use cereal grains such as rice or bulgur wheat (precooked and dried whole wheat) in combination with legumes and/or vegetables. Use small amounts of nuts and seeds in dishes.

2. *Use a wide variety of vegetables.* Steaming, stir-frying, or microwaving vegetables retains flavor, nutrients, and color.

3. *Offer entrées that are acceptable to lactovegetarians and lacto-ovovegetarians, and an entrée for vegans.* Although lactovegetarians and lacto-ovovegetarians will eat vegan entrees, vegans won't eat entrées with any dairy products or eggs.

4. *Choose low-fat and nonfat varieties of milk and milk products and limit the use of eggs.* This is important to prevent a high intake of saturated fat, found in whole milk, low-fat milk, regular cheeses, eggs, and other foods.

5. *Offer dishes made with soybean-based products, such as tofu and tempeh.* Soybeans are unique: They contain the only plant protein that is nutritionally equivalent to animal protein.

6. *Provide foods that contain nutrients of special importance to vegetarians.* Table 11-5 lists good sources of vitamin B_{12}, vitamin D, calcium, iron, zinc, and linolenic acid.

TABLE 11-5 Good Sources of Vitamin B$_{12}$, Vitamin D, Calcium, Iron, Zinc, and Linolenic Acid

Vitamin B$_{12}$:	Dairy products, eggs, fortified cereals, and meat analogs
Vitamin D:	Fortified milk, eggs, fortified cereals, and soy milk
Calcium:	Milk and milk products, canned salmon and sardines (with bones), oysters, calcium-fortified juice or soy milk, broccoli, collards, kale, greens
Iron:	Liver, meats, breads and cereals, green leafy vegetables, legumes, dried fruits
Zinc:	Whole grains, legumes, nuts and seeds, peanut butter
Linolenic acid:	Walnuts, walnut oil, canola oil, soybean oil, soybeans

■ **MINI-SUMMARY**

Legumes, grains, nuts, and seeds are important foods for vegetarians. Vegetarians can still follow the Food Pyramid concept by concentrating on meat substitutes in the Meat/Meat Substitute group and adding fortified soy milk and soy cheeses to the dairy group. Menu-planning guidelines are given.

Check-Out Quiz

1. The two medical conditions that lead to most cardiovascular diseases are atherosclerosis and high blood pressure.
   a. True   b. False
2. Atherosclerosis develops by a process that is totally silent.
   a. True   b. False
3. The most effective dietary method to lower the level of your blood cholesterol is to eat less cholesterol.
   a. True   b. False
4. The Step I diet is used to treat diabetes mellitus.
   a. True   b. False
5. Blood pressure increases with age.
   a. True   b. False
6. Both heart attacks and strokes are usually due to clots caught in arteries.
   a. True   b. False
7. Dietary protein probably promotes cancer.
   a. True   b. False

8. Beta-carotene, vitamin C, and vitamin D are antioxidants that may protect against cancer.
   a. True   b. False
9. Symptoms of diabetes include increased hunger and thirst.
   a. True   b. False
10. Vegetarians enjoy certain health benefits from their diets.
    a. True   b. False

Activities and Applications

1. A Diet for Disease Prevention

You have read about several diseases and the dietary means for preventing and treating each one. Using this information, write what you would consider to be an ideal diet for disease prevention (other than vegetarian). Focus on different groups of foods and what role each would play (and why) in disease prevention.

2. Vegetarian Meal Planning

With the help of vegetarian cookbooks and magazines or trade publications, suggest two vegetarian entrées and desserts for use at each of the following establishments: a casual-themed restaurant for younger people, a college cafeteria, and a dining room for an investment bank.

3. Vegetarian Meal

With your classmates, plan and carry out a vegetarian meal that uses foods you may not be familiar with. Plan the meal so there are items for both lacto-ovovegetarians and vegans.

Nutrition Web Explorer

American Heart Association   www.americanheart.org
Go to the website of the American Heart Association and click on "Interactive Risk Assessment" to find out your risk for heart disease.

American Cancer Society   www.cancer.org
On the home page for the American Cancer Society, click on "Prevention and Early Detection." Then click on "Nutrition and Prevention" and read the dietary guidelines for cancer prevention.

Center for Science in the Public Interest   www.cspinet.org
To prevent disease, you need to eat a healthy diet. Assess your diet at this website by clicking on "Rate Your Diet."

American Diabetes Association   www.diabetes.org
On the home page of the American Diabetes Association, click on "Nutrition." Then click on "Diabetes Nutrition Quiz." Take the quiz and get your score.

National Restaurant Association   www.restaurant.org
Read the article "Vegetarian Cuisine" that appeared in the National Restaurant Association's magazine, *Restaurants USA,* in January 1999 on this website.

Food Facts *Caffeine*

Check out how much you know, or don't know, about caffeine.

True or False?

1. Tea has more caffeine than coffee.
2. Brewed coffee has more caffeine than instant coffee.
3. Some nonprescription drugs contain caffeine.
4. Caffeine is a nervous-system stimulant.
5. Withdrawing from regular caffeine use causes physical symptoms.

Check your answers as you read on.

Caffeine, a stimulant, is the most widely used psychoactive substance in the world. Eighty percent of Americans drink at least one caffeine-containing beverage each day, and average caffeine consumption is about 280 milligrams. Caffeine is present in over 60 plant species in various parts of the world, such as the coffee bean in Arabia, the tea leaf in China, the kola nut in West Africa, and the cocoa bean in Mexico.

Coffee, tea, cola, and cocoa are the most common sources of caffeine in the American diet (Table 11-6), with coffee being the chief source. The caffeine content of coffee or tea depends on the variety of coffee bean or tea leaf, the particle size, the brewing method, and the length of brewing or steeping time. Brewed coffee always has more caffeine than instant coffee, and espresso coffee always

has more caffeine than brewed coffee. Espresso is made by forcing hot pressurized water through finely ground, dark-roast beans. Because it is brewed with less water, it contains more caffeine than regular coffee.

In soft drinks, caffeine is both a natural and an added ingredient. The Food and Drug Administration requires caffeine as an ingredient in colas and pepper-flavored beverages and allows it to be added to other soft drinks as well. About 5 percent of the caffeine in colas and pepper-flavored soft drinks is obtained naturally from cola nuts; the remaining 95 percent is added. Caffeine-free soft drinks contain virtually no caffeine and make up a small part of the soft-drink market.

Numerous prescription and nonprescription drugs also contain caffeine. It is often used in alertness or stay-awake tablets, headache and pain-relief remedies, cold products, and diuretics. When caffeine is an ingredient, it must be listed on the product label.

Caffeine is absorbed very well into the body and is rapidly absorbed into the bloodstream. For most people, caffeine raises blood pressure and heart rate, increases attentiveness and performance, and gives relief from fatigue. In high doses it can produce insomnia, nervousness, a racing heart, and other troublesome symptoms. Many people build a tolerance to caffeine's effects that may then lead to increased usage.

TABLE 11-6 Caffeine Content of Beverages, Foods, and Drugs

Item	Milligrams Caffeine Average	Range	Item	Milligrams Caffeine Average	Range
Coffee (5-ounce cup)			Cocoa		
Brewed, drip method	115	110–150	Cocoa beverage		
Instant	65	30–120	(5-ounce cup)	4	2–20
Decaffeinated, brewed	3	2–5	Chocolate milk beverage		
Decaffeinated, instant	2	1–5	(8 ounces)	5	2–7
Tea (5-ounce cup)			Milk chocolate (1 ounce)	6	1–15
Brewed, major U.S.			Dark chocolate,		
brands	40	20–90	semisweet (1 ounce)	20	5–35
Brewed, imported			Baker's chocolate		
brands	60	25–110	(1 ounce)	26	26
Instant	30	25–50	Chocolate-flavored syrup		
Iced (12-ounce			(1 ounce)	4	4
glass)	70	67–76			
Soft drinks (12-ounce			Prescription drugs:		
can)			(caffeine per tablet or capsule)		
Cola, pepper		30–46	Cafergot (migraine		
Decaffeinated cola,			headache)		100
pepper		0–2	Norgesic Forte (muscle		
Cherry cola		36–46	relaxant)		60
Lemon-lime		0	Fiorinal (tension headache)		40
Other citrus		0–64	Darvon (pain relief)		32
Root beer		0	Synalogos-DC (pain relief)		30
Ginger ale		0	Nonprescription drugs:		
Tonic water		0	(caffeine per tablet or capsule)		
Other regular soda		0–44	Alertness Tablets		
Juice added		0	No Doz		100
Diet cola, pepper		0.6	Pain Relief		
Decaffeinated diet			Anacin, Maximum		
cola, pepper		0–0.2	Strength Anacin		32
Diet cherry cola		0–46	Vanquish		33
Diet lemon-lime, diet			Excedrin		65
root beer		0	Midol		32
Other diets		0–70	Diuretics		
Club soda, seltzer,			Aqua-Ban		100
sparkling water		0	Cold/Allergy Remedies		
			Coryban-D capsules		30

Source: Lecos, Chris W. 1987–1988. "Caffeine jitters: Some safety questions remain." *FDA Consumer* 21(10):22–27.

It is easy to become dependent on caffeine. When caffeine is withdrawn, symptoms include headache, fatigue, irritability depression, and poor concentration. The symptoms peak on day 1 or 2 and progressively decrease over the course of a week. It has been shown that even moderate consumption—about one or two 10-ounce cups of coffee daily (or the equivalent of caffeine from other sources)—often causes people to experience these debilitating symptoms of caffeine withdrawal. To minimize withdrawal symptoms, experts recommend reducing one's caffeine intake by about 20 percent a week over four to five weeks.

Although caffeine use has stirred fears in the past (it reportedly caused pancreatic cancer and birth defects), moderate use probably confers the benefits of caffeine with few of the risks. There are, however, two groups of individuals who should abstain from caffeine. In some susceptible people with heart disease, caffeine can cause irregular heartbeat. Caffeine is also not recommended for individuals with peptic ulcers, because caffeine increases the production of stomach acid. Pregnant and lactating women should consume caffeine in moderation—less than two cups of coffee a day. Studies on caffeine and pregnancy have shown conflicting results, so moderation is recommended.

High intakes of caffeine may be linked to heart attacks and bone loss in women. Once an individual drinks more than two 10-ounce cups of coffee daily, there is added cardiac risk among both smokers and nonsmokers, and especially in people with high blood pressure. Women who consume caffeine and drink little or no milk lose more calcium in their urine and have less dense bones than do women who don't consume any caffeine. For optimum bone density, women should moderate their caffeine intake and get at least two to three servings from the milk group daily.

Hot Topic B Vitamins and Heart Disease

Back in the 1970s, a researcher in Boston thought that a derivative of an amino acid (called homocysteine) in the blood was damaging artery walls and therefore stimulating the process of atherosclerosis and causing heart disease. He also thought that certain vitamins—B_6, B_{12}, and folate—worked to maintain lower homocysteine levels in the blood. When the amino acid methionine is broken down to make other compounds in the body, it produces homocysteine. High blood levels of homocysteine have been associated with higher rates of cardiovascular deaths. Vitamin B_6 and folate appear to lower the amount of homocysteine in the blood by helping to convert it back into amino acids.

In 1998, a major study involving 80,000 nurses was the first to show a link between B_6 and folate and reduced risk of heart attacks. Nurses with higher intakes of folate and B_6 had lower rates of cardiovascular disease. The study involves nurses whose diets and health status were first evaluated in 1980 and who were periodically reevaluated for the following 14 years. Previous studies in women and men showed that folate and B_6 consumed through food and supplements reduced blood levels of homocysteine and protected against narrowing of the arteries that go into the brain. Homocysteine is believed to be involved in mechanisms that damage the arterial wall and/or foster blood clots that often start heart attacks and strokes.

The most cardiac protection in the nurses' study came from levels of folate, 400 micrograms, that are the same as the current Dietary Reference Intake (DRI) recommendation. However, the most protection came from a level of vitamin B_6, about 3 milligrams, that is double the current RDA recommendation. These levels were protective whether they came from foods or supplements.

It isn't difficult to get enough B_6 and folate from a variety of foods. Good sources of folate include orange juice, lentils, black-eyed peas, dried beans, leafy greens such as spinach, and chicken liver. Good sources of B_6 are bananas, chicken, dried beans, whole grains, tuna, and nuts. Food sources of both these vitamins are nutrient-rich foods that provide many other benefits.

Chapter 12

Weight Management and Exercise

Obesity—The condition of being 30 percent or more over one's ideal body weight.

Americans trying to lose weight have plenty of company. According to the Centers for Disease Control and Prevention, more than two-thirds of American adults are trying to lose weight or to keep from gaining weight. Tens of millions of Americans are dieting at any given time and spending more than $33 billion yearly on weight-reduction products, such as diet foods and drinks.

Why is the weight-loss industry such a giant? A growing obesity epidemic is threatening the health of millions of Americans. **Obesity** (defined as being over 30 percent above ideal body weight) in the population increased from 12 percent in 1991 to 17.9 percent in 1998, while physical inactivity did not change substantially. The 1996 Surgeon General's Report on Physical Activity and Health shows that more than 60 percent of adults are not participating in the recommended 30 minutes per day of moderate physical activity most days of the week. In addition, an important 1999 study of over 100,000 Americans found that only 21 percent of men and 19 percent of women report using the recommended combination of eating fewer calories and engaging in at least 150 minutes of leisure-time physical activity per week.

Obesity is a disease, not a moral failing, in which many factors are involved: genetic, environmental, psychological, and others. It occurs when energy intake exceeds the amount of energy expended over time. In a small number of cases obesity is caused by illnesses such as hypothyroidism or is the result of taking medications that can cause weight gain.

Because so many factors affect how much or how little food a person eats and how food is metabolized by the body, losing weight is not simple. This chapter discusses obesity's impact on health, theories about what causes obesity, and treatment to lose weight and maintain weight loss. Exercise, an important activity for obese and non-obese individuals, is also examined, along with nutrition for athletes. This chapter will help you to:

■ define obesity and overweight
■ give one advantage and one disadvantage of the three methods of measuring obesity
■ list the health implications of obesity
■ give possible causes of obesity
■ list the six components of a comprehensive weight-reduction program
■ describe seven basic concepts of nutrition education to keep in mind when planning diets
■ explain the relationship between exercise and weight loss
■ define behavior and attitude modification theory and discuss how they can be used to help someone lose weight
■ discuss five strategies that appear to support weight maintenance

- design menus for weight loss and maintenance
- give three tips for someone who is underweight
- identify the primary fuels for exercise and the most crucial nutrient for athletes
- design menus for athletes

"How Much Should I Weigh?"

Obesity can be measured using various methods. **Height-weight tables** are easy to use and understand. Table 12-1 shows suggested weights for adults from the 1995 *Dietary Guidelines for Americans*. One concern with height-weight tables is that the weights say nothing about body composition. For instance, a 250-pound linebacker at 6 feet 2 inches would be considered overweight according to the tables, yet his excess pounds are not fat, but muscle. Conversely, a person whose weight is within the appropriate range may indeed have excess fat stores.

A second method of measuring degree of obesity is **body mass index (BMI)**, a mathematical formula that correlates weight with body fat. The BMI is calculated as follows:

$$\text{BMI} = \frac{\text{body weight (in kilograms)}}{\text{height (in meters)}^2}$$

The BMI is a more sensitive indicator and a better measure of the amount of fat you have compared with the height-weight tables. Still, BMI is not

Height-weight tables— Tables that show an appropriate weight for a given height.

Body Mass Index— A method of measuring degree of obesity that is a more sensitive indicator than height-weight tables.

TABLE 12-1	**Suggested Weights for Adults**				
Height*	Weight in pounds†	Height*	Weight in pounds†	Height*	Weight in pounds*
4'10"	91–119	5'5"	114–150	6'0"	140–184
4'11"	94–124	5'6"	118–155	6'1"	144–189
5'0"	97–128	5'7"	121–160	6'2"	148–195
5'1"	101–132	5'8"	125–164	6'3"	152–200
5'2"	104–137	5'9"	129–169	6'4"	156–205
5'3"	107–141	5'10"	132–174	6'5"	160–211
5'4"	111–146	5'11"	136–179	6'6"	164–216

* Without shoes

† Without clothes

Source: Dietary Guidelines for Americans (1995).

useful for athletes and pregnant women, whose increased weight is not due to fat. On Table 12-2, your BMI is at the intersection of your height and weight.

Overweight—Having a Body Mass Index of 25 or higher.

The National Institutes of Health defines **overweight** as a BMI of 25–29.9. A BMI of 30 or greater is considered obese. These cut-off points were chosen because studies show that as BMI rises above 25, blood pressure and blood cholesterol rise, HDL levels fall, and maintaining normal blood-sugar levels becomes more difficult. Thus, individuals with a BMI of 25 or higher run a higher risk of coronary artery disease, high blood pressure, heart attacks, stroke, type 2 diabetes, and certain cancers, such as breast cancer.

Because BMI doesn't tell you how much of your excess weight is fat and where that fat is located, the National Institutes of Health has asked physicians to measure patients' waistlines. Studies show that excessive abdominal fat is more health-threatening than fat in the hips or thighs. If an individual has a BMI of 25 to 35, and a waist greater than 40 inches for men or 35 inches for women, this person is at an increased risk of developing serious health problems.

Another way to measure obesity is to examine the percentage of your body that is fat. For men, a desirable percentage of body fat is 13 to 25 percent, for women about 17 to 29 percent. When a man's body fat goes over 25 percent (or 29 percent for a woman), health risks increase. Body fat is most often measured by using special calipers to measure the skinfold thickness of the triceps and other parts of the body. Because half of all your fat is under the skin, this method is quite accurate when performed by an experienced professional. Other methods of estimating body fatness include underwater weighing and bioelectrical impedance.

> ■ **MINI-SUMMARY**
>
> Obesity can be measured using various methods. The BMI is a more sensitive indicator than height-weight tables and can be calculated easily using Table 12-2. A BMI of 25 or over is considered overweight, and health risks increase as BMI rises.

Health Implications of Obesity

Obesity is the second leading cause of preventable death in the United States (smoking is first). An obese individual is at increased risk for hypertension, high blood cholesterol levels, Type 2 diabetes, and coronary heart

TABLE 12-2 Body Mass Index

HEIGHT	100	105	110	115	120	125	130	135	140	145	150	155	160	165	170	175	180	185	190	195	200	205
											WEIGHT											
5'0"	20	21	21	22	23	24	25	26	27	28	29	30	31	32	33	34	35	36	37	38	39	40
5'1"	19	20	21	22	23	24	25	26	26	27	28	29	30	31	32	33	34	35	36	37	38	39
5'2"	18	19	20	21	22	23	24	25	26	27	27	28	29	30	31	32	33	34	35	36	37	37
5'3"	18	19	20	21	22	22	23	24	25	26	27	27	28	29	30	31	32	33	34	35	35	36
5'4"	17	18	19	20	21	22	23	23	24	25	26	27	27	28	29	30	31	32	33	33	34	35
5'5"	17	18	18	19	20	21	22	22	23	24	25	26	27	27	28	29	30	31	32	32	33	34
5'6"	16	17	18	19	19	20	21	22	23	23	24	25	26	27	27	28	29	30	31	31	32	33
5'7"	16	16	17	18	19	20	20	21	22	23	23	24	25	26	27	27	28	29	30	31	31	32
5'8"	15	16	17	17	18	19	20	21	21	22	23	23	24	25	26	27	27	28	29	30	30	31
5'9"	15	16	16	17	18	18	19	20	21	21	22	23	24	24	25	26	27	28	28	29	30	30
5'10"	14	15	16	16	17	18	19	19	20	21	22	22	23	24	24	25	26	27	27	28	29	29
5'11"	14	15	15	16	17	17	18	19	20	20	21	22	22	23	24	24	25	26	27	27	28	29
6'0"	14	14	15	16	16	17	18	18	19	20	20	21	22	22	23	24	24	25	25	26	27	28
6'1"	13	14	15	15	16	17	17	18	18	19	20	20	21	22	22	23	24	24	25	26	26	27
6'2"	13	13	14	15	15	16	17	17	18	19	19	20	21	21	22	23	23	24	24	25	26	26
6'3"	12	13	14	14	15	16	16	17	17	18	19	19	20	21	21	22	22	23	24	24	25	26
6'4"	12	13	13	14	15	15	16	16	17	18	18	19	19	20	21	21	22	23	23	24	24	25

Source: Shape Up America, National Institutes of Health.

389

disease. Obese people do have more hypertension, diabetes, and high blood cholesterol levels, as well as other cardiovascular problems.

Obese people are also at higher risk for developing certain types of cancer. Obese men are at higher risk for colon, rectal, and prostate cancer. Obese women are at higher risk for breast and uterine cancer.

Conditions aggravated by obesity include breathing problems, abdominal hernia, arthritis, varicose veins, gout, gallbladder disease, and pregnancy. In addition, obese individuals are higher surgical risks.

Obesity creates a psychological burden that, in terms of suffering, may be its greatest adverse effect. Obese people, particularly children, are ridiculed, teased, and excluded by their peers. They are also victims of discrimination. They are self-conscious about their weight and frequently blame themselves.

Weight reduction is strongly sought when excess fat is combined with any of the following conditions: non-insulin-dependent diabetes, family history of diabetes, high blood pressure, high blood lipid or cholesterol levels, coronary heart disease, or gout. Losing weight often decreases blood pressure and blood cholesterol levels, and brings diabetes under better control. Although obesity does not cause these medical conditions, losing weight can help to reduce some of their negative aspects.

■ **MINI-SUMMARY**

An obese individual is at increased risk for hypertension, high blood cholesterol levels, adult-onset diabetes, cancer (certain types), and coronary heart disease. Conditions aggravated by obesity include abdominal hernia, arthritis, varicose veins, gout, gallbladder disease, and pregnancy. In addition, obese individuals are higher surgical risks.

Theories of Obesity

As researchers try to figure out why some people get fat and others don't, it is becoming increasingly apparent that obesity has a variety of causes—heredity, environment, metabolism, and level of physical activity—and therefore no single cure.

The body has an almost limitless capacity to store fat. Not only can each fat cell balloon to more than 10 times its original size, but should the available cells get filled to the brim, new ones will propagate. As the body stores more fat, weight and girth increase.

A number of studies have shown that genetics may be the most important determinant of how much you weigh, perhaps explaining about half the problems with obesity. Some people are more prone to weight gain than others, even when caloric intake is the same. Identical twins show similar body weight, even when they have been raised apart. Researchers conclude that genetic factors, apart from diet or lifestyle, strongly influence how much a person weighs.

Obesity tends to run in families. If both parents are obese, their children have an 80 percent chance of being obese. If neither parent is obese, the likelihood of the children being obese is no more than 14 percent. This difference could reflect genetics and/or the family's tendency to teach poor eating and exercise habits.

Three interesting theories reflect the physiological school of thought concerning obesity: the fat cell, set point, and dietary obesity theories. Obese people usually have a larger than normal number of fat cells and/or enlarged fat cells. Fat cells can be added but never subtracted, so when an obese person loses weight, the cells reduce in size, not number. The fat cell theory states that weight loss will be met with internal resistance when cells are reduced below their normal size.

Set point theory is based on the supposition that the body strives to maintain a given range of weight, fat, muscle, or related factors. Therefore, obesity results from maintaining body weight and fat at a higher than normal level. An obese individual who diets encounters strong biological resistance. Many studies show a tendency for humans to keep their weight within a fairly constant weight range.

Dietary obesity refers to a tendency to prefer calorically dense foods. Most laboratory animals fed standard foods do not overeat even when food is abundant. When these animals were offered supermarket diets of chocolate chip cookies, milk chocolate, salami, marshmallows, cheese, peanut butter, bananas, and fat, however, they gained 269 percent more weight, even when they had the option of eating their standard foods. Human beings and animals may share an inborn biological preference for calorically dense foods as a leftover survival mechanism that served a purpose when food was scarce.

■ MINI-SUMMARY

As researchers try to figure out why some people get fat and others do not, it is becoming increasingly apparent that obesity has a variety of causes—heredity, environment, metabolism, and level of physical activity—and therefore, no single cure.

Treatment of Obesity

Before discussing the treatment of obesity, it is a good idea to discuss treatment goals. The goals of most weight-loss programs have focused on short-term weight loss. Critics of this type of goal feel that short-term weight loss is not a valid measure of success, because it is not associated with health benefits (as is long-term weight loss). Also, weight-loss goals tend to reinforce the American preoccupation with being slender, especially for women. Lastly, critics point out that weight-loss goals are often set too high, since even a small weight loss can reduce the risks of developing chronic diseases. Overweight individuals who lose even relatively small amounts of weight are likely to:

- lower their blood pressure, and thereby the risks of high blood pressure
- reduce abnormally high levels of blood glucose associated with diabetes
- bring down blood levels of cholesterol and triglycerides associated with cardiovascular disease
- reduce sleep apnea or irregular breathing during sleep
- decrease the risk of osteoarthritis of the weight-bearing joints
- decrease depression
- increase self-esteem

Current recommendations from the Food and Nutrition Board favor shifting from a weight-loss perspective to one of weight management, in which success is judged more by the program's effect on an individual's health status than on its effect on weight.

Treatment of obesity generally consists of one or more of the following: diet, exercise, behavior, modification, and drug therapy. Treatment programs often concentrate on two aspects of losing weight: diets and exercise. Individuals in treatment programs that just offer diets tend to regain any lost weight within two or three years.

The next section will discuss each of the following components of treatment: diet and nutrition education, exercise, behavior and attitude modification, social support, and maintenance support. But first, let's take a look at a newer approach to treating obesity that doesn't use diets.

More and more health professionals are adopting a nondieting approach to treating obesity that includes eating less fat and exercising. This new approach steers clear of dieting and emphasizes helping obese people adopt a healthier lifestyle. In many cases diets simply don't work. By restricting food intake, diets often cause dieters to become obsessed with food, which may then lead to binge eating. Eating fewer fat calories and exercising can help many obese people to lose some weight and keep it off.

Diet and Nutrition Education

Basic nutrition education is crucial for dieters. They need education about fat, carbohydrate, and protein in foods and about balancing them in a lower-calorie and lower-fat diet. They need to understand variety, moderation, and nutrient, density, particularly because they have to pack the same amount of nutrients into fewer calories.

Dieters need to understand seven basic concepts of nutrition education before planning their actual diets:

1. Calories should not be overly restricted during dieting because this practice decreases the likelihood of success. A dieter who normally eats 2,500 calories daily and goes on a 1,200-calorie diet is eating less than half of what he or she normally eats. Restricting calories by 500 each day amounts to about one pound lost in a week. In any case, calories should not be restricted below 1,200, because getting adequate nutrients is impossible below this level. A progressive weight loss of 1 to 2 pounds a week is considered safe.

2. The focus should be on decreasing fat calories, since they provide more than twice the calories of carbohydrate and protein. Fat calories also tend to be stored in fat cells. In obesity research, the Flatt hypotheses state that high-fat diets are related to the development and maintenance of obesity. By restricting the fat you eat, you can cut down on the calories you eat. This might be the only adjustment that a moderately overweight person needs to make in order to lose weight (hopefully, he or she will also exercise). With the emphasis on eating less fat, keep in mind that fat-free foods still contain calories.

3. No foods should be forbidden, as that only makes them more attractive.

4. Eating three meals and one or two snacks each day is crucial to minimizing the possibility of getting hungry. People tend to overeat when hungry.

5. Portion control is vital. Measuring and weighing foods is important, because "eyeballing" is not always accurate.

6. Variety, balance, and moderation are crucial to satisfying all nutrient needs.

7. Weighing is important but should not be done every day, because 1- to 2-pound weight gains and losses can occur on a daily basis due to fluid shifts. Weekly weigh-ins are more accurate and less likely to cause disappointment.

Exercise

Exercise is a vital component of any weight-control program. Research consistently shows that time spent exercising is a major predictor of long-term weight loss. Exercise not only facilitates weight loss through direct energy expenditure, but burns fat both during and after exercise. Regular exercise also helps control or suppress appetite, and builds and tones muscles, which in turn raises your basal metabolic rate. But regular exercise has many benefits beyond simply losing pounds and keeping them off. Indeed, the Centers for Disease Control and Prevention and the American College of Sports Medicine recommend that every American adult accumulate 30 minutes or more of moderate-intensity physical activity on most, preferably all, days of the week. Additional advantages of regular physical activity include:

1. improved functioning of the cardiovascular system
2. reduced levels of blood lipids associated with cardiovascular disease
3. increased ability to cope with stress, anxiety, and depression
4. increased stamina
5. increased resistance to fatigue
6. improved self-image

A consistent pattern of exercise is vital to achieving these beneficial results.

The key to a successful exercise program is choosing an enjoyable activity. Some questions obese people need to answer to find a good exercise include:

1. Do you like to exercise alone or with others?
2. Do you prefer to exercise outdoors or in your home?
3. Do you particularly like any activities?
4. How much money are you willing to spend for sports equipment or facilities if needed?
5. When can you best fit the activity into your schedule?

An obese person may resist starting an exercise program for several reasons. Many have had bad experiences in school physical education classes and want to avoid such activity. Some tend to be self-conscious and may not want to be seen exercising. Activity is also harder for obese people and requires more effort.

Aerobic activities such as walking, jogging, cycling, and swimming are ideal as the major component of an exercise program. Aerobic activities must be brisk enough to raise heart and breathing rates, and they must be sustained, meaning that they must be done at least 15 to 30 minutes without interruption. Activities such as baseball, bowling, and golf are not vigorous or sustained. They still have certain benefits—they can be enjoyable, help improve coordination and muscle tone, and help relieve tension—but they are not aerobic.

Any exercise program for sedentary and overweight people must be started slowly, with enjoyment and commitment as the major goals. A buildup in intensity and duration should be gradual and progressive, depending to a large extent on how overweight the individual is. Aerobic capacity improves when exercise increases the heart rate to a target zone of 65 to 80 percent of the maximum heart rate, which is the fastest the heart can beat. Maximum heart rate can be calculated by subtracting your age from 220. One goal of the exercise program should be to build up to this intensity of exercise. Exercise that increases the heart rate to between 65 and 80 percent of its maximum conditions the heart and lungs, besides burning calories (Table 12-3). To determine whether your heart rate is in the target zone, take your pulse as follows:

1. When you stop exercising, quickly place the tip of your third finger lightly over one of the blood vessels on your neck located to the left or

TABLE 12-3 Calories Burned per Hour by a 150-Pound Person (Reduce the Calories 1/3 for a 100-Pound Person and Multiply by 1-1/3 for a 200-Pound Person)	
Activity	Calories Burned per Hour
Bicycling at 6 mph	240
Bicycling at 12 mph	410
Cross-country skiing	700
Jogging at 5.5 mph	740
Jogging at 7 mph	920
Running in place	650
Jumping rope	750
Swimming 25 yards/minute	275
Swimming 50 yards/minute	500
Tennis, singles	400
Walking at 2 mph	240
Walking at 3 mph	320
Walking at 4.5 mph	440

Source: U.S. Department of Health and Human Services. 1983. *Exercise and Your Heart.* Washington, D.C.: U.S. Government Printing Office.

right of your Adam's apple. (Another convenient pulse spot is the inside of your wrist just below the base of your thumb.)

2. Count your pulse for 30 seconds and multiply by 2.
3. If your pulse is below your target zone, exercise a little harder the next time. If you are above your target zone, exercise a little easier. If it falls within the target zone, you are doing fine.
4. Once you are exercising within your target zone, you should check your pulse at least once each week.

Exercise should take place at least three times per week and include a warm-up period, the exercise itself, and then a cooling-down period. To warm up before exercising, do stretching exercises slowly and in a steady, rhythmic way. Start at a medium pace and gradually increase. Next, begin jumping rope or jogging in place slowly before starting any vigorous activities to ease the cardiovascular system into the aerobic exercise. The exercise part of the session should burn at least 300 calories, which can be achieved with 15 to 30 minutes of aerobic activity in the target zone or 40 to 60 minutes of lower-intensity activity, such as leisurely walking. After exercising, slowing down the exercise activity or changing to a less vigorous activity for 5 to 10 minutes is important to allow the body to relax gradually.

Beyond the exercise program, obese people should be encouraged to schedule more activity into their daily routines. For instance, they should use stairs, both up and down, instead of elevators or escalators. They can also park or get off public transportation farther away from their destination to allow more walking.

Behavior and Attitude Modification

Behavior modification deals with identifying and changing behaviors that affect weight gain, such as raiding the refrigerator at midnight. Elements of behavior and attitude modification can be grouped into several categories: self-monitoring, stimulus or cue control, eating behaviors, reinforcement or self-reward, self-control, and attitude modification.

Self-monitoring involves keeping a food diary or daily record of types and amounts of foods and beverages consumed, as well as time and place of eating, mood at the time, and degree of hunger felt. Its purpose is, of course, to increase awareness of what is actually being eaten and whether it is in response to hunger or other stimuli. Once harmful patterns that encourage overeating are identified, negative behaviors can be changed to more positive ones. Figure 12-1 contains a sample food diary page.

Through self-monitoring, cues or stimuli to overeating can be identified. For example, passing a bakery may be a cue for someone to stop and buy

Figure 12-1
Food diary form

Time?	What Was Eaten?	Where?	How Much?	Hungry?	With Whom?	Mood?

a dozen cookies. Examples of behavioral modification techniques for cue or stimulus control follow.

Food Purchasing, Storage, and Cooking

1. Plan meals a week or more ahead.
2. Make a shopping list.
3. Do food shopping after eating, on a full stomach.
4. Do not shop for food with someone who will pressure you to buy foods you do not need.
5. If you feel you must buy high-calorie foods for someone else in the family who can afford the calories, buy something you do not like or let them buy, store, and serve the particular food.
6. Store food out of sight and limit storage to the kitchen.
7. Keep low-calorie snacks on hand and ready to eat.
8. When cooking, keep a small spoon such as a half-teaspoon on hand to use if you must taste while cooking.

Mealtime

1. Do not serve food at the table or leave serving dishes on the table.
2. Leave the table immediately after eating.

Holidays and Parties

1. Eat and drink something before you go to a party.
2. Drink fewer alcoholic beverages.
3. Bring a low-calorie food to the party.

4. Decide what you will eat before the meal.

5. Stay away from the food as much as possible.

6. Concentrate on socializing.

7. Be polite but firm and persistent when refusing another portion or drink.

Eating behaviors need to be modified to discourage overeating. First, one or two eating areas, such as the kitchen and dining-room tables, need to be set up so that eating occurs in only these designated locations. Eat only while sitting at the table, and make the environment as attractive as possible. Do not read, watch television, or do anything else while you eat, because you can easily form associations between certain activities and food, such as television and snacking. Do not eat while standing at cabinets or refrigerators.

Second, plan three meals and at least two snacks daily, preferably at certain times of the day. Third, eat slowly by putting your fork down between bites, eating your favorite foods first, talking to others, eating a high-fiber food that requires time to chew, and drinking a no-calorie beverage to help you fill up. In addition, take smaller bites, savor each bite, use a smaller plate to make the food look bigger, and leave a bite or two on your plate. If you clean your plate, you are responding to the sight of food, not to real hunger. When you want a snack, postpone it for 10 minutes.

Reward yourself for positive steps taken to lose weight, but do not use food as a reward. Other rewards could include reading a good book or magazine, doing a hobby such as sewing, taking a long bubble bath, picking some fresh flowers, or calling a friend.

Overeating sometimes occurs in reaction to stressful situations, emotions, or cravings. The food diary is very useful in identifying these situations. Then you can handle these situations in new ways. You can express your feelings verbally if you are overeating in response to frustration or similar stressful emotions. You can exercise or use relaxation techniques to relieve stress, or you can switch to a new activity, such as taking a walk, knitting, reading, or engaging in a hobby to help you take your mind off food. If you allow yourself five minutes before getting something to eat, you often will go on to something else and forget about the food. Positive self-talk is important for good self-control. Instead of repeating a negative statement such as "I cannot resist that cookie," say, "I will resist that cookie."

The most common attitude problem obese people have is thinking of themselves as either on or off a diet. Being "on a diet" implies that at some point the diet will be over, resulting in weight gain if old habits are resumed. Dieting should not be so restrictive or have such unrealistic goals that the person cannot wait to get "off the diet." When combined with exercise, behavior and attitude modification, social support, and a maintenance plan, dieting is really a plan of sensible eating that allows for periodic indulgences.

Another attitude that needs modification revolves around using words such as *always, never,* or *every*. The following are examples of unrealistic statements using these terms.

I will always control my desire for chocolates.
I will never eat more than 1,500 calories each day.
I will exercise every day.

Goals stated in this manner decrease the likelihood that you will ever accomplish them, and thus result in discouragement and thoughts of failure.

Setting realistic goals, followed by monitoring and self-reward when appropriate, is crucial to the success of any weight-loss program. Through goal setting, complex behavior changes can be broken down into a series of small, successive steps. Goals need to be reasonable and stated in a positive, behavioral manner. For example, if a problem behavior is buying a chocolate bar every afternoon at work, a goal may be to bring an appropriate afternoon snack from home. If this goal is not truly attainable, perhaps one chocolate bar per week should be allowed and worked into the diet.

Even with reasonable goals, occasional lapses in behavior occur. This is when having a constructive attitude is critical. After eating and drinking too much at a party one night, for example, feelings of guilt and failure are common. However, they do nothing to help people get back on their feet. Instead, the dieter must stay calm, realize that what is done is done, and that no one is perfect.

Two other attitudes that often need correcting concern hunger and foods that are "bad for you." Hunger is a physiological need for food, whereas appetite is a psychological need. Eating should be in response to hunger, not to appetite. Although obese people frequently regard certain foods as "good" or "bad," they must realize that no food is inherently good or bad. Some foods do contain more nutrients per calorie, and some are mostly empty calories with few nutrients. However, no food is so bad that it can never be eaten.

Social Support

In general, obese people are more likely to lose weight when their families and friends are supportive and involved in their weight-loss plans. As social support increases, so do a person's chances of maintaining weight loss. When possible, obese people need to enlist the help of someone who is easy to talk to, understands and empathizes with the problems of losing weight, and is genuinely interested in helping. Supporters can model good eating habits and give praise and encouragement. The obese person needs to tell others exactly how to be supportive by, for example, not offering high-calorie snacks. Requests need to be specific and positive.

Maintenance Support

Not enough is known about factors associated with weight-maintenance success or what support is needed during the first few months of weight maintenance, when a majority of dieters begin to relapse. Being at a normal, or more normal, weight can bring about stress as adjustments are made. Food is no longer a focal point, and old friends and activities may not fit into the new lifestyle. Support and encouragement from significant others will probably diminish. A formal maintenance program can help deal with those issues as well as others.

Strategies that appear to support weight maintenance include:

1. Determining how many calories are needed for weight maintenance and working out a livable diet
2. Learning skills for dealing with high-risk situations when a lapse in eating behavior may occur, and what to do when a relapse occurs
3. Continued self-monitoring
4. Continued exercise
5. Continued social support
6. Continued use of other strategies that were useful during weight loss
7. Dealing with unrealistic expectations about being thin

These strategies can be used during formal maintenance programs and after treatment is terminated.

■ **MINI-SUMMARY**

A comprehensive treatment plan for obesity needs to include diet and nutrition education, exercise, behavior and attitude modification, social support, and maintenance support. A newer, nondieting approach de-emphasizes dieting because dieting works in so few cases. This approach emphasizes exercise and fewer fat calories.

Menu Planning for Weight Loss and Maintenance

1. Cut down on fats; they are loaded with calories. But be wary of low-fat and reduced-fat foods; they often have as many calories as their full-fat counterparts.

2. De-emphasize protein foods as the central part of a meal, especially as entrées. Instead, highlight complex carbohydrates that are high in fiber, such as legumes, grains, fruits, and vegetables.
3. Be sure to suggest snacks that are healthy and not too high in calories. Try fresh fruit and vegetables, popcorn without added fat and sprinkled with garlic or chili powder, regular and soft pretzels, bread sticks, muffins or quick breads, wholegrain cookies (preferably homemade), graham crackers, vanilla wafers, gingersnaps, small amounts of peanut butter or cheese on wholegrain crackers, unsweetened ready-to-eat cereals, puddings such as rice or bread made with skim milk, frozen yogurt, some commercial frozen fruit and pudding bars with less than 100 calories per serving, plain low-fat yogurt, skim or low-fat milk, diet soft drinks, bottled waters, fruit spritzers (fruit juice with bottled water), coffee, tea, and herbal tea.

The Problem of Underweight

A person is considered *underweight* if he or she weighs 10 percent below the listing in the height-weight tables. Although anyone who has seriously dieted may think the underweight person is problem free, this is hardly the case. Underweight persons who have trouble gaining weight have very real concerns. Just as some people cannot seem to lose weight, so some people have trouble putting on a few extra pounds. The cause could be genetics, metabolism, or environment. However, some thin people, if they were to gain weight, would feel uncomfortable. Anyone who is underweight due to wasting diseases such as cancer or eating disorders also has a problem: malnutrition.

The following list contains tips on gaining weight.

1. If you find you cannot eat large meals, do not get discouraged. You can increase your intake by eating smaller meals frequently.
2. Avoid drinking low-calorie beverages such as coffee, tea, or water, especially with meals. Try fruit juices, milk, and milkshakes for more calories.
3. Add additional calories to your meals by using margarine mayonnaise, oil, salad dressing, or other fats that do not contain much saturated fat. For example, spread margarine on bread or use oil-packed tuna fish.
4. Add skim milk power to soups, sauces, gravies, mixed casseroles, scrambled eggs, and hot cereals. It adds both calories and protein. It can also be blended with milk at the rate of 2 to 4 tablespoons of powder to 1 cup of milk.
5. Add cheese to favorite sandwiches. Use grated cheese on top of casseroles, salads, soups, sauces, and baked potatoes.

6. Try breaded foods.
7. Eat regular yogurt, peanut butter or cheese with crackers, nuts, milkshakes, and wholegrain cookies and muffins as snacks.
8. Add regular cottage cheese to casseroles or egg dishes such as quiche, scrambled eggs, and souffles. Add it to spaghetti or noodles.
9. Make every mouthful count!

Nutrition for the Athlete

The amount of energy required by the athlete depends on the type of activity and its duration, frequency, and intensity. In addition, the athlete's basal metabolic rate, body composition, age, and environment must be taken into account. Many athletes require between 3,000 and 6,000 calories daily.

Carbohydrates and fat are the primary fuel sources for exercise. Protein plays a minor role. An appropriate diet for many athletes consists of 60 to 65 percent of calories as carbohydrates, 30 percent or less as fat, and enough protein to provide 1 to 1.5 grams per kilogram body weight.

Although many athletes take vitamin and mineral supplements, these will not enhance performance unless there is a deficiency. Most athletes get plenty of vitamins and minerals in their regular diets, although young athletes and women need to pay special attention to iron and calcium.

Water is the most crucial nutrient for athletes. They need about 1 liter of water for every 1,000 calories consumed. For moderate exercise without extreme temperatures or duration, cold water is the choice for replacing fluids. Cold water both cools the body and empties more quickly from the stomach. Athletes need 2 cups of water about 15 to 20 minutes before endurance exercise and at regular intervals during exercise. A good way to determine how much fluid to replace after exercising is to weigh in before and after exercise and also the next morning. For every pound that is lost, the athlete needs to drink 2 cups, or 16 ounces, of water.

For endurance events, some carbohydrates in fluids taken before and during competition may be helpful in maintaining normal blood sugar levels.

Carbohydrate or glycogen loading is a regimen involving three or more days of decreasing amounts of exercise and increased consumption of carbohydrates before an event to increase glycogen stores. The theory is that increased glycogen stores (which increase 50 to 80 percent) will enhance performance by providing more energy during lengthy competition.

Carbohydrate or glycogen loading—A regimen involving both decreased exercise and increased consumption of carbohydrates before an athletic event to increase the amount of glycogen stores.

Here are some menu-planning guidelines for athletes.

1. Offer a variety of foods from all four food groups: meat, poultry, and seafood; milk and dairy products; fruits and vegetables; and grain products.
2. Good sources of complex carbohydrates to emphasize on menus include pasta, rice, other grain products such as breads and cereals, legumes, and fruits and vegetables. On the eve of the New York City Marathon each year, marathon officials typically host a pasta dinner for runners featuring spaghetti with marinara sauce and cold pasta primavera. Complex carbohydrates such as pasta also provide needed B vitamins, minerals, and fiber. Wholegrain products such as whole-wheat bread contain more nutrients than refined products such as white bread. If using refined products, be sure they are enriched (the thiamin, riboflavin, niacin, and iron have been replaced). Here are some ways to include complex carbohydrates in your menu:

   ■ At breakfast, offer a variety of pancakes, waffles, cold and hot cereals, breads, and rolls.
   ■ At lunch, make sandwiches with different types of bread, such as pita pockets, raisin bread, onion rolls, and brown bread. Also, have a variety of breads and rolls available for nonsandwich items.
   ■ Serve pasta and rice as a side or main dish with, for example, chicken and vegetables. Cold pasta and rice salads are great too.
   ■ Potatoes, whether baked, mashed, or boiled, are excellent sources of carbohydrates.
   ■ Always have available as many types of fresh fruits and salads as possible.
   ■ Don't forget to use beans and peas in soups, salads, entrées, and side dishes.
   ■ Nutritious desserts emphasizing carbohydrates are frozen yogurt with fruit toppings, oatmeal cookies, and fresh fruit.

3. Don't offer too much protein and fat in the belief that athletes need the extra calories. They do, but much of those extra calories should come from complex carbohydrates. The days of steak-and-egg dinners are over for athletes. The protein and fat present in these meals does nothing to improve performance. Here are some ways to moderate the amount of fat and protein:

   ■ Use lean, well-trimmed cuts of beef.
   ■ Offer chicken, turkey, and fish—all lower in fat than beef. Broiling, roasting, and grilling are preferred cooking methods, with frying being acceptable occasionally.

▪ Offer larger serving sizes of meat, poultry, and fish, perhaps 1 to 2 ounces more, but don't overdo it!
▪ Offer fried food in moderation.
▪ Offer low-fat and skim milk.
▪ Offer high-fat desserts, such as ice cream and many types of sweets, in moderation. Frozen yogurt and ice milk generally contain less fat than ice cream and can be topped with fruit or crushed oatmeal cookies. Fruit ice and sorbet contain no fat.

4. Offer a variety of fluids, not just soft drinks and other sugared drinks. Good beverage choices include fruit juices, iced tea, and iced coffee (preferably freshly brewed decaffeinated), and plain and flavored mineral and seltzer water, spritzers (fruit juice and mineral water), and milkshakes made with yogurt or ice milk and fruit. Soft drinks and juice drinks—both loaded with sugar—should be offered in moderation.

5. Make sure iodized salt is on the table.

6. Be sure to include sources of iron at each meal. Good iron sources include liver, red meats, legumes, and iron-fortified breakfast cereal. Moderate iron sources include raisins, dried fruit, bananas, nuts, and wholegrain and fortified grain products. Be sure to include good vitamin C sources at each meal, as vitamin C assists in iron absorption. Vitamin C sources include citrus fruits and juices, cantaloupe, strawberries, broccoli, potatoes, and brussels sprouts.

Precompetition meal— The meal closest to the time of a competition or event.

7. The most important meal is the one closest to the competition, commonly called the **precompetition meal.** The functions of this meal include getting the athlete fueled up, both physically and psychologically, helping to settle the stomach, and preventing hunger. The meal should consist of mostly complex carbohydrates (they are digested easier and faster and help maintain blood sugar levels), and should be low in fat. High-fat foods take longer to digest and can cause sluggishness. Substantial precompetition meals are usually served three to four hours prior to competition, in order to allow enough time for stomach emptying (to avoid cramping and discomfort during the competition). Menus might include cereals with low-fat or skim milk topped with fresh fruit, low-fat yogurt with muffins and juice, or one or two eggs with toast and jelly and juice. The meal should include 2 to 3 cups of fluid for hydration and typically provide 300 to 1,000 calories. Smaller precompetition meals may be served two to three hours before competition. Many athletes have specific "comfort" foods that they enjoy before competition.

8. After competition and workouts, again emphasize complex carbohydrates to ensure glycogen restoration. The sooner an athlete fuels up after exercising, the more glycogen will be stored in the muscle. Food is also important to restore the minerals lost in sweating.

Check-Out Quiz

1. Obesity is due simply to overeating.
   a. True   b. False
2. An obese person is at increased risk for coronary heart disease.
   a. True   b. False
3. Obesity is defined as being 25 percent or more over desirable weight.
   a. True   b. False
4. When trying to lose weight, omitting junk foods that you like from your diet is a good idea.
   a. True   b. False
5. Carbohydrate loading is a regimen involving three days of increasing exercise and increased intake of carbohydrates.
   a. True   b. False
6. Serious athletes need about 8.0 grams of protein per kilogram of body weight.
   a. True   b. False
7. Body mass index tells you about your weight more accurately than height/weight tables.
   a. True   b. False
8. An example of using stimulus control to lose weight is to leave the table immediately after eating.
   a. True   b. False
9. A person is considered underweight if he or she weighs 10 percent below his or her desirable weight.
   a. True   b. False
10. Fat that collects around the belly presents an increased risk of developing serious health problems.
    a. True   b. False

Activities and Applications

1. **Your Desirable Weight**

Using Tables 12-1 and 12-2, determine your desirable weight. What is the difference between the weights? Using body mass index (you can use the website noted below to determine your BMI), see if you are overweight (a BMI of 25 or more). If you are overweight, find out if you have a family history of any of the medical conditions discussed in the section on health and obesity.

2. Low-Calorie Menu Planning

A local steak and seafood restaurant has asked you to design lower-calorie menu items as follows: two appetizers, three entrées, and one dessert. Their emphasis is freshly made traditional American cooking. Provide recipes and calorie information if possible.

3. Using a Food Diary

Using the food diary form, complete a three-day food diary. Then examine it to increase your awareness of how much and how often you are eating and whether you are eating in response to moods, people, and/or activities. Write down five insights you gained about your eating habits.

4. Box Lunches

Design three complete box lunches for calorie- and fat-conscious customers of a gourmet take-out deli with complete kitchen facilities located in San Francisco. The meal must be well balanced and provide a main dish, side dish, dessert, and beverage.

5. Precompetition Meals

Devise a menu for a precompetition meal for long-distance runners on a university track team who will compete at 4 P.M. Have two selections for each category you choose.

Nutrition Web Explorer

National Heart, Lung, and Blood Institute www.nhlbi.nih.gov

At NHLBI's web page, click on "Achieving Your Healthy Weight." Calculate your BMI and get some tips on behavior changes and exercise.

Shape Up America www.shapeup.org/fitness/home.htm

This web page welcomes you to the Shape Up America Fitness Center. Click on "Improvement" to find out about the different ways to add physical activity to your daily life.

Sports Science News www.sportsci.org

On the home page, click on the index on the left for "Sports Nutrition." Read one of the articles available on carbohydrates or fluids. Summarize what you read in one paragraph.

Food Facts *Sports Drinks*

A topic of much interest to athletes is whether sports drinks, such as Gatorade or Exceed, are needed during an event or workout. Sports drinks contain a dilute mixture of carbohydrate and electrolytes. Most contain about 50 calories per cup (or about 12 grams of carbohydrate) and small amounts of sodium and potassium. Sports drinks are purposely made to be weak solutions so they can empty faster from the stomach, and the nutrients in it are therefore available to the body more quickly. They are primarily designed to be used during exercise, although there are some specially formulated sports drinks with slightly more sugar that can be used just prior to exercising.

During exercise lasting 90 minutes or more, sports drinks can help replace water and electrolytes and provide some carbohydrates for energy. During an endurance event or workout, you increasingly rely on blood sugar for energy as your muscle glycogen stores diminish. Carbohydrates taken during exercise can help you maintain a normal blood sugar level and enhance (as well as lengthen) performance. Athletes often consume 1/2 to 1 cup of sports fluids every 15 to 20 minutes during exercise.

Some sports drinks claim that they contain glucose polymers, which are chains of glucose. It was thought that sports drinks with glucose polymers emptied from the stomach faster than solutions with sucrose or glucose, but it turned out that each of these solutions empties at approximately the same rate and provides the same positive effects on performance as long as the carbohydrate concentration is between 5 and 8 percent (which it usually is).

Although sports drinks clearly can help the athlete in lengthy events or workouts, are there other products that can do much the same? Long before sports drinks were available, athletes had their own homemade sports drinks: diluted juices with a pinch of salt, tea with honey, and dilute lemonade. Which works best?—whichever satisfies you best both physically and psychologically.

Hot Topic Diet Pills

A study by the Food and Drug Administration (FDA) found that 5 percent of women and 2 percent of men trying to lose weight use diet pills. In 1992, the FDA banned 111 ingredients in over-the-counter (OTC) diet products—including amino acids, cellulose, and grapefruit extract—after manufacturers were unable to prove that they worked. A number of products were also recalled because they posed serious health risks. The products contained guar gum, which supposedly swelled in the stomach to provide a feeling of fullness. However, the swelling from the guar gum caused blockages in the throat and stomach.

Products considered by the FDA to be over-the-counter (OTC) weight-control drugs are primarily those containing the active ingredient phenyl-propanolamine (PPA), such as Dexatrim. PPA is an ingredient found not only in many OTC diet pills but in cough-cold and allergy products as well. The FDA is concerned that PPA may increase the risk of a type of stroke, as was suggested by some reports among PPA users, typically young women. This possible risk could be further increased if a person took more than the recommended dose of PPA, which might occur inadvertently if one also took a cough-cold product with PPA. While the FDA agrees that studies have not shown a definite link between PPA and stroke, the agency believes that data from a more comprehensive study are needed to confirm the ingredient's safety.

PPA is not recommended for use by teenagers, because they are still growing and if they suppress their appetite, they may not get proper nutrition. This is especially true of teens who don't need to lose weight but think they do. Diet pills containing PPA must include the warning label: "For use by people 18 years of age and older."

Besides OTC diet pills, there are also prescribed drugs. In 1999 the FDA approved orlistat, a new drug to treat obesity. Unlike other obesity drugs, orlistat prevents enzymes in the gastrointestinal tract from breaking down dietary fats into smaller molecules that can be absorbed by the body. Absorption of fat is decreased by about 30 percent. Since undigested triglycerides are not absorbed, the reduced caloric intake may have a positive effect on weight control.

The recommended dose of orlistat is one capsule with each main meal that includes fat. During treatment, the patient should be on a nutritionally balanced, reduced-calorie diet that contains no more than 30 percent of calories from fat. Orlistat is indicated for obese patients with a body mass index (BMI) of 30 or more, or for patients with a BMI of 27 or more who also have high blood pressure, high cholesterol, or diabetes.

Because orlistat reduces the absorption of some fat-soluble vitamins and beta-carotene, patients should take a supplement that contains these nutrients. The most common side effects of orlistat are oily spotting, gas with discharge, and frequent bowel movements.

There are pros and cons about the use of diet pills. Although they do help some people lose weight initially, what happens when they stop taking the diet pill? At best, diet pills are a temporary solution. After all, you can't take them forever. Most can be used for only a few months or a year.

Chapter 13
Nutrition Over the Life Cycle

Although much nutrition advice we hear on television or read about in magazines is for adults—and indeed, the first part of this book is mostly for adults—other groups have their own special nutrition needs and concerns too. What do pregnant women, babies, children, and even teenagers have in common? They are all growing, and growth demands more nutrients. Did you know that a one-month-old baby, compared with an adult, needs twice the amount, proportionally, of many vitamins and minerals? At the other end of the age spectrum, the fastest-growing age group in the United States is the over-85 group. As people age, many new factors affect their nutrition status: the aging process, onset of chronic diseases such as heart disease, living alone, dentures, inability to get out to shop for food, and so on.

This chapter takes you from pregnancy through infancy, childhood, and adolescence and on to the golden years. Along the way, we will explore the nutritional needs and factors affecting nutrition status for each group, along with menu-planning guides. This chapter will help you to:

- explain why good nutrition during pregnancy is so vital
- identify nutrients that must be supplemented during pregnancy
- plan menus for women during pregnancy and lactation
- describe what an infant is fed during the first year, including the progression of solid foods
- plan menus for preschool and school-age children
- describe influences on children's and adolescents' eating habits
- plan menus for adolescents
- distinguish between anorexia nervosa, bulimia nervosa, and binge eating disorder
- list factors that influence the nutrition status of adults and older adults
- plan menus for healthy older adults

Pregnancy

Fetus—The infant in the mother's uterus from the eighth week after conception until birth.
Amniotic sac—The protective bag, or sac, that cushions and protects the fetus during pregnancy.

From a modest one-cell beginning, an actual living and breathing baby is born after 40 weeks. From eight weeks after conception until birth, the infant in the mother's uterus is called a **fetus.** At eight weeks the fetus is about 1 inch long and has a beating heart and gastrointestinal and nervous systems. To cushion and protect the fetus, it floats in a protected bag, or sac, called the **amniotic sac.** From the second to eighth week after conception, the infant is called an **embryo.**

During the first month of pregnancy, an organ called the **placenta** develops to provide an exchange of nutrients and wastes between fetus and

Embryo—The name of the fertilized egg from conception to the eighth week.

Placenta—An organ that develops during the first month of pregnancy, which provides for exchange of nutrients and wastes between fetus and mother and secretes the hormones necessary to maintain pregnancy.

Low birth weight baby—A newborn who weighs less than 5-1/2 pounds; these infants are at higher risk for disease.

mother and to secrete the hormones necessary to maintain pregnancy. If a mother is not sufficiently nourished during early pregnancy (when she probably doesn't even know she's pregnant), the placenta will not perform properly and the fetus will not get optimal nourishment.

Nutrition During Pregnancy

The nutritional status of women before and during pregnancy influences both the mother's and the baby's health. Factors that place a woman at nutritional risk during pregnancy include an inadequate diet, smoking, and other influences described in Table 13-1.

Table 13-2 shows optimum weight gain during pregnancy. Underweight women (10 percent or more below their desirable weight) must either gain weight before or gain more weight during pregnancy. Overweight women need to lose weight before or gain less weight during pregnancy, otherwise they are at greater risk of developing gestational diabetes and hypertension, both of which can cause complications. Table 13-2 also shows that of the weight gained, about 8 pounds is actually baby, with the rest serving to support the baby's growth.

Both prepregnancy weight and weight gain during pregnancy directly influence infant birth weight. The newborn's weight is the number-one indicator of his or her future health status. A newborn who weighs less than 5-1/2 pounds is referred to as a **low birth weight baby.** These babies are at higher risk for complications and experience more difficulties surviving the first year. Often the mother of a low birth weight baby has a history

---

TABLE 13-1 Nutrition Risk Factors for Pregnant Women

- Pregnancy during adolescence.
- Inadequate diets.
- Multiple birth (twins, triplets, and so on)
- Use of cigarettes, alcohol, or illicit drugs
- Lactose intolerance
- Underweight or overweight at time of conception
- Gaining too few or too many pounds during pregnancy
- Health conditions such as diabetes mellitus, hypertension, and HIV infection

TABLE 13-2 Optimum Weight Gain in Pregnancy	
Weight at Conception	Optimum Weight Gain
Normal weight	25–35 pounds
Underweight	28–40 pounds
Overweight	15–25 pounds
Components of Weight Gain	Pounds
Fetus	8 pounds
Placenta	1.5 pounds
Amniotic fluid	2.0 pounds
Increase in size of uterus, breast, fluid and blood volume	12 pounds
Fat	2.0 to 8.0 pounds

of poor nutrition status before and/or during pregnancy. Other factors associated with low birth weight are smoking, alcohol use, drug use, and certain disease conditions.

For healthy babies, pregnant women need to eat more calories, but not a whole lot more. If a pregnant woman "eats for two" during pregnancy, she is probably asking for trouble! Pregnancy does increase calorie needs, but only by an additional 300 calories per day. Within that measly 300 calories, however, the pregnant woman must pack more protein and more of 13 different vitamins and minerals! See Appendix B for a comparison of the Dietary Reference Intakes for nonpregnant and pregnant women.

During the first 13 weeks of pregnancy (referred to as the **first trimester**), the total weight gain is between 2 and 4 pounds. Thereafter, about 1 pound per week is normal. Corresponding with the timing of weight gain, it makes sense that the greatest need for calories begins around the tenth week of pregnancy and continues until birth. The two major factors influencing calorie requirements during pregnancy are the woman's activity level and basal metabolic rate (BMR). The BMR increases to support the growth of the fetus. Pregnancy is no time to diet, or especially to follow a fad diet, which could have dangerous implications for the fetus.

Protein needs increase from 45–50 grams for nonpregnant women to 60 grams for pregnant women. The requirements for protein are generous during pregnancy and probably high for some women. Meeting protein needs is rarely a problem.

During pregnancy, calcium, phosphorus, and magnesium are necessary for the proper development of the skeleton and teeth. In adequate amounts, calcium may help reduce high blood pressure and the incidence of pre-eclampsia, a sometimes deadly disorder marked by high blood pressure after the sixth month of pregnancy. On the positive side, much more calcium (about double) is absorbed through the intestine. On the negative side, many women do not eat enough calcium-rich foods during pregnancy or lactation. The need for calcium is moderately high (1,000 milligrams) but can be met by having at least three servings from the dairy group each day. Magnesium is found in green leafy vegetables, nuts, seeds, legumes, and whole grains.

The need for folate increases 50 percent during pregnancy. This makes perfect sense when you realize that folate is needed to sustain the growth of new cells and the increased blood volume that occur during pregnancy. Folate is critical in the first four to six weeks of pregnancy (when most women don't even know they are pregnant) because this is when the **neural tube,** the tissue that develops into the brain and spinal cord, forms. Without enough folate, birth defects of the brain and spinal cord, such as **spina bifida,** can occur. In spina bifida, parts of the spinal cord are not properly fused, so gaps are present (Figure 13-1). Not every woman who has insufficient folate during early pregnancy will have a child with such a birth defect. However, if all women of childbearing age consumed enough folate, 50 to 70 percent of birth defects of the brain and spinal cord could be

Neural tube—The embryonic tissue that develops into the brain and spinal cord.

Spina bifida—A birth defect in which parts of the spinal cord are not fused together properly, so gaps are present where the spinal cord has little or no protection.

Figure 13-1
Spina bifida aperta

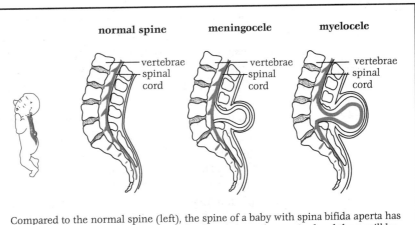

Compared to the normal spine (left), the spine of a baby with spina bifida aperta has a noticeable sac. When the sac is small (center), it can be repaired and there will be no muscle paralysis. But in 90 percent of cases, a portion of the nondeveloped spinal cord protrudes through the spine and into the sac (right), causing paralysis and incontinence.

prevented, according the the U.S. Centers for Disease Control and Prevention.

Folate is also critical during the entire pregnancy. A study of pregnant women showed that those women consuming less than 240 micrograms per day of folate had about a two- to threefold greater risk of preterm delivery and low birth weight.

Since vitamin B_{12} works with folate to make new cells, increased amounts of this vitamin are also needed. As long as animal products, such as meat and milk, are being consumed, vitamin B_{12} deficiency is not a concern.

Along with folate and vitamin B_{12}, iron also helps in the formation of blood—it is necessary for hemoglobin in both maternal and fetal red blood cells. After 34 weeks of pregnancy, a woman's blood volume has increased 50 percent from the time of conception. Although iron absorption does increase during pregnancy, whether the diet can supply enough iron is questionable. The National Academy of Sciences recommends iron supplements during the second and third trimesters.

In the past it was thought that sodium restriction was necessary for women with **edema,** or tissue swelling. It is now known that moderate swelling is normal during pregnancy and that sodium restriction is unnecessary and could actually be harmful for healthy pregnant women.

Table 13-3 lists the current supplementation recommendations for pregnant women. The only nutrient suggested for all pregnant women is iron.

Edema—Swelling due to an abnormal accumulation of fluid in the intercellular spaces.

Menu Planning During Pregnancy

To plan menus properly, certain diet-related concerns that occur during pregnancy need to be discussed. These include morning sickness, changes in taste and smell, constipation, heartburn, and the intake of alcohol, saccharin, aspartame, and mercury and polychlorinated biphenyls.

Nausea and vomiting, commonly referred to as morning sickness (although it can occur at any time of day), may be due to an increase in one or more of the 30 hormones that increase during pregnancy. The hallmarks of morning sickness are nausea and vomiting. Aversion to certain odors and fatigue also accompany the nausea and vomiting. Constipation may also be a concern. Each woman experiences it a little differently. Morning sickness lasts for an average of 17 weeks, but for some unlucky women it lasts until delivery. From 50 to 90 percent of women experience some gastrointestinal discomfort in early pregnancy. A major health concern with morning sickness is that it can cause dehydration, which in turn causes nausea.

Dietary advice in the past concentrated on small, carbohydrate-rich meals and tea and crackers. For many women, this dietary advice doesn't work.

TABLE 13-3 Supplementation Recommendations: National Academy of Sciences (1990)

Nutrient	Candidates for Supplementation	Level of Nutrient Supplementation
Iron	All pregnant women (2nd and 3rd trimesters)	30 mg ferrous iron daily
Folate	Pregnant women with suspected dietary inadequacy of folate	300 μg/day
Vitamin D	Complete vegetarians and others with low intake of vitamin D-fortified milk	10 μg/day
Calcium	Women under age 25 whose daily dietary calcium intake is less than 600 mg	600 mg/day
Vitamin B_{12}	Complete vegetarians	2 μg/day
Zinc/copper	Women under treatment with iron for iron deficiency anemia	15 mg Zn/day .2 mg Cu/day
Multivitamin-mineral supplements	Pregnant women with poor diets and for those who are considered high risk: multiple gestation, heavy smokers, alcohol/drug abusers, other	Preparation containing iron-30 mg zinc-15 mg copper-2 mg calcium-250 mg vitamin B_6-2 mg folate-300 μg vitamin C-50 mg vitamin D-5 μg

Source: Institute of Medicine. 1990. *Nutrition During Pregnancy.* Washington, DC: National Academy Press.

Recent advice centers on eating whatever foods you can keep down, even foods that aren't terribly nutritious, such as potato chips. The logic behind this recommendation is that tastes change when you are sick and you often crave something when you feel ready to eat. It's better to eat that food and keep it down than to eat something that is not appealing and throw it up. During pregnancy, women often develop a fine-tuned sense of smell, which often adds to their nausea. The smell of foods and cooking can make them sick.

Pregnant women commonly report changes in taste and smell. They may prefer saltier foods and crave sweets and dairy products such as ice cream. Certain foods they may have aversions to include alcohol, caffeinated drinks, and meats. Their cravings and aversions do not necessarily reflect actual physiological needs.

Constipation is not uncommon, due to the relaxation of gastrointestinal muscles. It can be counteracted by eating more high-fiber foods, drinking more fluids, and getting additional exercise.

Heartburn is a common complaint toward the end of pregnancy when the growing uterus crowds the stomach. This condition has nothing to do with the heart but is actually a painful burning sensation in the esophagus. It occurs when stomach contents, which are acidic, flow back into the lower esophagus. Possible solutions include eating small and frequent meals, eating slowly and in a relaxed atmosphere, avoiding caffeine, wearing comfortable clothes, and not lying down after eating.

Several food and beverage ingredients may affect the course and outcome of pregnancy, including alcohol, mercury, and polychlorinated biphenyls (PCBs).

If you look at the label on any bottle of beer, wine, or liquor, you will see the following:

Government Warning: According to the Surgeon General, women should not drink alcoholic beverages during pregnancy because of the risk of birth defects.

Alcohol and pregnancy don't mix. During pregnancy, alcohol crosses the placenta, and high alcohol levels can build up in the fetus. Alcohol can limit the amount of oxygen delivered to the fetus, oxygen that is vital to its development, as well as slow the growth of cells. It can also produce abnormal cells.

The heavy consumption of alcohol during pregnancy may cause a variety of symptoms called **fetal alcohol syndrome (FAS).** FAS children may show signs of mental retardation, growth retardation, brain damage, and facial deformities. Newborns with FAS are generally small in size and irritable because of alcohol withdrawal. The most serious concern with FAS infants is their impaired physical and mental development. They have problems gaining weight and are frequently mentally impaired.

You don't have to be a chronic alcoholic to have problems with FAS. Even moderate drinkers can have babies that show subtle signs of FAS. These women also have a higher rate of miscarriages and low birth weight babies. The American Academy of Pediatrics recommends that women stop drinking alcohol as soon as they plan to become pregnant, because harm can be done during the first six to eight weeks, when a woman doesn't yet know for sure whether she is pregnant.

Heartburn—A painful burning sensation in the esophagus caused by acidic stomach contents flowing back into the lower esophagus.

Fetal alcohol syndrome—A set of symptoms occurring in newborn babies that are due to alcohol use by the mother during pregnancy; symptoms may include mental retardation, brain damage, etc.

Studies reveal that neither saccharin nor aspartame are known to cause problems during pregnancy. However, use of both saccharin and aspartame should be moderated in pregnancy.

High levels of mercury have been found in certain large fish, including swordfish, large tuna, shark, halibut, and marlin. According to the Center for Science in the Public Interest, pregnant women should limit their intake of tuna to a half-pound per week and completely avoid the other fish.

TABLE 13-4	Daily Food Guide for Pregnancy and Lactation	
Food Group	Servings	Serving Size
Meat/meat alternative	3—pregnancy to include 1 serving legumes 3—lactation	2 ounces cooked lean meat, poultry, or fish 2 eggs 2 ounces cheese 1/2 cup cottage cheese 1 cup dried beans or peas 4 tablespoons peanut butter
Milk and Dairy	3—pregnancy and lactation 4—pregnant/lactating teenagers	1 cup milk, yogurt, pudding or custard 1-1/2 ounces cheese 1-1/2 to 2 cups cottage cheese
Vegetables*	4—5 during pregnancy and lactation	1/2 cup cooked or juice 1 cup raw
Fruits	3—4 during pregnancy and lactation	Portion commonly served, such as a medium apple or banana
Grain	7—11 during pregnancy and lactation	1 slice whole-grain or enriched bread 1 cup ready-to-eat cereal 1/2 cup cooked cereal or pasta 1/2 bagel or hamburger roll 6 crackers 1 small roll 1/2 cup rice or grits
Fats and sweets**		Includes butter, margarine, salad dressings, mayonnaise, oils, candy, sugar, jams, jellies, syrups, soft drinks, and any other fats and sweets

* A source rich in vitamin C (citrus, strawberries, melons, tomatoes) is needed daily; a source rich in vitamin A (dark green and deep yellow vegetables) is needed every other day.

** In general, the amount of these foods to use depends on the number of calories you require. Get your essential nutrients in the other food groups first before choosing foods from this group.

Pregnant women should also avoid eating fish contaminated with PCBs: freshwater carp, wild catfish, lake trout, whitefish, bluefish, mackerel, and striped bass.

Table 13-4 is a daily food guide for pregnancy. Problems can arise when an individual omits entire or substantial parts of certain groups. For instance, vegetarians, who do not eat any food of animal origin, need varied, adequate diets and supplements to obtain adequate vitamin B_{12}, calcium, zinc, and, unless getting adequate sunshine, vitamin D. Individuals who avoid the dairy group may need calcium and vitamin D supplements unless they eat foods fortified with these nutrients, such as calcium-fortified orange juice or calcium- and vitamin D-fortified soy milk.

The following are some menu-planning guidelines for pregnant and lactating women.

1. Offer a varied and balanced selection of nutrient-dense foods. Because energy needs increase less than nutrient needs, empty calories are rarely an acceptable choice.
2. In addition to traditional meat entrées, choose entrées based on legumes and/or grains and dairy products. Beans, peas, rice, pasta, and cheese can be used in many entrées. Chapter 11 covers vegetarianism and has much information on meatless entrées.
3. Be sure to offer dairy products made with skim or low-fat milk.
4. Use a variety of wholegrain and enriched breads, rolls, cereals, rice, pasta, and other grains.

TABLE 13-5	Sources of Problem Nutrients During Pregnancy
Nutrient	Food Sources
Fiber	Bran, dried beans and peas, wholegrain breads and cereals, fruits and vegetables, nuts, seeds
Folate	Organ meats, legumes, dark green leafy vegetables, orange juice, whole-wheat breads and cereals
Vitamin D	Vitamin D fortified milk
Iron	Red meat, liver, shellfish, poultry, dried fruit, beans, nuts, wholegrain and enriched breads and cereals
Calcium	Milk, dairy products, calcium-fortified orange juice
Magnesium	Green leafy vegetables, nuts, seeds, whole grains, and legumes
Zinc	Red meat, liver, poultry, fish, legumes

5. Use assorted fruits and vegetables in all areas of the menu, including appetizers, salads, entrées, side dishes, and desserts.
6. Be sure to have good sources of problem nutrients: fiber, vitamin B_6, folate, vitamin D, iron, calcium, magnesium, and zinc (See Table 13-5).
7. Be sure to use iodized salt.

■ **MINI-SUMMARY**

Women's nutritional status before and during pregnancy influences both the mother and baby's health. Both prepregnancy weight and weight gain during pregnancy directly influence infant birth weight, the most important indicator of the baby's future health status. Although a pregnant woman should consume only 300 additional calories per day, she must take in more nutrients such as iron, folate, zinc, and vitamin B_{12}. Iron supplements are likely to be prescribed because diet just isn't enough. When menu planning, keep in mind that pregnant women need nutrient-dense foods, an extra serving from the dairy and meat/meat-alternate groups, and plenty of fruits and vegetables. Recent advice for morning sickness is to let women eat whatever foods they feel they can keep down, without worrying excessively about how nutritious they are. Moderate use of saccharin and aspartame is advised during pregnancy, but it is probably best to stay away from alcoholic beverages.

Nutrition and Menu Planning During Lactation

Table 13-4 shows the daily food guide for breast-feeding mothers. During lactation, the period of milk production, 500 additional calories and more protein are necessary. Actually, more than 500 extra calories are needed daily, but some (about 150 each day) are supplied by extra fat stored during pregnancy. Lactating mothers, who normally produce about 25 ounces of milk a day, also need at least 2 to 3 quarts of water each day to prevent dehydration. Extra cups of coffee contain extra fluid but also contribute excess caffeine, which can cross to the baby and cause irritability.

A balanced, varied, and adequate diet (at least 1,800 calories per day) is critical to successful breast-feeding and infant health. If the mother is not eating properly, any nutritional deficiencies are more likely to affect the quantity of milk she makes, rather than the quality. Menu-planning guidelines for lactating women follow those for pregnant women, with emphasis on fluids, dairy products, fruits, and vegetables. Occasional consumption of small amounts of alcohol will probably have no consequences. The National Research Council suggests iron supplementation for the mother to replenish stores depleted during pregnancy. Lactating vegetarian mothers who eat no food of animal origin need to pay special attention to getting enough calories, iron, zinc, calcium, vitamin D, and vitamin B_{12}.

■ MINI-SUMMARY

During lactation, mothers need 500 extra calories a day, plenty of fluids, two extra servings from the dairy group, one extra serving from the meat/meat-alternate group, and a variety of nutrient-dense foods. If the mother is not eating right, the quantity of milk will be adversely affected. Small amounts of alcohol or caffeine are probably okay.

Infancy: The First Year of Life

The nutrient needs of infants are about double those of an adult when viewed in proportion to their weights. Little wonder, considering that infants generally double their birth weight in the first four to five months and then triple their birth weight by the first birthday. An infant will also grow 50 percent in length by the first birthday. (In other words, a baby who was 20 inches at birth grows to 30 inches in one year.)

Nutrition During Infancy

Newborns need a plentiful supply of all nutrients, especially those necessary for growth, such as vitamins C and D, folate, vitamin B_{12}, calcium, and iron. The DRI is set for infants from 0 to 6 months, and then from 7 to 12 months. By 6 months, growth occurs at a slower rate.

For the first 4 to 6 months of life, the source of all nutrients is breast milk or formula. Breast milk is recommended for all infants in the United States under ordinary circumstances from birth to 12 months. A baby needs breast milk for the first year of life, and as long as desired after that.

The number of women who are choosing to breast-feed is increasing. Current estimates are that more than 50 percent of American mothers breastfeed their babies in the hospital, but only 19 percent are still breast-feeding six months later. The reasons behind this increase include research findings that show definite health benefits of breast milk, as well as support and information groups that communicate breast-feeding guidelines and advantages. However, too few mothers breast-feed. Unfortunately, women who are young, unemployed, and on low incomes are the least likely to breast-feed. Their babies could greatly benefit from breast-feeding because they typically face the highest risk of health problems.

The following list shows the advantages of breast-feeding, compared to formula feeding.

1. Breast milk is nutritionally superior to any formula or other type of feeding. It provides exactly the right proportion and form of calories and nutrients needed for optimal growth, brain development, and digestion. Cow's milk contains a different type of protein than breast milk and infants can have difficulty digesting it. It also provides more of the essential fatty acids compared to formula. The composition of breast milk changes to meet the needs of the growing infant.

2. Newborns are less apt to be allergic to breast milk than to any other food.

3. Suckling promotes the development of the infant's jaw and teeth. It's harder work to get milk out of a breast than from a bottle; the exercise strengthens the jaw and encourages the growth of straight, healthy teeth. The baby at the breast can control the flow of milk by sucking and stopping. With a bottle, the baby must constantly suck or react to the pressure of the nipple in the mouth.

4. Breast-feeding promotes a close relationship—a bonding between mother and child. At birth, infants see only 12 to 15 inches, the distance between a nursing baby and its mother's face.

5. Breast milk is less likely to be mishandled. Some formulas require accurate dilutions, and all are much more apt to be mishandled, which can result in foodborne illness.

6. Breast milk helps the infant build up immunities to infectious disease, because it contains the mother's antibodies to disease. Breast-fed infants are much less likely to develop serious respiratory and gastrointestinal illnesses.

7. Breast milk also contains growth factors, thought to help in developing body tissues, and hormones and other substances that may subtly shape the newborn's brain and behavior.

8. Breast-feeding may reduce the risk of breast cancer for the mother.

9. Breast-feeding is less expensive.

10. Breast-fed infants have lower rates of hospital admissions, ear infections, diarrhea, rashes, allergies, and other medical problems than do bottle-fed babies.

Breast-feeding is not recommended if the mother uses addictive drugs, drinks more than a minimal amount of alcohol, is on certain medications, or is HIV (the virus that causes AIDS) positive.

If the infant or mother is not exposed regularly to sunlight or if the mother's intake of vitamin D is low, the breast-fed infant may need vitamin D. Vitamin D supplements are generally recommended for breast-fed infants.

Formula feeding is an acceptable substitute for breast-feeding and has some advantages. Some women find formula feeding more convenient (others find breast-feeding more convenient). Other family members can take part in formula feeding. For some women who are uncomfortable with breast-feeding, even after education, formula feeding is the method of choice.

All formulas must meet nutrient standards set by the American Academy of Pediatrics. The three forms of formulas on the market are ready-to-feed formula, liquid concentrate that needs to be mixed with equal amounts of water, and powdered formulas, which also need to be mixed with water. All formulas must be handled in a sanitary manner to prevent contamination and possible food poisoning.

Cow-milk-based formulas are normally used unless the baby is allergic to the protein or sugar in milk. In that case, a soy-based formula is used. For the baby who is allergic to both milk-based and soy-based formulas, predigested formulas are available. Symptoms of allergies usually include diarrhea and/or vomiting. Cow's milk has too much protein and minerals and too little essential fatty acids, vitamin C, and iron. Therefore it is not recommended until 12 months of age, when the baby is less likely to be allergic to it. Babies are normally switched slowly from formula to cow's milk.

Whether infants are breast-fed or formula fed, their iron stores are relatively depleted by 4 to 6 months, at which time they typically start to eat iron-fortified cereals. Fluoride supplements may also be prescribed for the formula-fed baby, unless the formula is made with fluoridated water.

Feeding the Infant

Successful infant feeding requires cooperative functioning between the mother and her baby. Feeding time should be a pleasurable period for both parent and child, so be sure to be comfortable and relaxed to better enjoy the experience. Ideally, the feeding schedule should be based on reasonable self-regulation by the baby. By the end of the first week of life, most infants want six to eight feedings a day. Formula-fed babies are fed about every four hours and breast-fed babies about every two or three hours.

Colostrum—A yellowish fluid that is the first secretion to come from the mother's breast a day or so after delivery of a baby; it is rich in proteins, antibodies, and other factors that protect against infectious disease.

Transitional milk—The type of breast milk produced from about the third to the tenth day after childbirth, when mature milk appears.

Milk letdown—The process by which milk comes out of the mother's breast to feed the baby; sucking causes the release of a hormone that allows milk letdown.

The mother must breast-feed the child as soon as possible after delivery to enhance success. **Colostrum,** a yellowish fluid, is the first secretion to come from the breast a day or so after delivery. It is rich in proteins, antibodies, and other factors that protect against infectious disease. Colostrum changes to **transitional milk** between the third and sixth days, and by the tenth day the major changes are finished.

The breast-feeding process begins with the infant using a sucking action that stimulates hormones to move milk into the ducts of the breast. This process is referred to as **milk letdown** and is hindered if the mother is tired or anxious. A baby will often empty a breast in about 10 minutes of nearly continuous sucking. The baby should empty at least one breast per feeding in order to stimulate the breast to produce more milk. To assure that the newborn is getting enough milk, he or she needs to be nursed frequently. In order to nurse the child successfully, the mother needs adequate rest, nutrition, and fluids, as well as education and support to decrease anxiety. Table 13-6 lists tips for breast-feeding success.

Babies are ready to eat semi-solid foods such as hot cereal when they can sit up and open their mouths. This usually occurs between five and seven months of age. Other signs that babies are ready for spoon feeding are when they:

■ have doubled their birth weight.
■ drink more than a quart of formula per day.
■ seem hungry often.
■ open their mouths in response to seeing food coming.

Although some parents think that feeding of solids will help the baby to sleep through the night, this is not often so. Feeding of solid food before a baby is ready can create problems, because the baby's digestive system is not ready for it. Feeding solids early also increases the risk of allergies and the chance of choking, and may encourage overfeeding.

Most babies can digest starchy foods at around four months of age. Once a baby starts on solid foods, it is important to make sure that the baby gets sufficient fluids. Up to this point, breast milk or formula met the baby's need for fluids. Now, however, drinking water or other fluid is needed to prevent dehydration. Proportionally, babies have more water in their bodies than adults do. They can become dehydrated very quickly due to hot weather, diarrhea, or vomiting, so fluids need to be offered at these times.

Although eating solid food is certainly simple for an adult, it involves a number of difficult steps for the baby. First the infant must have enough muscle control to close his or her mouth over the spoon, scrape the food from the spoon with the lips, and then move the food from the front to the

TABLE 13-6 Tips for Breast-Feeding Success

■ ***Get an early start:*** Nursing should begin within an hour after delivery if possible, when an infant is awake and the suckling instinct is strong. Even though the mother won't be producing milk yet, her breasts contain colostrum, a thin fluid that contains antibodies to disease.

■ ***Proper positioning:*** The baby's mouth should be wide open, with the nipple as far back into his or her mouth as possible. This minimizes soreness for the mother. A nurse, midwife, or other knowledgeable person can help her find a comfortable nursing position.

■ ***Nurse on demand:*** Newborns need to nurse frequently, at least every two hours, and not on any strict schedule. This will stimulate the mother's breasts to produce plenty of milk. Later, the baby can settle into a more predictable routine. But because breast milk is more easily digested than formula, breast-fed babies often eat more frequently than bottle-fed babies.

■ ***Delay artificial nipples:*** It's best to wait a week or two before introducing a pacifier, so that the baby doesn't get confused. Artificial nipples require a different sucking action than real ones. Sucking at a bottle could also confuse some babies in the early days. They, too, are learning how to breast-feed.

■ ***Air dry:*** In the early postpartum period or until her nipples toughen, the mother should air dry them after each nursing to prevent them from cracking, which can lead to infection. If her nipples do crack, the mother can coat them with breast milk or other natural moisturizers to help them heal. Vitamin E oil and lanolin are commonly used, although some babies may have allergic reactions to them. Proper positioning at the breast can help prevent sore nipples. If the mother's very sore, the baby may not have the nipple far enough back in his or her mouth.

■ ***Watch for infection:*** Symptoms of breast infection include fever and painful lumps and redness in the breast. These require immediate medical attention.

■ ***Expect engorgement:*** A new mother usually produces lots of milk, making her breasts big, hard, and painful for a few days. To relieve this engorgement, she should feed the baby frequently and on demand until her body adjusts and produces only what the baby needs. In the meantime, the mother can take over-the-counter pain relievers, apply warm, wet compresses to her breasts, and take warm baths to relieve the pain.

■ ***Eat right, get rest:*** To produce plenty of good milk, the nursing mother needs a balanced diet that includes 500 extra calories a day and six to eight glasses of fluid. She should also rest as much as possible to prevent breast infections, which are aggravated by fatigue.

back of the tongue. By about 16 weeks, a baby generally has these skills, but probably no teeth! The baby's first teeth will cut through the gums between six and ten months of age. If a baby can't swallow well enough to get the food from the back of the tongue into the pharynx, the baby will gag. The **gag reflex** prevents choking and sometimes results in vomiting.

Foods are generally introduced as follows. Keep in mind that the order of introducing different types and textures of foods is tied to the baby's developmental stages.

- 4 to 6 months: Iron-fortified baby cereals
- 5 to 7 months: Strained or puréed vegetables and fruits
- 7 to 9 months: Strained or soft protein foods (meat, chicken, fish,
  cheese, yogurt, beans, egg yolk)
  Finger foods such as crackers
  Fruit juice
- 9 to 12 months: Soft, chopped foods (finely chopped at first)
  Breads and grain products
- 12 months: Cut-up table foods
  Whole milk
  Whole eggs

The first solid food is iron-fortified baby cereal mixed with breast milk or formula. Usually, rice cereal is offered first because it is the least likely to cause an allergic reaction. Barley and oatmeal cereals follow. The iron found in these cereals is very important to meet the infant's high iron needs. Avoid putting cereals or any other solids into the infant's bottle.

Once the baby is used to various cereals, puréed or mashed vegetables and fruits can be tried at about five to six months. It is a good idea to start with vegetables so that the baby does not become accustomed to the sweet taste of fruits (babies like sweets) and then reject the vegetables. When adding new foods to the infant's diet, always do so one at a time (and in small quantities) so that, if there is an allergic reaction (such as hives or diarrhea), you will know which food caused it. Introduce new vegetables and fruits about three or four days apart. Babies adjust differently to new tastes and new textures. If the baby does not like a certain food, offer it a few days later. If you offer new foods when the baby is hungry, as at the beginning of a meal, he or she is more likely to eat them.

Fruit juice that is fortified with vitamin C can be started about the fifth to seventh month. Although some babies get two or more bottles a day of apple juice (or other type of juice), it is a good idea to limit juice to a half cup, or 4 fluid ounces, daily. More than a half cup of juice daily can result in growth failure if substituted for breast milk or formula. Another problem with fruit juice can occur when you let your baby go to sleep with a bottle in his or her mouth. The natural sugars in the juice can cause serious tooth decay, called **baby bottle tooth decay.** Letting a baby go to bed with a bottle of formula, cow's milk, or breast milk will also cause baby bottle tooth decay.

Before a baby can move on to finger foods, he or she has to be able to grab them. At about eight months, a baby discovers and starts to use the thumb and forefinger together to pick things up (called the **pincer grasp**). From about six months, the baby has been using the palm (called the **palmar**

Baby bottle tooth decay—Serious tooth decay in babies caused by letting a baby go to bed with a bottle of juice, formula, cow's milk, or breast milk.

Pincer grasp—The ability of a baby at about 8 months of age to use the thumb and forefinger together to pick things up.

Palmar grasp—The ability of a baby from about 6 months of age to grab objects with the palm of the hand.

grasp) to do this. Suitable finger foods include chopped ripe bananas, dry cereal, and pieces of cheese. About this time infants can also start eating protein foods. Poultry and fish must be very tender, and meat will have to be chopped or cut very fine and possibly moistened.

Between 10 and 12 months of age, babies may have four to six sharp teeth, and many are eating soft, chopped foods with the family. At this time it is appropriate to let your child begin drinking from a cup. It takes time, but sooner or later your child will get the idea.

By one year of age, infants can enjoy cut-up table foods as well as whole milk. By 12 months, a baby should be almost entirely self-feeding. Children should not be switched to low-fat milk until they are at least two years old, because they need the fat in whole milk for proper growth and development. Table 13-7 is a food guide for infants from birth to 12 months.

Several foods should be avoided during the first year. Because honey and liquid corn syrup may be contaminated with botulism, these foods may cause food poisoning or foodborne illness in children younger than one year. Certain foods are also more apt than others to cause choking. They include nuts, raisins, hot dogs, popcorn, whole grapes, peanut butter, chunks of apple or pear, and cherries with pits. Other foods are also more apt to cause allergies: milk, eggs, wheat, nuts, chocolate, and shellfish. Whole milk and eggs are usually introduced at about 10 to 12 months.

■ **MINI-SUMMARY**

The growth rate during the first year will never be duplicated again. The nutrient needs of infants are about double those of an adult when viewed in proportion to their weight. For the first four to six months, the only food an infant should get is either breast milk or formula. Breast milk is recommended for many reasons. It is nutritionally superior to formula, and newborns are less likely to be allergic to it. Breast milk contains antibodies, to help babies build up immunities, and growth factors, to help babies develop. Breast-fed babies are generally given vitamin D supplements. For formula-fed babies, the formula is generally cow-milk-based. If the baby is allergic to it, then a soy-based formula is used. A baby is generally ready to eat semi-solid foods between five and seven months. The progression of foods starts with iron-fortified baby cereals, strained or puréed vegetables and fruits, strained or soft protein foods, finger foods, and fruit juice, to soft chopped foods, and finally to cut-up table foods by age one. Whole milk and whole eggs are not recommended until 12 months because of possible allergic reactions. Infants start getting baby teeth after six months and many have four to six teeth by their first birthday. Foods that cause choking, such as nuts and raisins, should be avoided.

TABLE 13-7 Food Guide for Infants

Age	Food*	Amount
0–4 months	Breast milk or formula**	21–29 ounces, formula, 5–8 feedings daily. 6–8 nursings.
4–6 months	Breast milk or formula	27–39 ounces formula, 4–6 feedings daily. 4–5 nursings.
	Iron-fortified infant cereal (usually starts at 5 months)	Give 1 tablespoon with mother's milk/formula to start. Start with rice cereal. Give once to twice daily. Can work up to 1-1/2 tablespoons twice daily.
5–7 months	Strained vegetables and fruits	Give 1–2 teaspoons once to twice daily. First fruits can be applesauce, pears, peaches, and bananas. First vegetables can be carrots, squash, and sweet potatoes. Slowly increase to 2 tablespoons twice daily.
7–9 months	Breast milk or formula	30–32 ounces formula, 3–5 feedings daily. 3–5 nursings.
	Iron-fortified infant cereal	3 tablespoons plus mother's milk/formula twice daily.
	Strained fruits and vegetables	3 tablespoons twice daily.
	Strained plain meats	1 to 2 tablespoons twice daily.
	Crackers, plain toast, or teething biscuit	When baby has teeth, offer these foods after other foods are eaten.
	Fruit juice (vitamin C fortified, non-acid) (usually starts at 5 months)	Start with 2 ounces watered down juice, usually apple juice. Limit fruit juice to 1/2 cup daily.
9–12 months	Breast milk or formula (Your physician may suggest switching to whole milk at 10 months or after.)	24–32 ounces formula, 3–4 times daiy. 3–4 nursings.
	Fruit juice (vitamin C fortified)	1/2 cup daily.
	Iron-fortified infant cereal	3–4 tablespoons plus mother's milk/formula twice daily.
	Vegetables, cut up	3–4 tablespoons twice daily.
	Fruits, cut up	3–4 tablespoons twice daily.
	Meats, cut up	2–3 tablespoons twice daily.
	Egg (usually at 12 months)	1 egg = 1 serving of meat.
	Bread and grain products	1/2 slice four times daily.

* Avoid the following foods in the first year because of possible allergic reactions: chocolate, nuts, berries, tomatoes, shellfish.

** Physician may request iron-fortified formula by third or fourth month.

Childhood

Around age one, the baby's growth rate decreases markedly. Yearly weight gain now approximates 4 to 6 pounds per year. Children can expect to grow about 3 inches per year between ages one and seven, and then 2 inches per year until the pubertal growth spurt. Until adolescence, growth will come in spurts, during which the child will grow more and eat more.

After age one, children start to lose baby fat and become leaner, with muscle accounting for a larger percentage of body weight. The legs become longer, and the baby now starts to look like a child and to walk, run, and jump like a child. By age two brain growth is 75 percent complete. A child's head size in relation to body size starts to decrease and look more normal, and by age six to ten, the brain becomes adult size.

By about age one, the child has six to eight teeth, and by age two the baby teeth are almost all in. Between ages six and twelve, these are gradually replaced with permanent teeth. After the first birthday, as children's physical capabilities and desire for independence increase, they are more capable of feeding themselves. By age 18 months, many children can successfully use a spoon without too much spilling, and by 24 months many children can drink properly from a cup.

Nutrition During Childhood

Table 13-8 shows the RDA for calories and protein for children. A one-year-old needs roughly 1,000 calories a day, and a three-year-old about 1,300. By age 10, a child needs about 2,000 calories. Energy needs of children of similar age, sex, and size can vary due to differing BMRs (basal metabolic

TABLE 13-8 Recommended Dietary Allowances for Calories and Protein for Children

Age	Weight (lb.)	Height (in.)	Calories*	Protein (grams)
1–3	29	35	1300	16
4–6	44	44	1800	24
7–10	62	52	2000	28

* Energy allowances for children are based on median energy intakes of children of these ages followed in longitudinal growth studies.

Source: *Recommended Dietary Allowances*, 1989, by the National Academy of Sciences, National Academy Press, Washington, DC.

rates), growth rates, and activity levels. Energy and protein needs decline gradually per pound of body weight.

During growth spurts, the requirements for calories and nutrients are greatly increased. Appetite fluctuates tremendously, with a good appetite during **growth spurts** (periods of rapid growth) and a seemingly terrible appetite during periods of slow growth. Parents may worry and force a child to eat more than needed at such times, when the child appears to be "living on air." A decreased appetite in childhood is perfectly normal. As long as the child is choosing nutrient-dense calories, nutritional problems are unlikely. In preparation for the adolescent growth spurt, children accumulate stores of nutrients, such as calcium, that will be drawn upon later, as intake cannot meet all the demands of this intensive growth spurt.

Although calorie and protein intakes are rarely inadequate in American children, there are concerns about iron intake. Lack of iron can cause decreased energy and affect behavior, mood, and attention span. A balanced diet with adequate consumption of iron-rich foods such as lean meat (ground meat for younger children is easier to chew), enriched breads and cereals, and legumes is important to get enough iron. A source of vitamin C, such as citrus fruits, increases the amount of iron absorbed.

To prevent coronary heart disease early in life, medical authorities generally agree that by age five all healthy children should comply with the following guidelines for adults:

1. Reduce total fat intake to 30 percent of calories.
2. Reduce saturated fat intake to less than 10 percent of calories.
3. Consume no more than 300 milligrams of cholesterol daily.

Some parents have overzealously interpreted these guidelines and restricted the fat content of their childrens' diets to the point where the children received inadequate calories, which then interferes with normal growth. After the child reaches age two parents may want to limit fatty meats and cheeses and use 2 percent milk.

Preschoolers exhibit some food-related behaviors that drive their parents crazy, such as **food jags** (eating mostly one food for a period of time). Food jags usually don't last long enough to cause any harm. Preschoolers often pick at foods or refuse to eat vegetables or drink milk. Lack of variety, erratic appetites, and food jags are typical of this age group. Toddlers (ages one to three) tend to be pickier eaters than older preschoolers (ages four to five). Toddlers are just starting to assert their independence and love to say "no" to parental requests. They may wage a control war, and parents need to set limits without being too controlling or rigid. Here are several tactics for dealing with preschoolers' (and school-age children's) food habits.

Growth spurt—Period of rapid growth.
Food jag—A habit of young children in which they have favorite foods they want to eat frequently.

1. Make mealtime as relaxing and enjoyable as possible.
2. Don't nag, bribe, force, or even cajole a child to eat. Stay calm. Pushing or prodding children almost always backfires. Children learn to hate the foods they are encouraged to eat and to desire the foods used as rewards, such as cake and ice cream. Once children know that you won't allow eating to be made into an issue of control, they will eat when they're hungry and stop when they're full. Your child is the best judge of when he or she is full.
3. Allow children to choose what they will eat from two or more healthy choices. You are responsible for choosing which foods are offered, and the child is responsible for deciding how much he or she wants of those foods.
4. Let children participate in food selection and preparation. Table 13-9 lists cooking activities for children of various ages.

TABLE 13-9　Age-Appropriate Cooking Activities

2-1/2–3-Year-Olds
■ Wash fruits and vegetables
■ Peel bananas
■ Stir batters
■ Slice soft foods with table knife (cooked potatoes, bananas)
■ Pour
■ Fetch cans from low cabinets
■ Spread with a knife (soft onto firm)
■ Use rotary egg beater (for a short time)
■ Measure (e.g., chocolate chips into 1-cup measure)

4–5-Year-Olds
■ Grease pans
■ Open packages
■ Peel carrots
■ Set table (with instruction)
■ Shape dough for cookies/hamburger patties*
■ Snip fresh herbs for salads or cooking

■ Wash and tear lettuce for salad, separate broccoli, cauliflower
■ Place toppings on pizza or snacks

6–8-Year-Olds
■ Take part in planning part of or entire meal
■ Set table (with less supervision)
■ Make a salad
■ Find ingredients in cabinet or spice rack
■ Shred cheese or vegetables
■ Garnish food
■ Use microwave, blender, or toaster oven (with previous instruction)
■ Measure ingredients
■ Present prepared food to family at table
■ Roll and shape cookies

9–12-Year-Olds
■ Depending on previous experience, plan and prepare an entire meal

* Children should not put their hands in their mouths while handling raw hamburger meat or dough with eggs. It can carry harmful bacteria. They should wash hands after shaping patties.

5. Respect your child's preferences when planning meals, but don't make your child a quick peanut butter sandwich, for instance, if he or she rejects your dinner.

6. Make sure your child has appropriately sized utensils and can reach the table comfortably.

7. Preschoolers love rituals, so start them early with the habit of eating three meals plus snacks each day at fairly regular times. Also, eat with your preschooler and model good eating habits.

8. Expect preschoolers to reject new foods at least once, if not many times. Simply continue presenting the new food, perhaps prepared differently, and one day they will try it (usually after 12 to 15 exposures).

9. Let the child serve himself or herself small portions.

10. Do not use desserts as a reward for eating meals. Make dessert a normal part of the meal, and make it nutritious.

11. Ask children to try new foods (just a little bite!) and praise them when they try something different. Encourage them by telling them about someone who really likes the food or relating the food to something they think is fun. Realize, though, that some children are less likely to try new things, including new foods.

12. If all else fails, keep in mind that children under six have more taste buds (which may explain why youngsters are such picky eaters) and that this, too, will pass.

Luckily, school-age children are much better eaters. Although they generally have better appetites and will eat a wider range of foods, they often dislike vegetables and casserole dishes.

Both preschoolers and school-age children learn about eating by watching others: their parents, their siblings, their friends, and their teachers. Parents, siblings, and friends provide role models for children and influence children's developing food patterns. Parents' interactions with their children will also influence what foods they or will not accept.

When children go to school, their peers influence their eating behaviors as well as what they eat for lunch. Lunch for school-age children often consists of the school lunch or a packed lunch from home.

Having breakfast makes a difference in how kids perform at school. Breakfast also makes a significant contribution to the child's intake of calories and nutrients for the day. Children who skip breakfast usually don't make up for the calories at other meals. If a child gets both breakfast and lunch at school, these meals together typically contribute about 50 percent of the day's calories.

Preschoolers and school-age children also learn about food by watching television. Research shows that children who watch a lot of television are more apt to be overweight. It not only takes them away from more robust activities but exposes them to commercials that are often for sugared cereals, candy, and other empty-calorie foods. Both obesity and inactivity are currently on the rise in school-age children, especially among adolescents.

So what can parents do to make sure their children eat nutritious diets and get exercise? Be a good role model by eating a well-balanced and varied diet. Have nutritious food choices readily available at home and serve a regular, nutritious breakfast. Maintain regular family meals as much as possible. Family meals are an appropriate time to model healthy eating habits and try out new foods. Also, limit television watching and encourage physical activity.

Menu Planning for Children

By the time children are four years old, they can eat amounts that count as regular Food Guide Pyramid servings eaten by older family members— that is, 1/2 cup fruit or vegetable, 3/4 cup of juice, 1 slice of bread, and 2 to 3 ounces of cooked lean meat, poultry, or fish. Children two to three years of age need the same variety of foods as four- to six-year-olds but may need smaller portions, about two-thirds of what counts a regular Food Guide Pyramid serving (except for milk). Two- to six-year-old children need a total of two full servings from the milk group each day. Following are additional menu-planning guidelines.

Preschoolers

1. Offer simply prepared foods and avoid casseroles or any foods that are mixed together, as children need to identify what they are eating.
2. Offer at least one colorful food, such as carrot sticks.
3. Preschoolers like nutritious foods in all food groups but are often reluctant to eat vegetables. Part of this problem may be due to the difficulty involved in getting them onto a spoon or fork. Vegetables are more likely to be accepted if served raw and cut up as finger foods. However, when serving celery, be sure to take off the strings. Serve cooked vegetables somewhat undercooked, so they are a little crunchy. Brightly colored, mild-flavored vegetables such as peas and corn are more popular with kids.
4. Provide at least one soft or moist food that is easy to chew at each meal. A crisp or chewy food is important, too, to develop chewing skills.

5. Avoid strong-flavored and highly salted foods. Children have more taste buds than adults, so these foods taste too strong to them.

6. Preschoolers love carbohydrate foods, including cereals, breads, and crackers, as they are easy to hold and chew.

7. Smooth-textured foods such as pea soup or mashed potatoes should not have any lumps—children find this unusual.

8. Before age four, when food-cutting skills start to develop, the child needs food served in bite-size pieces that are either eaten as finger foods or with utensils. For example, cut meat into strips or use ground meat, cut fruit into wedges or slices, and serve pieces of raw vegetables instead of a mixed salad. Other good finger foods include cheese sticks, wedges of hard-boiled eggs, dry ready-to-eat cereal, fish sticks, arrowroot biscuits, and graham crackers.

9. Serve foods warm, not hot; a child's mouth is more sensitive to hot and cold than an adult's. Also, little children need little plates, utensils, and cups, as well as seats that allow them to reach the table comfortably.

10. Cut-up fruit and vegetables make good snacks. Let preschoolers spread peanut butter on crackers or use a spoon to eat yogurt. Snacks are important to preschoolers because they need to eat more often than adults.

11. To minimize choking hazards for children under four:

   ■ Slice hotdogs in quarters lengthwise.
   ■ Shred hard raw vegetables and fruits.
   ■ Remove pits from apples, cherries, plums, peaches, and other fruits.
   ■ Cut grapes in half lengthwise.
   ■ Spread peanut butter thin.
   ■ Chop nuts and seeds fine.
   ■ Check to make sure fish is boneless.
   ■ Avoid popcorn and hard candies.

12. Children learn to like new foods by being presented with them repeatedly.

School-Age Children

1. Serve a wide variety of foods, including children's favorites: tuna fish, pizza (use vegetable toppings), macaroni and cheese, hamburgers (use lean beef combined with ground turkey breast), hot dogs (use low-fat varieties), and peanut butter.

2. Good snack choices are important, as children do not always have the desire or the time to sit down and eat. Snacks can include fresh fruits and vegetables, dried fruits, fruit juices, breads, cold cereals, popcorn

(without excessive fat), pretzels, tortillas, muffins, milk, yogurt, cheese, pudding, sliced lean meats and poultry, and peanut butter.

3. Balance menu items that are higher in fat with those containing less fat.

4. Pay attention to serving sizes.

5. Children's most common nutritional problem is iron-deficiency anemia. Offer iron-rich foods such as meat in hamburgers or roast beef sandwiches, peanut butter, baked beans, chili, dried fruits, and fortified dry cereals.

6. As children grow, they need to eat more high-fiber foods, such as fruits, vegetables, beans and peas, and wholegrain foods. Whereas adults need at least 25 grams of fiber daily, children need a daily amount equal to or greater than their age plus 5 grams. In other words, a 12-year-old needs 17 grams of fiber daily.

All children up to age 10 need to eat every four to six hours to keep their blood glucose at a desirable level; therefore, snacking is necessary between meals. Nutritious snack choices for both preschoolers and school-age children are noted above.

Snacks as well as meals should provide good sources of calcium, such as dairy foods, and iron and zinc, such as meats and legumes. If a child drinks little or no milk, try adding flavorings to milk such as chocolate or strawberry, or make cocoa, milkshakes, puddings, and custards. Milk can be fortified with powdered milk by blending 2 cups of fluid milk with 1/3 cup powdered milk. One cup of fortified milk is equal to 1-1/2 cups of regular milk. Powdered milk can also be added in baking and to casseroles, soups, sauces, gravies, ground meats, mashed potatoes, and scrambled eggs. Cheese and yogurt, of course, are also good sources of calcium.

■ MINI-SUMMARY

By a child's first birthday, the growth rate decreases markedly and yearly weight gain until puberty is about 4 to 6 pounds. By the second birthday, the baby teeth are almost all in. During growth spurts, children's appetites are good; otherwise, their appetites may seem poor. Preschoolers can be fussy eaters, often have food jags, and can take the pleasure out of mealtime. Guidelines for eating with preschoolers and menu planning for them are detailed. Children's eating habits are influenced by family, friends, teachers, availability of school breakfast and lunch programs, and television. Menu-planning guidelines for school-age children are detailed. The most common nutritional problem of children is iron-deficiency anemia.

Adolescence

Puberty, the process of physically developing from a child to an adult, starts at about age 10 or 11 for girls and 12 or 13 for boys. In girls, it peaks at age 12 and is completed by age 15. In males, it peaks at age 14 and is completed by age 19. The timing of puberty and rates of growth show much individual variation. During the five to seven years of pubertal development, adolescents gain about 20 percent of adult height and 50 percent of adult weight. Most of the body organs double in size, and almost half of total bone growth occurs.

Whereas before puberty the proportion of fat and muscle was similar in males and females, males now put on twice as much muscle as females, and females gain proportionately more fat. In adolescent girls, an increasing amount of fat is being stored under the skin, particularly in the abdominal area. The male also experiences a greater increase in bone mass than does the female.

Nutrition During Adolescence

Table 13-10 compares the RDA for calories and protein for adolescent males and females to adult needs. A major limitation of the RDAs for adolescents is that they are categorized according to age groups and are not related to stages of physical maturity, which show individual variation. The highest levels of nutrients are for individuals growing at the fastest rate.

Males now need more calories, protein, calcium, iron, and zinc for muscle and bone development than females; however, females need increased iron due to the onset of menstruation. Owing to their big appetites and calorie needs, teenage boys are more likely than girls to get sufficient nutrients. Females have to pack nutrients into fewer calories, which can become difficult if they decrease their food intake in an effort to lose weight.

With their increased independence, adolescents assume responsibility for their own eating habits. Teenagers are not fed; they make most of their own food choices. They eat more meals away from home, such as at fast-food restaurants, and skip more meals than previously. Irregular meals and snacking are common due to busy social lives and after-school activities and jobs. Teenagers will tell you that they lack the time or discipline to eat right, although many are pretty well informed about good nutrition practices. Ready-to-eat foods such as cookies, chips, and soft drinks are readily available, and teenagers pick them up as snack foods. Studies show that snacks contribute one-quarter to one-third of total daily calories and make substantial nutrient contributions except for iron.

TABLE 13-10   Recommended Dietary Allowances for Adolescents for Calories and Protein

Age	Weight	Height	Calories*	Protein
Males				
11–14	99	62	2500	45
15–18	145	69	3000	59
19–24	160	70	2900	58
25–50	174	70	2900	63
Females				
11–14	101	62	2200	46
15–18	120	64	2200	44
19–24	128	65	2200	46
25–50	138	64	2200	50

* Energy allowances are based on median energy intakes of people of those ages followed in longitudinal growth studies.

Source: *Recommended Dietary Allowances:* ©1989 by the National Academy of Sciences, National Academy Press, Washington, DC.

Adolescents often have a variety of lunch options when they are at school. These choices may include leaving the school to buy lunch, eating a lunch from the National School Lunch Program, buying à la carte foods in the school cafeteria that do not qualify as a lunch in the National School Lunch Program, or buying food from a school store or vending machines. Although federal regulations prohibit the sale of carbonated beverages, chewing gum, water ices, and most hard candies in the foodservice area or cafeteria during mealtimes, vending machines with these foods are often found just outside of the cafeteria. State, local, or school rules may close vending machines during mealtime and other times during the school day. Two professional associations, the American Dietetic Association and the American School Food Service Association, have concerns that the foods sold in vending machines, school stores, and à la carte cafeterias discourage students from eating meals provided by the National School Lunch and Breakfast programs.

The media have a powerful influence on adolescents' eating patterns and behaviors. Advertising for not-so-nutritious foods and fast foods permeates television, radio, and billboards. In addition, questionable eating habits are portrayed on television shows. In a study of 12- to 17-year-old adolescents, the prevalence of obesity increased 2 percent for each additional hour of television watched.

A typical meal at a fast-food restaurant—a 4-ounce hamburger, French fries, and a regular soft drink—is high in calories, fat, and sodium. However, more nutritious choices are available at fast-food restaurants. Smaller hamburgers, milk, salads, and grilled chicken sandwiches are examples of more nutritious options.

Parents can positively influence adolescents' eating habits by being good role models and by having dinner and nutritious breakfast and snack foods available at home. Adolescents can become involved in food purchasing and preparation. Parents can also influence their children's fitness level by limiting sedentary activities and encouraging exercise.

Both adolescent boys and girls are influenced by their body image. Adolescent boys may take nutrition supplements and fill up on protein in hopes of becoming more muscular. Adolescent girls who feel they are overweight may skip meals and modify their food choices in hopes of losing weight. Teens who need to lose weight should limit the amount of high-fat food and/or substitute lower-fat choices, such as skim milk for whole milk or nonfat frozen yogurt instead of ice cream. High-fat foods such as French fries and candy bars that have no low-fat substitutes should be eaten only once in a while or in very small amounts. Whether overweight or not, teens need regular exercise.

Menu Planning for Adolescents

1. Emphasize complex carbohydrates such as assorted breads, rolls, cereals, fruits, vegetables, potatoes, pasta, rice, and dried beans and peas. These foods supply calories along with needed nutrients. Wholegrain products are preferred.
2. Offer well-trimmed lean beef, poultry, and fish. Don't think that just because adolescents need more calories that fatty meats are in order. Their fat calories should be less saturated.
3. Low-fat and skim milk need to be offered at all meals. Girls are more likely to need to select skim milk than males. Other forms of calcium also need to be available, such as pizza, macaroni and cheese and other entrées using cheese, yogurt, frozen yogurt, ice milk, puddings, and custards made with skim milk.
4. Offer margarine; many adolescents are probably used to eating it at home.
5. Have nutritious choices available for hungry on-the-run adolescents looking for a snack. Nutritious snack choices include fresh fruit, muffins and other quick breads, crackers or rolls with low-fat cheese or peanut butter, vegetable-stuffed pita pockets, yogurt or cottage cheese with fruit, or fig bars.

6. Emphasize quick and nutritious breakfasts, such as wholegrain pancakes or waffles with fruit, juices, wholegrain toast or muffins, with low-fat cheese, cereal topped with fresh fruits, or bagels with peanut butter.

7. The nutrients most often lacking in adolescent diets are iron, folate, and calcium. Significant iron sources include meats, poultry, fish, eggs, legumes, and dried fruits. Vitamin C aids the absorption of iron from legumes and dried fruits. Folate is found in leafy green vegetables, orange juice, and beans and peas. Calcium may be lacking for those who have an inadequate intake of milk and other dairy products. If teenagers frequently drink soft drinks instead of milk, they may not have enough calcium in their diets to support bone growth.

◾ **MINI-SUMMARY**

The pubertal growth spurt starts at about age 10 for girls and 12 for boys. During the five to seven years of pubertal development, the adolescent gains about 20 percent of adult height and 50 percent of adult weight. Boys gain twice as much muscle and more bone mass than girls, who gain proportionately more fat. Males now need more calories, protein, calcium, iron, and zinc for muscle and bone development than females, who need increased amounts of iron due to menstruation. Teenagers make most of their own food choices, which are influenced by their body image, peers, and the media. Menu-planning guidelines for adolescents are detailed.

Eating Disorders

Each year millions of Americans develop serious and sometimes life-threatening eating disorders. The vast majority—more than 90 percent—of those afflicted with eating disorders are adolescent and young adult women. One reason that women in this age group are particularly vulnerable to eating disorders is their tendency to go on strict diets to achieve an "ideal" figure. Researchers have found that such stringent dieting can play a key role in triggering eating disorders. The actual cause of eating disorders is not entirely understood, but many risk factors have been identified. Risk factors may include a high degree of perfectionism, low self-esteem, genetics, or family preoccupation with dieting and weight.

Eating-disorder patients deal with two sets of issues: those surrounding their eating behaviors and those surrounding their interactions with others and themselves. Eating disorders are considered a mental disorder, and both psychotherapy and medical nutrition therapy are cornerstones of treatment.

Anorexia nervosa—An
eating disorder most
prevalent in adolescent
females who starve
themselves.

Bulimia nervosa—An
eating disorder
characterized by a
destructive pattern of
excessive overeating
followed by vomiting or
other "purging" behaviors
to control weight.

**Binge eating
disorder**—An eating
disorder characterized by
episodes of uncontrolled
eating or binging.

Approximately 1 percent of adolescent girls develop **anorexia nervosa,** a dangerous condition in which they can literally starve themselves to death. Another 2 to 3 percent of young women develop **bulimia nervosa,** a destructive pattern of excessive overeating followed by vomiting or other "purging" behaviors to control their weight. The most recently recognized eating disorder, **binge eating disorder,** could turn out to be the most common. With this disorder, binges are not followed by purges, so these individuals often become overweight. Eating disorders also occur in men and older women, but much less frequently.

The consequences of eating disorders can be severe, with one in ten cases leading to death from starvation, cardiac arrest, or suicide over the course of ten years. The outlook is better for bulimia than for anorexia; anorexia patients tend to relapse more. Many patients with anorexia or bulimia also suffer from other psychiatric illnesses such as clinical depression, anxiety, obsessive-compulsive disorder, or substance abuse. Fortunately, increasing awareness of the dangers of eating disorders—sparked by medical studies and extensive media coverage of the illness—has led many people to seek help. Nevertheless, some people with eating disorders refuse to admit that they have a problem and do not get treatment. Family members and friends can help recognize the problem and encourage the person to seek treatment. The earlier treatment is started, the better the chance of a full recovery.

Anorexia Nervosa

People who intentionally starve themselves suffer from anorexia nervosa. This disorder, which usually begins in young people around the time of puberty, involves extreme weight loss—at least 15 percent below the individual's normal body weight. Many people with the disorder look emaciated but are convinced they are overweight. Sometimes they must be hospitalized to prevent starvation. Let's look at a typical case.

Deborah developed anorexia nervosa when she was 16. A rather shy, studious teenager, she tried hard to please everyone. She had an attractive appearance but was slightly overweight. Like many teenage girls, she was interested in boys but concerned that she wasn't pretty enough to get their attention. When her father jokingly remarked that she would never get a date if she didn't take off some weight, she took him seriously and began to diet relentlessly—never believing she was thin enough, even when she became extremely underweight.

Soon after the pounds started dropping off, Deborah's menstrual periods stopped. As anorexia tightened its grip, she became obsessed with dieting and food, and developed strange eating rituals. Every day she weighed all the food she would eat on a kitchen scale, cutting solids into minuscule pieces and precisely measuring liquids. She would then put her daily ration in small containers, lining them up in neat rows. She also exercised compulsively, even after she weakened and became faint.

No one could convince Deborah that she was in danger. Finally, her doctor insisted that she be hospitalized and carefully monitored for treatment of her illness. While in the hospital, she secretly continued her exercise regimen in the bathroom. It took several hospitalizations and a good deal of individual and family outpatient therapy for Deborah to face and solve her problems.

One of the most frightening aspects of the disorder is that people with anorexia continue to think they are overweight, even when they are bone-thin. Food and weight become obsessions. For some, the compulsiveness shows up strange eating rituals or the refusal to eat in front of others. It is not uncommon for anorexics to collect recipes and prepare gourmet feasts for family and friends but not partake in the meals themselves.

In patients with anorexia, starvation can damage vital organs such as the heart and brain. To protect itself, the body shifts into "slow gear": menstrual periods stop and breathing, pulse, and blood pressure rates drop. Nails and hair become brittle. The skin dries, yellows, and becomes covered with soft hair called **lanugo.** Reduced body fat leads to lowered body temperature and the inability to withstand cold.

Lanugo—Downy hair on the skin.

Mild anemia, swollen joints, reduced muscle mass, and lightheadedness are also common. If the disorder becomes severe, patients may lose calcium from the bones, making them brittle and prone to breakage. They may also experience irregular heart rhythms and heart failure.

Bulimia Nervosa

People with bulimia nervosa consume large amounts of food and then rid their bodies of excess calories by vomiting, abusing laxatives or diuretics, taking enemas, or exercising obsessively. Some use a combination of all these forms of purging. Because many individuals with bulimia "binge and purge" in secret and maintain normal or above-normal body weight, they can often successfully hide their problem for years. Let's take a look at Lisa.

Lisa developed bulimia at age 18. As with Deborah, her strange eating behavior began when she started to diet. She too dieted and exercised to lose weight, but unlike Deborah, she regularly ate huge amounts of food and maintained her normal weight by forcing herself to vomit. Lisa often felt like an emotional powder keg—angry, frightened, and depressed.

Unable to understand her own behavior, she thought no one else would either, so she felt isolated and lonely. Typically, when things were not going well, she would be overcome with an uncontrollable desire for sweets. She would eat pounds of candy and cake at a time and often not stop until she was exhausted or in severe pain. Then, overwhelmed with guilt and disgust, she would make herself vomit.

While recuperating in a hospital from a suicide attempt, she was referred to an eating disorders clinic, where she got into group therapy. She also received medications to treat the illness and the understanding and help she so desperately needed from others who had the same problem.

Individuals with this disorder may binge and purge once or twice a week or as much as several times a day. Dieting heavily between episodes of binging and purging is also common.

As with anorexia, bulimia typically begins during adolescence. The condition occurs most often in women but is also found in men. Many individuals with bulimia, ashamed of their strange habits, do not seek help until they reach their thirties or forties. By this time, their eating behavior is deeply ingrained and more difficult to change.

Bulimic patients—even those of normal weight—can severely damage their bodies by frequent binge eating and purging. Vomiting causes serious problems: The acid in vomit wears down the outer layer of the teeth and can cause scarring on the backs of hands when fingers are pushed down the throat to induce vomiting. Further, the esophagus becomes inflamed and the glands near the cheeks become swollen.

Binge Eating Disorder

Binge eating disorder resembles bulimia in that it is characterized by episodes of uncontrolled eating, or binging. However, binge eating disorder differs from bulimia in that its sufferers do not purge their bodies of excess food. Binge eating was recognized as a mental disorder in 1994. This is not to say that binge eating is a recent development—it's been around for a long time, but only lately has it been categorized as a mental disorder.

Binge eaters feel that they lose control of themselves when eating. They eat large quantities of food and do not stop until they are uncomfortably full. Usually, they have more difficulty losing weight and keeping it off than do people with other serious weight problems. Most people with this disorder are obese and have a history of weight fluctuations. Binge eating disorder is found in about 2 percent of the general population—more often in women than men. Binge eating disorder occurs in about 30 percent of people participating in medically supervised weight control programs.

Treatment

The sooner a disorder is diagnosed, the better the chances that treatment can work. The longer abnormal eating behaviors persist, the more difficult it is to overcome the disorder and its effects on the body. In some cases, long-term treatment is required.

Once an eating disorder is diagnosed, the clinician must determine whether the patient is in immediate medical danger and requires hospitalization. Although most patients can be treated as outpatients, some need hospital care, as in the case of severe purging or risk of suicide.

Eating-disorder patients commonly work with a treatment team that includes an internist, a nutritionist, an individual psychotherapist, and someone who is knowledgeable about psychoactive medications used in treating these disorders. Treatment usually includes individual psychotherapy, family therapy, cognitive-behavioral therapy, medical nutrition therapy, and possibly medications such as antidepressant drugs.

Eating disorders, unfortunately, have a very high death rate; one out of every ten patients will die. With that in mind, prevention of these diseases needs to be seriously examined. Research has identified the community groups most important to reach: junior high school students, coaches, and parents.

Table 13-11 lists questions to help individuals determine whether they have an eating disorder.

TABLE 13-11   Do You Have an Eating Disorder?

A positive answer to one or more of these questions may indicate an eating disorder.

1. Do you eat large amounts of food in a very short period while feeling out of control and by yourself?
2. Do you frequently eat a lot of food when you are not hungry and usually when you are alone?
3. Do you feel guilty after overeating?
4. Do you make yourself vomit or use laxatives or diuretics to purge yourself?
5. Do you carefully make sure you eat only a small number of calories each day, such as 500 calories or less, and exercise a lot?
6. Do you avoid going out to maintain your eating and exercise schedule?
7. Do you feel that food controls your life?

■ MINI-SUMMARY

As shown in Table 13-12, there are three different types of eating disorders. Most people afflicted with these problems are adolescent girls and young women. The sooner the disorder is diagnosed, the better the chances for successful treatment. Treatment usually includes individual psychotherapy, family therapy, cognitive-behavior therapy, medical nutrition therapy, and possibly medications. Of all types of mental illness, eating disorders have one of the highest death rates.

TABLE 13-12 Common Symptoms of Eating Disorders			
Symptoms	Anorexia Nervosa*	Bulimia Nervosa*	Binge Eating Disorder
Excessive weight loss in relatively short period of time	√		
Continuation of dieting although bone-thin	√		
Dissatisfaction with appearance; belief that body is fat, even though severely underweight	√		
Loss of monthly menstrual periods	√	√	
Unusual interest in food and development of strange eating rituals	√	√	
Eating in secret	√	√	√
Obsession with exercise	√	√	
Serious depression	√	√	√
Binging—consumption of large amounts of food		√	√
Vomiting or use of drugs to stimulate vomiting, bowel movements, and urination		√	
Binging but no noticeable weight gain		√	
Disappearance into bathroom for long periods of time to induce vomiting		√	
Abuse of drugs or alcohol		√	√

* Some individuals suffer from anorexia and bulimia and have symptoms of both disorders.

Older Adults

The fastest-growing age group in the United States comprises those over age 85! With the baby boom generation entering their fifties, the graying of America is in full swing. This trend is seen in the growing number of retirement communities and nursing facilities. Not only are there more elderly, but they are living longer. A 65-year-old woman can probably expect to live into her eighties. Her male counterpart still has about nine or ten more years.

Before looking at nutrition during aging, let's take a look at what happens when we age. Studies suggest that the maximum efficiency of many organ systems occurs between ages 20 and 35. After age 35, the functional capability of almost every organ system declines. Similar changes occur in adults as they age, but the rate of decline shows great individual variation. Both genetics and environmental factors such as nutrition affect the rate of aging. Conversely, changes brought about by the aging process affect nutrition status. Of particular importance are changes that affect digestion, absorption, and metabolism of nutrients.

The basal metabolic rate declines between 8 and 12 percent from age 30 to 70 and is accompanied by a 25 to 30 percent loss in muscle mass. Combined with a general decrease in activity level, these factors clearly indicate a need for decreased calorie intake, which generally does take place during aging. But the elderly need not lose all that muscle mass. Studies have shown that when the elderly do regular weight-training exercises, they increase their muscular strength, increase basal metabolism, improve appetite, and improve blood flow to the brain.

Overall, the functioning of the cardiovascular system declines with age. The workload of the heart increases due to atherosclerotic deposits and less elasticity in the arteries. The heart does not pump as hard as before, and cardiac output is reduced in elderly people who do not remain physically active. Blood pressure increases normally with age. Pulmonary capacity decreases by about 40 percent throughout life. This decrease does not restrict the normal activity of healthy older persons but may limit vigorous exercise. Kidney function deteriorates over time, and the aging kidney is less able to excrete waste. Adequate fluid intake is important, as is avoiding megadoses of water-soluble vitamins because they will put a strain on the kidney to excrete them. Last, loss of bone occurs normally during aging and osteoporosis is common (see Chapter 7).

Factors Affecting Nutrition Status

The nutrition status of an elderly person is greatly influenced by many variables, including physiological, psychosocial, and socioeconomic factors.

Physiological Factors

- **Disease.** The presence of disease, both acute and chronic, and use of modified diets can affect nutrition status. The elderly are major users of modified diets. The most prevalent nutrition-related problems of the elderly are chronic conditions that require modified diets. Certain chronic diseases are associated with **anorexia** (lack of appetite), such as gastrointestinal disease, congestive heart failure, renal disease, and cancer. Other diseases, such as stroke, are not associated with anorexia but can cause the individual to take in little food.

- **Less muscle mass.** With aging, there is less muscle mass, so the basal metabolic rate decreases. As the BMR slows down, the number of calories needed by the elderly decreases.

- **Activity level.** Because active individuals tend to eat more calories than their sedentary counterparts do, they are more likely to ingest more nutrients.

- **Dentition.** Approximately 50 percent of Americans have lost their teeth by age 65. Despite widespread use of dentures, chewing still presents problems for many of the elderly.

- **Functional disabilities.** Functional disabilities interfere with the ability of the elderly to perform daily tasks, such as the purchasing and preparation of food and eating. These disabilities may be due to arthritis or rheumatism, stroke, visual impairment, heart trouble, or dementia. One study reported that 39 percent of the elderly subjects needed help when food shopping and 26 percent needed help making meals.

- **Taste and smell.** Sensitivity to taste and smell decline slowly with age. The taste buds become less sensitive and the nasal nerves that register aromas need extra stimulation to detect smells. That's why seniors may find ordinarily seasoned foods too bland. Medications also may alter an individual's ability to taste.

- **Changes in the gastrointestinal tract.** The movement of food through the gastrointestinal tract slows down over time, causing problems such as constipation, a frequent complaint of older people. Constipation may also be related to low fiber and fluid intake, medications, or lack of exercise. Other frequent complaints include nausea, indigestion, and heartburn. (Heartburn, a burning sensation in the area of the throat, has nothing to do with the heart. It occurs when acidic stomach contents are pushed into the lower part of the esophagus or throat.)

Anorexia—Lack of appetite.

■ **Medications.** More than half of all seniors take at least one medication daily, and many take six or more a day. Medications may alter appetite or the digestion, absorption, and metabolism of nutrients (Table 13-13).

■ **Thirst.** Many of the elderly suffer a diminished perception of thirst—especially problematic when they are not feeling well. Because the aging kidney is less able to concentrate urine, more fluid is lost, setting the stage for dehydration.

Psychosocial Factors

■ **Cognitive functioning.** Poor cognitive functioning may affect nutrition, or perhaps poor nutritional status contributes to poor cognitive functioning.

■ **Social support.** An individual's nutritional health results in part from a series of social acts. The purchasing, preparing, and eating of foods are social events for most people. For example, elderly people may rely on one another for rides to the supermarket, cooking, and sharing meals. The benefits of social networks or support are largely due to the companionship and emotional support they provide. It is anticipated that this has a positive effect on appetite and dietary intake.

Socioeconomic Factors

■ **Education.** Higher levels of education are positively associated with increased nutrient intakes.

■ **Income.** In a large study of older Americans using data from the Nationwide Food Consumption Survey 1977–78, money spent on food was found to be a highly significant predictor of dietary quality.

TABLE 13-13 Nutrients Depleted by Selected Drugs		
Drug Group	Drug	Nutrients Depleted
Analgesics	Uncoated aspirin	Iron
Antacids	Aluminum or magnesium hydroxide	Phosphate, calcium, and folate
	Sodium bicarbonate	calcium, folate
Antiulcer drugs	Cimetidine	B_{12}
Chemotherapeutic agents	Methotrexate	Folate
Cholesterol-lowering agents	Cholestyramine	Fat, vitamins A and K
Diuretics	Lasix	Potassium

■ **Living arrangements.** The elderly, particularly women, are more likely to be widowed. The trend has been for widows and widowers in the United States to live alone after the spouse dies. Research focusing on the impact of living arrangements on dietary quality showed that living alone is a risk factor for dietary inadequacy for older men, especially those over age 75 years of age, and for women in the youngest age group (55 to 64).

■ **Availability of federally funded meals.** The availability of nutritious meals through federal programs such as Meals on Wheels, in which meals are delivered to the home, is crucial to the nutritional health of many elderly. Another popular elderly feeding program is the Congregate Meals Program, in which the elderly go to a senior center to eat.

Nutrition for Older Adults

A survey completed for the Nutrition Screening Initiative—targeted at improving the nutritional health status of the aging—shows that, although 85 percent of seniors surveyed believe that nutrition is important for their health and well-being, few act on their beliefs. Further, 30 percent admit to skipping at least one meal a day. These numbers may well soar as America continues to gray at an increasing rate. Studies of the elderly have shown that maintaining adequate calorie intake is vital to good nutrition.

Because the elderly consume fewer calories, this means there is less room in the diet for empty-calorie foods such as sweets, alcohol, and fats. At a time when good nutrition is so important to good health, there are many obstacles to fitting more nutrients into fewer calories, such as medical conditions, dentures, and medications.

Nutrients of concern to the elderly include the following:

■ **Water.** Due to decreased thirst sensation and other factors, fluid intake is more important for older adults than for younger people. It is also important to prevent constipation (as is fiber).

■ **Vitamin B_{12}.** The elderly have a problem with vitamin B_{12} even if they take in enough. The stomach of an elderly person secretes less gastric acid and pepsin, both of which are necessary to break vitamin B_{12} from its polypeptide linkages in food. The result is that less vitamin B_{12} is absorbed. Vitamin B_{12} is necessary to convert folate into its active form so that folate can do its job of making new cells, such as new red blood cells. Vitamin B_{12} also maintains the protective cover around nerve fibers. A deficiency in vitamin B_{12} can cause a type of anemia as well as nervous-system problems that can cause a poor sense of balance, numbness and tingling in the arms and legs, and mental confusion. If a vitamin B_{12} deficiency is due to problems in absorption, injections are recommended.

■ **Vitamin D.** Several factors adversely affect the vitamin D status of the elderly. First, the elderly tend to be outside less, so they make less vitamin D from exposure to the sun. Also, they have less of the vitamin D precursor in the skin necessary to make vitamin D, and older women absorb less vitamin D from food. Because milk is the only dairy product with vitamin D added to it, milk is important to getting enough vitamin D. If milk intake is low, supplements may be recommended.

■ **Calcium.** Current intakes for calcium are below the recommendations for individuals over 65 years of age (1,200 milligrams). To meet this recommendation, an elderly person would need to eat three servings of dairy products or other calcium-rich foods daily. Because this can be difficult, supplements may be recommended.

■ **Zinc.** Because the elderly are at risk for taking in less than the RDA for zinc, and due to the importance of zinc in cell production, wound healing, the immune system, and taste, attention needs to be placed on getting enough of this mineral.

Here are some ways in which the elderly can increase their chances of eating nutritiously.

1. Eat with other people. This usually makes mealtime more enjoyable and stimulates appetite. Taking a walk before eating also stimulates appetite.
2. Prepare larger amounts of food and freeze some for heating up at a later time. This saves cooking time and is helpful if you are reluctant to cook for yourself.
3. If big meals are too much, eat small amounts more frequently during the day. Eat regular meals.
4. If getting to the supermarket is a bother, go at a time when it is not busy or, if you have the money, engage a delivery service.
5. Use unit pricing and sales to cut back on the amount of money you spend on food.
6. Take advantage of community meal programs for the elderly, such as Meals on Wheels.
7. To perk up a sluggish appetite, increase your use of herbs, spices, lemon juice, vinegar, and garlic.

Menu Planning for Older Adults

Figure 13-2 is a modified Food Guide Pyramid for people over 70 developed by researchers at the USDA Human Nutrition Research Center on Aging at Tufts University. This pyramid is designed for healthy individuals over 70 years of age. You will notice the following differences.

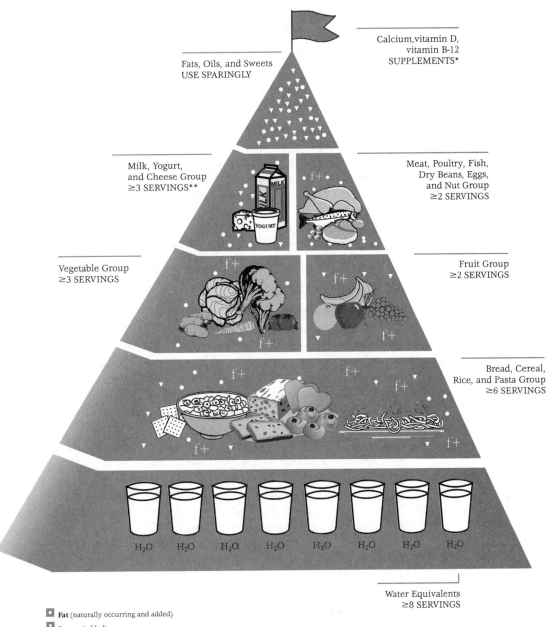

Calcium, vitamin D,
vitamin B-12
SUPPLEMENTS*

Fats, Oils, and Sweets
USE SPARINGLY

Milk, Yogurt,
and Cheese Group
≥3 SERVINGS**

Meat, Poultry, Fish,
Dry Beans, Eggs,
and Nut Group
≥2 SERVINGS

Vegetable Group
≥3 SERVINGS

Fruit Group
≥2 SERVINGS

Bread, Cereal,
Rice, and Pasta Group
≥6 SERVINGS

H₂O   H₂O   H₂O   H₂O   H₂O   H₂O   H₂O   H₂O

Water Equivalents
≥8 SERVINGS

■ **Fat** (naturally occurring and added)

▼ **Sugars** (added)

f+Fiber (should be present)

These symbols show fat, added sugars, and fiber in foods

*Not all individuals need supplements, consult your healthcare provider

**≥ Greater than or equal to

Figure 13-2

Modified food pyramid for 70+ adults

© Copyright 1999 Tufts University. Reprinted with permission.

1. The Pyramid is narrower than the traditional Pyramid to illustrate that older adults have decreased calorie needs and therefore fewer food selections. Their selections must be more nutrient dense than for other groups of people.

2. A small supplement flag at the top of the Pyramid suggests supplements for those nutrients (calcium, vitamin D, vitamin B_{12}) that may be insufficient in the diet due to smaller and fewer servings of food and medical conditions such as lactose intolerance.

3. At the base of the Pyramid you will find a suggestion to drink at least eight servings of water (fluid) daily.

4. A symbol for fiber within the Pyramid emphasizes the importance of high-fiber foods, which are especially important to avoid constipation.

When planning meals for older adults, use these guidelines.

1. Offer moderately sized meals. Older adults frequently complain when given too much food because they hate to see waste. Restaurants might reduce the size of their entrées by 15 to 25 percent.

2. Emphasize complex carbohydrates and high-fiber foods such as fruits, vegetables, grains, and beans. Older people requiring softer diets may have problems chewing some high-fiber foods. High-fiber foods that are soft in texture include cooked beans and peas, bran cereals soaked in milk, canned prunes and pears, and cooked vegetables such as potatoes, corn, green peas, and winter squash.

3. Moderate the use of fat. Many seniors don't like to see their entrée swimming in a pool of butter. Use lean meats, poultry, or fish and sauces prepared with vegetable or fruit purées. Have low-fat dairy products available, such as skim or 1-percent milk.

4. Dairy products are important sources of calcium, vitamin D, protein, potassium, vitamin B_{12}, and riboflavin.

5. Offer adequate but not too much protein. Use a variety of both animal and vegetable sources. Providing protein on a budget, as in a nursing home, need not be a problem. Lower-cost protein sources include beans and peas, cottage cheese, macaroni and cheese, eggs, liver, dried skim milk, chicken, and ground beef.

6. Moderate the use of salt. Many seniors are on low-sodium diets and recognize a salty soup when they taste it. Avoid highly salted soups, sauces, and other dishes. It is better to let them season food to taste.

7. Use herbs and spices to make foods flavorful. Seniors are looking for tasty foods just like anyone else, and they may need them more than ever!

8. Offer a variety of foods, including traditional menu items and cooking from other countries and regions of the United States.

9. Fluid intake is critical, so offer a variety of beverages. Diminished sensitivity to dehydration may cause older adults to drink less fluid than needed. Special attention must be paid to fluids, particularly for those who need assistance with eating and drinking. Beverages such as water, milk, juice, coffee, or tea, and foods such as soup contribute to fluid intake.

10. Intake of the following vitamins and minerals may be inadequate in older adults and needs to be considered in menu planning: vitamin B_{12}, vitamin D, calcium, and zinc. See chapters 6 and 7 for food sources.

11. If chewing is a problem, softer foods can be chosen to provide a well-balanced diet. Following are some guidelines for soft diets.

   ■ Use tender meats, and if necessary, chop or grind them. Ground meats can be used in soups, stews, and casseroles. Cooked beans and peas, soft cheeses, and eggs are additional softer protein sources.

   ■ Cook vegetables thoroughly and dice or chop them by hand if necessary after cooking.

   ■ Serve mashed potatoes or rice, with gravy if desired.

   ■ Serve chopped salads.

   ■ Soft fruits such as fresh or canned bananas, berries, peaches, pears, or melon, as well as applesauce, are some good choices.

   ■ Soft breads and rolls can be made even softer by dipping them briefly in milk.

   ■ Puddings and custard are good dessert choices.

   ■ Many foods that are not soft can be easily chopped by hand or blended in a blender or food processor to provide additional variety.

■ **MINI-SUMMARY**

The graying of America is in full swing. During aging, the functional capability of almost every organ declines, and muscle tissue and mass decreases. Along with a declining basal metabolic rate, the need for calories decreases. The nutrition status of an elderly person is greatly influenced by many variables: presence of disease, activity level, quality of dentition, functional disabilities, decline in taste and smell acuity, changes in the gastrointestinal tract, use of medications, diminished sense of thirst, level of cognitive functioning, available social support, level of education and income, living arrangements, and availability of federally funded meals. The modified Food Guide Pyramid for people over 70, as well as guidelines given, will help you plan menus for older adults.

Check–Out Quiz

1. Dieting during pregnancy is medically allowed.
   a. True   b. False
2. Morning sickness occurs only in the morning.
   a. True   b. False
3. Nutritional deficiencies during lactation are more likely to affect the quantity of milk the mother makes, rather than the quality.
   a. True   b. False
4. For the first year, the newborn's only source of nutrients is either breast milk or formula.
   a. True   b. False
5. A deficiency of iron during the first weeks of pregnancy can cause birth defects.
   a. True   b. False
6. Breast-feeding is considered to be healthier and more nutritious than formula feeding.
   a. True   b. False
7. Moderate drinking during pregnancy may cause fetal alcohol syndrome.
   a. True   b. False
8. Baby's first food is strained vegetables.
   a. True   b. False
9. School-age children tend to be better eaters than preschoolers are.
   a. True   b. False
10. After one year of age, children start to lose baby fat and become leaner.
    a. True   b. False
11. A good way to get a child to finish dinner is to promise dessert once all foods are eaten.
    a. True   b. False
12. A child will ask for and need more food during growth spurts than during slower periods of growth.
    a. True   b. False
13. Breast-fed babies need supplementation with vitamin A.
    a. True   b. False
14. Children's most common nutritional problem is iron-deficiency anemia.
    a. True   b. False
15. During adolescence, females put on proportionately more muscle than fat.
    a. True   b. False
16. The elderly are major users of modified diets.
    a. True   b. False

17. Certain drugs can impair absorption and metabolism of certain nutrients.
    a. True   b. False
18. After age 35, functional capability of almost every organ system declines.
    a. True   b. False
19. As you get older, your energy needs decrease.
    a. True   b. False
20. Two vitamins of concern to the elderly are thiamin and riboflavin.
    a. True   b. False

Activities and Applications

1. **Media Watch**

On a Saturday morning, watch children's television for one hour and record the name of each advertiser and what is being advertised. How many of the total number of advertisers were selling food? Were the majority of the advertised foods healthy foods or junk foods?

2. **Childhood Eating Habits**

Think back to when you were a child and teenager. What influenced what you put in your mouth? Consider influences such as home, school, friends, and relatives. Which positively influenced your eating style? Which negatively? Discuss this with someone else in your class who is close in age.

3. **Preschoolers' Eating Habits**

Visit a preschool or day-care center when a meal is being served. Observe the children while they eat. Ask the caregivers about the children's food preferences and eating habits. Ask how well the children accept new foods and how the caregivers introduce new foods.

4. **School Lunch Menu**

Write a five-day lunch menu for elementary, middle school, or high school students. The menu must provide one-third of the DRI/RDA for all nutrients. Only 30 percent of total calories are allowed from fat. Saturated fat can provide no more than 10 percent of total calories.

5. **Eldercare Menu**

Visit or phone the foodservice director of a local continuing-care retirement community, congregate meals feeding center, or nursing home. Ask about the type of menu being used (restaurant-style or cycle menu) and meals and foods being offered. What are the major meal-planning considerations used in planning meals for the elderly? What special circumstances come up that are unique to them?

Nutrition Web Explorer

Food and Nutrition Services, USDA www.fns.usda.gov/fns
On this home page, click on "Team Nutrition." Find out what Team Nutrition is, and what they do.

Michigan Aging Services www.mdch.state.mi.us/mass/Health/TOC.html
This web site offers a number of useful reading materials for older adults. Click on "Advent of the Solo Diner" and find tips about how older adults living alone can still eat well.

American Anorexia Bulimia Association www.aabainc.org
Once on the home page of this organization, click on "Information for Families and Friends." Find out what to do if someone you know is anorexic or bulimic.

Food Facts *Creative Puréed Foods*

In the past, puréed foods had the reputation of looking pretty miserable when served in most hospitals and nursing facilities. Puréed meat and vegetables were often scooped into small bowls (which for some reason are called "monkey" dishes), covered with gravy, and sent away to some unfortunate patient as supper.

Puréed diets are often necessary for individuals with chewing or swallowing disorders. But that doesn't mean they need runny, liquid foods. On the contrary, the hardest foods for these individuals to swallow are runny foods. The easiest texture to swallow resembles that of mashed potatoes.

Luckily, times have changed. Many cooks are using thickeners to help shape puréed foods so they look like the original foods (Figure 13-3). Thickeners are often powdered and can be mixed directly with liquids and puréed foods. Although there are several commercial thickeners available, such as Thick & Easy, some cooks use thickeners such as cornstarch or instant mashed-potato flakes.

Preparing puréed foods is a challenge not only because the foods must look good and be the right consistency for swallowing, but also because the volume of puréed foods differs from that of regular foods. Fruits and vegetables tend to decrease in volume when puréed, so half a cup of puréed peaches, for example, would contain more calories and nutrients than half a cup of regular peaches. On the other hand, meats almost double in volume because of the liquid required to purée them. Using standardized recipes ensures that puréed foods

■ are nutritionally adequate.
■ are the right consistency.
■ look and taste appropriate.
■ are not too expensive (recipes cut down on waste).

Figure 13-3

Creative and attractive puréed foods using thickeners. Puréed peaches, puréed ham on slurried pumpernickel with puréed lettuce and tomato wedges, garnished with puréed cantaloupe thickened with Menu Magic's Thicken Right.™

Courtesy: Menu Magic, Indianapolis, Indiana.

Hot Topic Food Allergies

Do you start itching whenever you eat peanuts? Does seafood cause your stomach to churn? Symptoms like those cause millions of Americans to suspect they have a **food allergy,** when indeed most probably have a **food intolerance.** True food allergies affect a relatively small percentage of people. Experts estimate that only 2 percent of adults, and from 4 to 8 percent of children are truly allergic to certain foods. So what's the difference between a food intolerance and a food allergy? A food allergy involves an abnormal immune-system response. If the response doesn't involve the immune system, it is called a food intolerance. Symptoms of food intolerance may include gas, bloating, constipation, dizziness, or difficulty sleeping.

Food allergy symptoms are quite specific **Food allergens,** the food components that cause allergic reactions, are usually proteins. When the allergen passes from the mouth into the stomach, the body recognizes it as a foreign substance and produces antibodies to halt the invasion. As the body fights off the invasion, symptoms begin to appear throughout the body. The most common sites (Figure 3-4) are the mouth (swelling of the lips or tongue, itching lips), digestive tract (stomach cramps, vomiting, diarrhea), the skin (hives, rashes, or eczema), and the airways (wheezing or breathing problems). Allergic reactions to foods usually begin within minutes to a few hours after eating.

Food intolerance may produce symptoms similar to those of food allergies, such as abdominal cramping. But whereas people with true

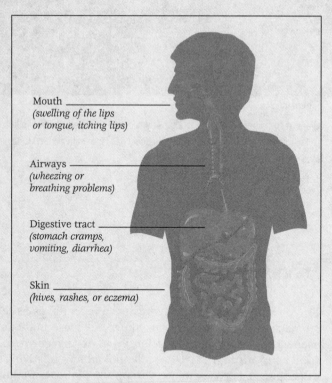

Mouth _____
(swelling of the lips
or tongue, itching lips)

Airways _____
(wheezing or
breathing problems)

Digestive tract _____
(stomach cramps,
vomiting, diarrhea)

Skin _____
(hives, rashes, or eczema)

Figure 13-4

Common sites for allergic reactions

food allergies must avoid offending foods altogether, people with food intolerance can often eat small amounts of the offending food without experiencing symptoms.

Food allergies are much more common in infants and young children, who often outgrow them later. Increased susceptibility of young infants to food allergic reactions is believed to be the result of immunologic immaturity and, to some extent, intestinal immaturity. Older children and adults may lose their sensitivity to certain foods if the responsible food allergen can be identified and completely eliminated from the diet, although some food allergies, such as those to peanuts and nuts, can last a lifetime.

Cow's milk, peanuts, eggs, wheat, and soy are the most common food allergies in children. In many cases, children outgrow these allergies later on in childhood. In general, the more severe the first allergic reaction, the longer it takes to outgrow. Adults are usually most affected by nuts, fish, shellfish, and peanuts.

Most cases of allergic reactions to foods are mild, but some are violent and life-threatening. The greatest danger in food allergy comes from **anaphylaxis,** a rare allergic reaction involving a number of body parts simultaneously. Anaphylaxis is also known as **anaphylactic shock.** Like less serious allergic reactions, anaphylaxis usually occurs after a person is exposed to an allergen to which he or she was sensitized by previous exposure. That is, it does not usually occur the first time a person eats a particular food. Anaphylaxis can produce severe symptoms in as little as five to fifteen minutes. Signs of such a reaction include difficulty breathing, swelling of the mouth and throat, drop in blood pressure, and loss of consciousness. The sooner anaphylaxis is treated, the greater the person's chance of surviving.

Although any food can trigger anaphylaxis, peanuts, nuts, shellfish, milk, eggs, and fish are the most common culprits. Peanuts are the leading cause of death from food allergies. As little as 1/5 to 1/5000 of a teaspoon of the offending food has caused death.

There is no specific test to predict the likelihood of anaphylaxis, although allergy testing may help determine which foods a person may be allergic to and provide some guidance as to the severity of the allergy. Experts advise people who are prone to anaphylaxis to carry medication—usually injectable epinephrine—with them at all times and to check the medicine's expiration date regularly.

Diagnosing a food allergy begins with a thorough medical history to identify the suspected food, the amount that must be eaten to cause a reaction, the amount of time between food consumption and development of symptoms, how often the reaction occurs, and other detailed information. A complete physical examination and selected laboratory tests are conducted to rule out underlying medical conditions not related to food allergy. Several tests, such as skin-testing and blood tests, are available to determine whether a person's immune system is sensitized to a specific food. In prick-skin testing, a diluted extract of the suspected food is placed on the skin, which is then scratched or punctured. If no reaction at the site occurs, then the skin test is negative and allergy to the food is unlikely. If a bump surrounded by redness (similar to a mosquito bite) forms within 15 minutes, then the skin test is positive and the person may be allergic to the tested food.

Once diagnosed, most food allergies are treated by avoidance of the food allergen. Newer approaches, such as drugs, are being explored. Avoiding allergens is relatively easy to do at home where you can read food labels and call food manufacturers, but this task become harder when you go out to eat. As a foodservice professional, you can do the following to help customers with food allergies.

1. Have recipe/ingredient information available for customers with food allergies. Some restaurants designate a person on each shift, who knows this information well, to be in charge of discussing allergy concerns with customers. To be useful, this information must be accurate and updated regularly. A manufacturer may change the ingredients in salsa, for example, and these changes are important to note.

2. If you are not sure what is in a menu item, don't give the customer false reassurances. It's better to say "I don't know—why not pick something else" than to give false information that could result in the customer's death and hundreds of thousands of dollars in fines and lawsuits.

3. Staff need training in the nature of food allergies, the foods commonly involved in anaphylactic shock, the restaurant's procedure on identifying and handling customers with food allergies, and emergency procedures.

4. Kitchen staff need training in avoiding ingredient substitutions. They also need to be trained to prepare and serve foods without contacting the foods most likely to cause anaphylactic shock: nuts, peanuts, fish, and shellfish. This means that all preparation and cooking equipment should be thoroughly cleaned after working with these foods. Remember that even minute amounts of the offending food, sometimes even a strong smell of the food, can cause anaphylactic reactions.

In short, take it seriously when a customer asks whether there are walnuts in your Waldorf salad. Some customers who ask such questions are probably just trying to avoid an upset stomach, but for some it's a much more serious, and possibly life-threatening, matter. Since you don't know which customer really suffers from food allergies, take every customer seriously.

Table 13-14 lists foods to omit for specific allergies.

TABLE 13-14　Foods to Omit for Specific Allergies

Food Allergen	Foods to Omit	Check Food Labels for
Milk	All fluid milk including buttermilk, evaporated or condensed milk, nonfat dry milk, all cheeses, all yogurts, ice cream and ice milk, butter, many margarines, most nondairy creamers and whipped toppings, hot cocoa mixes, creamed soups, many breads, crackers and cereals, pancakes, waffles, many baked goods such as cakes and cookies (check the label), fudge, instant potatoes, custards, puddings, some hot dogs and luncheon meats	Instant nonfat dry milk, nonfat milk, milk solids, whey, curds, casein, caseinate, milk, lactose-free milk, lactalbumin, lactoglobulin, sour cream, butter, cheese, cheese food, butter, milk chocolate, buttermilk
Eggs	All forms of eggs, most egg substitutes, eggnog, any baked goods made with eggs such as muffins and cookies or glazed with eggs such as sweet rolls, ice cream, sherbet, custards, meringues, cream pies, puddings, French toast, pancakes, waffles, some candies, some salad dressings and sandwich spreads such as mayonnaise, any sauce made with egg such as hollandaise, souffles, any meat or potato made with egg, all pastas unless egg free, soups made with eggs of noodles, soups made with stocks that were cleared with eggs, marshmallows	Eggs, albumin, globulin, livetin, ovalbumin, ovomucin, ovomucoid, ovoglobulin egg albumin, ovovitellin, vitellin
Gluten	All foods containing wheat, oats, barley or rye as flour or in any other form, salad dressings, gravies, malted beverages, postum, soy sauce, instant puddings, distilled vinegar, beer, ale, some wines, gin, whiskey, vodka	Flour* (unless from sources noted below), modified food starch, monosodium glutamate, hydrolyzed vegetable protein, cereals, malt or cereal extracts, food starch, vegetable gum, wheat germ, wheat bran, bran, semolina, malt flavoring, distilled vinegar, emulsifiers, stabilizers.

* Corn, rice, soy, arrowroot, tapioca, and potato do not contain gluten, so they are safe to use.

This table shows the nutritive value of 908 common foods. The foods are grouped under the following main headings:

Beverages
Dairy products
Eggs
Fats and oils
Fish and shellfish
Fruits and fruit juices
Grain products
Legumes, nuts, and seeds

Meat and meat products
Mixed dishes and fast foods
Poultry and poultry products
Soups, sauces, and gravies
Sugars and sweets
Vegetables and vegetable products
Miscellaneous items

Most of the foods are listed in ready-to-eat form. Some are basic products widely used in food preparation, such as flour, fat, and cornmeal.

Source: The Nutritive Value of Foods. USDA Home and Garden Bulletin no. 72, 1989.

(Tr indicates nutrient present in trace amount.)

Item No.	Foods, Approximate Measures, Units, and Weight (Weight of Edible Portion Only)		Water	Food Energy	Protein	Fat	Fatty Acids		
							Saturated	Mono-unsaturated	Poly-unsaturated
		Grams	Percent	Calories	Grams	Grams	Grams	Grams	Grams
	Beverages								
	Alcoholic:								
	Beer:								
1	Regular- - - - - - - - - - - - - 12 fl oz - - - - -	360	92	150	1	0	0.0	0.0	0.0
2	Light - - - - - - - - - - - - - - 12 fl oz - - - - -	355	95	95	1	0	0.0	0.0	0.0
	Gin, rum, vodka, whiskey:								
3	80-proof - - - - - - - - - - - - 1-1/2 fl oz - - -	42	67	95	0	0	0.0	0.0	0.0
4	86-proof - - - - - - - - - - - - 1-1/2 fl oz - - -	42	64	105	0	0	0.0	0.0	0.0
5	90-proof - - - - - - - - - - - - 1-1/2 fl oz - - -	42	62	110	0	0	0.0	0.0	0.0
	Wines:								
6	Dessert- - - - - - - - - - - - - 3-1/2 fl oz - - -	103	77	140	Tr	0	0.0	0.0	0.0
	Table:								
7	Red - - - - - - - - - - - - - - 3-1/2 fl oz - - -	102	88	75	Tr	0	0.0	0.0	0.0
8	White - - - - - - - - - - - - - 3-1/2 fl oz - - -	102	87	80	Tr	0	0.0	0.0	0.0
	Carbonated:[2]								
9	Club soda - - - - - - - - - - - - 12 fl oz - - - - -	355	100	0	0	0	0.0	0.0	0.0
	Cola type:								
10	Regular- - - - - - - - - - - - - 12 fl oz - - - - -	369	89	160	0	0	0.0	0.0	0.0
11	Diet, artificially sweetened 12 fl oz - - - - -	355	100	Tr	0	0	0.0	0.0	0.0
12	Ginger ale - - - - - - - - - - - 12 fl oz - - - - -	366	91	125	0	0	0.0	0.0	0.0
13	Grape - - - - - - - - - - - - - - 12 fl oz - - - - -	372	88	180	0	0	0.0	0.0	0.0
14	Lemon-lime - - - - - - - - - - - 12 fl oz - - - - -	372	89	155	0	0	0.0	0.0	0.0
15	Orange - - - - - - - - - - - - - 12 fl oz - - - - -	372	88	180	0	0	0.0	0.0	0.0
16	Pepper type - - - - - - - - - - - 12 fl oz - - - - -	369	89	160	0	0	0.0	0.0	0.0
17	Root beer- - - - - - - - - - - - 12 fl oz - - - - -	370	89	165	0	0	0.0	0.0	0.0
	Cocoa and chocolate-flavored beverages. See Dairy Products (items 95–98).								
	Coffee:								
18	Brewed - - - - - - - - - - - - - 6 fl oz - - - - -	180	100	Tr	Tr	Tr	Tr	Tr	Tr
19	Instant, prepared (2 tsp powder plus 6 fl oz water) - - - - - - - 6 fl oz - - - - -	182	99	Tr	Tr	Tr	Tr	Tr	Tr
	Fruit drinks, noncarbonated:								
	Canned:								
20	Fruit punch drink - - - - - - - - 6 fl oz - - - - -	190	88	85	Tr	0	0.0	0.0	0.0
21	Grape drink - - - - - - - - - - - 6 fl oz - - - - -	187	86	100	Tr	0	0.0	0.0	0.0
22	Pineapple-grapefruit juice drink - - - - - - - - - - - - - - 6 fl oz - - - - -	187	87	90	Tr	Tr	Tr	Tr	Tr
	Frozen:								
	Lemonade concentrate:								
23	Undiluted - - - - - - - - - - - 6-fl-oz can- - -	219	49	425	Tr	Tr	Tr	Tr	Tr
24	Diluted with 4-1/3 parts water by volume - - - - - 6 fl oz - - - - -	185	89	80	Tr	Tr	Tr	Tr	Tr
	Limeade concentrate:								
25	Undiluted - - - - - - - - - - - 6-fl-oz can- - -	218	50	410	Tr	Tr	Tr	Tr	Tr
26	Diluted with 4-1/3 parts water by volume - - - - - 6 fl oz - - - - -	185	89	75	Tr	Tr	Tr	Tr	Tr
	Fruit Juices. See type under Fruits and Fruit Juices.								
	Milk beverages. See Dairy Products (items 92–105).								

[1] Value not determined.

[2] Mineral content varies depending on water source.

Choles-terol	Carbo-hydrate	Calcium	Phos-phorus	Iron	Potas-sium	Sodium	Vitamin A Value (IU)	Vitamin A Value (RE)	Thiamin	Ribo-flavin	Niacin	Ascorbic Acid	Item No.
Milli-grams	Grams	Milli-grams	Milli-grams	Milli-grams	Milli-grams	Milli-grams	Inter-national units	Retinol equiva-lents	Milli-grams	Milli-grams	Milli-grams	Milli-grams	
0	13	14	50	0.1	115	18	0	0	0.02	0.09	1.8	0	1
0	5	14	43	0.1	64	11	0	0	0.03	0.11	1.4	0	2
0	Tr	Tr	Tr	Tr	1	Tr	0	0	Tr	Tr	Tr	0	3
0	Tr	Tr	Tr	Tr	1	Tr	0	0	Tr	Tr	Tr	0	4
0	Tr	Tr	Tr	Tr	1	Tr	0	0	Tr	Tr	Tr	0	5
0	8	8	9	0.2	95	9	(1)	(1)	0.01	0.02	0.2	0	6
0	3	8	18	0.4	113	5	(1)	(1)	0.00	0.03	0.1	0	7
0	3	9	14	0.3	83	5	(1)	(1)	0.00	0.01	0.1	0	8
0	0	18	0	Tr	0	78	0	0	0.00	0.00	0.0	0	9
0	41	11	52	0.2	7	18	0	0	0.00	0.00	0.0	0	10
0	Tr	14	39	0.2	7	[3]32	0	0	0.00	0.00	0.0	0	11
0	32	11	0	0.1	4	29	0	0	0.00	0.00	0.0	0	12
0	46	15	0	0.4	4	48	0	0	0.00	0.00	0.0	0	13
0	39	7	0	0.4	4	33	0	0	0.00	0.00	0.0	0	14
0	46	15	4	0.3	7	52	0	0	0.00	0.00	0.0	0	15
0	41	11	41	0.1	4	37	0	0	0.00	0.00	0.0	0	16
0	42	15	0	0.2	4	48	0	0	0.00	0.00	0.0	0	17
0	Tr	4	2	Tr	124	2	0	0	0.00	0.02	0.4	0	18
0	1	2	6	0.1	71	Tr	0	0	0.00	0.03	0.6	0	19
0	22	15	2	0.4	48	15	20	2	0.03	0.04	Tr	[4]61	20
0	26	2	2	0.3	9	11	Tr	Tr	0.01	0.01	Tr	[4]64	21
0	23	13	7	0.9	97	24	60	6	0.06	0.04	0.5	[4]110	22
0	112	9	13	0.4	153	4	40	4	0.04	0.07	0.7	66	23
0	21	2	2	0.1	30	1	10	1	0.01	0.02	0.2	13	24
0	108	11	13	0.2	129	Tr	Tr	Tr	0.02	0.02	0.2	26	25
0	20	2	2	Tr	24	Tr	Tr	Tr	Tr	Tr	Tr	4	26

[3] Blend of aspartame and saccharin; if only sodium saccharin is used, sodium is 75 mg: if only aspartame is used, sodium is 23 mg.

[4] With added ascorbic acid.

(Tr indicates nutrient present in trace amount.)

Item No.	Foods, Approximate Measures, Units, and Weight (Weight of Edible Portion Only)			Water	Food Energy	Protein	Fat	Saturated	Fatty Acids Mono-unsaturated	Poly-unsaturated
			Grams	Percent	Calories	Grams	Grams	Grams	Grams	Grams
	Beverages *(continued)*									
	Tea:									
27	Brewed - - - - - - - - - - - - - -	8 fl oz - - - - -	240	100	Tr	Tr	Tr	Tr	Tr	Tr
	Instant, powder, prepared:									
28	Unsweetened (1 tsp powder									
	plus 8 fl oz water) - - - - -	8 fl oz - - - - -	241	100	Tr	Tr	Tr	Tr	Tr	Tr
29	Sweetened (3 tsp powder									
	plus 8 fl oz water) - - - - - -	8 fl oz - - - - -	262	91	85	Tr	Tr	Tr	Tr	Tr
	Dairy Products									
	Butter. See Fats and Oils (items 128–130).									
	Cheese:									
	Natural:									
30	Blue - - - - - - - - - - - - - -	1 oz - - - - - - -	28	42	100	6	8	5.3	2.2	0.2
31	Camembert (3 wedges per									
	4-oz container) - - - - - -	1 wedge - - - -	38	52	115	8	9	5.8	2.7	0.3
	Cheddar:									
32	Cut pieces - - - - - - - - - -	1 oz - - - - - - -	28	37	115	7	9	6.0	2.7	0.3
33		1 in³ - - - - - -	17	37	70	4	6	3.6	1.6	0.2
34	Shredded - - - - - - - - - - -	1 cup - - - - -	113	37	455	28	37	23.8	10.6	1.1
	Cottage (curd not pressed down):									
	Creamed (cottage cheese, 4% fat):									
35	Large curd - - - - - - - - -	1 cup - - - - -	225	79	235	28	10	6.4	2.9	0.3
36	Small curd - - - - - - - - -	1 cup - - - - -	210	79	215	26	9	6.0	2.7	0.3
37	With fruit - - - - - - - - - -	1 cup - - - - -	226	72	280	22	8	4.9	2.2	0.2
38	Lowfat (2%) - - - - - - - - -	1 cup - - - - -	226	79	205	31	4	2.8	1.2	0.1
39	Uncreamed (cottage cheese dry curd, less than 1/2% fat) - - - - - -	1 cup - - - - -	145	80	125	25	1	0.4	0.2	Tr
40	Cream - - - - - - - - - - - - -	1 oz - - - - - - -	28	54	100	2	10	6.2	2.8	0.4
41	Feta - - - - - - - - - - - - - -	1 oz - - - - - - -	28	55	75	4	6	4.2	1.3	0.2
	Mozzarella, made with:									
42	Whole milk - - - - - - - - -	1 oz - - - - - - -	28	54	80	6	6	3.7	1.9	0.2
43	Part skim milk (low moisture) - - - - - - - - -	1 oz - - - - - - -	28	49	80	8	5	3.1	1.4	0.1
44	Muenster - - - - - - - - - - - -	1 oz - - - - - - -	28	42	105	7	9	5.4	2.5	0.2
	Parmesan, grated:									
45	Cup, not pressed down - -	1 cup - - - - -	100	18	455	42	30	19.1	8.7	0.7
46	Tablespoon - - - - - - - - -	1 tbsp - - - - -	5	18	25	2	2	1.0	0.4	Tr
47	Ounce - - - - - - - - - - - -	1 oz - - - - - - -	28	18	130	12	9	5.4	2.5	0.2
48	Provolone - - - - - - - - - - -	1 oz - - - - - - -	28	41	100	7	8	4.8	2.1	0.2
	Ricotta, made with:									
49	Whole milk - - - - - - - - -	1 cup - - - - -	246	72	430	28	32	20.4	8.9	0.9
50	Part skim milk - - - - - - -	1 cup - - - - -	246	74	340	28	19	12.1	5.7	0.6
51	Swiss - - - - - - - - - - - - -	1 oz - - - - - - -	28	37	105	8	8	5.0	2.1	0.3
	Pasteurized process cheese:									
52	American - - - - - - - - - - -	1 oz - - - - - - -	28	39	105	6	9	5.6	2.5	0.3
53	Swiss - - - - - - - - - - - - -	1 oz - - - - - - -	28	42	95	7	7	4.5	2.0	0.2
54	Pasteurized process cheese food, American - - - - - - -	1 oz - - - - - - -	28	43	95	6	7	4.4	2.0	0.2
55	Pasteurized process cheese spread, American - - - - - - -	1 oz - - - - - - -	28	48	80	5	6	3.8	1.8	0.2
	Cream, sweet:									
56	Half-and-half (cream and milk)	1 cup - - - - -	242	81	315	7	28	17.3	8.0	1.0
57		1 tbsp - - - - -	15	81	20	Tr	2	1.1	0.5	0.1

464

Choles-terol	Carbo-hydrate	Calcium	Phos-phorus	Iron	Potas-sium	Sodium	Vitamin A Value (IU)	(RE)	Thiamin	Ribo-flavin	Niacin	Ascorbic Acid	Item No.
Milli-grams	Grams	Milli-grams	Milli-grams	Milli-grams	Milli-grams	Milli-grams	Inter-national units	Retinol equiva-lents	Milli-grams	Milli-grams	Milli-grams	Milli-grams	
0	Tr	0	2	Tr	36	1	0	0	0.00	0.03	Tr	0	27
0	1	1	4	Tr	61	1	0	0	0.00	0.02	0.1	0	28
0	22	1	3	Tr	49	Tr	0	0	0.00	0.04	0.1	0	29
21	1	150	110	0.1	73	396	200	65	0.01	0.11	0.3	0	30
27	Tr	147	132	0.1	71	320	350	96	0.01	0.19	0.2	0	31
30	Tr	204	145	0.2	28	176	300	86	0.01	0.11	Tr	0	32
18	Tr	123	87	0.1	17	105	180	52	Tr	0.06	Tr	0	33
119	1	815	579	0.8	111	701	1,200	342	0.03	0.42	0.1	0	34
34	6	135	297	0.3	190	911	370	108	0.05	0.37	0.3	Tr	35
31	6	126	277	0.3	177	850	340	101	0.04	0.34	0.3	Tr	36
25	30	108	236	0.2	151	915	280	81	0.04	0.29	0.2	Tr	37
19	8	155	340	0.4	217	918	160	45	0.05	0.42	0.3	Tr	38
10	3	46	151	0.3	47	19	40	12	0.04	0.21	0.2	0	39
31	1	23	30	0.3	34	84	400	124	Tr	0.06	Tr	0	40
25	1	140	96	0.2	18	316	130	36	0.04	0.24	0.3	0	41
22	1	147	105	0.1	19	106	220	68	Tr	0.07	Tr	0	42
15	1	207	149	0.1	27	150	180	54	0.01	0.10	Tr	0	43
27	Tr	203	133	0.1	38	178	320	90	Tr	0.09	Tr	0	44
79	4	1,376	807	1.0	107	1,861	700	173	0.05	0.39	0.3	0	45
4	Tr	69	40	Tr	5	93	40	9	Tr	0.02	Tr	0	46
22	1	390	229	0.3	30	528	200	49	0.01	0.11	0.1	0	47
20	1	214	141	0.1	39	248	230	75	0.01	0.09	Tr	0	48
124	7	509	389	0.9	257	207	1,210	330	0.03	0.48	0.3	0	49
76	13	669	449	1.1	307	307	1,060	278	0.05	0.46	0.2	0	50
26	1	272	171	Tr	31	74	240	72	0.01	0.10	Tr	0	51
27	Tr	174	211	0.1	46	406	340	82	0.01	0.10	Tr	0	52
24	1	219	216	0.2	61	388	230	65	Tr	0.08	Tr	0	53
18	2	163	130	0.2	79	337	260	62	0.01	0.13	Tr	0	54
16	2	159	202	0.1	69	381	220	54	0.01	0.12	Tr	0	55
89	10	254	230	0.2	314	98	1,050	259	0.08	0.36	0.2	2	56
6	1	16	14	Tr	19	6	70	16	0.01	0.02	Tr	Tr	57

(Tr indicates nutrient present in trace amount.)

Item No.	Foods, Approximate Measures, Units, and Weight (Weight of Edible Portion Only)			Water	Food Energy	Protein	Fat	Fatty Acids		
								Saturated	Mono-unsaturated	Poly-unsaturated
				Grams	Percent	Calories	Grams	Grams	Grams	Grams	Grams

Item No.	Foods	Measure		Grams	Percent	Calories	Grams	Grams	Grams	Grams	Grams
	Dairy Products *(continued)*										
58	Light, coffee, or table - - - - -	1 cup - - - - - -		240	74	470	6	46	28.8	13.4	1.7
59		1 tbsp - - - - -		15	74	30	Tr	3	1.8	0.8	0.1
	Whipping, unwhipped (volume about double when whipped):										
60	Light - - - - - - - - - - - - - -	1 cup - - - - - -		239	64	700	5	74	46.2	21.7	2.1
61		1 tbsp - - - - -		15	64	45	Tr	5	2.9	1.4	0.1
62	Heavy - - - - - - - - - - - - -	1 cup - - - - - -		238	58	820	5	88	54.8	25.4	3.3
63		1 tbsp - - - - -		15	58	50	Tr	6	3.5	1.6	0.2
64	Whipped topping, (pressurized)	1 cup - - - - - -		60	61	155	2	13	8.3	3.9	0.5
65		1 tbsp - - - - -		3	61	10	Tr	1	0.4	0.2	Tr
66	Cream, sour - - - - - - - - - - -	1 cup - - - - - -		230	71	495	7	48	30.0	13.9	1.8
67		1 tbsp - - - - -		12	71	25	Tr	3	1.6	0.7	0.1
	Cream products, imitation (made with vegetable fat):										
	Sweet:										
	Creamers:										
68	Liquid (frozen) - - - - - - - -	1 tbsp - - - - -		15	77	20	Tr	1	1.4	Tr	Tr
69	Powdered - - - - - - - - - - -	1 tsp - - - - -		2	2	10	Tr	1	0.7	Tr	Tr
	Whipped topping:										
70	Frozen - - - - - - - - - - - - -	1 cup - - - - - -		75	50	240	1	19	16.3	1.2	0.4
71		1 tbsp - - - - -		4	50	15	Tr	1	0.9	0.1	Tr
72	Powdered, made with whole milk - - - - - - - -	1 cup - - - - - -		80	67	150	3	10	8.5	0.7	0.2
73		1 tbsp - - - - -		4	67	10	Tr	Tr	0.4	Tr	Tr
74	Pressurized - - - - - - - - - -	1 cup - - - - - -		70	60	185	1	16	13.2	1.3	0.2
75		1 tbsp - - - - -		4	60	10	Tr	1	0.8	0.1	Tr
76	Sour dressing (filled cream type product, nonbutterfat) - - - - - - - - - -	1 cup - - - - - -		235	75	415	8	39	31.2	4.6	1.1
77		1 tbsp - - - - -		12	75	20	Tr	2	1.6	0.2	0.1
	Ice cream. See Milk desserts, frozen (items 106–111).										
	Ice milk. See Milk desserts, frozen (items 112–114).										
	Milk:										
	Fluid:										
78	Whole (3.3% fat) - - - - - - - -	1 cup - - - - - -		244	88	150	8	8	5.1	2.4	0.3
	Lowfat (2%):										
79	No milk solids added - - - -	1 cup - - - - - -		244	89	120	8	5	2.9	1.4	0.2
80	Milk solids added, label claim less than 10 g of protein per cup - - - - - - - - - - - - -	1 cup - - - - - -		245	89	125	9	5	2.9	1.4	0.2
	Lowfat (1%):										
81	No milk solids added - - - -	1 cup - - - - - -		244	90	100	8	3	1.6	0.7	0.1
82	Milk solids added, label claim less than 10 g of protein per cup - - - - - - - - - - - - -	1 cup - - - - - -		245	90	105	9	2	1.5	0.7	0.1
	Nonfat (skim):										
83	No milk solids added - - - -	1 cup - - - - - -		245	91	85	8	Tr	0.3	0.1	Tr

Choles-terol	Carbo-hydrate	Calcium	Phos-phorus	Iron	Potas-sium	Sodium	Vitamin A Value		Thiamin	Ribo-flavin	Niacin	Ascorbic Acid	Item No.
							(IU)	(RE)					
Milli-grams	Grams	Milli-grams	Milli-grams	Milli-grams	Milli-grams	Milli-grams	Inter-national units	Retinol equiva-lents	Milli-grams	Milli-grams	Milli-grams	Milli-grams	
159	9	231	192	0.1	292	95	1,730	437	0.08	0.36	0.1	2	58
10	1	14	12	Tr	18	6	110	27	Tr	0.02	Tr	Tr	59
265	7	166	146	0.1	231	82	2,690	705	0.06	0.30	0.1	1	60
17	Tr	10	9	Tr	15	5	170	44	Tr	0.02	Tr	Tr	61
326	7	154	149	0.1	179	89	3,500	1,002	0.05	0.26	0.1	1	62
21	Tr	10	9	Tr	11	6	220	63	Tr	0.02	Tr	Tr	63
46	7	61	54	Tr	88	78	550	124	0.02	0.04	Tr	0	64
2	Tr	3	3	Tr	4	4	30	6	Tr	Tr	Tr	0	65
102	10	268	195	0.1	331	123	1,820	448	0.08	0.34	0.2	2	66
5	1	14	10	Tr	17	6	90	23	Tr	0.02	Tr	Tr	67
0	2	1	10	Tr	29	12	[5]10	[5]1	0.00	0.00	0.0	0	68
0	1	Tr	8	Tr	16	4	Tr	Tr	0.00	Tr	0.0	0	69
0	17	5	6	0.1	14	19	[5]650	[5]65	0.00	0.00	0.0	0	70
0	1	Tr	Tr	Tr	1	1	[5]30	[5]3	0.00	0.00	0.0	0	71
8	13	72	69	Tr	121	53	[5]290	[5]39	0.02	0.09	Tr	1	72
Tr	1	4	3	Tr	6	3	[5]10	[5]2	Tr	Tr	Tr	Tr	73
0	11	4	13	Tr	13	43	[5]330	[5]33	0.00	0.00	0.0	0	74
0	1	Tr	1	Tr	1	2	[5]20	[5]2	0.00	0.00	0.0	0	75
13	11	266	205	0.1	380	113	[20]	5	0.09	0.38	0.2	2	76
1	1	14	10	Tr	19	6	Tr	Tr	Tr	0.02	Tr	Tr	77
33	11	291	228	0.1	370	120	310	76	0.09	0.40	0.2	2	78
18	12	297	232	0.1	377	122	500	139	0.10	0.40	0.2	2	79
18	12	313	245	0.1	397	128	500	140	0.10	0.42	0.2	2	80
10	12	300	235	0.1	381	123	500	144	0.10	0.41	0.2	2	81
10	12	313	245	0.1	397	128	500	145	0.10	0.42	0.2	2	82
4	12	302	247	0.1	406	126	500	149	0.09	0.34	0.2	2	83

[5] Vitamin A value is largely from beta-carotene used for coloring.

(Tr indicates nutrient present in trace amount.)

Item No.	Foods, Approximate Measures, Units, and Weight (Weight of Edible Portion Only)		Water	Food Energy	Protein	Fat	Fatty Acids Saturated	Mono-unsaturated	Poly-unsaturated
			Grams	Percent	Calories	Grams	Grams	Grams	Grams
	Dairy Products *(continued)*								
84	Milk solids added, label claim less than 10 g of protein per cup - - - - - - - - - - - -	1 cup - - - - - -	245	90	90	9	1	0.4	0.2	Tr
85	Buttermilk - - - - - - - - - - - -	1 cup - - - - - -	245	90	100	8	2	1.3	0.6	0.1
	Canned:									
86	Condensed, sweetened - - - -	1 cup - - - - - -	306	27	980	24	27	16.8	7.4	1.0
	Evaporated:									
87	Whole milk - - - - - - - - - -	1 cup - - - - - -	252	74	340	17	19	11.6	5.9	0.6
88	Skim milk - - - - - - - - - - -	1 cup - - - - - -	255	79	200	19	1	0.3	0.2	Tr
	Dried:									
89	Buttermilk - - - - - - - - - - -	1 cup - - - - - -	120	3	465	41	7	4.3	2.0	0.3
	Nonfat, instantized:									
90	Envelope, 3.2 oz, net wt.[6] - - - - - - - - - - - -	1 envelope - -	91	4	325	32	1	0.4	0.2	Tr
91	Cup - - - - - - - - - - - - - -	1 cup - - - - - -	68	4	245	24	Tr	0.3	0.1	Tr
	Milk beverages:									
	Chocolate milk (commercial):									
92	Regular - - - - - - - - - - - - -	1 cup - - - - - -	250	82	210	8	8	5.3	2.5	0.3
93	Lowfat (2%) - - - - - - - - - -	1 cup - - - - - -	250	84	180	8	5	3.1	1.5	0.2
94	Lowfat (1%) - - - - - - - - - -	1 cup - - - - - -	250	85	160	8	3	1.5	0.8	0.1
	Milk beverages:									
	Cocoa and chocolate-flavored beverages:									
95	Powder containing nonfat dry milk - - - - - -	1 oz - - - - - - -	28	1	100	3	1	0.6	0.3	Tr
96	Prepared (6 oz water plus 1 oz powder) - - - -	1 serving - - - -	206	86	100	3	1	0.6	0.3	Tr
97	Powder without nonfat dry milk - - - - - - - - -	3/4 oz - - - - -	21	1	75	1	1	0.3	0.2	Tr
98	Prepared (8 oz whole milk plus 3/4 oz powder) - - - - - - - - - -	1 serving - - - -	265	81	225	9	9	5.4	2.5	0.3
99	Eggnog (commercial) - - - - - -	1 cup - - - - - -	254	74	340	10	19	11.3	5.7	0.9
	Malted milk:									
	Chocolate:									
100	Powder - - - - - - - - - - - -	3/4 oz - - - - -	21	2	85	1	1	0.5	0.3	0.1
101	Prepared (8 oz whole milk plus 3/4 oz powder) - - - -	1 serving - - - -	265	81	235	9	9	5.5	2.7	0.4
	Natural:									
102	Powder - - - - - - - - - - - -	3/4 oz - - - - -	21	3	85	3	2	0.9	0.5	0.3
103	Prepared (8 oz whole milk plus 3/4 oz powder) - - - -	1 serving - - - -	265	81	235	11	10	6.0	2.9	0.6
	Shakes, thick:									
104	Chocolate - - - - - - - - - - -	10-oz container - -	283	72	335	9	8	4.8	2.2	0.3
105	Vanilla - - - - - - - - - - - - -	10-oz container - -	283	74	315	11	9	5.3	2.5	0.3

[6] Yields 1 qt of fluid milk when reconstituted according to package directions.

Choles-terol	Carbo-hydrate	Calcium	Phos-phorus	Iron	Potas-sium	Sodium	Vitamin A Value (IU)	Vitamin A Value (RE)	Thiamin	Ribo-flavin	Niacin	Ascorbic Acid	Item No.
Milli-grams	Grams	Milli-grams	Milli-grams	Milli-grams	Milli-grams	Milli-grams	Inter-national units	Retinol equiva-lents	Milli-grams	Milli-grams	Milli-grams	Milli-grams	
5	12	316	255	0.1	418	130	500	149	0.10	0.43	0.2	2	84
9	12	285	219	0.1	371	257	80	20	0.08	0.38	0.1	2	85
104	166	868	775	0.6	1,136	389	1,000	248	0.28	1.27	0.6	8	86
74	25	657	510	0.5	764	267	610	136	0.12	0.80	0.5	5	87
9	29	738	497	0.7	845	293	1,000	298	0.11	0.79	0.4	3	88
83	59	1,421	1,119	0.4	1,910	621	260	65	0.47	1.89	1.1	7	89
17	47	1,120	896	0.3	1,552	499	[7]2,160	[7]646	0.38	1.59	0.8	5	90
12	35	837	670	0.2	1,160	373	[7]1,610	[7]483	0.28	1.19	0.6	4	91
31	26	280	251	0.6	417	149	300	73	0.09	0.41	0.3	2	92
17	26	284	254	0.6	422	151	500	143	0.09	0.41	0.3	2	93
7	26	287	256	0.6	425	152	500	148	0.10	0.42	0.3	2	94
1	22	90	88	0.3	223	139	Tr	Tr	0.03	0.17	0.2	Tr	95
1	22	90	88	0.3	223	139	Tr	Tr	0.03	0.17	0.2	Tr	96
0	19	7	26	0.7	136	56	Tr	Tr	Tr	0.03	0.1	Tr	97
33	30	298	254	0.9	508	176	310	76	0.10	0.43	0.3	3	98
149	34	330	278	0.5	420	138	890	203	0.09	0.48	0.3	4	99
1	18	13	37	0.4	130	49	20	5	0.04	0.04	0.4	0	100
34	29	304	265	0.5	500	168	330	80	0.14	0.43	0.7	2	101
4	15	56	79	0.2	159	96	70	17	0.11	0.14	1.1	0	102
37	27	347	307	0.3	529	215	380	93	0.20	0.54	1.3	2	103
30	60	374	357	0.9	634	314	240	59	0.13	0.63	0.4	0	104
33	50	413	326	0.3	517	270	320	79	0.08	0.55	0.4	0	105

[7] With added vitamin A.

(Tr indicates nutrient present in trace amount.)

Item No.	Foods, Approximate Measures, Units, and Weight (Weight of Edible Portion Only)		Water	Food Energy	Protein	Fat	Fatty Acids Saturated	Mono-unsaturated	Poly-unsaturated
		Grams	Percent	Calories	Grams	Grams	Grams	Grams	Grams
	Dairy Products *(continued)*								
	Milk desserts, frozen:								
	Ice cream, vanilla:								
	Regular (about 11% fat):								
106	Hardened - - - - - - - - - - - 1/2 gal - - - - -	1,064	61	2,155	38	115	71.3	33.1	4.3
107	1 cup- - - - - -	133	61	270	5	14	8.9	4.1	0.5
108	3 fl oz - - - - -	50	61	100	2	5	3.4	1.6	0.2
109	Soft serve (frozen custard) - - - - - - - - - - 1 cup- - - - - -	173	60	375	7	23	13.5	6.7	1.0
110	Rich (about 16% fat), hardened - - - - - - - - - - - 1/2 gal - - - - -	1,188	59	2,805	33	190	118.3	54.9	7.1
111	1 cup- - - - - -	148	59	350	4	24	14.7	6.8	0.9
	Ice milk, vanilla:								
112	Hardened (about 4% fat) - - - - - - - - - - - 1/2 gal - - - - -	1,048	69	1,470	41	45	28.1	13.0	1.7
113	1 cup- - - - - -	131	69	185	5	6	3.5	1.6	0.2
114	Soft serve (about 3% fat) - - - - - - - - - - 1 cup- - - - - -	175	70	225	8	5	2.9	1.3	0.2
115	Sherbet (about 2% fat)- - - - - 1/2 gal - - - - -	1,542	66	2,160	17	31	19.0	8.8	1.1
116	1 cup- - - - - -	193	66	270	2	4	2.4	1.1	0.1
	Yogurt:								
	With added milk solids:								
	Made with lowfat milk:								
117	Fruit-flavored[8] - - - - - - - - 8-oz container	227	74	230	10	2	1.6	0.7	0.1
118	Plain - - - - - - - - - - - - - - 8-oz container	227	85	145	12	4	2.3	1.0	0.1
119	Made with nonfat milk- - - - - 8-oz container	227	85	125	13	Tr	0.3	0.1	Tr
	Without added milk solids:								
120	Made with whole milk - - - - - 8-oz container	227	88	140	8	7	4.8	2.0	0.2
	Eggs								
	Eggs, large (24 oz per dozen):								
	Raw:								
121	Whole, without shell - - - - - 1 egg- - - - - -	50	75	80	6	6	1.7	2.2	0.7
122	White - - - - - - - - - - - - - - 1 white - - - - -	33	88	15	3	Tr	0.0	0.0	0.0
123	Yolk- - - - - - - - - - - - - - - 1 yolk- - - - - -	17	49	65	3	6	1.7	2.2	0.7
	Cooked:								
124	Fried in butter- - - - - - - - - - 1 egg- - - - - -	46	68	95	6	7	2.7	2.7	0.8
125	Hard-cooked, shell removed - - - - - - - - - - - 1 egg- - - - - -	50	75	80	6	6	1.7	2.2	0.7
126	Poached - - - - - - - - - - - - - 1 egg- - - - - -	50	74	80	6	6	1.7	2.2	0.7
127	Scrambled (milk added) in butter. Also omelet- - - - 1 egg- - - - - -	64	73	110	7	8	3.2	2.9	0.8
	Fats and Oils								
	Butter (4 sticks per lb):								
128	Stick- - - - - - - - - - - - - - - 1/2 cup - - - -	113	16	810	1	92	57.1	26.4	3.4
129	Tablespoon (1/8 stick) - - - - - 1 tbsp - - - - -	14	16	100	Tr	11	7.1	3.3	0.4
130	Pat (1 in square, 1/3 in high; 90 per lb) - - - - - - - - - - - 1 pat - - - - - -	5	16	35	Tr	4	2.5	1.2	0.2
131	Fats, cooking (vegetable shortenings) - - - - - - - - - - - 1 cup- - - - - -	205	0	1,810	0	205	51.3	91.2	53.5
132	1 tbsp - - - - -	13	0	115	0	13	3.3	5.8	3.4
133	Lard - - - - - - - - - - - - - - - 1 cup- - - - - -	205	0	1,850	0	205	80.4	92.5	23.0
134	1 tbsp - - - - -	13	0	115	0	13	5.1	5.9	1.5

[8] Carbohydrate content varies widely because of amount of sugar added and amount and solids content of added flavoring.

Choles-terol	Carbo-hydrate	Calcium	Phos-phorus	Iron	Potas-sium	Sodium	Vitamin A Value		Thiamin	Ribo-flavin	Niacin	Ascorbic Acid	Item No.
							(IU)	(RE)					
Milli-grams	Grams	Milli-grams	Milli-grams	Milli-grams	Milli-grams	Milli-grams	Inter-national units	Retinol equiva-lents	Milli-grams	Milli-grams	Milli-grams	Milli-grams	
476	254	1,406	1,075	1.0	2,052	929	4,340	1,064	0.42	2.63	1.1	6	106
59	32	176	134	0.1	257	116	540	133	0.05	0.33	0.1	1	107
22	12	66	51	Tr	96	44	200	50	0.02	0.12	0.1	Tr	108
153	38	236	199	0.4	338	153	790	199	0.08	0.45	0.2	1	109
703	256	1,213	927	0.8	1,771	868	7,200	1,758	0.36	2.27	0.9	5	110
88	32	151	115	0.1	221	108	990	219	0.04	0.28	0.1	1	111
146	232	1,409	1,035	1.5	2,117	836	1,710	419	0.61	2.78	0.9	6	112
18	29	176	129	0.2	265	105	210	52	0.08	0.35	0.1	1	113
13	38	274	202	0.3	412	163	175	44	0.12	0.54	0.2	1	114
113	469	827	594	2.5	1,585	706	1,480	308	0.26	0.71	1.0	31	115
14	59	103	74	0.3	198	88	190	39	0.03	0.09	0.1	4	116
10	43	345	271	0.2	442	133	100	25	0.08	0.40	0.2	1	117
14	16	415	326	0.2	531	159	150	36	0.10	0.49	0.3	2	118
4	17	452	355	0.2	579	174	20	5	0.11	0.53	0.3	2	119
29	11	274	215	0.1	351	105	280	68	0.07	0.32	0.2	1	120
274	1	28	90	1.0	65	69	260	78	0.04	0.15	Tr	0	121
0	Tr	4	4	Tr	45	50	0	0	Tr	0.09	Tr	0	122
272	Tr	26	86	0.9	15	8	310	94	0.04	0.07	Tr	0	123
278	1	29	91	1.1	66	162	320	94	0.04	0.14	Tr	0	124
274	1	28	90	1.0	65	69	260	78	0.04	0.14	Tr	0	125
273	1	28	90	1.0	65	146	260	78	0.03	0.13	Tr	0	126
282	2	54	109	1.0	97	176	350	102	0.04	0.18	Tr	Tr	127
247	Tr	27	26	0.2	29	[9]933	[10]3,460	[10]852	0.01	0.04	Tr	0	128
31	Tr	3	3	Tr	4	[9]116	[10]430	[10]106	Tr	Tr	Tr	0	129
11	Tr	1	1	Tr	1	[9]41	[10]150	[10]38	Tr	Tr	Tr	0	130
0	0	0	0	0.0	0	0	0	0	0.00	0.00	0.0	0	131
0	0	0	0	0.0	0	0	0	0	0.00	0.00	0.0	0	132
195	0	0	0	0.0	0	0	0	0	0.00	0.00	0.0	0	133
12	0	0	0	0.0	0	0	0	0	0.00	0.00	0.0	0	134

[9] For salted butter; unsalted butter contains 12 mg sodium per stick, 2 mg per tbsp, or 1 mg per pat.

[10] Values for vitamin A are year-round average.

Item No.	Foods, Approximate Measures, Units, and Weight (Weight of Edible Portion Only)		Water	Food Energy	Protein	Fat	Fatty Acids		
							Saturated	Mono-unsaturated	Poly-unsaturated
			Grams	Percent	Calories	Grams	Grams	Grams	Grams	Grams

Fats and Oils *(continued)*

Margarine:

| 135 | Imitation (about 40% fat), soft | 8-oz container | 227 | 58 | 785 | 1 | 88 | 17.5 | 35.6 | 31.3 |
| 136 | | 1 tbsp - - - - - | 14 | 58 | 50 | Tr | 5 | 1.1 | 2.2 | 1.9 |

Regular (about 80% fat):
Hard (4 sticks per lb):

137	Stick - - - - - - - - - - - - -	1/2 cup - - - -	113	16	810	1	91	17.9	40.5	28.7
138	Tablespoon (1/8 stick) - - -	1 tbsp - - - - -	14	16	100	Tr	11	2.2	5.0	3.6
139	Pat (1 in square, 1/3 in high; 90 per lb) - - - - - - -	1 pat - - - - - -	5	16	35	Tr	4	0.8	1.8	1.3
140	Soft - - - - - - - - - - - - - - - -	8-oz container	227	16	1,625	2	183	31.3	64.7	78.5
141		1 tbsp - - - - -	14	16	100	Tr	11	1.9	4.0	4.8

Spread (about 60% fat):
Hard (4 sticks per lb):

142	Stick - - - - - - - - - - - - -	1/2 cup - - - -	113	37	610	1	69	15.9	29.4	20.5
143	Tablespoon (1/8 stick) - - -	1 tbsp - - - - -	14	37	75	Tr	9	2.0	3.6	2.5
144	Pat (1 in square, 1/3 in high; 90 per lb) - - - - - - -	1 pat - - - - - -	5	37	25	Tr	3	0.7	1.3	0.9
145	Soft - - - - - - - - - - - - - - - -	8-oz container	227	37	1,225	1	138	29.1	71.5	31.3
146		1 tbsp - - - - -	14	37	75	Tr	9	1.8	4.4	1.9

Oils, salad or cooking:

147	Corn - - - - - - - - - - - - - - -	1 cup - - - - -	218	0	1,925	0	218	27.7	52.8	128.0
148		1 tbsp - - - - -	14	0	125	0	14	1.8	3.4	8.2
149	Olive - - - - - - - - - - - - - - -	1 cup - - - - -	216	0	1,910	0	216	29.2	159.2	18.1
150		1 tbsp - - - - -	14	0	125	0	14	1.9	10.3	1.2
151	Peanut - - - - - - - - - - - - - -	1 cup - - - - -	216	0	1,910	0	216	36.5	99.8	69.1
152		1 tbsp - - - - -	14	0	125	0	14	2.4	6.5	4.5
153	Safflower - - - - - - - - - - - -	1 cup - - - - -	218	0	1,925	0	218	19.8	26.4	162.4
154		1 tbsp - - - - -	14	0	125	0	14	1.3	1.7	10.4
155	Soybean oil, hydrogenated (partially hardened) - - - - - -	1 cup - - - - -	218	0	1,925	0	218	32.5	93.7	82.0
156		1 tbsp - - - - -	14	0	125	0	14	2.1	6.0	5.3
157	Soybean-cottonseed oil blend, hydrogenated - - - - -	1 cup - - - - -	218	0	1,925	0	218	39.2	64.3	104.9
158		1 tbsp - - - - -	14	0	125	0	14	2.5	4.1	6.7
159	Sunflower - - - - - - - - - - - - -	1 cup - - - - -	218	0	1,925	0	218	22.5	42.5	143.2
160		1 tbsp - - - - -	14	0	125	0	14	1.4	2.7	9.2

Salad dressings:
Commercial:

| 161 | Blue cheese - - - - - - - - - - - | 1 tbsp - - - - - | 15 | 32 | 75 | 1 | 8 | 1.5 | 1.8 | 4.2 |

French:

| 162 | Regular - - - - - - - - - - - - | 1 tbsp - - - - - | 16 | 35 | 85 | Tr | 9 | 1.4 | 4.0 | 3.5 |
| 163 | Low calorie - - - - - - - - - - | 1 tbsp - - - - - | 16 | 75 | 25 | Tr | 2 | 0.2 | 0.3 | 1.0 |

Italian:

| 164 | Regular - - - - - - - - - - - - | 1 tbsp - - - - - | 15 | 34 | 80 | Tr | 9 | 1.3 | 3.7 | 3.2 |
| 165 | Low calorie - - - - - - - - - - | 1 tbsp - - - - - | 15 | 86 | 5 | Tr | Tr | Tr | Tr | Tr |

Mayonnaise:

166	Regular - - - - - - - - - - - -	1 tbsp - - - - -	14	15	100	Tr	11	1.7	3.2	5.8
167	Imitation - - - - - - - - - - -	1 tbsp - - - - -	15	63	35	Tr	3	0.5	0.7	1.6
168	Mayonnaise type - - - - - - - -	1 tbsp - - - - -	15	40	60	Tr	5	0.7	1.4	2.7
169	Tartar sauce - - - - - - - - - - -	1 tbsp - - - - -	14	34	75	Tr	8	1.2	2.6	3.9

Thousand island:

| 170 | Regular - - - - - - - - - - - - | 1 tbsp - - - - - | 16 | 46 | 60 | Tr | 6 | 1.0 | 1.3 | 3.2 |
| 171 | Low calorie - - - - - - - - - | 1 tbsp - - - - - | 15 | 69 | 25 | Tr | 2 | 0.2 | 0.4 | 0.9 |

Cholesterol Milligrams	Carbohydrate Grams	Calcium Milligrams	Phosphorus Milligrams	Iron Milligrams	Potassium Milligrams	Sodium Milligrams	Vitamin A Value (IU) International units	Vitamin A Value (RE) Retinol equivalents	Thiamin Milligrams	Riboflavin Milligrams	Niacin Milligrams	Ascorbic Acid Milligrams	Item No.
0	1	40	31	0.0	57	[11]2,178	[12]7,510	[12]2,254	0.01	0.05	Tr	Tr	135
0	Tr	2	2	0.0	4	[11]134	[12]460	[12]139	Tr	Tr	Tr	Tr	136
0	1	34	26	0.1	48	[11]1,066	[12]3,740	[12]1,122	0.01	0.04	Tr	Tr	137
0	Tr	4	3	Tr	6	[11]132	[12]460	[12]139	Tr	0.01	Tr	Tr	138
0	Tr	1	1	Tr	2	[11]47	[12]170	[12]50	Tr	Tr	Tr	Tr	139
0	1	60	46	0.0	86	[11]2,449	[12]7,510	[12]2,254	0.02	0.07	Tr	Tr	140
0	Tr	4	3	0.0	5	[11]151	[12]460	[12]139	Tr	Tr	Tr	Tr	141
0	0	24	18	0.0	34	[11]1,123	[12]3,740	[12]1,122	0.01	0.03	Tr	Tr	142
0	0	3	2	0.0	4	[11]139	[12]460	[12]139	Tr	Tr	Tr	Tr	143
0	0	1	1	0.0	1	[11]50	[12]170	[12]50	Tr	Tr	Tr	Tr	144
0	0	47	37	0.0	68	[11]2,256	[12]7,510	[12]2,254	0.02	0.06	Tr	Tr	145
0	0	3	2	0.0	4	[11]139	[12]460	[12]139	Tr	Tr	Tr	Tr	146
0	0	0	0	0.0	0	0	0	0	0.00	0.00	0.0	0	147
0	0	0	0	0.0	0	0	0	0	0.00	0.00	0.0	0	148
0	0	0	0	0.0	0	0	0	0	0.00	0.00	0.0	0	149
0	0	0	0	0.0	0	0	0	0	0.00	0.00	0.0	0	150
0	0	0	0	0.0	0	0	0	0	0.00	0.00	0.0	0	151
0	0	0	0	0.0	0	0	0	0	0.00	0.00	0.0	0	152
0	0	0	0	0.0	0	0	0	0	0.00	0.00	0.0	0	153
0	0	0	0	0.0	0	0	0	0	0.00	0.00	0.0	0	154
0	0	0	0	0.0	0	0	0	0	0.00	0.00	0.0	0	155
0	0	0	0	0.0	0	0	0	0	0.00	0.00	0.0	0	156
0	0	0	0	0.0	0	0	0	0	0.00	0.00	0.0	0	157
0	0	0	0	0.0	0	0	0	0	0.00	0.00	0.0	0	158
0	0	0	0	0.0	0	0	0	0	0.00	0.00	0.0	0	159
0	0	0	0	0.0	0	0	0	0	0.00	0.00	0.0	0	160
3	1	12	11	Tr	6	164	30	10	Tr	0.02	Tr	Tr	161
0	1	2	1	Tr	2	188	Tr	Tr	Tr	Tr	Tr	Tr	162
0	2	6	5	Tr	3	306	Tr	Tr	Tr	Tr	Tr	Tr	163
0	1	1	1	Tr	5	162	30	3	Tr	Tr	Tr	Tr	164
0	2	1	1	Tr	4	136	Tr	Tr	Tr	Tr	Tr	Tr	165
8	Tr	3	4	0.1	5	80	40	12	0.00	0.00	Tr	0	166
4	2	Tr	Tr	0.0	2	75	0	0	0.00	0.00	0.0	0	167
4	4	2	4	Tr	1	107	30	13	Tr	Tr	Tr	0	168
4	1	3	4	0.1	11	182	30	9	Tr	Tr	0.0	Tr	169
4	2	2	3	0.1	18	112	50	15	Tr	Tr	Tr	0	170
2	2	2	3	0.1	17	150	50	14	Tr	Tr	Tr	0	171

[11] For Salted margarine.
[12] Based on average vitamin A content of fortified margarine. Federal specifications for fortified margarine require a minium of 15,000 IU per pound.

Item No.	Foods, Approximate Measures, Units, and Weight (Weight of Edible Portion Only)		Water	Food Energy	Protein	Fat	Fatty Acids		
							Saturated	Mono-unsaturated	Poly-unsaturated
		Grams	Percent	Calories	Grams	Grams	Grams	Grams	Grams
	Fats and Oils *(continued)*								
	Salad dressings:								
	Prepared from home recipe:								
172	Cooked type[13] - - - - - - - - - 1 tbsp - - - - -	16	69	25	1	2	0.5	0.6	0.3
173	Vinegar and oil - - - - - - - - - 1 tbsp - - - - -	16	47	70	0	8	1.5	2.4	3.9
	Fish and Shellfish								
	Clams:								
174	Raw, meat only - - - - - - - - - 3 oz - - - - - - -	85	82	65	11	1	0.3	0.3	0.3
175	Canned, drained solids - - - - - 3 oz - - - - - - -	85	77	85	13	2	0.5	0.5	0.4
176	Crabmeat, canned - - - - - - - - - 1 cup - - - - - -	135	77	135	23	3	0.5	0.8	1.4
177	Fish sticks, frozen, reheated, (stick, 4 by 1 by 1/2 in) - - - - - 1 fish stick - - -	28	52	70	6	3	0.8	1.4	0.8
	Flounder or Sole, baked, with lemon juice:								
178	With butter - - - - - - - - - - - - - 3 oz - - - - - - -	85	73	120	16	6	3.2	1.5	0.5
179	With margarine - - - - - - - - - - 3 oz - - - - - - -	85	73	120	16	6	1.2	2.3	1.9
180	Without added fat - - - - - - - - - 3 oz - - - - - - -	85	78	80	17	1	0.3	0.2	0.4
181	Haddock, breaded, fried[14] - - - - - 3 oz - - - - - - -	85	61	175	17	9	2.4	3.9	2.4
182	Halibut, broiled, with butter and lemon juice - - - - - - - - - - - - - 3 oz - - - - - - -	85	67	140	20	6	3.3	1.6	0.7
183	Herring, pickled - - - - - - - - - - - 3 oz - - - - - - -	85	59	190	17	13	4.3	4.6	3.1
184	Ocean perch, breaded, fried[14] - - - 1 fillet - - - - - -	85	59	185	16	11	2.6	4.6	2.8
	Oysters:								
185	Raw, meat only (13–19 medium Selects) - - - - - - - - 1 cup - - - - - -	240	85	160	20	4	1.4	0.5	1.4
186	Breaded, fried[14] - - - - - - - - - - 1 oyster - - - -	45	65	90	5	5	1.4	2.1	1.4
	Salmon:								
187	Canned (pink), solids and liquid 3 oz - - - - - - -	85	71	120	17	5	0.9	1.5	2.1
188	Baked (red) - - - - - - - - - - - - - 3 oz - - - - - - -	85	67	140	21	5	1.2	2.4	1.4
189	Smoked - - - - - - - - - - - - - - - 3 oz - - - - - - -	85	59	150	18	8	2.6	3.9	0.7
190	Sardines, Atlantic, canned in oil, drained solids - - - - - - - - - - - 3 oz - - - - - - -	85	62	175	20	9	2.1	3.7	2.9
191	Scallops, breaded, frozen, reheated - - - - - - - - - - - - - - 6 scallops - - -	90	59	195	15	10	2.5	4.1	2.5
	Shrimp:								
192	Canned, drained solids - - - - - 3 oz - - - - - - -	85	70	100	21	1	0.2	0.2	0.4
193	French fried (7 medium)[16] - - - - 3 oz - - - - - - -	85	55	200	16	10	2.5	4.1	2.6
194	Trout, broiled, with butter and lemon juice - - - - - - - - - - - - - 3 oz - - - - - - -	85	63	175	21	9	4.1	2.9	1.6
	Tuna, canned, drained solids:								
195	Oil pack, chunk light - - - - - - - 3 oz - - - - - - -	85	61	165	24	7	1.4	1.9	3.1
196	Water pack, solid white - - - - - 3 oz - - - - - - -	85	63	135	30	1	0.3	0.2	0.3
197	Tuna Salad[17] - - - - - - - - - - - - - 1 cup - - - - - -	205	63	375	33	19	3.3	4.9	9.2
	Fruits and Fruit Juices								
	Apples:								
	Raw:								
	Unpeeled, without cores:								
198	2-3/4-in diam. (about 3 per lb with cores) - - - - - 1 apple - - - - -	138	84	80	Tr	Tr	0.1	Tr	0.1
199	3-1/4-in diam. (about 2 per lb with cores) - - - - - 1 apple - - - - -	212	84	125	Tr	1	0.1	Tr	0.2

[13] Fatty acid values apply to product made with regular margarine.

[14] Dipped in egg, and breadcrumbs; fried in vegetable shortening.

[15] If bones are discarded, value for calcium will be greatly reduced.

Choles-terol	Carbo-hydrate	Calcium	Phos-phorus	Iron	Potas-sium	Sodium	Vitamin A Value (IU)	Vitamin A Value (RE)	Thiamin	Ribo-flavin	Niacin	Ascorbic Acid	Item No.
Milli-grams	Grams	Milli-grams	Milli-grams	Milli-grams	Milli-grams	Milli-grams	Inter-national units	Retinol equiva-lents	Milli-grams	Milli-grams	Milli-grams	Milli-grams	
9	2	13	14	0.1	19	117	70	20	0.01	0.02	Tr	Tr	172
0	Tr	0	0	0.0	1	Tr	0	0	0.00	0.00	0.0	0	173
43	2	59	138	2.6	154	102	90	26	0.09	0.15	1.1	9	174
54	2	47	116	3.5	119	102	90	26	0.01	0.09	0.9	3	175
135	1	61	246	1.1	149	1,350	50	14	0.11	0.11	2.6	0	176
26	4	11	58	0.3	94	53	20	5	0.03	0.05	0.6	0	177
68	Tr	13	187	0.3	272	145	210	54	0.05	0.08	1.6	1	178
55	Tr	14	187	0.3	273	151	230	69	0.05	0.08	1.6	1	179
59	Tr	13	197	0.3	286	101	30	10	0.05	0.08	1.7	1	180
75	7	34	183	1.0	270	123	70	20	0.06	0.10	2.9	0	181
62	Tr	14	206	0.7	441	103	610	174	0.06	0.07	7.7	1	182
85	0	29	128	0.9	85	850	110	33	0.04	0.18	2.8	0	183
66	7	31	191	1.2	241	138	70	20	0.10	0.11	2.0	0	184
120	8	226	343	15.6	290	175	740	223	0.34	0.43	6.0	24	185
35	5	49	73	3.0	64	70	150	44	0.07	0.10	1.3	4	186
34	0	[15]167	243	0.7	307	443	60	18	0.03	0.15	6.8	0	187
60	0	26	269	0.5	305	55	290	87	0.18	0.14	5.5	0	188
51	0	12	208	0.8	327	1,700	260	77	0.17	0.17	6.8	0	189
85	0	[15]371	424	2.6	349	425	190	56	0.03	0.17	4.6	0	190
70	10	39	203	2.0	369	298	70	21	0.11	0.11	1.6	0	191
128	1	98	224	1.4	1	1,955	50	15	0.01	0.03	1.5	0	192
168	11	61	154	2.0	189	384	90	26	0.06	0.09	2.8	0	193
71	Tr	26	259	1.0	297	122	230	60	0.07	0.07	2.3	1	194
55	0	7	199	1.6	298	303	70	20	0.04	0.09	10.1	0	195
48	0	17	202	0.6	255	468	110	32	0.03	0.10	13.4	0	196
80	19	31	281	2.5	531	877	230	53	0.06	0.14	13.3	6	197
0	21	10	10	0.2	159	Tr	70	7	0.02	0.02	0.1	8	198
0	32	15	15	0.4	244	Tr	110	11	0.04	0.03	0.2	12	199

[16] Dipped in egg, breadcrumbs, and flour; fried in vegetable shortening.

[17] Made with drained chunk light tuna, celery, onion, pickle relish, and mayonnaise-type salad dressing.

(Tr indicates nutrient present in trace amount.)

Item No.	Foods, Approximate Measures, Units, and Weight (Weight of Edible Portion Only)		Water	Food Energy	Protein	Fat	Fatty Acids		
							Saturated	Mono-unsaturated	Poly-unsaturated
		Grams	Percent	Calories	Grams	Grams	Grams	Grams	Grams

Fruits and Fruit Juices *(continued)*

Item No.	Food	Measure	Grams	Percent	Calories	Protein Grams	Fat Grams	Saturated Grams	Mono-unsaturated Grams	Poly-unsaturated Grams
	Apples:									
	Raw:									
200	Peeled, sliced	1 cup	110	84	65	Tr	Tr	0.1	Tr	0.1
201	Dried, sulfured	10 rings	64	32	155	1	Tr	Tr	Tr	0.1
202	Apple juice, bottled or canned[19]	1 cup	248	88	115	Tr	Tr	Tr	Tr	0.1
	Applesauce, canned:									
203	Sweetened	1 cup	255	80	195	Tr	Tr	0.1	Tr	0.1
204	Unsweetened	1 cup	244	88	105	Tr	Tr	Tr	Tr	Tr
	Apricots:									
205	Raw, without pits (about 12 per lb with pits)	3 apricots	106	86	50	1	Tr	Tr	0.2	0.1
	Canned (fruit and liquid):									
206	Heavy syrup pack	1 cup	258	78	215	1	Tr	Tr	0.1	Tr
207		3 halves	85	78	70	Tr	Tr	Tr	Tr	Tr
208	Juice pack	1 cup	248	87	120	2	Tr	Tr	Tr	Tr
209		3 halves	84	87	40	1	Tr	Tr	Tr	Tr
	Dried:									
210	Uncooked (28 large or 37 medium halves per cup)	1 cup	130	31	310	5	1	Tr	0.3	0.1
211	Cooked, unsweetened, fruit and liquid	1 cup	250	76	210	3	Tr	Tr	0.2	0.1
212	Apricot nectar, canned	1 cup	251	85	140	1	Tr	Tr	0.1	Tr
	Avocados, raw, whole, without skin and seed:									
213	California (about 2 per lb with skin and seed)	1 avocado	173	73	305	4	30	4.5	19.4	3.5
214	Florida (about 1 per lb with skin and seed)	1 avocado	304	80	340	5	27	5.3	14.8	4.5
	Bananas, raw, without peel:									
215	Whole (about 2-1/2 per lb with peel)	1 banana	114	74	105	1	1	0.2	Tr	0.1
216	Sliced	1 cup	150	74	140	2	1	0.3	0.1	0.1
217	Blackberries, raw	1 cup	144	86	75	1	1	0.2	0.1	0.1
	Blueberries:									
218	Raw	1 cup	145	85	80	1	1	Tr	0.1	0.3
219	Frozen, sweetened	10-oz container	284	77	230	1	Tr	Tr	0.1	0.2
220		1 cup	230	77	185	1	Tr	Tr	Tr	0.1
	Cantaloupe. See Melons (item 251).									
	Cherries:									
221	Sour, red, pitted, canned, water pack	1 cup	244	90	90	2	Tr	0.1	0.1	0.1
222	Sweet, raw, without pits and stems	10 cherries	68	81	50	1	1	0.1	0.2	0.2
223	Cranberry juice cocktail, bottled, sweetened	1 cup	253	85	145	Tr	Tr	Tr	Tr	0.1
224	Cranberry sauce, sweetened, canned, strained	1 cup	277	61	420	1	Tr	Tr	0.1	0.2

[18] Sodium bisulfite used to preserve color; unsulfited product would contain less sodium.

[19] Also applies to pasteurized apple cider.

Choles-terol	Carbo-hydrate	Calcium	Phos-phorus	Iron	Potas-sium	Sodium	Vitamin A Value (IU)	Vitamin A Value (RE)	Thiamin	Ribo-flavin	Niacin	Ascorbic Acid	Item No.
Milli-grams	Grams	Milli-grams	Milli-grams	Milli-grams	Milli-grams	Milli-grams	Inter-national units	Retinol equiva-lents	Milli-grams	Milli-grams	Milli-grams	Milli-grams	
0	16	4	8	0.1	124	Tr	50	5	0.02	0.01	0.1	4	200
0	42	9	24	0.9	288	[18]56	0	0	0.00	0.10	0.6	2	201
0	29	17	17	0.9	295	7	Tr	Tr	0.05	0.04	0.2	[20]2	202
0	51	10	18	0.9	156	8	30	3	0.03	0.07	0.5	[20]4	203
0	28	7	17	0.3	183	5	70	7	0.03	0.06	0.5	[20]3	204
0	12	15	20	0.6	314	1	2,770	277	0.03	0.04	0.6	11	205
0	55	23	31	0.8	361	10	3,170	317	0.05	0.06	1.0	8	206
0	18	8	10	0.3	119	3	1,050	105	0.02	0.02	0.3	3	207
0	31	30	50	0.7	409	10	4,190	419	0.04	0.05	0.9	12	208
0	10	10	17	0.3	139	3	1,420	142	0.02	0.02	0.3	4	209
0	80	59	152	6.1	1,791	13	9,410	941	0.01	0.20	3.9	3	210
0	55	40	103	4.2	1,222	8	5,910	591	0.02	0.08	2.4	4	211
0	36	18	23	1.0	286	8	3,300	330	0.02	0.04	0.7	[20]2	212
0	12	19	73	2.0	1,097	21	1,060	106	0.19	0.21	3.3	14	213
0	27	33	119	1.6	1,484	15	1,860	186	0.33	0.37	5.8	24	214
0	27	7	23	0.4	451	1	90	9	0.05	0.11	0.6	10	215
0	35	9	30	0.5	594	2	120	12	0.07	0.15	0.8	14	216
0	18	46	30	0.8	282	Tr	240	24	0.04	0.06	0.6	30	217
0	20	9	15	0.2	129	9	150	15	0.07	0.07	0.5	19	218
0	62	17	20	1.1	170	3	120	12	0.06	0.15	0.7	3	219
0	50	14	16	0.9	138	2	100	10	0.05	0.12	0.6	2	220
0	22	27	24	3.3	239	17	1,840	184	0.04	0.10	0.4	5	221
0	11	10	13	0.3	152	Tr	150	15	0.03	0.04	0.3	5	222
0	38	8	3	0.4	61	10	10	1	0.01	0.04	0.1	[21]108	223
0	108	11	17	0.6	72	80	60	6	0.04	0.06	0.3	6	224

[20] Without added ascorbic acid. For value with added ascorbic acid, refer to label.

[21] With added ascorbic acid.

(Tr indicates nutrient present in trace amount.)

Item No.	Foods, Approximate Measures, Units, and Weight (Weight of Edible Portion Only)		Water	Food Energy	Protein	Fat	Fatty Acids		
							Saturated	Mono-unsaturated	Poly-unsaturated
		Grams	Percent	Calories	Grams	Grams	Grams	Grams	Grams
	Fruits and Fruit Juices *(continued)*								
	Dates:								
225	Whole, without pits - - - - - - - - 10 dates - - - -	83	23	230	2	Tr	0.1	0.1	Tr
226	Chopped - - - - - - - - - - - - - - 1 cup - - - - - -	178	23	490	4	1	0.3	0.2	Tr
227	Figs, dried - - - - - - - - - - - - - 10 figs - - - - -	187	28	475	6	2	0.4	0.5	1.0
	Fruit cocktail, canned, fruit and liquid:								
228	Heavy syrup pack - - - - - - - - - 1 cup - - - - - -	255	80	185	1	Tr	Tr	Tr	0.1
229	Juice pack - - - - - - - - - - - - - 1 cup - - - - - -	248	87	115	1	Tr	Tr	Tr	Tr
	Grapefruit:								
	Raw, without peel, membrane and seeds (3-3/4-in diam., 1 lb								
230	1 oz, whole, with refuse) - - - 1/2 grapefruit	120	91	40	1	Tr	Tr	Tr	Tr
	Canned, sections with								
231	syrup - - - - - - - - - - - - - - - 1 cup - - - - - -	254	84	150	1	Tr	Tr	Tr	0.1
	Grapefruit juice:								
232	Raw - - - - - - - - - - - - - - - - 1 cup - - - - - -	247	90	95	1	Tr	Tr	Tr	0.1
	Canned:								
233	Unsweetened - - - - - - - - - - 1 cup - - - - - -	247	90	95	1	Tr	Tr	Tr	0.1
234	Sweetened - - - - - - - - - - - 1 cup - - - - - -	250	87	115	1	Tr	Tr	Tr	0.1
	Frozen concentrate, unsweetened								
235	Undiluted - - - - - - - - - - - - - 6-fl-oz can - - -	207	62	300	4	1	0.1	0.1	0.2
236	Diluted with 3 parts water by volume - - - - - - - - 1 cup - - - - - -	247	89	100	1	Tr	Tr	Tr	0.1
	Grapes, European type (adherent skin), raw:								
237	Thompson Seedless - - - - - - - 10 grapes - - -	50	81	35	Tr	Tr	0.1	Tr	0.1
238	Tokay and Emperor, seeded types 10 grapes - - -	57	81	40	Tr	Tr	0.1	Tr	0.1
	Grape juice:								
239	Canned or bottled - - - - - - - - 1 cup - - - - - -	253	84	155	1	Tr	0.1	Tr	0.1
	Frozen concentrate, sweetened:								
240	Undiluted - - - - - - - - - - - - 6-fl-oz can - - -	216	54	385	1	1	0.2	Tr	0.2
241	Diluted with 3 parts water by volume - - - - - - - - - - - 1 cup - - - - - -	250	87	125	Tr	Tr	0.1	Tr	0.1
242	Kiwifruit, raw, without skin (about 5 per lb with skin) - - - - - - - - 1 kiwifruit - - -	76	83	45	1	Tr	Tr	0.1	0.1
243	Lemons, raw, without peel and seeds (about 4 per lb with peel and seeds) - - - - - - - - - - - - 1 lemon - - - -	58	89	15	1	Tr	Tr	Tr	0.1
	Lemon juice:								
244	Raw - - - - - - - - - - - - - - - - 1 cup - - - - - -	244	91	60	1	Tr	Tr	Tr	Tr
245	Canned or bottled, unsweetened 1 cup - - - - - -	244	92	50	1	1	0.1	Tr	0.2
246	1 tbsp - - - - -	15	92	5	Tr	Tr	Tr	Tr	Tr
247	Frozen, single-strength, unsweetened - - - - - - - - - - 6-fl-oz can - - -	244	92	55	1	1	0.1	Tr	0.2
	Lime juice:								
248	Raw - - - - - - - - - - - - - - - - 1 cup - - - - - -	246	90	65	1	Tr	Tr	Tr	0.1

[20] Without added ascorbic acid. For value with added ascorbic acid, refer to label.

[21] With added ascorbic acid.

Choles-terol	Carbo-hydrate	Calcium	Phos-phorus	Iron	Potas-sium	Sodium	Vitamin A Value		Thiamin	Ribo-flavin	Niacin	Ascorbic Acid	Item No.
							(IU)	(RE)					
Milli-grams	Grams	Milli-grams	Milli-grams	Milli-grams	Milli-grams	Milli-grams	Inter-national units	Retinol equiva-lents	Milli-grams	Milli-grams	Milli-grams	Milli-grams	
0	61	27	33	1.0	541	2	40	4	0.07	0.08	1.8	0	225
0	131	57	71	2.0	1,161	5	90	9	0.16	0.18	3.9	0	226
0	122	269	127	4.2	1,331	21	250	25	0.13	0.16	1.3	1	227
0	48	15	28	0.7	224	15	520	52	0.05	0.05	1.0	5	228
0	29	20	35	0.5	236	10	760	76	0.03	0.04	1.0	7	229
0	10	14	10	0.1	167	Tr	[22]10	[22]1	0.04	0.02	0.3	41	230
0	39	36	25	1.0	328	5	Tr	Tr	0.10	0.05	0.6	54	231
0	23	22	37	0.5	400	2	20	2	0.10	0.05	0.5	94	232
0	22	17	27	0.5	378	2	20	2	0.10	0.05	0.6	72	233
0	28	20	28	0.9	405	5	20	2	0.10	0.06	0.8	67	234
0	72	56	101	1.0	1,002	6	60	6	0.30	0.16	1.6	248	235
0	24	20	35	0.3	336	2	20	2	0.10	0.05	0.5	83	236
0	9	6	7	0.1	93	1	40	4	0.05	0.03	0.2	5	237
0	10	6	7	0.1	105	1	40	4	0.05	0.03	0.2	6	238
0	38	23	28	0.6	334	8	20	2	0.07	0.09	0.7	[20]Tr	239
0	96	28	32	0.8	160	15	60	6	0.11	0.20	0.9	[21]179	240
0	32	10	10	0.3	53	5	20	2	0.04	0.07	0.3	[21]60	241
0	11	20	30	0.3	252	4	130	13	0.02	0.04	0.4	74	242
0	5	15	9	0.3	80	1	20	2	0.02	0.01	0.1	31	243
0	21	17	15	0.1	303	2	50	5	0.07	0.02	0.2	112	244
0	16	27	22	0.3	249	[23]51	40	4	0.10	0.02	0.5	61	245
0	1	2	1	Tr	15	[23]3	Tr	Tr	0.01	Tr	Tr	4	246
0	16	20	20	0.3	217	2	30	3	0.14	0.03	0.3	77	247
0	22	22	17	0.1	268	2	20	2	0.05	0.02	0.2	72	248

[22] For white grapefruit; pink grapefruit have about 310 IU or 31 RE.

[23] Sodium benzoate and sodium bisulfite added as preservatives.

(Tr indicates nutrient present in trace amount.)

Item No.	Foods, Approximate Measures, Units, and Weight (Weight of Edible Portion Only)		Water	Food Energy	Protein	Fat	Fatty Acids		
							Saturated	Mono-unsaturated	Poly-unsaturated
		Grams	Percent	Calories	Grams	Grams	Grams	Grams	Grams
	Fruits and Fruit Juices *(continued)*								
	Lime juice:								
249	Canned, unsweetened - - - - - 1 cup - - - - - -	246	93	50	1	1	0.1	0.1	0.2
250	Mangos, raw, without skin and seed (about 1-1/2 per lb with skin and seed) - - - - - - - - - - 1 mango - - - -	207	82	135	1	1	0.1	0.2	0.1
	Melons, raw, without rind and cavity contents:								
251	Cantaloupe, orange-fleshed (5-in diam., 2-1/3 lb, whole, with rind and cavity contents) - - - - - - - - - - - 1/2 melon - - -	267	90	95	2	1	0.1	0.1	0.3
252	Honeydew (6-1/2-1 diam., 5-1/4 lb, whole, with rind and cavity contents) - - - - - - 1/10 melon - -	129	90	45	1	Tr	Tr	Tr	0.1
253	Nectarines, raw, without pits (about 3 per lb with pits) - - - - 1 nectarine - -	136	86	65	1	1	0.1	0.2	0.3
	Oranges, raw:								
254	Whole, without peel and seeds (2-5/8-in diam., about 2-1/2 per lb, with peel and seeds) - - - - - - - - - - - - - 1 orange - - - -	131	87	60	1	Tr	Tr	Tr	Tr
255	Sections without membranes - - 1 cup - - - - - -	180	87	85	2	Tr	Tr	Tr	Tr
	Orange juice:								
256	Raw, all varieties - - - - - - - - - 1 cup - - - - - -	248	88	110	2	Tr	0.1	0.1	0.1
257	Canned, unsweetened - - - - - - 1 cup - - - - - -	249	89	105	1	Tr	Tr	0.1	0.1
258	Chilled - - - - - - - - - - - - - 1 cup - - - - - -	249	88	110	2	1	0.1	0.1	0.2
	Frozen concentrate:								
259	Undiluted - - - - - - - - - - - - 6-fl-oz can - -	213	58	340	5	Tr	0.1	0.1	0.1
260	Diluted with 3 parts water by volume - - - - - - - - - - - - 1 cup - - - - - -	249	88	110	2	Tr	Tr	Tr	Tr
261	Orange and grapefruit juice, canned - - - - - - - - - - - - - 1 cup - - - - - -	247	89	105	1	Tr	Tr	Tr	Tr
262	Papayas, raw, 1/2-in cubes - - - 1 cup - - - - - -	140	86	65	1	Tr	0.1	0.1	Tr
	Peaches:								
	Raw:								
263	Whole, 2-1/2-in diam., peeled, pitted (about 4 per lb with peels and pits) - - - 1 peach - - - -	87	88	35	1	Tr	Tr	Tr	Tr
264	Sliced - - - - - - - - - - - - - - 1 cup - - - - - -	170	88	75	1	Tr	Tr	0.1	0.1
	Canned, fruit and liquid:								
265	Heavy syrup pack - - - - - - - 1 cup - - - - - -	256	79	190	1	Tr	Tr	0.1	0.1
266	1 half - - - - - -	81	79	60	Tr	Tr	Tr	Tr	Tr
267	Juice pack - - - - - - - - - - - 1 cup - - - - - -	248	87	110	2	Tr	Tr	Tr	Tr
268	1 half - - - - - -	77	87	35	Tr	Tr	Tr	Tr	Tr
	Peaches:								
	Dried:								
269	Uncooked - - - - - - - - - - - - 1 cup - - - - - -	160	32	380	6	1	0.1	0.4	0.6
270	Cooked, unsweetened, fruit and liquid - - - - - - - - - - - 1 cup - - - - - -	258	78	200	3	1	0.1	0.2	0.3
271	Frozen, sliced, sweetened - - - - 10-oz container - - - -	284	75	265	2	Tr	Tr	0.1	0.2
272	1 cup - - - - - -	250	75	235	2	Tr	Tr	0.1	0.2

Cholesterol	Carbohydrate	Calcium	Phosphorus	Iron	Potassium	Sodium	Vitamin A Value		Thiamin	Riboflavin	Niacin	Ascorbic Acid	Item No.
							(IU)	(RE)					
Milligrams	Grams	Milligrams	Milligrams	Milligrams	Milligrams	Milligrams	International units	Retinol equivalents	Milligrams	Milligrams	Milligrams	Milligrams	
0	16	30	25	0.6	185	[23]39	40	4	0.08	0.01	0.4	16	249
0	35	21	23	0.3	323	4	8,060	806	0.12	0.12	1.2	57	250
0	22	29	45	0.6	825	24	8,610	861	0.10	0.06	1.5	113	251
0	12	8	13	0.1	350	13	50	5	0.10	0.02	0.8	32	252
0	16	7	22	0.2	288	Tr	1,000	100	0.02	0.06	1.3	7	253
0	15	52	18	0.1	237	Tr	270	27	0.11	0.05	0.4	70	254
0	21	72	25	0.2	326	Tr	370	37	0.16	0.07	0.5	96	255
0	26	27	42	0.5	496	2	500	50	0.22	0.07	1.0	124	256
0	25	20	35	1.1	436	5	440	44	0.15	0.07	0.8	86	257
0	25	25	27	0.4	473	2	190	19	0.28	0.05	0.7	82	258
0	81	68	121	0.7	1,436	6	590	59	0.60	0.14	1.5	294	259
0	27	22	40	0.2	473	2	190	19	0.20	0.04	0.5	97	260
0	25	20	35	1.1	390	7	290	29	0.14	0.07	0.8	72	261
0	17	35	12	0.3	247	9	400	40	0.04	0.04	0.5	92	262
0	10	4	10	0.1	171	Tr	470	47	0.01	0.04	0.9	6	263
0	19	9	20	0.2	335	Tr	910	91	0.03	0.07	1.7	11	264
0	51	8	28	0.7	236	15	850	85	0.03	0.06	1.6	7	265
0	16	2	9	0.2	75	5	270	27	0.01	0.02	0.5	2	266
0	29	15	42	0.7	317	10	940	94	0.02	0.04	1.4	9	267
0	9	5	13	0.2	99	3	290	29	0.01	0.01	0.4	3	268
0	98	45	190	6.5	1,594	11	3,460	346	Tr	0.34	7.0	8	269
0	51	23	98	3.4	826	5	510	51	0.01	0.05	3.9	10	270
0	68	9	31	1.1	369	17	810	81	0.04	0.10	1.9	[21]268	271
0	60	8	28	0.9	325	15	710	71	0.03	0.09	1.6	[21]236	272

[21] With added ascorbic acid.

(Tr indicates nutrient present in trace amount.)

Item No.	Foods, Approximate Measures, Units, and Weight (Weight of Edible Portion Only)		Water	Food Energy	Protein	Fat	Saturated	Fatty Acids Mono- unsaturated	Poly- unsaturated
			Grams	Percent	Calories	Grams	Grams	Grams	Grams	Grams

			Grams	Percent	Calories	Grams	Grams	Grams	Grams
	Fruits and Fruit Juices *(continued)*								
	Pears:								
	Raw, with skin, cored:								
273	Bartlett, 2-1/2-in diam. (about 2-1/2 per lb with cores and stems) - - - - - -	1 pear - - - - -	166	84	100	1	1	Tr	0.1	0.2
274	Bosc, 2-1/2-in diam. (about 3 per lb with cores and stems) - - - - - - - - - - -	1 pear - - - - -	141	84	85	1	1	Tr	0.1	0.1
275	D'Anjou, 3-in diam. (about 2 per lb with cores and stems) - - - - - - - - - - -	1 pear - - - - -	200	84	120	1	1	Tr	0.2	0.2
	Canned, fruit and liquid:								
276	Heavy syrup pack - - - - - - -	1 cup- - - - - -	255	80	190	1	Tr	Tr	0.1	0.1
277		1 half - - - - - -	79	80	60	Tr	Tr	Tr	Tr	Tr
278	Juice pack- - - - - - - - - - - -	1 cup- - - - - -	248	86	125	1	Tr	Tr	Tr	Tr
279		1 half - - - - - -	77	86	40	Tr	Tr	Tr	Tr	Tr
	Pineapple:								
280	Raw, diced- - - - - - - - - - - -	1 cup- - - - - -	155	87	75	1	1	Tr	0.1	0.2
	Canned, fruit and liquid:								
	Heavy syrup pack:								
281	Crushed, chunks, tidbits - - -	1 cup- - - - - -	255	79	200	1	Tr	Tr	Tr	0.1
282	Slices - - - - - - - - - - - - - -	1 slice - - - - -	58	79	45	Tr	Tr	Tr	Tr	Tr
	Juice pack:								
283	Chunks or tidbits - - - - - - - -	1 cup- - - - - -	250	84	150	1	Tr	Tr	Tr	0.1
284	Slices- - - - - - - - - - - - - -	1 slice - - - - -	58	84	35	Tr	Tr	Tr	Tr	Tr
285	Pineapple juice, unsweetened, canned - - - - - - - - - - - - - -	1 cup- - - - - -	250	86	140	1	Tr	Tr	Tr	0.1
	Plantains, without peel:								
286	Raw - - - - - - - - - - - - - - - -	1 plantain - - -	179	65	220	2	1	0.3	0.1	0.1
287	Cooked, boiled, sliced - - - - - -	1 cup- - - - - -	154	67	180	1	Tr	0.1	Tr	0.1
	Plums, without pits:								
	Raw:								
288	2-1/8-in diam. (about 6-1/2 per lb with pits) - - - - - - -	1 plum - - - - -	66	85	35	1	Tr	Tr	0.3	0.1
289	1-1/2-in diam. (about 15 per lb with pits) - - - - - - - - - -	1 plum - - - - -	28	85	15	Tr	Tr	Tr	0.1	Tr
	Canned, purple, fruit and liquid:								
290	Heavy syrup pack - - - - - - -	1 cup- - - - - -	258	76	230	1	Tr	Tr	0.2	0.1
291		3 plums - - - -	133	76	120	Tr	Tr	Tr	0.1	Tr
292	Juice pack- - - - - - - - - - - -	1 cup- - - - - -	252	84	145	1	Tr	Tr	Tr	Tr
293		3 plums - - - -	95	84	55	Tr	Tr	Tr	Tr	Tr
	Prunes, dried:								
294	Uncooked - - - - - - - - - - - - -	4 extra large or 5 large prunes	49	32	115	1	Tr	Tr	0.2	0.1
295	Cooked, unsweetened, fruit and liquid - - - - - - - - - - - -	1 cup- - - - - -	212	70	225	2	Tr	Tr	0.3	0.1
296	Prune juice, canned or bottled - -	1 cup- - - - - -	256	81	180	2	Tr	Tr	0.1	Tr
	Raisins, seedless:								
297	Cup, not pressed down - - - - -	1 cup- - - - - -	145	15	435	5	1	0.2	Tr	0.2
298	Packet, 1/2 or (1-1/2 tbsp) - -	1 packet - - - -	14	15	40	Tr	Tr	Tr	Tr	Tr
	Raspberries:								
299	Raw - - - - - - - - - - - - - - - -	1 cup- - - - - -	123	87	60	1	1	Tr	0.1	0.4
300	Frozen, sweetened - - - - - - -	10-oz container	284	73	295	2	Tr	Tr	Tr	0.3
301		1 cup- - - - - -	250	73	255	2	Tr	Tr	Tr	0.2

Choles-terol	Carbo-hydrate	Calcium	Phos-phorus	Iron	Potas-sium	Sodium	Vitamin A Value (IU)	Vitamin A Value (RE)	Thiamin	Ribo-flavin	Niacin	Ascorbic Acid	Item No.
Milli-grams	Grams	Milli-grams	Milli-grams	Milli-grams	Milli-grams	Milli-grams	Inter-national units	Retinol equiva-lents	Milli-grams	Milli-grams	Milli-grams	Milli-grams	
0	25	18	18	0.4	208	Tr	30	3	0.03	0.07	0.2	7	273
0	21	16	16	0.4	176	Tr	30	3	0.03	0.06	0.1	6	274
0	30	22	22	0.5	250	Tr	40	4	0.04	0.08	0.2	8	275
0	49	13	18	0.6	166	13	10	1	0.03	0.06	0.6	3	276
0	15	4	6	0.2	51	4	Tr	Tr	0.01	0.02	0.2	1	277
0	32	22	30	0.7	238	10	10	1	0.03	0.03	0.5	4	278
0	10	7	9	0.2	74	3	Tr	Tr	0.01	0.01	0.2	1	279
0	19	11	11	0.6	175	2	40	4	0.14	0.06	0.7	24	280
0	52	36	18	1.0	265	3	40	4	0.23	0.06	0.7	19	281
0	12	8	4	0.2	60	1	10	1	0.05	0.01	0.2	4	282
0	39	35	15	0.7	305	3	100	10	0.24	0.05	0.7	24	283
0	9	8	3	0.2	71	1	20	2	0.06	0.01	0.2	6	284
0	34	43	20	0.7	335	3	10	1	0.14	0.06	0.6	27	285
0	57	5	61	1.1	893	7	2,020	202	0.09	0.10	1.2	33	286
0	48	3	43	0.9	716	8	1,400	140	0.07	0.08	1.2	17	287
0	9	3	7	0.1	114	Tr	210	21	0.03	0.06	0.3	6	288
0	4	1	3	Tr	48	Tr	90	9	0.01	0.03	0.1	3	289
0	60	23	34	2.2	235	49	670	67	0.04	0.10	0.8	1	290
0	31	12	17	1.1	121	25	340	34	0.02	0.05	0.4	1	291
0	38	25	38	0.9	388	3	2,540	254	0.06	0.15	1.2	7	292
0	14	10	14	0.3	146	1	960	96	0.02	0.06	0.4	3	293
0	31	25	39	1.2	365	2	970	97	0.04	0.08	1.0	2	294
0	60	49	74	2.4	708	4	650	65	0.05	0.21	1.5	6	295
0	45	31	64	3.0	707	10	10	1	0.04	0.18	2.0	10	296
0	115	71	141	3.0	1,089	17	10	1	0.23	0.13	1.2	5	297
0	11	7	14	0.3	105	2	Tr	Tr	0.02	0.01	0.1	Tr	298
0	14	27	15	0.7	187	Tr	160	16	0.04	0.11	1.1	31	299
0	74	43	48	1.8	324	3	170	17	0.05	0.13	0.7	47	300
0	65	38	43	1.6	285	3	150	15	0.05	0.11	0.6	41	301

Item No.	Foods, Approximate Measures, Units, and Weight (Weight of Edible Portion Only)		Water	Food Energy	Protein	Fat	Fatty Acids		
							Saturated	Mono-unsaturated	Poly-unsaturated
		Grams	Percent	Calories	Grams	Grams	Grams	Grams	Grams
	Fruits and Fruit Juices *(continued)*								
302	Rhubarb, cooked, added sugar - -	1 cup - - - - -	240	68	280	1	Tr	Tr	Tr	0.1
	Strawberries:								
303	Raw, capped, whole - - - - - - -	1 cup - - - - -	149	92	45	1	1	Tr	0.1	0.3
304	Frozen, sweetened, sliced - - - -	10-oz container	284	73	275	2	Tr	Tr	0.1	0.2
305		1 cup - - - - - -	255	73	245	1	Tr	Tr	Tr	0.2
	Tangerines:								
306	Raw, without peel and seeds (2-3/8-in diam., about 4 per lb, with peel and seeds) - - - - - - - - - - - - -	1 tangerine - -	84	88	35	1	Tr	Tr	Tr	Tr
307	Canned, light syrup, fruit and liquid - - - - - - - - - - - - - -	1 cup - - - - - -	252	83	155	1	Tr	Tr	Tr	0.1
308	Tangerine juice, canned, sweetened - - - - - - - - - - - - -	1 cup - - - - - -	249	87	125	1	Tr	Tr	Tr	0.1
	Watermelon, raw, without rind and seeds:								
309	Piece (4 by 8 in wedge with rind and seeds; 1/16 of 32-2/3-lb melon, 10 by 16 in) - - - - - - - - - - - - - -	1 piece - - - - -	482	92	155	3	2	0.3	0.2	1.0
310	Diced - - - - - - - - - - - - - - -	1 cup - - - - -	160	92	50	1	1	0.1	0.1	0.3
	Grain Products								
311	Bagels, plain or water, enriched, 3-1/2-in diam.[24]	1 bagel - - - - -	68	29	200	7	2	0.3	0.5	0.7
312	Barley, pearled, light, uncooked	1 cup - - - - - -	200	11	700	16	2	0.3	0.2	0.9
	Biscuits, baking powder, 2-in diam. (enriched flour, vegetable shortening):								
313	From home recipe - - - - - - - -	1 biscuit - - - -	28	28	100	2	5	1.2	2.0	1.3
314	From mix - - - - - - - - - - - - -	1 biscuit - - - -	28	29	95	2	3	0.8	1.4	0.9
315	From refrigerated dough - - - - -	1 biscuit - - - -	20	30	65	1	2	0.6	0.9	0.6
	Breadcrumbs, enriched:								
316	Dry, grated - - - - - - - - - - - -	1 cup - - - - - -	100	7	390	13	5	1.5	1.6	1.0
	Soft. See White bread (item 351).								
	Breads:								
317	Boston brown bread, canned, slice, 3-1/4 in by 1/2 in[25] - -	1 slice - - - - -	45	45	95	2	1	0.3	0.1	0.1
	Cracked-wheat bread (3/4 enriched flour, 1/4 cracked wheat flour):[25]								
318	Loaf, 1 lb - - - - - - - - - - -	1 loaf - - - - - -	454	35	1,190	42	16	3.1	4.3	5.7
319	Slice (18 per loaf) - - - - - - -	1 slice - - - - -	25	35	65	2	1	0.2	0.2	0.3
320	Toasted - - - - - - - - - - - -	1 slice - - - - -	21	26	65	2	1	0.2	0.2	0.3
	French or vienna bread, enriched:[25]								
321	Loaf, 1 lb - - - - - - - - - - -	1 loaf - - - - - -	454	34	1,270	43	18	3.8	5.7	5.9
	Slice:								
322	French, 5 by 2-1/2 by 1 in	1 slice - - - - -	35	34	100	3	1	0.3	0.4	0.5

[24] Egg bagels have 44 mg cholesterol and 22 IU or 7 RE vitamin A per bagel.

[25] Made with vegetable shortening.

Choles-terol	Carbo-hydrate	Calcium	Phos-phorus	Iron	Potas-sium	Sodium	Vitamin A Value		Thiamin	Ribo-flavin	Niacin	Ascorbic Acid	Item No.
							(IU)	(RE)					
Milli-grams	Grams	Milli-grams	Milli-grams	Milli-grams	Milli-grams	Milli-grams	Inter-national units	Retinol equiva-lents	Milli-grams	Milli-grams	Milli-grams	Milli-grams	
0	75	348	19	0.5	230	2	170	17	0.04	0.06	0.5	8	302
0	10	21	28	0.6	247	1	40	4	0.03	0.10	0.3	84	303
0	74	31	37	1.7	278	9	70	7	0.05	0.14	1.1	118	304
0	66	28	33	1.5	250	8	60	6	0.04	0.13	1.0	106	305
0	9	12	8	0.1	132	1	770	77	0.09	0.02	0.1	26	306
0	41	18	25	0.9	197	15	2,120	212	0.13	0.11	1.1	50	307
0	30	45	35	0.5	443	2	1,050	105	0.15	0.05	0.2	55	308
0	35	39	43	0.8	559	10	1,760	176	0.39	0.10	1.0	46	309
0	11	13	14	0.3	186	3	590	59	0.13	0.03	0.3	15	310
0	38	29	46	1.8	50	245	0	0	0.26	0.20	2.4	0	311
0	158	32	378	4.2	320	6	0	0	0.24	0.10	6.2	0	312
Tr	13	47	36	0.7	32	195	10	3	0.08	0.08	0.8	Tr	313
Tr	14	58	128	0.7	56	262	20	4	0.12	0.11	0.8	Tr	314
1	10	4	79	0.5	18	249	0	0	0.08	0.05	0.7	0	315
5	73	122	141	4.1	152	736	0	0	0.35	0.35	4.8	0	316
3	21	41	72	0.9	131	113	[26]0	[26]0	0.06	0.04	0.7	0	317
0	227	295	581	12.1	608	1,966	Tr	Tr	1.73	1.73	15.3	Tr	318
0	12	16	32	0.7	34	106	Tr	Tr	0.10	0.09	0.8	Tr	319
0	12	16	32	0.7	34	106	Tr	Tr	0.07	0.09	0.8	Tr	320
0	230	499	386	14.0	409	2,633	Tr	Tr	2.09	1.59	18.2	Tr	321
0	18	39	30	1.1	32	203	Tr	Tr	0.16	0.12	1.4	Tr	322

[26] Made with white cornmeal. If made with yellow cornmeal, value is 32 IU or 3 RE.

(Tr indicates nutrient present in trace amount.)

Item No.	Foods, Approximate Measures, Units, and Weight (Weight of Edible Portion Only)		Water	Food Energy	Protein	Fat	Saturated	Fatty Acids Mono- unsaturated	Poly- unsaturated
			Grams	Percent	Calories	Grams	Grams	Grams	Grams	Grams

Item No.	Foods, Approximate Measures, Units, and Weight (Weight of Edible Portion Only)		Water (Grams)	Water (Percent)	Food Energy (Calories)	Protein (Grams)	Fat (Grams)	Saturated (Grams)	Mono-unsaturated (Grams)	Poly-unsaturated (Grams)
	Grain Products *(continued)*									
	French or vienna bread, enriched:[25]									
	Slice:									
323	Vienna, 4-3/4 by 4 by 1/2 in - - - - - - - - - - -	1 slice - - - - -	25	34	70	2	1	0.2	0.3	0.3
	Italian bread, enriched:									
324	Loaf, 1 lb - - - - - - - - - - -	1 loaf - - - - - -	454	32	1,255	41	4	0.6	0.3	1.6
325	Slice, 4-1/2 by 3-1/4 by 3/4 in - - - - - - - - - - -	1 slice - - - - -	30	32	85	3	Tr	Tr	Tr	0.1
	Mixed grain bread, enriched:[25]									
326	Loaf, 1 lb - - - - - - - - - - -	1 loaf - - - - - -	454	37	1,165	45	17	3.2	4.1	6.5
327	Slice (18 per loaf) - - - - - - -	1 slice - - - - -	25	37	65	2	1	0.2	0.2	0.4
328	Toasted - - - - - - - - - - - -	1 slice - - - - -	23	27	65	2	1	0.2	0.2	0.4
	Oatmeal bread, enriched:[25]									
329	Loaf, 1 lb - - - - - - - - - - -	1 loaf - - - - - -	454	37	1,145	38	20	3.7	7.1	8.2
330	Slice (18 per loaf) - - - - - - -	1 slice - - - - -	25	37	65	2	1	0.2	0.4	0.5
331	Toasted - - - - - - - - - - - -	1 slice - - - - -	23	30	65	2	1	0.2	0.4	0.5
332	Pita bread, enriched, white, 6-1/2-in diam. - - - -	1 pita - - - - - -	60	31	165	6	1	0.1	0.1	0.4
	Pumpernickel (2/3 rye flour, 1/3 enriched wheat flour):[25]									
333	Loaf, 1 lb - - - - - - - - - - -	1 loaf - - - - - -	454	37	1,160	42	16	2.6	3.6	6.4
334	Slice, 5 by 4 by 3/8 in - - - -	1 slice - - - - -	32	37	80	3	1	0.2	0.3	0.5
335	Toasted - - - - - - - - - - - -	1 slice - - - - -	29	28	80	3	1	0.2	0.3	0.5
	Raisin bread, enriched:[25]									
336	Loaf, 1 lb - - - - - - - - - - -	1 loaf - - - - - -	454	33	1,260	37	18	4.1	6.5	6.7
337	Slice (18 per loaf) - - - - - - -	1 slice - - - - -	25	33	65	2	1	0.2	0.3	0.4
338	Toasted - - - - - - - - - - - -	1 slice - - - - -	21	24	65	2	1	0.2	0.3	0.4
	Rye bread, light (2/3 enriched wheat flour, 1/3 rye flour):[25]									
339	Loaf, 1 lb - - - - - - - - - - -	1 loaf - - - - - -	454	37	1,190	38	17	3.3	5.2	5.5
340	Slice, 4-3/4 by 3-3/4 by 7/16 in - - - - - - - - - - -	1 slice - - - - -	25	37	65	2	1	0.2	0.3	0.3
341	Toasted - - - - - - - - - - - -	1 slice - - - - -	22	28	65	2	1	0.2	0.3	0.3
	Wheat bread, enriched:[25]									
342	Loaf, 1 lb - - - - - - - - - - -	1 loaf - - - - - -	454	37	1,160	43	19	3.9	7.3	4.5
343	Slice (18 per loaf) - - - - - - -	1 slice - - - - -	25	37	65	2	1	0.2	0.4	0.3
344	Toasted - - - - - - - - - - - -	1 slice - - - - -	23	28	65	3	1	0.2	0.4	0.3
	White bread, enriched:[25]									
345	Loaf, 1 lb - - - - - - - - - - -	1 loaf - - - - - -	454	37	1,210	38	18	5.6	6.5	4.2
346	Slice (18 per loaf) - - - - - - -	1 slice - - - - -	25	37	65	2	1	0.3	0.4	0.2
347	Toasted - - - - - - - - - - -	1 slice - - - - -	22	28	65	2	1	0.3	0.4	0.2
348	Slice (22 per loaf) - - - - - -	1 slice - - - - -	20	37	55	2	1	0.2	0.3	0.2
349	Toasted - - - - - - - - - -	1 slice - - - - -	17	28	55	2	1	0.2	0.3	0.2
350	Cubes - - - - - - - - - - - -	1 cup - - - - - -	30	37	80	2	1	0.4	0.4	0.3
351	Crumbs, soft - - - - - - - - - -	1 cup - - - - - -	45	37	120	4	2	0.6	0.6	0.4
	Whole-wheat bread:[25]									
352	Loaf, 1 lb - - - - - - - - - - -	1 loaf - - - - - -	454	38	1,110	44	20	5.8	6.8	5.2
353	Slice (16 per loaf) - - - - - - -	1 slice - - - - -	28	38	70	3	1	0.4	0.4	0.3
354	Toasted - - - - - - - - - - - -	1 slice - - - - -	25	29	70	3	1	0.4	0.4	0.3

[25] Made with vegetable shortening.

Choles-terol	Carbo-hydrate	Calcium	Phos-phorus	Iron	Potas-sium	Sodium	Vitamin A Value (IU)	(RE)	Thiamin	Ribo-flavin	Niacin	Ascorbic Acid	Item No.
Milli-grams	Grams	Milli-grams	Milli-grams	Milli-grams	Milli-grams	Milli-grams	Inter-national units	Retinol equiva-lents	Milli-grams	Milli-grams	Milli-grams	Milli-grams	
0	13	28	21	0.8	23	145	Tr	Tr	0.12	0.09	1.0	Tr	323
0	256	77	350	12.7	336	2,656	0	0	1.80	1.10	15.0	0	324
0	17	5	23	0.8	22	176	0	0	0.12	0.07	1.0	0	325
0	212	472	962	14.8	990	1,870	Tr	Tr	1.77	1.73	18.9	Tr	326
0	12	27	55	0.8	56	106	Tr	Tr	0.10	0.10	1.1	Tr	327
0	12	27	55	0.8	56	106	Tr	Tr	0.08	0.10	1.1	Tr	328
0	212	267	563	12.0	707	2,231	0	0	2.09	1.20	15.4	0	329
0	12	15	31	0.7	39	124	0	0	0.12	0.07	0.9	0	330
0	12	15	31	0.7	39	124	0	0	0.09	0.07	0.9	0	331
0	33	49	60	1.4	71	339	0	0	0.27	0.12	2.2	0	332
0	218	322	990	12.4	1,966	2,461	0	0	1.54	2.36	15.0	0	333
0	16	23	71	0.9	141	177	0	0	0.11	0.17	1.1	0	334
0	16	23	71	0.9	141	177	0	0	0.09	0.17	1.1	0	335
0	239	463	395	14.1	1,058	1,657	Tr	Tr	1.50	2.81	18.6	Tr	336
0	13	25	22	0.8	59	92	Tr	Tr	0.08	0.15	1.0	Tr	337
0	13	25	22	0.8	59	92	Tr	Tr	0.06	0.15	1.0	Tr	338
0	218	363	658	12.3	926	3,164	0	0	1.86	1.45	15.0	0	339
0	12	20	36	0.7	51	175	0	0	0.10	0.08	0.8	0	340
0	12	20	36	0.7	51	175	0	0	0.08	0.08	0.8	0	341
0	213	572	835	15.8	627	2,447	Tr	Tr	2.09	1.45	20.5	Tr	342
0	12	32	47	0.9	35	138	Tr	Tr	0.12	0.08	1.2	Tr	343
0	12	32	47	0.9	35	138	Tr	Tr	0.10	0.08	1.2	Tr	344
0	222	572	490	12.9	508	2,334	Tr	Tr	2.13	1.41	17.0	Tr	345
0	12	32	27	0.7	28	129	Tr	Tr	0.12	0.08	0.9	Tr	346
0	12	32	27	0.7	28	129	Tr	Tr	0.09	0.08	0.9	Tr	347
0	10	25	21	0.6	22	101	Tr	Tr	0.09	0.06	0.7	Tr	348
0	10	25	21	0.6	22	101	Tr	Tr	0.07	0.06	0.7	Tr	349
0	15	38	32	0.9	34	154	Tr	Tr	0.14	0.09	1.1	Tr	350
0	22	57	49	1.3	50	231	Tr	Tr	0.21	0.14	1.7	Tr	351
0	206	327	1,180	15.5	799	2,887	Tr	Tr	1.59	0.95	17.4	Tr	352
0	13	20	74	1.0	50	180	Tr	Tr	0.10	0.06	1.1	Tr	353
0	13	20	74	1.0	50	180	Tr	Tr	0.08	0.06	1.1	Tr	354

(Tr indicates nutrient present in trace amount.)

Item No.	Foods, Approximate Measures, Units, and Weight (Weight of Edible Portion Only)		Water	Food Energy	Protein	Fat	Fatty Acids Saturated	Mono-unsaturated	Poly-unsaturated
			Grams	Percent	Calories	Grams	Grams	Grams	Grams	Grams
	Grain Products *(continued)*									
	Bread stuffing (from enriched bread), prepared from mix:									
355	Dry type	1 cup	140	33	500	9	31	6.1	13.3	9.6
356	Moist type	1 cup	203	61	420	9	26	5.3	11.3	8.0
	Breakfast cereals:									
	Hot type, cooked:									
	Corn (hominy) grits:									
357	Regular and quick, enriched	1 cup	242	85	145	3	Tr	Tr	0.1	0.2
358	Instant, plain	1 pkt	137	85	80	2	Tr	Tr	Tr	0.1
	Cream of Wheat ®:									
359	Regular, quick, instant	1 cup	244	86	140	4	Tr	0.1	Tr	0.2
360	Mix'n Eat, Plain	1 pkt	142	82	100	3	Tr	Tr	Tr	0.1
361	Malt-O-Meal ®	1 cup	240	88	120	4	Tr	Tr	Tr	0.1
	Oatmeal or rolled oats:									
362	Regular, quick, instant, nonfortified	1 cup	234	85	145	6	2	0.4	0.8	1.0
	Instant, fortified:									
363	Plain	1 pkt	177	86	105	4	2	0.3	0.6	0.7
364	Flavored	1 pkt	164	76	160	5	2	0.3	0.7	0.8
	Ready to eat:									
365	All-Bran® (about 1/3 cup)	1 oz	28	3	70	4	1	0.1	0.1	0.3
366	Cap'n Crunch® (about 3/4 cup)	1 oz	28	3	120	1	3	1.7	0.3	0.4
367	Cheerios® (about 1-1/4 cup)	1 oz	28	5	110	4	2	0.3	0.6	0.7
	Corn Flakes (about 1-1/4 cup):									
368	Kellogg's®	1 oz	28	3	110	2	Tr	Tr	Tr	Tr
369	Toasties®	1 oz	28	3	110	2	Tr	Tr	Tr	Tr
	40% Bran Flakes:									
370	Kellogg's® (about 3/4 cup)	1 oz	28	3	90	4	1	0.1	0.1	0.3
371	Post® (about 2/3 cup)	1 oz	28	3	90	3	Tr	0.1	0.1	0.2
372	Froot Loops® (about 1 cup)	1 oz	28	3	110	2	1	0.2	0.1	0.1
373	Golden Grahams® (about 3/4 cup)	1 oz	28	2	110	2	1	0.7	0.1	0.2
374	Grape-Nuts® (about 1/4 cup)	1 oz	28	3	100	3	Tr	Tr	Tr	0.1
375	Honey Nut Cheerios® (about 3/4 cup)	1 oz	28	3	105	3	1	0.1	0.3	0.3
376	Lucky Charms® (about 1 cup)	1 oz	28	3	110	3	1	0.2	0.4	0.4
377	Nature Valley® Granola (about 1/3 cup)	1 oz	28	4	125	3	5	3.3	0.7	0.7
378	100% Natural Cereal (about 1/4 cup)	1 oz	28	2	135	3	6	4.1	1.2	0.5
379	Product 19® (about 3/4 cup)	1 oz	28	3	110	3	Tr	Tr	Tr	0.1

Cholesterol	Carbohydrate	Calcium	Phosphorus	Iron	Potassium	Sodium	Vitamin A Value		Thiamin	Riboflavin	Niacin	Ascorbic Acid	Item No.
							(IU)	(RE)					
Milligrams	Grams	Milligrams	Milligrams	Milligrams	Milligrams	Milligrams	International units	Retinol equivalents	Milligrams	Milligrams	Milligrams	Milligrams	
0	50	92	136	2.2	126	1,254	910	273	0.17	0.20	2.5	0	355
67	40	81	134	2.0	118	1,023	850	256	0.10	0.18	1.6	0	356
0	31	0	29	[27]1.5	53	[28]0	[29]0	[29]0	[27]0.24	[27]0.15	[27]2.0	0	357
0	18	7	16	[27]1.0	29	343	0	0	[27]0.18	[27]0.08	[27]1.3	0	358
0	29	[30]54	[31]43	[30]10.9	46	[31,32]5	0	0	[30]0.24	[30]0.07	[30]1.5	0	359
0	21	[30]20	[34]20	[30]8.1	38	241	[30]1,250	[30]376	[30]0.43	[30]0.28	[30]5.0	0	360
0	26	5	[30]24	[30]9.6	31	[33]2	0	0	[30]0.48	[30]0.24	[30]5.8	0	361
0	25	19	178	1.6	131	[34]2	40	4	0.26	0.05	0.3	0	362
0	18	[27]163	133	[27]6.3	99	[27]285	[27]1,510	[27]453	[27]0.53	[27]0.28	[27]5.5	0	363
0	31	[27]168	148	[27]6.7	137	[27]254	[27]1,530	[27]460	[27]0.53	[27]0.38	[27]5.9	Tr	364
0	21	23	264	[30]4.5	350	320	[30]1,250	[30]375	[30]0.37	[30]0.43	[30]5.0	[30]15	365
0	23	5	36	[27]7.5	37	213	40	4	[27]0.50	[27]0.55	[27]6.6	0	366
0	20	48	134	[30]4.5	101	307	[30]1,250	[30]375	[30]0.37	[30]0.43	[30]5.0	[30]15	367
0	24	1	18	[30]1.8	26	351	[30]1,250	[30]375	[30]0.37	[30]0.43	[30]5.0	[30]15	368
0	24	1	12	[27]0.7	33	297	[30]1,250	[30]375	[30]0.37	[30]0.43	[30]5.0	0	369
0	22	14	139	[30]8.1	180	264	[30]1,250	[30]375	[30]0.37	[30]0.43	[30]5.0	0	370
0	22	12	179	[30]4.5	151	260	[30]1,250	[30]375	[30]0.37	[30]0.43	[30]5.0	0	371
0	25	3	24	[30]4.5	26	145	[30]1,250	[30]375	[30]0.37	[30]0.43	[30]5.0	[30]15	372
Tr	24	17	41	[30]4.5	63	346	[30]1,250	[30]375	[30]0.37	[30]0.43	[30]5.0	[30]15	373
0	23	11	71	1.2	95	197	[30]1,250	[30]375	[30]0.37	[30]0.43	[30]5.0	0	374
0	23	20	105	[30]4.5	99	257	[30]1,250	[30]375	[30]0.37	[30]0.43	[30]5.0	[30]15	375
0	23	32	79	[30]4.5	59	201	[30]1,250	[30]375	[30]0.37	[30]0.43	[30]5.0	[30]15	376
0	19	18	89	0.9	98	58	20	2	0.10	0.05	0.2	0	377
Tr	18	49	104	0.8	140	12	20	2	0.09	0.15	0.6	0	378
0	24	3	40	[30]18.0	44	325	[30]5,000	[30]1,501	[30]1.50	[30]1.70	[30]20.0	[30]60	379

[27] Nutrient added.

[28] Cooked without salt. If salt is added according to label recommendations, sodium content is 540 mg.

[29] For white corn grits. Cooked yellow grits contain 145 IU or 14 RE.

[30] Value based on label declaration for added nutrients.

(Tr indicates nutrient present in trace amount.)

Item No.	Foods, Approximate Measures, Units, and Weight (Weight of Edible Portion Only)		Water	Food Energy	Protein	Fat	Fatty Acids		
							Saturated	Mono-unsaturated	Poly-unsaturated
		Grams	Percent	Calories	Grams	Grams	Grams	Grams	Grams
	Grain Products *(continued)*								
	Breakfast cereals:								
	Ready to eat:								
	Raisin Bran:								
380	Kellogg's® (about 3/4 cup) - - - - - - - - - - - 1 oz - - - - - -	28	8	90	3	1	0.1	0.1	0.3
381	Post® (about 1/2 cup) - - - 1 oz - - - - - -	28	9	85	3	1	0.1	0.1	0.3
382	Rice Krispies® (about 1 cup) - - - - - - - - - - - - 1 oz - - - - - -	28	2	110	2	Tr	Tr	Tr	0.1
383	Shredded Wheat (about 2/3 cup) - - - - - - - - - - - 1 oz - - - - - -	28	5	100	3	1	0.1	0.1	0.3
384	Special K® (about 1-1/3 cup) 1 oz - - - - - -	28	2	110	6	Tr	Tr	Tr	Tr
385	Super Sugar Crisp® (about 7/8 cup) - - - - - - - - - - - - 1 oz - - - - - -	28	2	105	2	Tr	Tr	Tr	0.1
386	Sugar Frosted Flakes, Kellogg's® (about 3/4 cup) - - - - - - - - - - - - 1 oz - - - - - -	28	3	110	1	Tr	Tr	Tr	Tr
387	Sugar Smacks® (about 3/4 cup) - - - - - - - - - - - - 1 oz - - - - - -	28	3	105	2	1	0.1	0.1	0.2
388	Total® (about 1 cup) - - - - - - 1 oz - - - - - -	28	4	100	3	1	0.1	0.1	0.3
389	Trix® (about 1 cup) - - - - - - 1 oz - - - - - -	28	3	110	2	Tr	0.2	0.1	0.1
390	Wheaties® (about 1 cup) - - - 1 oz - - - - - -	28	5	100	3	Tr	0.1	Tr	0.2
391	Buckwheat flour, light, sifted - - - - 1 cup - - - - - -	98	12	340	6	1	0.2	0.4	0.4
392	Bulgur, uncooked - - - - - - - - - - 1 cup - - - - - -	170	10	600	19	3	1.2	0.3	1.2
	Cakes prepared from cake mixes with enriched flour:[35]								
	Angelfood:								
393	Whole cake, 9-3/4-in diam. tube cake - - - - - - - 1 cake - - - - -	635	38	1,510	38	2	0.4	0.2	1.0
394	Piece, 1/12 of cake - - - - - - 1 piece - - - - -	53	38	125	3	Tr	Tr	Tr	0.1
	Coffeecake, crumb:								
395	Whole cake, 7-3/4 by 5-5/8 by 1-1/4 in - - - - - - 1 cake - - - - -	430	30	1,385	27	41	11.8	16.7	9.6
396	Piece, 1/6 of cake - - - - - - - 1 piece - - - - -	72	30	230	5	7	2.0	2.8	1.6
	Devil's food with chocolate frosting:								
397	Whole, 2-layer cake, 8- or 9-in diam. - - - - - - - - 1 cake - - - - -	1,107	24	3,755	49	136	55.6	51.4	19.7
398	Piece, 1/16 of cake - - - - - - 1 piece - - - - -	69	24	235	3	8	3.5	3.2	1.2
399	Cupcake, 2-1/2-in diam. - - 1 cupcake - - -	35	24	120	2	4	1.8	1.6	0.6
	Gingerbread:								
400	Whole cake, 8 in square - - - 1 cake - - - - -	570	37	1,575	18	39	9.6	16.4	10.5
401	Piece, 1/9 of cake - - - - - - - 1 piece - - - - -	63	37	175	2	4	1.1	1.8	1.2
	Cakes prepared from cake mixes with enriched flour:[35]								
	Yellow with chocolate frosting:								
402	Whole, 2-layer cake, 8- or 9-in diam. - - - - - - - - - - 1 cake - - - - -	1,108	26	3,735	45	125	47.8	48.0	21.8
403	Piece, 1/16 of cake - - - - - - 1 piece - - - - -	69	26	235	3	8	3.0	3.0	1.4

[31] For regular and instant cereal. For quick cereal, phosphorus is 102 mg and sodium is 142 mg.

[32] Cooked without salt. If salt is added according to label recommendations, sodium content is 390 mg.

[33] Cooked without salt. If salt is added according to label recommendations, sodium content is 324 mg.

Choles-terol	Carbo-hydrate	Calcium	Phos-phorus	Iron	Potas-sium	Sodium	Vitamin A Value (IU)	Vitamin A Value (RE)	Thiamin	Ribo-flavin	Niacin	Ascorbic Acid	Item No.
Milli-grams	Grams	Milli-grams	Milli-grams	Milli-grams	Milli-grams	Milli-grams	Inter-national units	Retinol equiva-lents	Milli-grams	Milli-grams	Milli-grams	Milli-grams	
0	21	10	105	[30]3.5	147	207	[30]960	[30]288	[30]0.28	[30]0.34	[30]3.9	0	380
0	21	13	119	[30]4.5	175	185	[30]1,250	[30]375	[30]0.37	[30]0.43	[30]5.0	0	381
0	25	4	34	[30]1.8	29	340	[30]1,250	[30]375	[30]0.37	[30]0.43	[30]5.0	[30]15	382
0	23	11	100	1.2	102	3	0	0	0.07	0.08	1.5	0	383
Tr	21	8	55	[30]4.5	49	265	[30]1,250	[30]375	[30]0.37	[30]0.43	[30]5.0	[30]15	384
0	26	6	52	[30]1.8	105	25	[30]1,250	[30]375	[30]0.37	[30]0.43	[30]5.0	0	385
0	26	1	21	[30]1.8	18	230	[30]1,250	[30]375	[30]0.37	[30]0.43	[30]5.0	[30]15	386
0	25	3	31	[30]1.8	42	75	[30]1,250	[30]375	[30]0.37	[30]0.43	[30]5.0	[30]15	387
0	22	48	118	[30]18.0	106	352	[30]5,000	[30]1,501	[30]1.50	[30]1.70	[30]20.0	[30]60	388
0	25	6	19	[30]4.5	27	181	[30]1,250	[30]375	[30]0.37	[30]0.43	[30]5.0	[30]15	389
0	23	43	98	[30]4.5	106	354	[30]1,250	[30]375	[30]0.37	[30]0.43	[30]5.0	[30]15	390
0	78	11	86	1.0	314	2	0	0	0.08	0.04	0.4	0	391
0	129	49	575	9.5	389	7	0	0	0.48	0.24	7.7	0	392
0	342	527	1,086	2.7	845	3,226	0	0	0.32	1.27	1.6	0	393
0	29	44	91	0.2	71	269	0	0	0.03	0.11	0.1	0	394
279	225	262	748	7.3	469	1,853	690	194	0.82	0.90	7.7	1	395
47	38	44	125	1.2	78	310	120	32	0.14	0.15	1.3	Tr	396
598	645	653	1,162	22.1	1,439	2,900	1,660	498	1.11	1.66	10.0	1	397
37	40	41	72	1.4	90	181	100	31	0.07	0.10	0.6	Tr	398
19	20	21	37	0.7	46	92	50	16	0.04	0.05	0.3	Tr	399
6	291	513	570	10.8	1,562	1,733	0	0	0.86	1.03	7.4	1	400
1	32	57	63	1.2	173	192	0	0	0.09	0.11	0.8	Tr	401
576	638	1,008	2,017	15.5	1,208	2,515	1,550	465	1.22	1.66	11.1	1	402
36	40	63	126	1.0	75	157	100	29	0.08	0.10	0.7	Tr	403

[34] Cooked without salt. If salt is added according to label recommendations, sodium content is 374 mg.

[35] Excepting angel food cake, cakes were made from mixes containing vegetable shortening and frostings were made with margarine.

Item No.	Foods, Approximate Measures, Units, and Weight (Weight of Edible Portion Only)		Water	Food Energy	Protein	Fat	Fatty Acids Saturated	Mono-unsaturated	Poly-unsaturated
			Grams	Percent	Calories	Grams	Grams	Grams	Grams
	Grain Products *(continued)*								
	Cakes prepared from home recipes using enriched flour:								
	Carrot, with cream cheese frosting:[36]								
404	Whole cake, 10-in diam. tube cake	1 cake	1,536	23	6,175	63	328	66.0	135.2	107.5
405	Piece, 1/16 of cake	1 piece	96	23	385	4	21	4.1	8.4	6.7
	Fruitcake, dark:[36]								
406	Whole cake, 7-1/2-in diam., 2-1/4-in high tube cake	1 cake	1,361	18	5,185	74	228	47.6	113.0	51.7
407	Piece, 1/32 of cake, 2/3-in arc	1 piece	43	18	165	2	7	1.5	3.6	1.6
	Plain sheet cake:[37]								
	Without frosting:								
408	Whole cake, 9-in square	1 cake	777	25	2,830	35	108	29.5	45.1	25.6
409	Piece, 1/9 of cake	1 piece	86	25	315	4	12	3.3	5.0	2.8
	With uncooked white frosting:								
410	Whole cake, 9-in square	1 cake	1,096	21	4,020	37	129	41.6	50.4	26.3
411	Piece, 1/9 of cake	1 piece	121	21	445	4	14	4.6	5.6	2.9
	Pound:[38]								
412	Loaf, 8-1/2 by 3-1/2 by 3-1/4 in	1 loaf	514	22	2,025	33	94	21.1	40.9	26.7
413	Slice, 1/17 of loaf	1 slice	30	22	120	2	5	1.2	2.4	1.6
	Cakes, commercial, made with enriched flour:								
	Pound:								
414	Loaf, 8-1/2 by 3-1/2 by 3 in	1 loaf	500	24	1,935	26	94	52.0	30.0	4.0
415	Slice, 1/17 of loaf	1 slice	29	24	110	2	5	3.0	1.7	0.2
	Snack cakes:								
416	Devil's food with creme filling (2 small cakes per pkg)	1 small cake	28	20	105	1	4	1.7	1.5	0.6
417	Sponge with creme filling (2 small cakes per pkg)	1 small cake	42	19	155	1	5	2.3	2.1	0.5
	White with white frosting:								
418	Whole, 2-layer cake, 8- or 9-in diam.	1 cake	1,140	24	4,170	43	148	33.1	61.6	42.2
419	Piece, 1/16 of cake	1 piece	71	24	260	3	9	2.1	3.8	2.6
	Yellow with chocolate frosting:								
420	Whole, 2-layer cake, 8- or 9-in diam.	1 cake	1,108	23	3,895	40	175	92.0	58.7	10.0
421	Piece, 1/16 of cake	1 piece	69	23	245	2	11	5.7	3.7	0.6
	Cheesecake:								
422	Whole cake, 9-in diam.	1 cake	1,110	46	3,350	60	213	119.9	65.5	14.4
423	Piece, 1/12 of cake	1 piece	92	46	280	5	18	9.9	5.4	1.2
	Cookies made with enriched flour:								
	Brownies with nuts:								
424	Commercial, with frosting, 1-1/2 by 1-3/4 by 7/8 in	1 brownie	25	13	100	1	4	1.6	2.0	0.6

[36] Made with vegetable oil.

[37] Cake made with vegetable shortening; frosting with margarine.

[38] Made with margarine.

Choles-terol	Carbo-hydrate	Calcium	Phos-phorus	Iron	Potas-sium	Sodium	Vitamin A Value (IU)	(RE)	Thiamin	Ribo-flavin	Niacin	Ascorbic Acid	Item No.
Milli-grams	Grams	Milli-grams	Milli-grams	Milli-grams	Milli-grams	Milli-grams	Inter-national units	Retinol equiva-lents	Milli-grams	Milli-grams	Milli-grams	Milli-grams	
1183	775	707	998	21.0	1,720	4,470	2,240	246	1.83	1.97	14.7	23	404
74	48	44	62	1.3	108	279	140	15	0.11	0.12	0.9	1	405
640	783	1,293	1,592	37.6	6,138	2,123	1,720	422	2.41	2.55	17.0	504	406
20	25	41	50	1.2	194	67	50	13	0.08	0.08	0.5	16	407
552	434	497	793	11.7	614	2,331	1,320	373	1.24	1.40	10.1	2	408
61	48	55	88	1.3	68	258	150	41	0.14	0.15	1.1	Tr	409
636	694	548	822	11.0	669	2,488	2,190	647	1.21	1.42	9.9	2	410
70	77	61	91	1.2	74	275	240	71	0.13	0.16	1.1	Tr	411
555	265	339	473	9.3	483	1,645	3,470	1,033	0.93	1.08	7.8	1	412
32	15	20	28	0.5	28	96	200	60	0.05	0.06	0.5	Tr	413
1100	257	146	517	8.0	443	1,857	2,820	715	0.96	1.12	8.1	0	414
64	15	8	30	0.5	26	108	160	41	0.06	0.06	0.5	0	415
15	17	21	26	1.0	34	105	20	4	0.06	0.09	0.7	0	416
7	27	14	44	0.6	37	155	30	9	0.07	0.06	0.6	0	417
46	670	536	1,585	15.5	832	2,827	640	194	3.19	2.05	27.6	0	418
3	42	33	99	1.0	52	176	40	12	0.20	0.13	1.7	0	419
609	620	366	1,884	19.9	1,972	3,080	1,850	488	0.78	2.22	10.0	0	420
38	39	23	117	1.2	123	192	120	30	0.05	0.14	0.6	0	421
2053	317	622	977	5.3	1,088	2,464	2,820	833	0.33	1.44	5.1	56	422
170	26	52	81	0.4	90	204	230	69	0.03	0.12	0.4	5	423
14	16	13	26	0.6	50	59	70	18	0.08	0.07	0.3	Tr	424

493

(Tr indicates nutrient present in trace amount.)

Item No.	Foods, Approximate Measures, Units, and Weight (Weight of Edible Portion Only)		Water	Food Energy	Protein	Fat	Fatty Acids		
							Saturated	Mono-unsaturated	Poly-unsaturated
			Grams	Percent	Calories	Grams	Grams	Grams	Grams	Grams

Item No.	Foods, Approximate Measures, Units, and Weight (Weight of Edible Portion Only)		Water (Percent)	Food Energy (Calories)	Protein (Grams)	Fat (Grams)	Saturated (Grams)	Mono-unsaturated (Grams)	Poly-unsaturated (Grams)
	Grain Products *(continued)*								
	Cookies made with enriched flour:								
	Brownies with nuts:								
425	From home recipe, 1-3/4 by 1-3/4 by 7/8 in[36]	1 brownie - - -	20	10	95	1	6	1.4	2.8	1.2
	Chocolate chip:								
426	Commercial, 2-1/4-in diam., 3/8 in thick - - - - - -	4 cookies - - -	42	4	180	2	9	2.9	3.1	2.6
	Cookies made with enriched flour:								
	Chocolate chip:								
427	From home recipe, 2-1/3-in diam.[25] - - - - - - - - - - - -	4 cookies - - -	40	3	185	2	11	3.9	4.3	2.0
428	From refrigerated dough, 2-1/4-in diam., 3/8 in thick	4 cookies - - -	48	5	225	2	11	4.0	4.4	2.0
429	Fig bars, square, 1-5/8 by 1-5/8 by 3/8 in or rectangular, 1-1/2 by 1-3/4 by 1/2 in	4 cookies - - -	56	12	210	2	4	1.0	1.5	1.0
430	Oatmeal with raisins, 2-5/8-in diam., 1/4 in thick - - - - - - -	4 cookies - - -	52	4	245	3	10	2.5	4.5	2.8
431	Peanut butter cookie, from home recipe, 2-5/8-in diam.[25] - - - - - - - - - - - - -	4 cookies - - -	48	3	245	4	14	4.0	5.8	2.8
432	Sandwich type (chocolate or vanilla), 1-3/4-in diam., 3/8 in thick - - - - - - - - - - - - -	4 cookies - - -	40	2	195	2	8	2.0	3.6	2.2
	Shortbread:								
433	Commercial - - - - - - - - - - -	4 small cookies - - - - -	32	6	155	2	8	2.9	3.0	1.1
434	From home recipe[38] - - - - - -	2 large cookies - - - - -	28	3	145	2	8	1.3	2.7	3.4
435	Sugar cookie, from refrigerated dough, 2-1/2-in diam., 1/4 in thick - - - - - - - - - - - - -	4 cookies - - -	48	4	235	2	12	2.3	5.0	3.6
436	Vanilla wafers, 1-3/4-in diam., 1/4 in thick - - - - - - - - - - -	10 cookies - -	40	4	185	2	7	1.8	3.0	1.8
437	Corn chips - - - - - - - - - - - - -	1-oz package - - -	28	1	155	2	9	1.4	2.4	3.7
	Cornmeal:								
438	Whole-ground, unbolted, dry form - - - - - - - - - - - - - -	1 cup - - - - - -	122	12	435	11	5	0.5	1.1	2.5
439	Bolted (nearly whole-grain), dry form - - - - - - - - - - - - -	1 cup - - - - - -	122	12	440	11	4	0.5	0.9	2.2
	Degermed, enriched:								
440	Dry form - - - - - - - - - - - - -	1 cup - - - - - -	138	12	500	11	2	0.2	0.4	0.9
441	Cooked - - - - - - - - - - - - -	1 cup - - - - - -	240	88	120	3	Tr	Tr	0.1	0.2
	Crackers:[39]								
	Cheese:								
442	Plain, 1 in square - - - - - - -	10 crackers - -	10	4	50	1	3	0.9	1.2	0.3
443	Sandwich type (peanut butter) - - - - - - - - - - - - -	1 sandwich - -	8	3	40	1	2	0.4	0.8	0.3
444	Graham, plain, 2-3/2 in square - - - - - - - - - - - - -	2 crackers - - -	14	5	60	1	1	0.4	0.6	0.4
445	Melba toast, plain - - - - - - - - -	1 piece - - - - -	5	4	20	1	Tr	0.1	0.1	0.1

[25] Made with vegetable shortening.

[38] Made with margarine.

Choles-terol	Carbo-hydrate	Calcium	Phos-phorus	Iron	Potas-sium	Sodium	Vitamin A Value		Thiamin	Ribo-flavin	Niacin	Ascorbic Acid	Item No.
							(IU)	(RE)					
Milli-grams	Grams	Milli-grams	Milli-grams	Milli-grams	Milli-grams	Milli-grams	Inter-national units	Retinol equiva-lents	Milli-grams	Milli-grams	Milli-grams	Milli-grams	
18	11	9	26	0.4	35	51	20	6	0.05	0.05	0.3	Tr	425
5	28	13	41	0.8	68	140	50	15	0.10	0.23	1.0	Tr	426
18	26	13	34	1.0	82	82	20	5	0.06	0.06	0.6	0	427
22	32	13	34	1.0	62	173	30	8	0.06	0.10	0.9	0	428
27	42	40	34	1.4	162	180	60	6	0.08	0.07	0.7	Tr	429
2	36	18	58	1.1	90	148	40	12	0.09	0.08	1.0	0	430
22	28	21	60	1.1	110	142	20	5	0.07	0.07	1.9	0	431
0	29	12	40	1.4	66	189	0	0	0.09	0.07	0.8	0	432
27	20	13	39	0.8	38	123	30	8	0.10	0.09	0.9	0	433
0	17	6	31	0.6	18	125	300	89	0.08	0.06	0.7	Tr	434
29	31	50	91	0.9	33	261	40	11	0.09	0.06	1.1	0	435
25	29	16	36	0.8	50	150	50	14	0.07	0.10	1.0	0	436
0	16	35	52	0.5	52	233	110	11	0.04	0.05	0.4	1	437
0	90	24	312	2.2	346	1	620	62	0.46	0.13	2.4	0	438
0	91	21	272	2.2	303	1	590	59	0.37	0.10	2.3	0	439
0	108	8	137	5.9	166	1	610	61	0.61	0.36	4.8	0	440
0	26	2	34	1.4	38	0	140	14	0.14	0.10	1.2	0	441
6	6	11	17	0.3	17	112	20	5	0.05	0.04	0.4	0	442
1	5	7	25	0.3	17	90	Tr	Tr	0.04	0.03	0.6	0	443
0	11	6	20	0.4	36	86	0	0	0.02	0.03	0.6	0	444
0	4	6	10	0.1	11	44	0	0	0.01	0.01	0.1	0	445

[39] Crackers made with enriched flour except for rye wafers and whole-wheat wafers.

Item No.	Foods, Approximate Measures, Units, and Weight (Weight of Edible Portion Only)		Water	Food Energy	Protein	Fat	Fatty Acids Saturated	Mono-unsaturated	Poly-unsaturated
			Grams	Percent	Calories	Grams	Grams	Grams	Grams	Grams

Item No.	Foods, Approximate Measures, Units, and Weight (Weight of Edible Portion Only)	Measure	Water (Grams)	Water (Percent)	Food Energy (Calories)	Protein (Grams)	Fat (Grams)	Saturated (Grams)	Mono-unsaturated (Grams)	Poly-unsaturated (Grams)
	Grain Products *(continued)*									
446	Rye wafers, whole-grain, 1-7/8 by 3-1/2 in	2 wafers	14	5	55	1	1	0.3	0.4	0.3
447	Saltines[40]	4 crackers	12	4	50	1	1	0.5	0.4	0.2
448	Snack-type, standard	1 round cracker	3	3	15	Tr	1	0.2	0.4	0.1
449	Wheat, thin	4 crackers	8	3	35	1	1	0.5	0.5	0.4
450	Whole-wheat wafers	2 crackers	8	4	35	1	2	0.5	0.6	0.4
451	Croissants, made with enriched flour, 4-1/2 by 4 by 1-3/4 in	1 croissant	57	22	235	5	12	3.5	6.7	1.4
	Danish pastry, made with enriched flour:									
	Plain without fruit or nuts:									
452	Packaged ring, 12 oz	1 ring	340	27	1,305	21	71	21.8	28.6	15.6
453	Round piece, about 4-1/4-in diam., 1 in high	1 pastry	57	27	220	4	12	3.6	4.8	2.6
454	Ounce	1 oz	28	27	110	2	6	1.8	2.4	1.3
455	Fruit, round piece	1 pastry	65	30	235	4	13	3.9	5.2	2.9
	Doughnuts, made with enriched flour:									
456	Cake type, plain, 3-1/4-in diam., 1 in high	1 doughnut	50	21	210	3	12	2.8	5.0	3.0
457	Yeast-leavened, glazed, 3-3/4-in diam., 1-1/4 in high	1 doughnut	60	27	235	4	13	5.2	5.5	0.9
458	English muffins, plain, enriched	1 muffin	57	42	140	5	1	0.3	0.2	0.3
459	Toasted	1 muffin	50	29	140	5	1	0.3	0.2	0.3
460	French toast, from home recipe	1 slice	65	53	155	6	7	1.6	2.0	1.6
	Macaroni, enriched, cooked (cut lengths, elbows, shells):									
461	Firm stage (hot)	1 cup	130	64	190	7	1	0.1	0.1	0.3
	Tender stage:									
462	Cold	1 cup	105	72	115	4	Tr	0.1	0.1	0.2
463	Hot	1 cup	140	72	155	5	1	0.1	0.1	0.2
	Muffins made with enriched flour, 2-1/2-in diam., 1-1/2 in high:									
	From home recipe:									
464	Blueberry[25]	1 muffin	45	37	135	3	5	1.5	2.1	1.2
465	Bran[36]	1 muffin	45	35	125	3	6	1.4	1.6	2.3
466	Corn (enriched, degermed cornmeal and flour)[25]	1 muffin	45	33	145	3	5	1.5	2.2	1.4
	From commercial mix (egg and water added):									
467	Blueberry	1 muffin	45	33	140	3	5	1.4	2.0	1.2
468	Bran	1 muffin	45	28	140	3	4	1.3	1.6	1.0
469	Corn	1 muffin	45	30	145	3	6	1.7	2.3	1.4
470	Noodles (egg noodles), enriched, cooked	1 cup	160	70	200	7	2	0.5	0.6	0.6
471	Noodles, chow mein, canned	1 cup	45	11	220	6	11	2.1	7.3	0.4
	Pancakes, 4-in diam.:									
472	Buckwheat, from mix (with buck-wheat and enriched flours), egg and milk added	1 pancake	27	58	55	2	2	0.9	0.9	0.5

[25] Made with vegetable shortening.

[36] Made with vegetable oil.

[40] Made with lard.

Choles-terol	Carbo-hydrate	Calcium	Phos-phorus	Iron	Potas-sium	Sodium	Vitamin A Value		Thiamin	Ribo-flavin	Niacin	Ascorbic Acid	Item No.
							(IU)	(RE)					
Milli-grams	Grams	Milli-grams	Milli-grams	Milli-grams	Milli-grams	Milli-grams	Inter-national units	Retinol equiva-lents	Milli-grams	Milli-grams	Milli-grams	Milli-grams	
0	10	7	44	0.5	65	115	0	0	0.06	0.03	0.5	0	446
4	9	3	12	0.5	17	165	0	0	0.06	0.05	0.6	0	447
0	2	3	6	0.1	4	30	Tr	Tr	0.01	0.01	0.1	0	448
0	5	3	15	0.3	17	69	Tr	Tr	0.04	0.03	0.4	0	449
0	5	3	22	0.2	31	59	0	0	0.02	0.03	0.4	0	450
13	27	20	64	2.1	68	452	50	13	0.17	0.13	1.3	0	451
292	152	360	347	6.5	316	1,302	360	99	0.95	1.02	8.5	Tr	452
49	26	60	58	1.1	53	218	60	17	0.16	0.17	1.4	Tr	453
24	13	30	29	0.5	26	109	30	8	0.08	0.09	0.7	Tr	454
56	28	17	80	1.3	57	233	40	11	0.16	0.14	1.4	Tr	455
20	24	22	111	1.0	58	192	20	5	0.12	0.12	1.1	Tr	456
21	26	17	55	1.4	64	222	Tr	Tr	0.28	0.12	1.8	0	457
0	27	96	67	1.7	331	378	0	0	0.26	0.19	2.2	0	458
0	27	96	67	1.7	331	378	0	0	0.23	0.19	2.2	0	459
112	17	72	85	1.3	86	257	110	32	0.12	0.16	1.0	Tr	460
0	39	14	85	2.1	103	1	0	0	0.23	0.13	1.8	0	461
0	24	8	53	1.3	64	1	0	0	0.15	0.08	1.2	0	462
0	32	11	70	1.7	85	1	0	0	0.20	0.11	1.5	0	463
19	20	54	46	0.9	47	198	40	9	0.10	0.11	0.9	1	464
24	19	60	125	1.4	99	189	230	30	0.11	0.13	1.3	3	465
23	21	66	59	0.9	57	169	80	15	0.11	0.11	0.9	Tr	466
45	22	15	90	0.9	54	225	50	11	0.10	0.17	1.1	Tr	467
28	24	27	182	1.7	50	385	100	14	0.08	0.12	1.9	0	468
42	22	30	128	1.3	31	291	90	16	0.09	0.09	0.8	Tr	469
50	37	16	94	2.6	70	3	110	34	0.22	0.13	1.9	0	470
5	26	14	41	0.4	33	450	0	0	0.05	0.03	0.6	0	471
20	6	59	91	0.4	66	125	60	17	0.04	0.05	0.2	Tr	472

(Tr indicates nutrient present in trace amount.)

Item No.	Foods, Approximate Measures, Units, and Weight (Weight of Edible Portion Only)		Water	Food Energy	Protein	Fat	Fatty Acids		
							Saturated	Mono-unsaturated	Poly-unsaturated
			Grams	Percent	Calories	Grams	Grams	Grams	Grams	Grams

			Grams	Percent	Calories	Grams	Grams	Grams	Grams
	Grain Products *(continued)*								
	Pancakes, 4-in diam.:								
	Plain:								
473	From home recipe using enriched flour - - - - - - - -	1 pancake - - -	27	50	60	2	2	0.5	0.8	0.5
474	From mix (with enriched flour), egg, milk, and oil added - - - - - - - - - - - - -	1 pancake - - -	27	54	60	2	2	0.5	0.9	0.5
	Piecrust, made with enriched flour and vegetable shortening, baked:								
475	From home recipe, 9-in diam. - - - - - - - - - - - - -	1 pie shell - - -	180	15	900	11	60	14.8	25.9	15.7
476	From mix, 9-in diam. - - - - -	Piecrust for 2-crust pie	320	19	1,485	20	93	22.7	41.0	25.0
	Pies, piecrust made with enriched flour, vegetable shortening, 9-in diam.:								
	Apple:								
477	Whole - - - - - - - - - - - - -	1 pie - - - - - -	945	48	2,420	21	105	27.4	44.4	26.5
478	Piece, 1/6 of pie - - - - - - - -	1 piece - - - - -	158	48	405	3	18	4.6	7.4	4.4
	Blueberry:								
479	Whole - - - - - - - - - - - - -	1 pie - - - - - -	945	51	2,285	23	102	25.5	44.4	27.4
480	Piece, 1/6 of pie - - - - - - - -	1 piece - - - - -	158	51	380	4	17	4.3	7.4	4.6
	Cherry:								
481	Whole - - - - - - - - - - - - -	1 pie - - - - - -	945	47	2,465	25	107	28.4	46.3	27.4
482	Piece, 1/6 of pie - - - - - - - -	1 piece - - - - -	158	47	410	4	18	4.7	7.7	4.6
	Creme:								
483	Whole - - - - - - - - - - - - -	1 pie - - - - - -	910	43	2,710	20	139	90.1	23.7	6.4
484	Piece, 1/6 of pie - - - - - - - -	1 piece - - - - -	152	43	455	3	23	15.0	4.0	1.1
	Custard:								
485	Whole - - - - - - - - - - - - -	1 pie - - - - - -	910	58	1,985	56	101	33.7	40.0	19.1
486	Piece, 1/6 of pie - - - - - - - -	1 piece - - - - -	152	58	330	9	17	5.6	6.7	3.2
	Lemon meringue:								
487	Whole - - - - - - - - - - - - -	1 pie - - - - - -	840	47	2,140	31	86	26.0	34.4	17.6
488	Piece, 1/6 of pie - - - - - - - -	1 piece - - - - -	140	47	355	5	14	4.3	5.7	2.9
	Peach:								
489	Whole - - - - - - - - - - - - -	1 pie - - - - - -	945	48	2,410	24	101	24.6	43.5	26.5
490	Piece, 1/6 of pie - - - - - - - -	1 piece - - - - -	158	48	405	4	17	4.1	7.3	4.4
	Pecan:								
491	Whole - - - - - - - - - - - - -	1 pie - - - - - -	825	20	3,450	42	189	28.1	101.5	47.0
492	Piece, 1/6 of piece- - - - - - -	1 piece - - - - -	138	20	575	7	32	4.7	17.0	7.9
	Pumpkin:								
493	Whole - - - - - - - - - - - - -	1 pie - - - - - -	910	59	1,920	36	102	38.2	40.0	18.2
494	Piece, 1/6 of pie - - - - - - - -	1 piece - - - - -	152	59	320	6	17	6.4	6.7	3.0
	Pies, fried:								
495	Apple - - - - - - - - - - - - - -	1 pie - - - - - -	85	43	255	2	14	5.8	6.6	0.6
496	Cherry- - - - - - - - - - - - - -	1 pie - - - - - -	85	42	250	2	14	5.8	6.7	0.6
	Popcorn, popped:								
497	Air-popped, unsalted- - - - - - -	1 cup - - - - - -	8	4	30	1	Tr	Tr	0.1	0.2
498	Popped in vegetable oil, salted	1 cup - - - - - -	11	3	55	1	3	0.5	1.4	1.2
499	Sugar syrup coated - - - - - - -	1 cup - - - - - -	35	4	135	2	1	0.1	0.3	0.6
	Pretzels, made with enriched flour:								
500	Stick, 2-1/4 in long - - - - - - -	10 pretzels - -	3	3	10	Tr	Tr	Tr	Tr	Tr
501	Twisted, dutch, 2-3/4 by 2-5/8 in - - - - - - - - - - -	1 pretzel - - - -	16	3	65	2	1	0.1	0.2	0.2

Choles-terol	Carbo-hydrate	Calcium	Phos-phorus	Iron	Potas-sium	Sodium	Vitamin A Value		Thiamin	Ribo-flavin	Niacin	Ascorbic Acid	Item No.
							(IU)	(RE)					
Milli-grams	Grams	Milli-grams	Milli-grams	Milli-grams	Milli-grams	Milli-grams	Inter-national units	Retinol equiva-lents	Milli-grams	Milli-grams	Milli-grams	Milli-grams	
16	9	27	38	0.5	33	115	30	10	0.06	0.07	0.5	Tr	473
16	8	36	71	0.7	43	160	30	7	0.09	0.12	0.8	Tr	474
0	79	25	90	4.5	90	1,000	0	0	0.54	0.40	5.0	0	475
0	141	131	272	9.3	179	2,602	0	0	1.06	0.80	9.9	0	476
0	360	76	208	9.5	756	2,844	280	28	1.04	0.76	9.5	9	477
0	60	13	35	1.6	126	476	50	5	0.17	0.13	1.6	2	478
0	330	104	217	12.3	945	2,533	850	85	1.04	0.85	10.4	38	479
0	55	17	36	2.1	158	423	140	14	0.17	0.14	1.7	6	480
0	363	132	236	9.5	992	2,873	4,160	416	1.13	0.85	9.5	0	481
0	61	22	40	1.6	166	480	700	70	0.19	0.14	1.6	0	482
46	351	273	919	6.8	796	2,207	1,250	391	0.36	0.89	6.4	0	483
8	59	46	154	1.1	133	369	210	65	0.06	0.15	1.1	0	484
1010	213	874	1,028	9.1	1,247	2,612	2,090	573	0.82	1.91	5.5	0	485
169	36	146	172	1.5	208	436	350	96	0.14	0.32	0.9	0	486
857	317	118	412	8.4	420	2,369	1,430	395	0.59	0.84	5.0	25	487
143	53	20	69	1.4	70	395	240	66	0.10	0.14	0.8	4	488
0	361	95	274	11.3	1,408	2,533	6,900	690	1.04	0.95	14.2	28	489
0	60	16	46	1.9	235	423	1,150	115	0.17	0.16	2.4	5	490
569	423	388	850	27.2	1,015	1,823	1,320	322	1.82	0.99	6.6	0	491
95	71	65	142	4.6	170	305	220	54	0.30	0.17	1.1	0	492
655	223	464	628	8.2	1,456	1,947	22,480	2,493	0.82	1.27	7.3	0	493
109	37	78	105	1.4	243	325	3,750	416	0.14	0.21	1.2	0	494
14	31	12	34	0.9	42	326	30	3	0.09	0.06	1.0	1	495
13	32	11	41	0.7	61	371	190	19	0.06	0.06	0.6	1	496
0	6	1	22	0.2	20	Tr	10	1	0.03	0.01	0.2	0	497
0	6	3	31	0.3	19	86	20	2	0.01	0.02	0.1	0	498
0	30	2	47	0.5	90	Tr	30	3	0.13	0.02	0.4	0	499
0	2	1	3	0.1	3	48	0	0	0.01	0.01	0.1	0	500
0	13	4	15	0.3	16	258	0	0	0.05	0.04	0.7	0	501

(Tr indicates nutrient present in trace amount.)

Item No.	Foods, Approximate Measures, Units, and Weight (Weight of Edible Portion Only)			Water	Food Energy	Protein	Fat	Fatty Acids		
								Saturated	Mono-unsaturated	Poly-unsaturated
				Grams	Percent	Calories	Grams	Grams	Grams	Grams	Grams

Grain Products *(continued)*

Pretzels, made with enriched flour:

Item No.	Foods	Measure	Grams	Percent	Calories	Protein (g)	Fat (g)	Saturated	Mono	Poly
502	Twisted, thin, 3-1/4 by 2-1/4 by 1/4 in	10 pretzels	60	3	240	6	2	0.4	0.8	0.6
	Rice:									
503	Brown, cooked, served hot	1 cup	195	70	230	5	1	0.3	0.3	0.4
	White, enriched: Commercial varieties, all types:									
504	Raw	1 cup	185	12	670	12	1	0.2	0.2	0.3
505	Cooked, served hot	1 cup	205	73	225	4	Tr	0.1	0.1	0.1
506	Instant, ready-to-serve, hot	1 cup	165	73	180	4	0	0.1	0.1	0.1
	Parboiled:									
507	Raw	1 cup	185	10	685	14	1	0.1	0.1	0.2
508	Cooked, served hot	1 cup	175	73	185	4	Tr	Tr	Tr	0.1
	Rolls, enriched: Commercial:									
509	Dinner, 2-1/2-in diam., 2 in high	1 roll	28	32	85	2	2	0.5	0.8	0.6
510	Frankfurter and hamburger (8 per 11-1/2-oz pkg.)	1 roll	40	34	115	3	2	0.5	0.8	0.6
511	Hard, 3-3/4-in diam., 2 in high	1 roll	50	25	155	5	2	0.4	0.5	0.6
512	Hoagie or submarine, 11-1/2 by 3 by 2-1/2 in	1 roll	135	31	400	11	8	1.8	3.0	2.2
	From home recipe:									
513	Dinner, 2-1/2-in diam., 2 in high	1 roll	35	26	120	3	3	0.8	1.2	0.9
	Spaghetti, enriched, cooked:									
514	Firm stage, "al dente," served hot	1 cup	130	64	190	7	1	0.1	0.1	0.3
515	Tender stage, served hot	1 cup	140	73	155	5	1	0.1	0.1	0.2
516	Toaster pastries	1 pastry	54	13	210	2	6	1.7	3.6	0.4
517	Tortillas, corn	1 tortilla	30	45	65	2	1	0.1	0.3	0.6
	Waffles, made with enriched flour, 7-in diam.:									
518	From home recipe	1 waffle	75	37	245	7	13	4.0	4.9	2.6
519	From mix, egg and milk added	1 waffle	75	42	205	7	8	2.7	2.9	1.5
	Wheat flours: All-purpose or family flour, enriched:									
520	Sifted, spooned	1 cup	115	12	420	12	1	0.2	0.1	0.5
521	Unsifted, spooned	1 cup	125	12	455	13	1	0.2	0.1	0.5
522	Cake or pastry flour, enriched, sifted, spooned	1 cup	96	12	350	7	1	0.1	0.1	0.3
523	Self-rising, enriched, unsifted, spooned	1 cup	125	12	440	12	1	0.2	0.1	0.5
524	Whole-wheat, from hard wheats, stirred	1 cup	120	12	400	16	2	0.3	0.3	1.1
	Legumes, Nuts, and Seeds									
	Almonds, shelled:									
525	Slivered, packed	1 cup	135	4	795	27	70	6.7	45.8	14.8
526	Whole	1 oz	28	4	165	6	15	1.4	9.6	3.1

Choles-terol	Carbo-hydrate	Calcium	Phos-phorus	Iron	Potas-sium	Sodium	Vitamin A Value		Thiamin	Ribo-flavin	Niacin	Ascorbic Acid	Item No.
							(IU)	(RE)					
Milli-grams	Grams	Milli-grams	Milli-grams	Milli-grams	Milli-grams	Milli-grams	Inter-national units	Retinol equiva-lents	Milli-grams	Milli-grams	Milli-grams	Milli-grams	
0	48	16	55	1.2	61	966	0	0	0.19	0.15	2.6	0	502
0	50	23	142	1.0	137	0	0	0	0.18	0.04	2.7	0	503
0	149	44	174	5.4	170	9	0	0	0.81	0.06	6.5	0	504
0	50	21	57	1.8	57	0	0	0	0.23	0.02	2.1	0	505
0	40	5	31	1.3	0	0	0	0	0.21	0.02	1.7	0	506
0	150	111	370	5.4	278	17	0	0	0.81	0.07	6.5	0	507
0	41	33	100	1.4	75	0	0	0	0.19	0.02	2.1	0	508
Tr	14	33	44	0.8	36	155	Tr	Tr	0.14	0.09	1.1	Tr	509
Tr	20	54	44	1.2	56	241	Tr	Tr	0.20	0.13	1.6	Tr	510
Tr	30	24	46	1.4	49	313	0	0	0.20	0.12	1.7	0	511
Tr	72	100	115	3.8	128	683	0	0	0.54	0.33	4.5	0	512
12	20	16	36	1.1	41	98	30	8	0.12	0.12	1.2	0	513
0	39	14	85	2.0	103	1	0	0	0.23	0.13	1.8	0	514
0	32	11	70	1.7	85	1	0	0	0.20	0.11	1.5	0	515
0	38	104	104	2.2	91	248	520	52	0.17	0.18	2.3	4	516
0	13	42	55	0.6	43	1	80	8	0.05	0.03	0.4	0	517
102	26	154	135	1.5	129	445	140	39	0.18	0.24	1.5	Tr	518
59	27	179	257	1.2	146	515	170	49	0.14	0.23	0.9	Tr	519
0	88	18	100	5.1	109	2	0	0	0.73	0.46	6.1	0	520
0	95	20	109	5.5	119	3	0	0	0.80	0.50	6.6	0	521
0	76	16	70	4.2	91	2	0	0	0.58	0.38	5.1	0	522
0	93	331	583	5.5	113	1,349	0	0	0.80	0.50	6.6	0	523
0	85	49	446	5.2	444	4	0	0	0.66	0.14	5.2	0	524
0	28	359	702	4.9	988	15	0	0	0.28	1.05	4.5	1	525
0	6	75	147	1.0	208	3	0	0	0.06	0.22	1.0	Tr	526

(Tr indicates nutrient present in trace amount.)

Item No.	Foods, Approximate Measures, Units, and Weight (Weight of Edible Portion Only)		Water	Food Energy	Protein	Fat	Fatty Acids		
							Saturated	Mono-unsaturated	Poly-unsaturated
		Grams	Percent	Calories	Grams	Grams	Grams	Grams	Grams
	Legumes, Nuts, and Seeds *(continued)*								
	Beans, dry:								
	Cooked, drained:								
527	Black - - - - - - - - - - - - - - - 1 cup - - - - - -	171	66	225	15	1	0.1	0.1	0.5
528	Great Northern - - - - - - - - - 1 cup - - - - - -	180	69	210	14	1	0.1	0.1	0.6
529	Lima - - - - - - - - - - - - - - 1 cup - - - - - -	190	64	260	16	1	0.2	0.1	0.5
530	Pea (navy) - - - - - - - - - - - - 1 cup - - - - - -	190	69	225	15	1	0.1	0.1	0.7
531	Pinto - - - - - - - - - - - - - - - 1 cup - - - - - -	180	65	265	15	1	0.1	0.1	0.5
	Canned, solids and liquid:								
	White with:								
532	Frankfurters (sliced) - - - - - 1 cup - - - - - -	255	71	365	19	18	7.4	8.8	0.7
533	Pork and tomato sauce - - 1 cup - - - - - -	255	71	310	16	7	2.4	2.7	0.7
534	Pork and sweet sauce - - - 1 cup - - - - - -	255	66	385	16	12	4.3	4.9	1.2
535	Red kidney - - - - - - - - - - - 1 cup - - - - - -	255	76	230	15	1	0.1	0.1	0.6
536	Black-eyed peas, dry, cooked (with residual cooking liquid) 1 cup - - - - - -	250	80	190	13	1	0.2	Tr	0.3
537	Brazil nuts, shelled - - - - - - - - - - 1 oz - - - - - -	28	3	185	4	19	4.6	6.5	6.8
538	Carob flour - - - - - - - - - - - - - 1 cup - - - - - -	140	3	255	6	Tr	Tr	0.1	0.1
	Cashew nuts, salted:								
539	Dry roasted - - - - - - - - - - - - 1 cup - - - - - -	137	2	785	21	63	12.5	37.4	10.7
540	1 oz - - - - - -	28	2	165	4	13	2.6	7.7	2.2
541	Roasted in oil - - - - - - - - - - - 1 cup - - - - - -	130	4	750	21	63	12.4	36.9	10.6
542	1 oz - - - - - -	28	4	165	5	14	2.7	8.1	2.3
543	Chestnuts, European (Italian), roasted, shelled - - - - - - - - - - 1 cup - - - - - -	143	40	350	5	3	0.6	1.1	1.2
544	Chickpeas, cooked, drained - - - - 1 cup - - - - - -	163	60	270	15	4	0.4	0.9	1.9
	Coconut:								
	Raw:								
545	Piece, about 2 by 2 by 1/2 in - - - - - - - - - - - - 1 piece - - - - -	45	47	160	1	15	13.4	0.6	0.2
546	Shredded or grated - - - - - - 1 cup - - - - - -	80	47	285	3	27	23.8	1.1	0.3
547	Dried, sweetened, shredded - - 1 cup - - - - - -	93	13	470	3	33	29.3	1.4	0.4
548	Filberts (hazelnuts), chopped - - - 1 cup - - - - - -	115	5	725	15	72	5.3	56.5	6.9
549	1 oz - - - - - -	28	5	180	4	18	1.3	13.9	1.7
550	Lentils, dry, cooked - - - - - - - - - 1 cup - - - - - -	200	72	215	16	1	0.1	0.2	0.5
551	Macadamia nuts, roasted in oil, salted - - - - - - - - - - - - - - - - 1 cup - - - - - -	134	2	960	10	103	15.4	80.9	1.8
552	1 oz - - - - - -	28	2	205	2	22	3.2	17.1	0.4
	Mixed nuts, with peanuts, salted:								
553	Dry roasted - - - - - - - - - - - - 1 oz - - - - - -	28	2	170	5	15	2.0	8.9	3.1
554	Roasted in oil - - - - - - - - - - - 1 oz - - - - - -	28	2	175	5	16	2.5	9.0	3.8
555	Peanuts, roasted in oil, salted - - - 1 cup - - - - - -	145	2	840	39	71	9.9	35.5	22.6
556	1 oz - - - - - -	28	2	165	8	14	1.9	6.9	4.4
557	Peanut butter - - - - - - - - - - - - 1 tbsp - - - - -	16	1	95	5	8	1.4	4.0	2.5
558	Peas, split, dry, cooked - - - - - - 1 cup - - - - - -	200	70	230	16	1	0.1	0.1	0.3
559	Pecans, halves - - - - - - - - - - - 1 cup - - - - - -	108	5	720	8	73	5.9	45.5	18.1
560	1 oz - - - - - -	28	5	190	2	19	1.5	12.0	4.7
561	Pine nuts (pinyons), shelled - - - - 1 oz - - - - - -	28	6	160	3	17	2.7	6.5	7.3
562	Pistachio nuts, dried shelled - - - - 1 oz - - - - - -	28	4	165	6	14	1.7	9.3	2.1
563	Pumpkin and squash kernels, dry, hulled - - - - - - - - - - - - 1 oz - - - - - -	28	7	155	7	13	2.5	4.0	5.9
564	Refried beans, canned - - - - - - - 1 cup - - - - - -	290	72	295	18	3	0.4	0.6	1.4

[41] Cashews without salt contain 21 mg sodium per cup or 4 mg per oz.

[42] Cashews without salt contain 22 mg sodium per cup or 5 mg per oz.

[43] Macadamia nuts without salt contain 9 mg sodium per cup or 2 mg per oz.

Choles-terol	Carbo-hydrate	Calcium	Phos-phorus	Iron	Potas-sium	Sodium	Vitamin A Value		Thiamin	Ribo-flavin	Niacin	Ascorbic Acid	Item No.
							(IU)	(RE)					
Milli-grams	Grams	Milli-grams	Milli-grams	Milli-grams	Milli-grams	Milli-grams	Inter-national units	Retinol equiva-lents	Milli-grams	Milli-grams	Milli-grams	Milli-grams	
0	41	47	239	2.9	608	1	Tr	Tr	0.43	0.05	0.9	0	527
0	38	90	266	4.9	749	13	0	0	0.25	0.13	1.3	0	528
0	49	55	293	5.9	1,163	4	0	0	0.25	0.11	1.3	0	529
0	40	95	281	5.1	790	13	0	0	0.27	0.13	1.3	0	530
0	49	86	296	5.4	882	3	Tr	Tr	0.33	0.16	0.7	0	531
30	32	94	303	4.8	668	1,374	330	33	0.18	0.15	3.3	Tr	532
10	48	138	235	4.6	536	1,181	330	33	0.20	0.08	1.5	5	533
10	54	161	291	5.9	536	969	330	33	0.15	0.10	1.3	5	534
0	42	74	278	4.6	673	968	10	1	0.13	0.10	1.5	0	535
0	35	43	238	3.3	573	20	30	3	0.40	0.10	1.0	0	536
0	4	50	170	1.0	170	1	Tr	Tr	0.28	0.03	0.5	Tr	537
0	126	390	102	5.7	1,275	24	Tr	Tr	0.07	0.07	2.2	Tr	538
0	45	62	671	8.2	774	[41]877	0	0	0.27	0.27	1.9	0	539
0	9	13	139	1.7	160	[41]181	0	0	0.06	0.06	0.4	0	540
0	37	53	554	5.3	689	[42]814	0	0	0.55	0.23	2.3	0	541
0	8	12	121	1.2	150	[42]177	0	0	0.12	0.05	0.5	0	542
0	76	41	153	1.3	847	3	30	3	0.35	0.25	1.9	37	543
0	45	80	273	4.9	475	11	Tr	Tr	0.18	0.09	0.9	0	544
0	7	6	51	1.1	160	9	0	0	0.03	0.01	0.2	1	545
0	12	11	90	1.9	285	16	0	0	0.05	0.02	0.4	3	546
0	44	14	99	1.8	313	244	0	0	0.03	0.02	0.4	1	547
0	18	216	359	3.8	512	3	80	8	0.58	0.13	1.3	1	548
0	4	53	88	0.9	126	1	20	2	0.14	0.03	0.3	Tr	549
0	38	50	238	4.2	498	26	40	4	0.14	0.12	1.2	0	550
0	17	60	268	2.4	441	[43]348	10	1	0.29	0.15	2.7	0	551
0	4	13	57	0.5	93	[43]74	Tr	Tr	0.06	0.03	0.6	0	552
0	7	20	123	1.0	169	[44]190	Tr	Tr	0.06	0.06	1.3	0	553
0	6	31	131	0.9	165	[44]185	10	1	0.14	0.06	1.4	Tr	554
0	27	125	734	2.8	1,019	[45]626	0	0	0.42	0.15	21.5	0	555
0	5	24	143	0.5	199	[45]122	0	0	0.08	0.03	4.2	0	556
0	3	5	60	0.3	110	75	0	0	0.02	0.02	2.2	0	557
0	42	22	178	3.4	592	26	80	8	0.30	0.18	1.8	0	558
0	20	39	314	2.3	423	1	140	14	0.92	0.14	1.0	2	559
0	5	10	83	0.6	111	Tr	40	4	0.24	0.04	0.3	1	560
0	5	2	10	0.9	178	20	10	1	0.35	0.06	1.2	1	561
0	7	38	143	1.9	310	2	70	7	0.23	0.05	0.3	Tr	562
0	5	12	333	4.2	229	5	110	11	0.06	0.09	0.5	Tr	563
0	51	141	245	5.1	1,141	1,228	0	0	0.14	0.16	1.4	17	564

[44] Mixed nuts without salt contain 3 mg sodium per oz.

[45] Peanuts without salt contain 22 mg sodium per cup or 4 mg per oz.

Item No.	Foods, Approximate Measures, Units, and Weight (Weight of Edible Portion Only)		Water	Food Energy	Protein	Fat	Fatty Acids		
							Saturated	Mono-unsaturated	Poly-unsaturated
		Grams	Percent	Calories	Grams	Grams	Grams	Grams	Grams
	Legumes, Nuts, and Seeds *(continued)*								
565	Sesame seeds, dry, hulled - - - - - 1 tbsp - - - - -	8	5	45	2	4	0.6	1.7	1.9
566	Soybeans, dry, cooked, drained - 1 cup - - - - - -	180	71	235	20	10	1.3	1.9	5.3
	Soy products:								
567	Miso - - - - - - - - - - - - - - - - 1 cup - - - - - -	276	53	470	29	13	1.8	2.6	7.3
568	Tofu, piece 2-1/2 by 2-3/4 by 1 in - - - - - - - - - - - - - 1 piece - - - - -	120	85	85	9	5	0.7	1.0	2.9
569	Sunflower seeds, dry, hulled - - - - 1 oz - - - - - - -	28	5	160	6	14	1.5	2.7	9.3
570	Tahini - - - - - - - - - - - - - - - - - 1 tbsp - - - - -	15	3	90	3	8	1.1	3.0	3.5
	Walnuts:								
571	Black, chopped - - - - - - - - - - 1 cup - - - - - -	125	4	760	30	71	4.5	15.9	46.9
572	1 oz - - - - - - -	28	4	170	7	16	1.0	3.6	10.6
573	English or Persian, pieces or chips - - - - - - - - - - - - - - - 1 cup - - - - - -	120	4	770	17	74	6.7	17.0	47.0
574	1 oz - - - - - - -	28	4	180	4	18	1.6	4.0	11.1
	Meat and Meat Products								
	Beef, cooked:[46]								
	Cuts braised, simmered, or pot roasted:								
	Relatively fat such as chuck blade:								
575	Lean and fat, piece, 2-1/2 by 2-1/2 by 3/4 in - - - - - - - - - - - 3 oz - - - - - - -	85	43	325	22	26	10.8	11.7	0.9
576	Lean only from item 575 - - 2.2 oz - - - - -	62	53	170	19	9	3.9	4.2	0.3
	Relatively lean, such as bottom round:								
577	Lean and fat, piece, 4-1/8 by 2-1/4 by 1/2 in - - - - - - - - - - - 3 oz - - - - - - -	85	54	220	25	13	4.8	5.7	0.5
578	Lean only from item 577 - - - - - 2.8 oz - - - - -	78	57	175	25	8	2.7	3.4	0.3
	Ground beef, broiled, patty, 3 by 5/8 in:								
579	Lean - - - - - - - - - - - - - - - 3 oz - - - - - - -	85	56	230	21	16	6.2	6.9	0.6
580	Regular - - - - - - - - - - - - - 3 oz - - - - - - -	85	54	245	20	18	6.9	7.7	0.7
581	Heart, lean, braised - - - - - - - 3 oz - - - - - - -	85	65	150	24	5	1.2	0.8	1.6
582	Liver, fried, slice, 6-1/2 by 2-3/8 by 3/8 in[47] - - - - - - - 3 oz - - - - - - -	85	56	185	23	7	2.5	3.6	1.3
	Roast, oven cooked, no liquid added:								
	Relatively fat, such as rib:								
583	Lean and fat, 2 pieces, 4-1/8 by 2-1/4 by 1/4 in - - - - - - - - - - - 3 oz - - - - - - -	85	46	315	19	26	10.8	11.4	0.9
584	Lean only from item 583 - - 2.2 oz - - - - -	61	57	150	17	9	3.6	3.7	0.3
	Relatively lean, such as eye of round:								
585	Lean and fat, 2 pieces, 2-1/2 by 2-1/2 by 3/8 in - - - - - - - - - - - 3 oz - - - - - - -	85	57	205	23	12	4.9	5.4	0.5
586	Lean only from item 585 - - 2.6 oz - - - - -	75	63	135	22	5	1.9	2.1	0.2

[46] Outer layer of fat was removed to within approximately 1/2 inch of the lean. Deposits of fat within the cut were not removed.

[47] Fried in vegetable shortening.

Choles-terol	Carbo-hydrate	Calcium	Phos-phorus	Iron	Potas-sium	Sodium	Vitamin A Value (IU)	(RE)	Thiamin	Ribo-flavin	Niacin	Ascorbic Acid	Item No.
Milli-grams	Grams	Milli-grams	Milli-grams	Milli-grams	Milli-grams	Milli-grams	Inter-national units	Retinol equiva-lents	Milli-grams	Milli-grams	Milli-grams	Milli-grams	
0	1	11	62	0.6	33	3	10	1	0.06	0.01	0.4	0	565
0	19	131	322	4.9	972	4	50	5	0.38	0.16	1.1	0	566
0	65	188	853	4.7	922	8,142	110	11	0.17	0.28	0.8	0	567
0	3	108	151	2.3	50	8	0	0	0.07	0.04	0.1	0	568
0	5	33	200	1.9	195	1	10	1	0.65	0.07	1.3	Tr	569
0	3	21	119	0.7	69	5	10	1	0.24	0.02	0.8	1	570
0	15	73	580	3.8	655	1	370	37	0.27	0.14	0.9	Tr	571
0	3	16	132	0.9	149	Tr	80	8	0.06	0.03	0.2	Tr	572
0	22	113	380	2.9	602	12	150	15	0.46	0.18	1.3	4	573
0	5	27	90	0.7	142	3	40	4	0.11	0.04	0.3	1	574
87	0	11	163	2.5	163	53	Tr	Tr	0.06	0.19	2.0	0	575
66	0	8	146	2.3	163	44	Tr	Tr	0.05	0.17	1.7	0	576
81	0	5	217	2.8	248	43	Tr	Tr	0.06	0.21	3.3	0	577
75	0	4	212	2.7	240	40	Tr	Tr	0.06	0.20	3.0	0	578
74	0	9	134	1.8	256	65	Tr	Tr	0.04	0.18	4.4	0	579
76	0	9	144	2.1	248	70	Tr	Tr	0.03	0.16	4.9	0	580
164	0	5	213	6.4	198	54	Tr	Tr	0.12	1.31	3.4	5	581
410	7	9	392	5.3	309	90	[48]30,690	[48]9,120	0.18	3.52	12.3	23	582
72	0	8	145	2.0	246	54	Tr	Tr	0.06	0.16	3.1	0	583
49	0	5	127	1.7	218	45	Tr	Tr	0.05	0.13	2.7	0	584
62	0	5	177	1.6	308	50	Tr	Tr	0.07	0.14	3.0	0	585
52	0	3	170	1.5	297	46	Tr	Tr	0.07	0.13	2.8	0	586

[48] Value varies widely.

Item No.	Foods, Approximate Measures, Units, and Weight (Weight of Edible Portion Only)		Water	Food Energy	Protein	Fat	Fatty Acids Saturated	Mono-unsaturated	Poly-unsaturated
			Grams	Percent	Calories	Grams	Grams	Grams	Grams
			Grams	Percent	Calories	Grams	Grams	Grams	Grams
	Meat and Meat Products *(continued)*								
	Steak:								
	Sirloin, broiled:								
587	Lean and fat, piece, 2-1/2 by 2-1/2 by 3/4 in	3 oz	85	53	240	23	15	6.4	6.9	0.6
588	Lean only from item 587	2.5 oz	72	59	150	22	6	2.6	2.8	0.3
589	Beef, canned, corned	3 oz	85	59	185	22	10	4.2	4.9	0.4
590	Beef, dried, chipped	2.5 oz	72	48	145	24	4	1.8	2.0	0.2
	Lamb, cooked:								
	Chops, (3 per lb with bone):								
	Arm, braised:								
591	Lean and fat	2.2 oz	63	44	220	20	15	6.9	6.0	0.9
592	Lean only from item 591	1.7 oz	48	49	135	17	7	2.9	2.6	0.4
	Loin, broiled:								
593	Lean and fat	2.8 oz	80	54	235	22	16	7.3	6.4	1.0
594	Lean only from item 593	2.3 oz	64	61	140	19	6	2.6	2.4	0.4
	Leg, roasted:								
595	Lean and fat, 2 pieces, 4-1/8 by 2-1/4 by 1/4 in	3 oz	85	59	205	22	13	5.6	4.9	0.8
596	Lean only from item 595	2.6 oz	73	64	140	20	6	2.4	2.2	0.4
	Rib, roasted:								
597	Lean and fat, 3 pieces, 2-1/2 by 2-1/2 by 1/4 in	3 oz	85	47	315	18	26	12.1	10.6	1.5
598	Lean only from item 597	2 oz	57	60	130	15	7	3.2	3.0	0.5
	Pork, cured, cooked:								
	Bacon:								
599	Regular	3 medium slices	19	13	110	6	9	3.3	4.5	1.1
600	Canadian-style	2 slices	46	62	85	11	4	1.3	1.9	0.4
	Ham, light cure, roasted:								
601	Lean and fat, 2 pieces, 4-1/8 by 2-1/4 by 1/4 in	3 oz	85	58	205	18	14	5.1	6.7	1.5
602	Lean only from item 601	2.4 oz	68	66	105	17	4	1.3	1.7	0.4
603	Ham, canned, roasted, 2 pieces, 4-1/8 by 2-1/4 by 1/4 in	3 oz	85	67	140	18	7	2.4	3.5	0.8
	Luncheon meat:								
604	Canned, spiced or unspiced, slice, 3 by 2 by 1/2 in	2 slices	42	52	140	5	13	4.5	6.0	1.5
605	Chopped ham (8 slices per 6 oz pkg)	2 slices	42	64	95	7	7	2.4	3.4	0.9
	Cooked ham (8 slices per 8-oz pkg):								
606	Regular	2 slices	57	65	105	10	6	1.9	2.8	0.7
607	Extra lean	2 slices	57	71	75	11	3	0.9	1.3	0.3
	Pork, fresh, cooked:								
	Chop, loin (cut 3 per lb with bone):								
	Broiled:								
608	Lean and fat	3.1 oz	87	50	275	24	19	7.0	8.8	2.2
609	Lean only from item 608	2.5 oz	72	57	165	23	8	2.6	3.4	0.9

Choles-terol	Carbo-hydrate	Calcium	Phos-phorus	Iron	Potas-sium	Sodium	Vitamin A Value		Thiamin	Ribo-flavin	Niacin	Ascorbic Acid	Item No.
							(IU)	(RE)					
Milli-grams	Grams	Milli-grams	Milli-grams	Milli-grams	Milli-grams	Milli-grams	Inter-national units	Retinol equiva-lents	Milli-grams	Milli-grams	Milli-grams	Milli-grams	
77	0	9	186	2.6	306	53	Tr	Tr	0.10	0.23	3.3	0	587
64	0	8	176	2.4	290	48	Tr	Tr	0.09	0.22	3.1	0	588
80	0	17	90	3.7	51	802	Tr	Tr	0.02	0.20	2.9	0	589
46	0	14	287	2.3	142	3,053	Tr	Tr	0.05	0.23	2.7	0	590
77	0	16	132	1.5	195	46	Tr	Tr	0.04	0.16	4.4	0	591
59	0	12	111	1.3	162	36	Tr	Tr	0.03	0.13	3.0	0	592
78	0	16	162	1.4	272	62	Tr	Tr	0.09	0.21	5.5	0	593
60	0	12	145	1.3	241	54	Tr	Tr	0.08	0.18	4.4	0	594
78	0	8	162	1.7	273	57	Tr	Tr	0.09	0.24	5.5	0	595
65	0	6	150	1.5	247	50	Tr	Tr	0.08	0.20	4.6	0	596
77	0	19	139	1.4	224	60	Tr	Tr	0.08	0.18	5.5	0	597
50	0	12	111	1.0	179	46	Tr	Tr	0.05	0.13	3.5	0	598
16	Tr	2	64	0.3	92	303	0	0	0.13	0.05	1.4	6	599
27	1	5	136	0.4	179	711	0	0	0.38	0.09	3.2	10	600
53	0	6	182	0.7	243	1,009	0	0	0.51	0.19	3.8	0	601
37	0	5	154	0.6	215	902	0	0	0.46	0.17	3.4	0	602
35	Tr	6	188	0.9	298	908	0	0	0.82	0.21	4.3	[49]19	603
26	1	3	34	0.3	90	541	0	0	0.15	0.08	1.3	Tr	604
21	0	3	65	0.3	134	576	0	0	0.27	0.09	1.6	[49]8	605
32	2	4	141	0.6	189	751	0	0	0.49	0.14	3.0	[49]16	606
27	1	4	124	0.4	200	815	0	0	0.53	0.13	2.8	[49]15	607
84	0	3	184	0.7	312	61	10	3	0.87	0.24	4.3	Tr	608
71	0	4	176	0.7	302	56	10	1	0.83	0.22	4.0	Tr	609

Item No.	Foods, Approximate Measures, Units, and Weight (Weight of Edible Portion Only)		Water	Food Energy	Protein	Fat	Fatty Acids		
							Saturated	Mono-unsaturated	Poly-unsaturated
		Grams	Percent	Calories	Grams	Grams	Grams	Grams	Grams

Meat and Meat Products *(continued)*

Pork, fresh, cooked:

Chop, loin (cut 3 per lb with bone):

Pan fried:

Item No.	Food	Measure	Grams	Percent	Calories	Protein	Fat	Saturated	Mono	Poly
610	Lean and fat - - - - - - - -	3.1 oz - - - - -	89	45	335	21	27	9.8	12.5	3.1
611	Lean only from item 610- -	2.4 oz - - - - -	67	54	180	19	11	3.7	4.8	1.3
	Ham (leg), roasted:									
612	Lean and fat, piece, 2-1/2 by 2-1/2 by 3/4 in - - - -	3 oz- - - - - -	85	53	250	21	18	6.4	8.1	2.0
613	Lean only from item 612 - - -	2.5 oz - - - - -	72	60	160	20	8	2.7	3.6	1.0
	Rib, roasted:									
614	Lean and fat, piece, 2-1/2 by 3/4 in - - - - - - - - - - -	3 oz- - - - - -	85	51	270	21	20	7.2	9.2	2.3
615	Lean only from item 614 - - -	2.5 oz - - - - -	71	57	175	20	10	3.4	4.4	1.2
	Shoulder cut, braised:									
616	Lean and fat, 3 pieces, 2-1/2 by 2-1/2 by 1/4 in - -	3 oz- - - - - - -	85	47	295	23	22	7.9	10.0	2.4
617	Lean only from item 616 - - -	2.4 oz - - - - -	67	54	165	22	8	2.8	3.7	1.0
	Sausages (See also Luncheon meats, items 604–607):									
618	Bologna, slice (8 per 8-oz pkg)- - - - - - - - - - - - - - -	2 slices- - - - -	57	54	180	7	16	6.1	7.6	1.4
619	Braunschweiger, slice (6 per 6-oz pkg) - - - - - - - - - - -	2 slices- - - - -	57	48	205	8	18	6.2	8.5	2.1
620	Brown and serve (10–11 per 8-oz pkg), browned - - - - -	1 link - - - - - -	13	45	50	2	5	1.7	2.2	0.5
621	Frankfurter (10 per 1-lb pkg), cooked (reheated) - - - - - - -	1 frankfurter - -	45	54	145	5	13	4.8	6.2	1.2
622	Pork link (16 per 1-lb pkg), cooked[50] - - - - - - - - - - - - -	1 link - - - - - -	13	45	50	3	4	1.4	1.8	0.5
	Salami:									
623	Cooked type, slice (8 per 8-oz pkg) - - - - - - - - - - -	2 slices- - - - -	57	60	145	8	11	4.6	5.2	1.2
624	Dry type, slice (12 per 4-oz pkg) - - - - - - - - - - -	2 slices- - - - -	20	35	85	5	7	2.4	3.4	0.6
625	Sandwich spread (pork, beef) -	1 tbsp - - - - -	15	60	35	1	3	0.9	1.1	0.4
626	Vienna sausage (7 per 4-oz can)- - - - - - - - - - - - - - -	1 sausage - - -	16	60	45	2	4	1.5	2.0	0.3
	Veal, medium fat, cooked, bone removed:									
627	Cutlet, 4-1/8 by 2-1/4 by 1/2 in, braised or broiled - - -	3 oz- - - - - - -	85	60	185	23	9	4.1	4.1	0.6
628	Rib, 2 pieces, 4-1/8 by 2-1/4 by 1/4 in, roasted - -	3 oz- - - - - - -	85	55	230	23	14	6.0	6.0	1.0
	Mixed Dishes and Fast Foods									
	Mixed dishes:									
629	Beef and vegetable stew, from home recipe- - - - - - - - - - -	1 cup- - - - - -	245	82	220	16	11	4.4	4.5	0.5
630	Beef potpie, from home recipe, baked, piece, 1/3 of 9-in diam. pie[51] - - - - - - - -	1 piece- - - - -	210	55	515	21	30	7.9	12.9	7.4

[49] Contains added sodium ascorbate. If sodium ascorbate is not added, ascorbic acid content is negligible.

[50] One patty (8 per pound) of bulk sausage is equivalent to 2 links.

Cholesterol	Carbohydrate	Calcium	Phosphorus	Iron	Potassium	Sodium	Vitamin A Value		Thiamin	Riboflavin	Niacin	Ascorbic Acid	Item No.
							(IU)	(RE)					
Milligrams	Grams	Milligrams	Milligrams	Milligrams	Milligrams	Milligrams	International units	Retinol equivalents	Milligrams	Milligrams	Milligrams	Milligrams	
92	0	4	190	0.7	323	64	10	3	0.91	0.24	4.6	Tr	610
72	0	3	178	0.7	305	57	10	1	0.84	0.22	4.0	Tr	611
79	0	5	210	0.9	280	50	10	2	0.54	0.27	3.9	Tr	612
68	0	5	202	0.8	269	46	10	1	0.50	0.25	3.6	Tr	613
69	0	9	190	0.8	313	37	10	3	0.50	0.24	4.2	Tr	614
56	0	8	182	0.7	300	33	10	2	0.45	0.22	3.8	Tr	615
93	0	6	162	1.4	286	75	10	3	0.46	0.26	4.4	Tr	616
76	0	5	151	1.3	271	68	10	1	0.40	0.24	4.0	Tr	617
31	2	7	52	0.9	103	581	0	0	0.10	0.08	1.5	[49]12	618
89	2	5	96	5.3	113	652	8,010	2,405	0.14	0.87	4.8	[49]6	619
9	Tr	1	14	0.1	25	105	0	0	0.05	0.02	0.4	0	620
23	1	5	39	0.5	75	504	0	0	0.09	0.05	1.2	[49]12	621
11	Tr	4	24	0.2	47	168	0	0	0.10	0.03	0.6	Tr	622
37	1	7	66	1.5	113	607	0	0	0.14	0.21	2.0	[49]7	623
16	1	2	28	0.3	76	372	0	0	0.12	0.06	1.0	[49]5	624
6	2	2	9	0.1	17	152	10	1	0.03	0.02	0.3	0	625
8	Tr	2	8	0.1	16	152	0	0	0.01	0.02	0.3	0	626
109	0	9	196	0.8	258	56	Tr	Tr	0.06	0.21	4.6	0	627
109	0	10	211	0.7	259	57	Tr	Tr	0.11	0.26	6.6	0	628
71	15	29	184	2.9	613	292	5,690	568	0.15	0.17	4.7	17	629
42	39	29	149	3.8	334	596	4,220	517	0.29	0.29	4.8	6	630

[51] Crust made with vegetable shortening and enriched flour.

Item No.	Foods, Approximate Measures, Units, and Weight (Weight of Edible Portion Only)		Water	Food Energy	Protein	Fat	Fatty Acids		
							Saturated	Mono-unsaturated	Poly-unsaturated
		Grams	Percent	Calories	Grams	Grams	Grams	Grams	Grams
	Mixed Dishes and Fast Foods								
	Mixed dishes:								
631	Chicken a la king, cooked, from home recipe · · · · · · · 1 cup · · · · ·	245	68	470	27	34	12.9	13.4	6.2
632	Chicken and noodles, cooked, from home recipe · · · · · · · 1 cup · · · · · ·	240	71	365	22	18	5.1	7.1	3.9
	Chicken chow mein:								
633	Canned · · · · · · · · · · · · · 1 cup · · · · ·	250	89	95	7	Tr	0.1	0.1	0.8
634	From home recipe · · · · · · · 1 cup · · · · ·	250	78	255	31	10	4.1	4.9	3.5
635	Chicken potpie, from home recipe, baked, piece, 1/3 of 9-in diam. pie[51] · · · · · · · · 1 piece · · · · ·	232	57	545	23	31	10.3	15.5	6.6
636	Chili con carne with beans, canned · · · · · · · · · · · · · · 1 cup · · · · · ·	255	72	340	19	16	5.8	7.2	1.0
637	Chop suey with beef and pork, from home recipe · · · · 1 cup · · · · · ·	250	75	300	26	17	4.3	7.4	4.2
	Macaroni (enriched) and cheese:								
638	Canned[52] · · · · · · · · · · · 1 cup · · · · · ·	240	80	230	9	10	4.7	2.9	1.3
639	From home recipe[38] · · · · · · 1 cup · · · · · ·	200	58	430	17	22	9.8	7.4	3.6
640	Quiche Lorraine, 1/8 of 8-in diam. quiche[51] · · · · · · · · 1 slice · · · · ·	176	47	600	13	48	23.2	17.8	4.1
	Spaghetti (enriched) in tomato sauce with cheese:								
641	Canned · · · · · · · · · · · · · 1 cup · · · · · ·	250	80	190	6	2	0.4	0.4	0.5
642	From home recipe · · · · · · · 1 cup · · · · · ·	250	77	260	9	9	3.0	3.6	1.2
	Spaghetti (enriched) with meatballs and tomato sauce:								
643	Canned · · · · · · · · · · · · · 1 cup · · · · · ·	250	78	260	12	10	2.4	3.9	3.1
644	From home recipe · · · · · · · 1 cup · · · · · ·	248	70	330	19	12	3.9	4.4	2.2
	Fast food entrées:								
	Cheeseburger:								
645	Regular · · · · · · · · · · · · · 1 sandwich · ·	112	46	300	15	15	7.3	5.6	1.0
646	4 oz patty · · · · · · · · · · · 1 sandwich · ·	194	46	525	30	31	15.1	12.2	1.4
	Chicken, fried. See Poultry and Poultry Products (items 656–659).								
647	Enchilada · · · · · · · · · · · · 1 enchilada · ·	230	72	235	20	16	7.7	6.7	0.6
648	English muffin, egg, cheese, and bacon · · · · · · · · · · · 1 sandwich · ·	138	49	360	18	18	8.0	8.0	0.7
	Fish sandwich:								
649	Regular, with cheese · · · · · 1 sandwich · ·	140	43	420	16	23	6.3	6.9	7.7
650	Large, without cheese · · · · · 1 sandwich · ·	170	48	470	18	27	6.3	8.7	9.5
	Hamburger:								
651	Regular · · · · · · · · · · · · · 1 sandwich · ·	98	46	245	12	11	4.4	5.3	0.5
652	4 oz patty · · · · · · · · · · · 1 sandwich · ·	174	50	445	25	21	7.1	11.7	0.6
653	Pizza, cheese, 1/8 of 15-in diam. pizza[51] · · · · · · · · 1 slice · · · · ·	120	46	290	15	9	4.1	2.6	1.3
654	Roast beef sandwich · · · · · · 1 sandwich · ·	150	52	345	22	13	3.5	6.9	1.8
655	Taco · · · · · · · · · · · · · · · · 1 taco · · · · ·	81	55	195	9	11	4.1	5.5	0.8

[38] Made with margarine.

[51] Crust made with vegetable shortening and enriched flour.

[52] Made with corn oil.

Cholesterol	Carbohydrate	Calcium	Phosphorus	Iron	Potassium	Sodium	Vitamin A Value (IU)	Vitamin A Value (RE)	Thiamin	Riboflavin	Niacin	Ascorbic Acid	Item No.
Milligrams	Grams	Milligrams	Milligrams	Milligrams	Milligrams	Milligrams	International units	Retinol equivalents	Milligrams	Milligrams	Milligrams	Milligrams	
221	12	127	358	2.5	404	760	1,130	272	0.10	0.42	5.4	12	631
103	26	26	247	2.2	149	600	430	130	0.05	0.17	4.3	Tr	632
8	18	45	85	1.3	418	725	150	28	0.05	0.10	1.0	13	633
75	10	58	293	2.5	473	718	280	50	0.08	0.23	4.3	10	634
56	42	70	232	3.0	343	594	7,220	735	0.32	0.32	4.9	5	635
28	31	82	321	4.3	594	1,354	150	15	0.08	0.18	3.3	8	636
68	13	60	248	4.8	425	1,053	600	60	0.28	0.38	5.0	33	637
24	26	199	182	1.0	139	730	260	72	0.12	0.24	1.0	Tr	638
44	40	362	322	1.8	240	1,086	860	232	0.20	0.40	1.8	1	639
285	29	211	276	1.0	283	653	1,640	454	0.11	0.32	Tr	Tr	640
3	39	40	88	2.8	303	955	930	120	0.35	0.28	4.5	10	641
8	37	80	135	2.3	408	955	1,080	140	0.25	0.18	2.3	13	642
23	29	53	113	3.3	245	1,220	1,000	100	0.15	0.18	2.3	5	643
89	39	124	236	3.7	665	1,009	1,590	159	0.25	0.30	4.0	22	644
44	28	135	174	2.3	219	672	340	65	0.26	0.24	3.7	1	645
104	40	236	320	4.5	407	1,224	670	128	0.33	0.48	7.4	3	646
19	24	322	662	11.0	2,180	4,451	2,720	352	0.18	0.26	Tr	Tr	647
213	31	197	290	3.1	201	832	650	160	0.46	0.50	3.7	1	648
56	39	132	223	1.8	274	667	160	25	0.32	0.26	3.3	2	649
91	41	61	246	2.2	375	621	110	15	0.35	0.23	3.5	1	650
32	28	56	107	2.2	202	463	80	14	0.23	0.24	3.8	1	651
71	38	75	225	4.8	404	763	160	28	0.38	0.38	7.8	1	652
56	39	220	216	1.6	230	699	750	106	0.34	0.29	4.2	2	653
55	34	60	222	4.0	338	757	240	32	0.40	0.33	6.0	2	654
21	15	109	134	1.2	263	456	420	57	0.09	0.07	1.4	1	655

(Tr indicates nutrient present in trace amount.)

Item No.	Foods, Approximate Measures, Units, and Weight (Weight of Edible Portion Only)		Water	Food Energy	Protein	Fat	Fatty Acids		
							Saturated	Mono-unsaturated	Poly-unsaturated
			Grams	Percent	Calories	Grams	Grams	Grams	Grams	Grams
	Poultry and Poultry Products								
	Chicken:								
	Fried, flesh, with skin:[53]								
	Batter dipped:								
656	Breast, 1/2 breast (5.6 oz with bones) - - - - - - - -	4.9 oz - - - - -	140	52	365	35	18	4.9	7.6	4.3
657	Drumstick (3.4 oz with bones) - - - - - - - - - - -	2.5 oz - - - - -	72	53	195	16	11	3.0	4.6	2.7
	Flour coated:								
658	Breast, 1/2 breast (4.2 oz with bones) - - - - - - - -	3.5 oz - - - - -	98	57	220	31	9	2.4	3.4	1.9
659	Drumstick (2.6 oz with bones) - - - - - - - - - - -	1.7 oz - - - - -	49	57	120	13	7	1.8	2.7	1.6
	Roasted, flesh only:								
660	Breast, 1/2 breast (4.2 oz withbones and skin) - - - -	3.0 oz - - - - -	86	65	140	27	3	0.9	1.1	0.7
661	Drumstick, (2.9 oz with bonesand skin) - - - - - - -	1.6 oz - - - - -	44	67	75	12	2	0.7	0.8	0.6
662	Stewed, flesh only, light and dark meat, chopped or diced - - - - - - - - - - - - - - -	1 cup - - - - -	140	67	250	38	9	2.6	3.3	2.2
663	Chicken liver, cooked - - - - - - - -	1 liver - - - - - -	20	68	30	5	1	0.4	0.3	0.2
664	Duck, roasted, flesh only - - - - - -	1/2 duck - - - -	221	64	445	52	25	9.2	8.2	3.2
	Turkey, roasted, flesh only:								
665	Dark meat, piece, 2-1/2 by 1-5/8 by 1/4 in - - - - - - - -	4 pieces - - - -	85	63	160	24	6	2.1	1.4	1.8
666	Light meat, piece, 4 by 2 by 1/4 in - - - - - - - - - - - - - -	2 pieces - - - -	85	66	135	25	3	0.9	0.5	0.7
	Light and dark meat:								
667	Chopped or diced - - - - - - -	1 cup - - - - - -	140	65	240	41	7	2.3	1.4	2.0
668	Pieces (1 slice white meat, 4 by 2 by 1/4 in and 2 slices dark meat, 2-1/2 by 1-5/8 by 1/4 in) - - - - - -	3 pieces - - - -	85	65	145	25	4	1.4	0.9	1.2
	Poultry food products:								
	Chicken:								
669	Canned, boneless - - - - - - -	5 oz - - - - - - -	142	69	235	31	11	3.1	4.5	2.5
670	Frankfurter (10 per 1-lb pkg) - - - - - - - - - - - - - -	1 frankfurter - -	45	58	115	6	9	2.5	3.8	1.8
671	Roll, light (6 slices per 6 oz pkg) - - - - - - - - - - - - - -	2 slices - - - - -	57	69	90	11	4	1.1	1.7	0.9
	Turkey:								
672	Gravy and turkey, frozen - - -	5-oz package - - - -	142	85	95	8	4	1.2	1.4	0.7
673	Ham, Cured turkey thigh meat (8 slices per 8-oz pkg) - - - - - - - - - - - - - -	2 slices - - - - -	57	71	75	11	3	1.0	0.7	0.9
674	Loaf, breast meat (8 slices per 6-oz pkg) - - - - - -	2 slices - - - - -	42	72	45	10	1	0.2	0.2	0.1
675	Patties, breaded, battered, fried (2.25 oz) - - - - - - - -	1 patty - - - - -	64	50	180	9	12	3.0	4.8	3.0
676	Roast, boneless, frozen, seasoned, light and dark meat, cooked - - - - - - - -	3 oz - - - - - - -	85	68	130	18	5	1.6	1.0	1.4

[53] Fried in vegetable shortening.

Choles-terol	Carbo-hydrate	Calcium	Phos-phorus	Iron	Potas-sium	Sodium	Vitamin A Value (IU)	(RE)	Thiamin	Ribo-flavin	Niacin	Ascorbic Acid	Item No.
Milli-grams	Grams	Milli-grams	Milli-grams	Milli-grams	Milli-grams	Milli-grams	Inter-national units	Retinol equiva-lents	Milli-grams	Milli-grams	Milli-grams	Milli-grams	
119	13	28	259	1.8	281	385	90	28	0.16	0.20	14.7	0	656
62	6	12	106	1.0	134	194	60	19	0.08	0.15	3.7	0	657
87	2	16	228	1.2	254	74	50	15	0.08	0.13	13.5	0	658
44	1	6	86	0.7	112	44	40	12	0.04	0.11	3.0	0	659
73	0	13	196	0.9	220	64	20	5	0.06	0.10	11.8	0	660
41	0	5	81	0.6	108	42	30	8	0.03	0.10	2.7	0	661
116	0	20	210	1.6	252	98	70	21	0.07	0.23	8.6	0	662
126	Tr	3	62	1.7	28	10	3,270	983	0.03	0.35	0.9	3	663
197	0	27	449	6.0	557	144	170	51	0.57	1.04	11.3	0	664
72	0	27	173	2.0	246	67	0	0	0.05	0.21	3.1	0	665
59	0	16	186	1.1	259	54	0	0	0.05	0.11	5.8	0	666
106	0	35	298	2.5	417	98	0	0	0.09	0.25	7.6	0	667
65	0	21	181	1.5	253	60	0	0	0.05	0.15	4.6	0	668
88	0	20	158	2.2	196	714	170	48	0.02	0.18	9.0	3	669
45	3	43	48	0.9	38	616	60	17	0.03	0.05	1.4	0	670
28	1	24	89	0.6	129	331	50	14	0.04	0.07	3.0	0	671
26	7	20	115	1.3	87	787	60	18	0.03	0.18	2.6	0	672
32	Tr	6	108	1.6	184	565	0	0	0.03	0.14	2.0	0	673
17	0	3	97	0.2	118	608	0	0	0.02	0.05	3.5	[54]0	674
40	10	9	173	1.4	176	512	20	7	0.06	0.12	1.5	0	675
45	3	4	207	1.4	253	578	0	0	0.04	0.14	5.3	0	676

[54] If sodium ascorbate is added, product contains 11 mg ascorbic acid.

(Tr indicates nutrient present in trace amount.)

Item No.	Foods, Approximate Measures, Units, and Weight (Weight of Edible Portion Only)		Water	Food Energy	Protein	Fat	Fatty Acids Saturated	Mono- unsaturated	Poly- unsaturated
		Grams	Percent	Calories	Grams	Grams	Grams	Grams	Grams
	Soups, Sauces, and Gravies								
	Soups:								
	Canned, condensed:								
	Prepared with equal volume of milk:								
677	Clam chowder, New England - - - - - - - - - 1 cup - - - - -	248	85	165	9	7	3.0	2.3	1.1
678	Cream of chicken - - - - - - 1 cup - - - - -	248	85	190	7	11	4.6	4.5	1.6
679	Cream of mushroom - - - - 1 cup - - - - -	248	85	205	6	14	5.1	3.0	4.6
680	Tomato - - - - - - - - - - - 1 cup - - - - -	248	85	160	6	6	2.9	1.6	1.1
	Soups:								
	Canned, condensed:								
	Prepared with equal volume of water:								
681	Bean with bacon - - - - - - 1 cup - - - - -	253	84	170	8	6	1.5	2.2	1.8
682	Beef broth, bouillon, consomme - - - - - - - - - 1 cup - - - - -	240	98	15	3	1	0.3	0.2	Tr
683	Beef noodle - - - - - - - - - 1 cup - - - - -	244	92	85	5	3	1.1	1.2	0.5
684	Chicken noodle - - - - - - - 1 cup - - - - -	241	92	75	4	2	0.7	1.1	0.6
685	Chicken rice - - - - - - - - - 1 cup - - - - -	241	94	60	4	2	0.5	0.9	0.4
686	Clam chowder, Manhattan 1 cup - - - - -	244	90	80	4	2	0.4	0.4	1.3
687	Cream of chicken - - - - - - 1 cup - - - - -	244	91	115	3	7	2.1	3.3	1.5
688	Cream of mushroom - - - - 1 cup - - - - -	244	90	130	2	9	2.4	1.7	4.2
689	Minestrone - - - - - - - - - - 1 cup - - - - -	241	91	80	4	3	0.6	0.7	1.1
690	Pea, green - - - - - - - - - - 1 cup - - - - -	250	83	165	9	3	1.4	1.0	0.4
691	Tomato - - - - - - - - - - - - 1 cup - - - - -	244	90	85	2	2	0.4	0.4	1.0
692	Vegetable beef - - - - - - - 1 cup - - - - -	244	92	80	6	2	0.9	0.8	0.1
693	Vegetarian - - - - - - - - - - 1 cup - - - - -	241	92	70	2	2	0.3	0.8	0.7
	Dehydrated:								
	Unprepared:								
694	Bouillon - - - - - - - - - - - 1 pkt - - - - - -	6	3	15	1	1	0.3	0.2	Tr
695	Onion - - - - - - - - - - - - - 1 pkt - - - - - -	7	4	20	1	Tr	0.1	0.2	Tr
	Prepared with water:								
696	Chicken noodle - - - - - - - 1 pkt (6-fl-oz)	188	94	40	2	1	0.2	0.4	0.3
697	Onion - - - - - - - - - - - - - 1 pkt (6-fl-oz)	184	96	20	1	Tr	0.1	0.2	0.1
698	Tomato vegetable - - - - - - 1 pkt (6-fl-oz)	189	94	40	1	1	0.3	0.2	0.1
	Sauces:								
	From dry mix:								
699	Cheese, prepared with milk - 1 cup - - - - - -	279	77	305	16	17	9.3	5.3	1.6
700	Hollandaise, prepared with water - - - - - - - - - - - - - 1 cup - - - - - -	259	84	240	5	20	11.6	5.9	0.9
701	White sauce, prepared with milk - - - - - - - - - - - - - - 1 cup - - - - - -	264	81	240	10	13	6.4	4.7	1.7
	From home recipe:								
702	White sauce, medium[55] - - - - 1 cup - - - - - -	250	73	395	10	30	9.1	11.9	7.2
	Ready to serve:								
703	Barbecue - - - - - - - - - - - 1 tbsp - - - - -	16	81	10	Tr	Tr	Tr	0.1	0.1
704	Soy - - - - - - - - - - - - - - 1 tbsp - - - - -	18	68	10	2	0	0.0	0.0	0.0
	Gravies:								
	Canned:								
705	Beef - - - - - - - - - - - - - 1 cup - - - - - -	233	87	125	9	5	2.7	2.3	0.2
706	Chicken - - - - - - - - - - - - 1 cup - - - - - -	238	85	190	5	14	3.4	6.1	3.6
707	Mushroom - - - - - - - - - - 1 cup - - - - - -	238	89	120	3	6	1.0	2.8	2.4
	From dry mix:								
708	Brown - - - - - - - - - - - - - 1 cup - - - - - -	261	91	80	3	2	0.9	0.8	0.1

[55] Made with enriched flour, margarine, and whole milk.

Choles-terol	Carbo-hydrate	Calcium	Phos-phorus	Iron	Potas-sium	Sodium	Vitamin A Value (IU)	(RE)	Thiamin	Ribo-flavin	Niacin	Ascorbic Acid	Item No.
Milli-grams	Grams	Milli-grams	Milli-grams	Milli-grams	Milli-grams	Milli-grams	Inter-national units	Retinol equiva-lents	Milli-grams	Milli-grams	Milli-grams	Milli-grams	
22	17	186	156	1.5	300	992	160	40	0.07	0.24	1.0	3	677
27	15	181	151	0.7	273	1,047	710	94	0.07	0.26	0.9	1	678
20	15	179	156	0.6	270	1,076	150	37	0.08	0.28	0.9	2	679
17	22	159	149	1.8	449	932	850	109	0.13	0.25	1.5	68	680
3	23	81	132	2.0	402	951	890	89	0.09	0.03	0.6	2	681
Tr	Tr	14	31	0.4	130	782	0	0	Tr	0.05	1.9	0	682
5	9	15	46	1.1	100	952	630	63	0.07	0.06	1.1	Tr	683
7	9	17	36	0.8	55	1,106	710	71	0.05	0.06	1.4	Tr	684
7	7	17	22	0.7	101	815	660	66	0.02	0.02	1.1	Tr	685
2	12	34	59	1.9	261	1,808	920	92	0.06	0.05	1.3	3	686
10	9	34	37	0.6	88	986	560	56	0.03	0.06	0.8	Tr	687
2	9	46	49	0.5	100	1,032	0	0	0.05	0.09	0.7	1	688
2	11	34	55	0.9	313	911	2,340	234	0.05	0.04	0.9	1	689
0	27	28	125	2.0	190	988	200	20	0.11	0.07	1.2	2	690
0	17	12	34	1.8	264	871	690	69	0.09	0.05	1.4	66	691
5	10	17	41	1.1	173	956	1,890	189	0.04	0.05	1.0	2	692
0	12	22	34	1.1	210	822	3,010	301	0.05	0.05	0.9	1	693
1	1	4	19	0.1	27	1,019	Tr	Tr	Tr	0.01	0.3	0	694
Tr	4	10	23	0.1	47	627	Tr	Tr	0.02	0.04	0.4	Tr	695
2	6	24	24	0.4	23	957	50	5	0.05	0.04	0.7	Tr	696
0	4	9	22	0.1	48	635	Tr	Tr	0.02	0.04	0.4	Tr	697
0	8	6	23	0.5	78	856	140	14	0.04	0.03	0.6	5	698
53	23	569	438	0.3	552	1,565	390	117	0.15	0.56	0.3	2	699
52	14	124	127	0.9	124	1,564	730	220	0.05	0.18	0.1	Tr	700
34	21	425	256	0.3	444	797	310	92	0.08	0.45	0.5	3	701
32	24	292	238	0.9	381	888	1,190	340	0.15	0.43	0.8	2	702
0	2	3	3	0.1	28	130	140	14	Tr	Tr	0.1	1	703
0	2	3	38	0.5	64	1,029	0	0	0.01	0.02	0.6	0	704
7	11	14	70	1.6	189	117	0	0	0.07	0.08	1.5	0	705
5	13	48	69	1.1	259	1,373	880	264	0.04	0.10	1.1	0	706
0	13	17	36	1.6	252	1,357	0	0	0.08	0.15	1.6	0	707
2	14	66	47	0.2	61	1,147	0	0	0.04	0.09	0.9	0	708

Item No.	Foods, Approximate Measures, Units, and Weight (Weight of Edible Portion Only)		Water	Food Energy	Protein	Fat	Fatty Acids		
							Saturated	Mono-unsaturated	Poly-unsaturated
		Grams	Percent	Calories	Grams	Grams	Grams	Grams	Grams
	Soups, Sauces, and Gravies *(continued)*								
	Gravies:								
	From dry mix:								
709	Chicken - - - - - - - - - - - - - 1 cup - - - - -	260	91	85	3	2	0.5	0.9	0.4
	Sugars and Sweets								
	Candy:								
710	Caramels, plain or chocolate - - 1 oz - - - - - -	28	8	115	1	3	2.2	0.3	0.1
	Chocolate:								
711	Milk, plain - - - - - - - - - - - 1 oz - - - - - -	28	1	145	2	9	5.4	3.0	0.3
712	Milk, with almonds - - - - - - - 1 oz - - - - - -	28	2	150	3	10	4.8	4.1	0.7
713	Milk, with peanuts - - - - - - - 1 oz - - - - - -	28	1	155	4	11	4.2	3.5	1.5
714	Milk, with rice cereal - - - - - - 1 oz - - - - - -	28	2	140	2	7	4.4	2.5	0.2
715	Semisweet, small pieces (60 per oz) - - - - - - - - - - - 1 cup or 6 oz	170	1	860	7	61	36.2	19.9	1.9
716	Sweet (dark) - - - - - - - - - - 1 oz - - - - - -	28	1	150	1	10	5.9	3.3	0.3
717	Fondant, uncoated (mints, candy corn, other) - - - - - - 1 oz - - - - - -	28	3	105	Tr	0	0.0	0.0	0.0
718	Fudge, chocolate, plain - - - - - 1 oz - - - - - -	28	8	115	1	3	2.1	1.0	0.1
719	Gum drops - - - - - - - - - - - - 1 oz - - - - - -	28	12	100	Tr	Tr	Tr	Tr	0.1
	Candy:								
720	Hard - - - - - - - - - - - - - - - 1 oz - - - - - -	28	1	110	0	0	0.0	0.0	0.0
721	Jelly beans - - - - - - - - - - - - 1 oz - - - - - -	28	6	105	Tr	Tr	Tr	Tr	0.1
722	Marshmallows - - - - - - - - - - - 1 oz - - - - - -	28	17	90	1	0	0.0	0.0	0.0
723	Custard, baked - - - - - - - - - - - 1 cup - - - - - -	265	77	305	14	15	6.8	5.4	0.7
724	Gelatin dessert prepared with gelatin dessert powder and water - - - - - - - - - - - - - 1/2 cup - - - -	120	84	70	2	0	0.0	0.0	0.0
725	Honey, strained or extracted - - - 1 cup - - - - - -	339	17	1,030	1	0	0.0	0.0	0.0
726	1 tbsp - - - - -	21	17	65	Tr	0	0.0	0.0	0.0
727	Jams and preserves - - - - - - - - 1 tbsp - - - - -	20	29	55	Tr	Tr	0.0	Tr	Tr
728	1 packet - - - -	14	29	40	Tr	Tr	0.0	Tr	Tr
729	Jellies - - - - - - - - - - - - - - - 1 tbsp - - - - -	18	28	50	Tr	Tr	Tr	Tr	Tr
730	1 packet - - - -	14	28	40	Tr	Tr	Tr	Tr	Tr
731	Popsicle, 3-fl-oz size - - - - - - - 1 popsicle - - -	95	80	70	0	0	0.0	0.0	0.0
	Puddings:								
	Canned:								
732	Chocolate - - - - - - - - - - - 5-oz can - - -	142	68	205	3	11	9.5	0.5	0.1
733	Tapioca - - - - - - - - - - - - 5-oz can - - -	142	74	160	3	5	4.8	Tr	Tr
734	Vanilla - - - - - - - - - - - - - 5-oz can - - -	142	69	220	2	10	9.5	0.2	0.1
	Dry mix, prepared with whole milk:								
	Chocolate:								
735	Instant - - - - - - - - - - - - 1/2 cup - - - -	130	71	155	4	4	2.3	1.1	0.2
736	Regular (cooked) - - - - - - 1/2 cup - - - -	130	73	150	4	4	2.4	1.1	0.1
737	Rice - - - - - - - - - - - - - - 1/2 cup - - - -	132	73	155	4	4	2.3	1.1	0.1
738	Tapioca - - - - - - - - - - - - 1/2 cup - - - -	130	75	145	4	4	2.3	1.1	0.1
	Vanilla:								
739	Instant - - - - - - - - - - - - 1/2 cup - - - -	130	73	150	4	4	2.2	1.1	0.2
740	Regular (cooked) - - - - - - 1/2 cup - - - -	130	74	145	4	4	2.3	1.0	0.1
	Sugars:								
741	Brown, pressed down - - - - - - 1 cup - - - - - -	220	2	820	0	0	0.0	0.0	0.0
	White:								
742	Granulated - - - - - - - - - - - 1 cup - - - - - -	200	1	770	0	0	0.0	0.0	0.0
743	1 tbsp - - - - -	12	1	45	0	0	0.0	0.0	0.0
744	1 packet - - - -	6	1	25	0	0	0.0	0.0	0.0

Choles-terol	Carbo-hydrate	Calcium	Phos-phorus	Iron	Potas-sium	Sodium	Vitamin A Value		Thiamin	Ribo-flavin	Niacin	Ascorbic Acid	Item No.
							(IU)	(RE)					
Milli-grams	Grams	Milli-grams	Milli-grams	Milli-grams	Milli-grams	Milli-grams	Inter-national units	Retinol equiva-lents	Milli-grams	Milli-grams	Milli-grams	Milli-grams	
3	14	39	47	0.3	62	1,134	0	0	0.05	0.15	0.8	3	709
1	22	42	35	0.4	54	64	Tr	Tr	0.01	0.05	0.1	Tr	710
6	16	50	61	0.4	96	23	30	10	0.02	0.10	0.1	Tr	711
5	15	65	77	0.5	125	23	30	8	0.02	0.12	0.2	Tr	712
5	13	49	83	0.4	138	19	30	8	0.07	0.07	1.4	Tr	713
6	18	48	57	0.2	100	46	30	8	0.01	0.08	0.1	Tr	714
0	97	51	178	5.8	593	24	30	3	0.10	0.14	0.9	Tr	715
0	16	7	41	0.6	86	5	10	1	0.01	0.04	0.1	Tr	716
0	27	2	Tr	0.1	1	57	0	0	Tr	Tr	Tr	0	717
1	21	22	24	0.3	42	54	Tr	Tr	0.01	0.03	0.1	Tr	718
0	25	2	Tr	0.1	1	10	0	0	0.00	Tr	Tr	0	719
0	28	Tr	2	0.1	1	7	0	0	0.10	0.00	0.0	0	720
0	26	1	1	0.3	11	7	0	0	0.00	Tr	Tr	0	721
0	23	1	2	0.5	2	25	0	0	0.00	Tr	Tr	0	722
278	29	297	310	1.1	387	209	530	146	0.11	0.50	0.3	1	723
0	17	2	23	Tr	Tr	55	0	0	0.00	0.00	0.0	0	724
0	279	17	20	1.7	173	17	0	0	0.02	0.14	1.0	3	725
0	17	1	1	0.1	11	1	0	0	Tr	0.01	0.1	Tr	726
0	14	4	2	0.2	18	2	Tr	Tr	Tr	0.01	Tr	Tr	727
0	10	3	1	0.1	12	2	Tr	Tr	Tr	Tr	Tr	Tr	728
0	13	2	Tr	0.1	16	5	Tr	Tr	Tr	0.01	Tr	1	729
0	10	1	Tr	Tr	13	4	Tr	Tr	Tr	Tr	Tr	1	730
0	18	0	0	Tr	4	11	0	0	0.00	0.00	0.0	0	731
1	30	74	117	1.2	254	285	100	31	0.04	0.17	0.6	Tr	732
Tr	28	119	113	0.3	212	252	Tr	Tr	0.03	0.14	0.4	Tr	733
1	33	79	94	0.2	155	305	Tr	Tr	0.03	0.12	0.6	Tr	734
14	27	130	329	0.3	176	440	130	33	0.04	0.18	0.1	1	735
15	25	146	120	0.2	190	167	140	34	0.05	0.20	0.1	1	736
15	27	133	110	0.5	165	140	140	33	0.10	0.18	0.6	1	737
15	25	131	103	0.1	167	152	140	34	0.04	0.18	0.1	1	738
15	27	129	273	0.1	164	375	140	33	0.04	0.17	0.1	1	739
15	25	132	102	0.1	166	178	140	34	0.04	0.18	0.1	1	740
0	212	187	56	4.8	757	97	0	0	0.02	0.07	0.2	0	741
0	199	3	Tr	0.1	7	5	0	0	0.00	0.00	0.0	0	742
0	12	Tr	Tr	Tr	Tr	Tr	0	0	0.00	0.00	0.0	0	743
0	6	Tr	Tr	Tr	Tr	Tr	0	0	0.00	0.00	0.0	0	744

Item No.	Foods, Approximate Measures, Units, and Weight (Weight of Edible Portion Only)		Water	Food Energy	Protein	Fat	Fatty Acids Saturated	Mono-unsaturated	Poly-unsaturated
			Grams	Percent	Calories	Grams	Grams	Grams	Grams	Grams
	Sugars and Sweets *(continued)*								
	Sugars:								
	White:								
745	Powdered, sifted, spooned into cup - - - - - - - - - - - -	1 cup - - - - -	100	1	385	0	0	0.0	0.0	0.0
	Syrups:								
	Chocolate-flavored syrup or topping:								
746	Thin type- - - - - - - - - - - - -	2 tbsp - - - - -	38	37	85	1	Tr	0.2	0.1	0.1
747	Fudge type - - - - - - - - - - -	2 tbsp - - - - -	38	25	125	2	5	3.1	1.7	0.2
748	Molasses, cane, blackstrap - - -	2 tbsp - - - - -	40	24	85	0	0	0.0	0.0	0.0
749	Table syrup (corn and maple)- -	2 tbsp - - - - -	42	25	122	0	0	0.0	0.0	0.0
	Vegetables and Vegetable Products								
750	Alfalfa seeds, sprouted, raw - - - -	1 cup - - - - - -	33	91	10	1	Tr	Tr	Tr	0.1
751	Artichokes, globe or French, cooked, drained- - - - - - - - - -	1 artichoke - -	120	87	55	3	Tr	Tr	Tr	0.1
	Asparagus, green:								
	Cooked, drained:								
	From raw:								
752	Cuts and tips - - - - - - - -	1 cup - - - - - -	180	92	45	5	1	0.1	Tr	0.2
753	Spears, 1/2-in diam. at base - - - - - - - - - - -	4 spears - - - -	60	92	15	2	Tr	Tr	Tr	0.1
	From frozen:								
754	Cuts and tips - - - - - - - -	1 cup - - - - - -	180	91	50	5	1	0.2	Tr	0.3
755	Spears, 1/2-in diam. at base - - - - - - - - - - -	4 spears - - - -	60	91	15	2	Tr	0.1	Tr	0.1
756	Canned, spears, 1/2-in diam. at base - - - - - - - - - - -	4 spears - - - -	80	95	10	1	Tr	Tr	Tr	0.1
757	Bamboo shoots, canned, drained	1 cup - - - - - -	131	94	25	2	1	0.1	Tr	0.2
	Beans:								
	Lima, immature seeds, frozen, cooked, drained:								
758	Thick-seeded types (Ford-hooks) - - - - - - - - - - - -	1 cup - - - - - -	170	74	170	10	1	0.1	Tr	0.3
759	Thin-seeded types (baby limas) - - - - - - - - - - - -	1 cup - - - - - -	180	72	190	12	1	0.1	Tr	0.3
	Snap:								
	Cooked, drained:								
760	From raw (cut and French style) - - - - - - - - - - - -	1 cup - - - - - -	125	89	45	2	Tr	0.1	Tr	0.2
761	From frozen (cut) - - - - - -	1 cup - - - - - -	135	92	35	2	Tr	Tr	Tr	0.1
762	Canned, drained solids (cut)	1 cup - - - - - -	135	93	25	2	Tr	Tr	Tr	0.1
	Beans, mature. See Beans, dry (items 527–535) and Black-eyed peas, dry (item 536).								
	Bean sprouts (mung):								
763	Raw - - - - - - - - - - - - - - - -	1 cup - - - - - -	104	90	30	3	Tr	Tr	Tr	0.1
764	Cooked, drained - - - - - - - -	1 cup - - - - - -	124	93	25	3	Tr	Tr	Tr	Tr
	Beets:								
	Cooked, drained:								
765	Diced or sliced - - - - - - - -	1 cup - - - - - -	170	91	55	2	Tr	Tr	Tr	Tr
766	Whole beets, 2-in diam. - -	2 beets - - - - -	100	91	30	1	Tr	Tr	Tr	Tr

[56] For regular pack; special dietary pack contains 3 mg sodium.

[57] For green varieties: yellow varieties contain 101 IU or 10 RE.

[58] For green varieties: yellow varieties contain 151 IU or 15 RE.

Choles-terol	Carbo-hydrate	Calcium	Phos-phorus	Iron	Potas-sium	Sodium	Vitamin A Value (IU)	Vitamin A Value (RE)	Thiamin	Ribo-flavin	Niacin	Ascorbic Acid	Item No.
Milli-grams	Grams	Milli-grams	Milli-grams	Milli-grams	Milli-grams	Milli-grams	Inter-national units	Retinol equiva-lents	Milli-grams	Milli-grams	Milli-grams	Milli-grams	
0	100	1	Tr	Tr	4	2	0	0	0.00	0.00	0.0	0	745
0	22	6	49	0.8	85	36	Tr	Tr	Tr	0.02	0.1	0	746
0	21	38	60	0.5	82	42	40	13	0.02	0.08	0.1	0	747
0	22	274	34	10.1	1,171	38	0	0	0.04	0.08	0.8	0	748
0	32	1	4	Tr	7	19	0	0	0.00	0.00	0.0	0	749
0	1	11	23	0.3	26	2	50	5	0.03	0.04	0.2	3	750
0	12	47	72	1.6	316	79	170	17	0.07	0.06	0.7	9	751
0	8	43	110	1.2	558	7	1,490	149	0.18	0.22	1.9	49	752
0	3	14	37	0.4	186	2	500	50	0.06	0.07	0.6	16	753
0	9	41	99	1.2	392	7	1,470	147	0.12	0.19	1.9	44	754
0	3	14	33	0.4	131	2	490	49	0.04	0.06	0.6	15	755
0	2	11	30	0.5	122	[56]278	380	38	0.04	0.07	0.7	13	756
0	4	10	33	0.4	105	9	10	1	0.03	0.03	0.2	1	757
0	32	37	107	2.3	694	90	320	32	0.13	0.10	1.8	22	758
0	35	50	202	3.5	740	52	300	30	0.13	0.10	1.4	10	759
0	10	58	49	1.6	374	4	[57]830	[57]83	0.09	0.12	0.8	12	760
0	8	61	32	1.1	151	18	[58]710	[58]71	0.06	0.10	0.6	11	761
0	6	35	26	1.2	147	[59]339	[60]470	[60]47	0.02	0.08	0.3	6	762
0	6	14	56	0.9	155	6	20	2	0.09	0.13	0.8	14	763
0	5	15	35	0.8	125	12	20	2	0.06	0.13	1.0	14	764
0	11	19	53	1.1	530	83	20	2	0.05	0.02	0.5	9	765
0	7	11	31	0.6	312	49	10	1	0.03	0.01	0.3	6	766

[59] For regular pack; special dietary pack contains 3 mg sodium.

[60] For green varieties; yellow varieties contain 142 IU or 14 RE.

Item No.	Foods, Approximate Measures, Units, and Weight (Weight of Edible Portion Only)		Water	Food Energy	Protein	Fat	Fatty Acids		
							Saturated	Mono-unsaturated	Poly-unsaturated
		Grams	Percent	Calories	Grams	Grams	Grams	Grams	Grams
	Vegetables and Vegetable Products *(continued)*								
	Beets:								
767	Canned, drained solids, diced or sliced - - - - - - - - - - - - - 1 cup - - - - -	170	91	55	2	Tr	Tr	Tr	0.1
768	Beet greens, leaves and stems, cooked, drained - - - - - - - - - - 1 cup - - - - -	144	89	40	4	Tr	Tr	0.1	0.1
	Black-eyed peas, immature seeds, cooked and drained:								
769	From raw - - - - - - - - - - - - - 1 cup - - - - -	165	72	180	13	1	0.3	0.1	0.6
770	From frozen - - - - - - - - - - - - 1 cup - - - - -	170	66	225	14	1	0.3	0.1	0.5
	Broccoli:								
771	Raw - - - - - - - - - - - - - - - - 1 spear - - - - -	151	91	40	4	1	0.1	Tr	0.3
	Cooked, drained:								
	From raw:								
772	Spear, medium - - - - - - - 1 spear - - - -	180	90	50	5	1	0.1	Tr	0.2
773	Spears, cut into 1/2-in pieces - - - - - - - - - - - 1 cup - - - - -	155	90	45	5	Tr	0.1	Tr	0.2
	From frozen:								
774	Piece, 4-1/2 to 5 in long 1 piece - - - - -	30	91	10	1	Tr	Tr	Tr	Tr
775	Chopped - - - - - - - - - - - 1 cup - - - - -	185	91	50	6	Tr	Tr	Tr	0.1
	Brussels sprouts, cooked, drained:								
776	From raw, 7–8 sprouts, 1-1/4 to 1-1/2-in diam. - - - - 1 cup - - - - -	155	87	60	4	1	0.2	0.1	0.4
777	From frozen - - - - - - - - - - - 1 cup - - - - -	155	87	65	6	1	0.1	Tr	0.3
	Cabbage, common varieties:								
778	Raw, coarsely shredded or sliced - - - - - - - - - - - - - - 1 cup - - - - -	70	93	15	1	Tr	Tr	Tr	0.1
779	Cooked, drained - - - - - - - - - 1 cup - - - - -	150	94	30	1	Tr	Tr	Tr	0.2
	Cabbage, Chinese:								
780	Pak-choi, cooked, drained - - - 1 cup - - - - -	170	96	20	3	Tr	Tr	Tr	0.1
781	Pe-tsai, raw, 1-in pieces - - - - 1 cup - - - - -	76	94	10	1	Tr	Tr	Tr	0.1
782	Cabbage, red, raw, coarsely shredded or sliced - - - - - - - - 1 cup - - - - -	70	92	20	1	Tr	Tr	Tr	0.1
783	Cabbage, savoy, raw, coarsely shredded or sliced - - - - - - - - 1 cup - - - - -	70	91	20	1	Tr	Tr	Tr	Tr
	Carrots:								
	Raw, without crowns and tips, scraped:								
784	Whole, 7-1/2 by 1-1/8 in, or strips, 2-1/2 to 3 in 1 carrot or 18 long strips - - - - - -	72	88	30	1	Tr	Tr	Tr	0.1
785	Grated - - - - - - - - - - - - - - 1 cup - - - - -	110	88	45	1	Tr	Tr	Tr	0.1
	Cooked, sliced, drained:								
786	From raw - - - - - - - - - - - - 1 cup - - - - -	156	87	70	2	Tr	0.1	Tr	0.1
787	From frozen - - - - - - - - - - - 1 cup - - - - -	146	90	55	2	Tr	Tr	Tr	0.1
788	Canned, sliced, drained solids 1 cup - - - - -	146	93	35	1	Tr	0.1	Tr	0.1
	Cauliflower:								
789	Raw, (flowerets) - - - - - - - - - - 1 cup - - - - -	100	92	25	2	Tr	Tr	Tr	0.1
	Cooked, drained:								
790	From raw (flowerets) - - - - - - 1 cup - - - - -	125	93	30	2	Tr	Tr	Tr	0.1

[61] For regular pack; special dietary pack contains 78 mg sodium.

[62] For regular pack; special dietary pack contains 61 mg sodium.

Choles-terol	Carbo-hydrate	Calcium	Phos-phorus	Iron	Potas-sium	Sodium	Vitamin A Value		Thiamin	Ribo-flavin	Niacin	Ascorbic Acid	Item No.
							(IU)	(RE)					
Milli-grams	Grams	Milli-grams	Milli-grams	Milli-grams	Milli-grams	Milli-grams	Inter-national units	Retinol equiva-lents	Milli-grams	Milli-grams	Milli-grams	Milli-grams	
0	12	26	29	3.1	252	[61]466	20	2	0.02	0.07	0.3	7	767
0	8	164	59	2.7	1,309	347	7,340	734	0.17	0.42	0.7	36	768
0	30	46	196	2.4	693	7	1,050	105	0.11	0.18	1.8	3	769
0	40	39	207	3.6	638	9	130	13	0.44	0.11	1.2	4	770
0	8	72	100	1.3	491	41	2,330	233	0.10	0.18	1.0	141	771
0	10	205	86	2.1	293	20	2,540	254	0.15	0.37	1.4	113	772
0	9	177	74	1.8	253	17	2,180	218	0.13	0.32	1.2	97	773
0	2	15	17	0.2	54	7	570	57	0.02	0.02	0.1	12	774
0	10	94	102	1.1	333	44	3,500	350	0.10	0.15	0.8	74	775
0	13	56	87	1.9	491	33	1,110	111	0.17	0.12	0.9	96	776
0	13	37	84	1.1	504	36	910	91	0.16	0.18	0.8	71	777
0	4	33	16	0.4	172	13	90	9	0.04	0.02	0.2	33	778
0	7	50	38	0.6	308	29	130	13	0.09	0.08	0.3	36	779
0	3	158	49	1.8	631	58	4,370	437	0.05	0.11	0.7	44	780
0	2	59	22	0.2	181	7	910	91	0.03	0.04	0.3	21	781
0	4	36	29	0.3	144	8	30	3	0.04	0.02	0.2	40	782
0	4	25	29	0.3	161	20	700	70	0.05	0.02	0.2	22	783
0	7	19	32	0.4	233	25	20,250	2,025	0.07	0.04	0.7	7	784
0	11	30	48	0.6	355	39	30,940	3,094	0.11	0.06	1.0	10	785
0	16	48	47	1.0	354	103	38,300	3,830	0.05	0.09	0.8	4	786
0	12	41	38	0.7	231	86	25,850	2,585	0.04	0.05	0.6	4	787
0	8	37	35	0.9	261	[62]352	20,110	2,011	0.03	0.04	0.8	4	788
0	5	29	46	0.6	355	15	20	2	0.08	0.06	0.6	72	789
0	6	34	44	0.5	404	8	20	2	0.08	0.07	0.7	69	790

Item No.	Foods, Approximate Measures, Units, and Weight (Weight of Edible Portion Only)		Water	Food Energy	Protein	Fat	Fatty Acids		
							Saturated	Mono-unsaturated	Poly-unsaturated
		Grams	Percent	Calories	Grams	Grams	Grams	Grams	Grams
	Vegetables and Vegetable Products *(continued)*								
	Cauliflower:								
	Cooked, drained:								
791	From frozen (flowerets) - - - - 1 cup - - - - -	180	94	35	3	Tr	0.1	Tr	0.2
	Celery, pascal type, raw:								
	Stalk, large outer, 8 b 1-1/2								
	in (at root end) - - - - - - - - - 1 stalk - - - - -	40	95	5	Tr	Tr	Tr	Tr	Tr
793	Pieces, diced - - - - - - - - - - - - 1 cup - - - - -	120	95	20	1	Tri	Tr	Tr	0.1
	Collards, cooked, drained:								
794	From raw (leaves without stems) 1 cup - - - - - -	190	96	25	2	Tr	0.1	Tr	0.2
795	From frozen (chopped)- - - - - - 1 cup - - - - - -	170	88	60	5	1	0.1	0.1	0.4
	Corn, sweet:								
	Cooked drained:								
796	From raw, ear 5 by 1-3/4 in. 1 ear - - - - - -	77	70	85	3	1	0.2	0.3	0.5
	From frozen:								
797	Ear, trimmed to about								
	3-1/2 in long - - - - - - - 1 ear - - - - - -	63	73	60	2	Tr	0.1	0.1	0.2
798	Kernels - - - - - - - - - - - - 1 cup - - - - -	165	76	135	5	Tr	Tr	Tr	0.1
	Canned:								
799	Cream style - - - - - - - - - - - 1 cup - - - - -	256	79	185	4	1	0.2	0.3	0.5
800	Whole kernel, vacuum pack - 1 cup - - - - - -	210	77	165	5	1	0.2	0.3	0.5
	Cowpeas. See Black-eyed peas, immature (items 769,770), mature (item 536).								
801	Cucumber, with peel, slices, 1/8 in thick (large, 2-1/8-in diam.; small, 1-3/4-in diam.) 6 large or 8 small slices - -	28	96	5	Tr	Tr	Tr	Tr	Tr
802	Dandelion greens, cooked, drained 1 cup - - - - - -	105	90	35	2	1	0.1	Tr	0.3
803	Eggplant, cooked, steamed - - - - 1 cup - - - - - -	96	92	25	1	Tr	Tr	Tr	0.1
804	Endive, curly (including escarole), raw, small pieces - - - - - - - - - 1 cup - - - - - -	50	94	10	1	Tr	Tr	Tr	Tr
805	Jerusalem-artichoke, raw, sliced 1 cup - - - - - -	150	78	115	3	Tr	0.0	Tr	Tr
	Kale, cooked, drained:								
806	From raw, chopped - - - - - - - 1 cup - - - - - -	130	91	40	2	1	0.1	Tr	0.3
807	From frozen, chopped - - - - - - 1 cup - - - - - -	130	91	40	4	1	0.1	Tr	0.3
808	Kohlrabi, thickened bulb-like stems, cooked, drained, diced 1 cup - - - - - -	165	90	50	3	Tr	Tr	Tr	0.1
	Lettuce, raw:								
	Butterhead, as Boston types:								
809	Head, 5-in diam - - - - - - - 1 head - - - - -	163	96	20	2	Tr	Tr	Tr	0.2
810	Leaves - - - - - - - - - - - - - - 1 outer or 2 inner leaves - -	15	96	Tr	Tr	Tr	Tr	Tr	Tr
	Crisphead, as iceberg:								
811	Head, 6-in diam - - - - - - - 1 head - - - - -	539	96	70	5	1	0.1	Tr	0.5
812	Wedge, 1/4 of head - - - - - - 1 wedge - - - -	135	96	20	1	Tr	Tr	Tr	0.1
813	Pieces, chopped or shredded - - - - - - - - - - - - 1 cup - - - - - -	55	96	5	1	Tr	Tr	Tr	0.1
814	Looseleaf (bunching varieties including romaine or cos), chopped or shredded pieces - - - - - - - - - - - - - 1 cup - - - - - -	56	94	10	1	Tr	Tr	Tr	0.1

[63] For yellow varieties; white varieties contain only a trace of vitamin A.

Choles-terol	Carbo-hydrate	Calcium	Phos-phorus	Iron	Potas-sium	Sodium	Vitamin A Value		Thiamin	Ribo-flavin	Niacin	Ascorbic Acid	Item No.
							(IU)	(RE)					
Milli-grams	Grams	Milli-grams	Milli-grams	Milli-grams	Milli-grams	Milli-grams	Inter-national units	Retinol equiva-lents	Milli-grams	Milli-grams	Milli-grams	Milli-grams	
0	7	31	43	0.7	250	32	40	4	0.07	0.10	0.6	56	791
0	1	14	10	0.2	114	35	50	5	0.01	0.01	0.1	3	792
0	4	43	31	0.6	341	106	150	15	0.04	0.04	0.4	8	793
0	5	148	19	0.8	177	36	4,220	422	0.03	0.08	0.4	19	794
0	12	357	46	1.9	427	85	10,170	1,017	0.08	0.20	1.1	45	795
0	19	2	79	0.5	192	13	[63]170	[63]17	0.17	0.06	1.2	5	796
0	14	2	47	0.4	158	3	[63]130	[63]13	0.11	0.04	1.0	3	797
0	34	3	78	0.5	229	8	[63]410	[63]41	0.11	0.12	2.1	4	798
0	46	8	131	1.0	343	[64]730	[63]250	[63]25	0.06	0.14	2.5	12	799
0	41	11	134	0.9	391	[65]571	[63]510	[63]51	0.09	0.15	2.5	17	800
0	1	4	5	0.1	42	1	10	1	0.01	0.01	0.1	1	801
0	7	147	44	1.9	244	46	12,290	1,229	0.14	0.18	0.5	19	802
0	6	6	21	0.3	238	3	60	6	0.07	0.02	0.6	1	803
0	2	26	14	0.4	157	11	1,030	103	0.04	0.04	0.2	3	804
0	26	21	117	5.1	644	6	30	3	0.30	0.09	2.0	6	805
0	7	94	36	1.2	296	30	9,620	962	0.07	0.09	0.7	53	806
0	7	179	36	1.2	417	20	8,260	826	0.06	0.15	0.9	33	807
0	11	41	74	0.7	561	35	60	6	0.07	0.03	0.6	89	808
0	4	52	38	0.5	419	8	1,580	158	0.10	0.10	0.5	13	809
0	Tr	5	3	Tr	39	1	150	15	0.01	0.01	Tr	1	810
0	11	102	108	2.7	852	49	1,780	178	0.25	0.16	1.0	21	811
0	3	26	27	0.7	213	12	450	45	0.06	0.04	0.3	5	812
0	1	10	11	0.3	87	5	180	18	0.03	0.02	0.1	2	813
0	2	38	14	0.8	148	5	1,060	106	0.03	0.04	0.2	10	814

[64] For regular pack; special dietary pack contains 8 mg sodium.

[65] For regular pack; special dietary pack contains 6 mg sodium.

Item No.	Foods, Approximate Measures, Units, and Weight (Weight of Edible Portion Only)		Water	Food Energy	Protein	Fat	Fatty Acids		
							Saturated	Mono-unsaturated	Poly-unsaturated
			Grams	Percent	Calories	Grams	Grams	Grams	Grams
	Vegetables and Vegetable Products *(continued)*								
	Mushrooms:								
815	Raw, sliced or chopped - - - - -	1 cup - - - - - -	70	92	20	1	Tr	Tr	Tr	0.1
816	Cooked, drained - - - - - - - - -	1 cup - - - - - -	156	91	40	3	1	0.1	Tr	0.3
817	Canned, drained solids - - - - -	1 cup - - - - - -	156	91	35	3	Tr	0.1	Tr	0.2
818	Mustard greens, without stems and midribs, cooked, drained - - - - - - - - - - - - - -	1 cup - - - - - -	140	94	20	3	Tr	Tr	0.2	0.1
819	Okra pods, 3 by 5/8 in, cooked -	8 pods - - - - -	85	90	25	2	Tr	Tr	Tr	Tr
	Onions:								
	Raw:								
820	Chopped - - - - - - - - - - - -	1 cup - - - - - -	160	91	55	2	Tr	0.1	0.1	0.2
821	Sliced - - - - - - - - - - - - - -	1 cup - - - - - -	115	91	40	1	Tr	0.1	Tr	0.1
822	Cooked (whole or sliced), drained	1 cup - - - - - -	210	92	60	2	Tr	0.1	Tr	0.1
823	Onions, spring, raw, bulb (3/8-in diam.) and white portion of top	6 onions - - - -	30	92	10	1	Tr	Tr	Tr	Tr
824	Onion rings, breaded, par-fried, frozen, prepared - - - - - - - - -	2 rings - - - - -	20	29	80	1	5	1.7	2.2	1.0
	Parsley:								
825	Raw - - - - - - - - - - - - - - - -	10 sprigs - - -	10	88	5	Tr	Tr	Tr	Tr	Tr
826	Freeze-dried - - - - - - - - - - -	1 tbsp - - - - -	0.4	2	Tr	Tr	Tr	Tr	Tr	Tr
827	Parsnips, cooked (diced or 2 in lengths), drained - - - - - - - - -	1 cup - - - - - -	156	78	125	2	Tr	0.1	0.2	0.1
828	Peas, edible pod, cooked, drained	1 cup - - - - - -	160	89	65	5	Tr	0.1	Tr	0.2
	Peas, green:								
829	Canned, drained solids - - - - -	1 cup - - - - - -	170	82	115	8	1	0.1	0.1	0.3
830	Frozen, cooked, drained - - - - -	1 cup - - - - - -	160	80	125	8	Tr	0.1	Tr	0.2
	Peppers:								
831	Hot chili, raw - - - - - - - - - - -	1 pepper - - - -	45	88	20	1	Tr	Tr	Tr	Tr
	Sweet (about 5 per 1b, whole), stem and seeds removed:								
832	Raw - - - - - - - - - - - - - - - -	1 pepper - - - -	74	93	20	1	Tr	Tr	Tr	0.2
833	Cooked, drained - - - - - - - - -	1 pepper - - - -	73	95	15	Tr	Tr	Tr	Tr	0.1
	Potatoes, cooked:								
	Baked (about 2 per 1b, raw):								
834	With skin - - - - - - - - - - - -	1 potato - - - -	202	71	220	5	Tr	0.1	Tr	0.1
835	Flesh only - - - - - - - - - - - -	1 potato - - - -	156	75	145	3	Tr	Tr	Tr	0.1
	Boiled (about 3 per 1b, raw):								
836	Peeled after boiling - - - - - -	1 potato - - - -	136	77	120	3	Tr	Tr	Tr	0.1
837	Peeled before boiling - - - - -	1 potato - - - -	135	77	115	2	Tr	Tr	Tr	0.1
	French fried, strip, 2 to 3-1/2 in long, frozen:								
838	Oven heated - - - - - - - - - -	10 strips - - - -	50	53	110	2	4	2.1	1.8	0.3
839	Fried in vegetable oil - - - - - -	10 strips - - - -	50	38	160	2	8	2.5	1.6	3.8
	Potato products, prepared:								
	Au gratin:								
840	From dry mix - - - - - - - - - -	1 cup - - - - - -	245	79	230	6	10	6.3	2.9	0.3
841	From home recipe - - - - - - -	1 cup - - - - - -	245	74	325	12	19	11.6	5.3	0.7
842	Hashed brown, from frozen - - -	1 cup - - - - - -	156	56	340	5	18	7.0	8.0	2.1

[66] For regular pack; special dietary pack contains 3 mg sodium.

[67] For red peppers; green peppers contain 350 IU or 35 RE.

[68] For green peppers; red peppers contain 4,220 IU or 422 RE.

Choles-terol	Carbo-hydrate	Calcium	Phos-phorus	Iron	Potas-sium	Sodium	Vitamin A Value		Thiamin	Ribo-flavin	Niacin	Ascorbic Acid	Item No.
							(IU)	(RE)					
Milli-grams	Grams	Milli-grams	Milli-grams	Milli-grams	Milli-grams	Milli-grams	Inter-national units	Retinol equiva-lents	Milli-grams	Milli-grams	Milli-grams	Milli-grams	
0	3	4	73	0.9	259	3	0	0	0.07	0.31	2.9	2	815
0	8	9	136	2.7	555	3	0	0	0.11	0.47	7.0	6	816
0	8	17	103	1.2	201	663	0	0	0.13	0.03	2.5	0	817
0	3	104	57	1.0	283	22	4,240	424	0.06	0.09	0.6	35	818
0	6	54	48	0.4	274	4	490	49	0.11	0.05	0.7	14	819
0	12	40	46	0.6	248	3	0	0	0.10	0.02	0.2	13	820
0	8	29	33	0.4	178	2	0	0	0.07	0.01	0.1	10	821
0	13	57	48	0.4	319	17	0	0	0.09	0.02	0.2	12	822
0	2	18	10	0.6	77	1	1,500	150	0.02	0.04	0.1	14	823
0	8	6	16	0.3	26	75	50	5	0.06	0.03	0.7	Tr	824
0	1	13	4	0.6	54	4	520	52	0.01	0.01	0.1	9	825
0	Tr	1	2	0.2	25	2	250	25	Tr	0.01	Tr	1	826
0	30	58	108	0.9	573	16	0	0	0.13	0.08	1.1	20	827
0	11	67	88	3.2	384	6	210	21	0.20	0.12	0.9	77	828
0	21	34	114	1.6	294	[66]372	1,310	131	0.21	0.13	1.2	16	829
0	23	38	144	2.5	269	139	1,070	107	0.45	0.16	2.4	16	830
0	4	8	21	0.5	153	3	[67]4,840	[67]484	0.04	0.04	0.4	109	831
0	4	4	16	0.9	144	2	[68]390	[68]39	0.06	0.04	0.4	[69]95	832
0	3	3	11	0.6	94	1	[70]280	[70]28	0.04	0.03	0.3	[71]81	833
0	51	20	115	2.7	844	16	0	0	0.22	0.07	3.3	26	834
0	34	8	78	0.5	610	8	0	0	0.16	0.03	2.2	20	835
0	27	7	60	0.4	515	5	0	0	0.14	0.03	2.0	18	836
0	27	11	54	0.4	443	7	0	0	0.13	0.03	1.8	10	837
0	17	5	43	0.7	229	16	0	0	0.06	0.02	1.2	5	838
0	20	10	47	0.4	366	108	0	0	0.09	0.01	1.6	5	839
12	31	203	233	0.8	537	1,076	520	76	0.05	0.20	2.3	8	840
56	28	292	277	1.6	970	1,061	650	93	0.16	0.28	2.4	24	841
0	44	23	112	2.4	680	53	0	0	0.17	0.03	3.8	10	842

[69] For green peppers; red peppers contain 141 mg ascorbic acid.

[70] For green peppers; red peppers contain 2,740 IU or 274 RE.

[71] For green peppers; red peppers contain 121 mg ascorbic acid.

(Tr indicates nutrient present in trace amount.)

Item No.	Foods, Approximate Measures, Units, and Weight (Weight of Edible Portion Only)		Water	Food Energy	Protein	Fat	Fatty Acids		
							Saturated	Mono-unsaturated	Poly-unsaturated
		Grams	Percent	Calories	Grams	Grams	Grams	Grams	Grams
	Vegetables and Vegetable Products								
	(continued)								
	Potato products, prepared:								
	Mashed:								
	From home recipe:								
843	Milk added - - - - - - - - - 1 cup - - - - - -	210	78	160	4	1	0.7	0.3	0.1
844	Milk and margarine added 1 cup - - - - - -	210	76	225	4	9	2.2	3.7	2.5
845	From dehydrated flakes (without milk), water, milk, butter, and salt added - - - 1 cup - - - - - -	210	76	235	4	12	7.2	3.3	0.5
846	Potato salad, made with mayonnaise - - - - - - - - - - - 1 cup - - - - - -	250	76	360	7	21	3.6	6.2	9.3
	Scalloped:								
847	From dry mix - - - - - - - - - - 1 cup - - - - - -	245	79	230	5	11	6.5	3.0	0.5
848	From home recipe - - - - - - - 1 cup - - - - - -	245	81	210	7	9	5.5	2.5	0.4
849	Potato chips - - - - - - - - - - - - - 10 chips - - - -	20	3	105	1	7	1.8	1.2	3.6
	Pumpkin:								
850	Cooked from raw, mashed - - - 1 cup - - - - - -	245	94	50	2	Tr	0.1	Tr	Tr
851	Canned - - - - - - - - - - - - - - 1 cup - - - - - -	245	90	85	3	1	0.4	0.1	Tr
852	Radishes, raw, stem ends, rootletscut off - - - - - - - - - - - 4 radishes - - -	18	95	5	Tr	Tr	Tr	Tr	Tr
853	Sauerkraut, canned, solids and liquid - - - - - - - - - - - - - - - 1 cup - - - - - -	236	93	45	2	Tr	0.1	Tr	0.1
	Seaweed:								
854	Kelp, raw - - - - - - - - - - - - - - 1 oz - - - - - - -	28	82	10	Tr	Tr	0.1	Tr	Tr
855	Spirulina, dried - - - - - - - - - - 1 oz - - - - - - -	28	5	80	16	2	0.8	0.2	0.6
	Southern peas. See Black-eyed peas, immature (items 769, 770), mature (item 536).								
	Spinach:								
856	Raw, chopped - - - - - - - - - - - 1 cup - - - - - -	55	92	10	2	Tr	Tr	Tr	0.1
	Cooked, drained:								
857	From raw - - - - - - - - - - - - 1 cup - - - - - -	180	91	40	5	Tr	0.1	Tr	0.2
858	From frozen (leaf) - - - - - - - 1 cup - - - - - -	190	90	55	6	Tr	0.1	Tr	0.2
859	Canned, drained solids - - - - - 1 cup - - - - - -	214	92	50	6	1	0.2	Tr	0.4
860	Spinach souffle - - - - - - - - - - - 1 cup - - - - - -	136	74	220	11	18	7.1	6.8	3.1
	Squash, cooked:								
861	Summer (all varieties), sliced, drained - - - - - - - - - - - - - 1 cup - - - - - -	180	94	35	2	1	0.1	Tr	0.2
862	Winter (all varieties), baked, cubes - - - - - - - - - - - - - - - 1 cup - - - - - -	205	89	80	2	1	0.3	0.1	0.5
	Sunchoke. See Jerusalem-artichoke (item 805).								
	Sweet potatoes:								
	Cooked (raw, 5 by 2 in; about 2-1/2 per lb):								
863	Baked in skin, peeled - - - - - 1 potato - - - -	114	73	115	2	Tr	Tr	Tr	0.1
864	Boiled, without skin - - - - - - 1 potato - - - -	151	73	160	2	Tr	0.1	Tr	0.2
865	Candied, 2-1/2 by 2-in piece - - - - - - - - - - - - - 1 piece - - - - -	105	67	145	1	3	1.4	0.7	0.2
	Canned:								
866	Solid pack (mashed) - - - - - - 1 cup - - - - - -	255	74	260	5	1	0.1	Tr	0.2

[1] Value not determined.

[72] With added salt; if none is added, sodium content is 58 mg.

Choles-terol	Carbo-hydrate	Calcium	Phos-phorus	Iron	Potas-sium	Sodium	Vitamin A Value (IU)	(RE)	Thiamin	Ribo-flavin	Niacin	Ascorbic Acid	Item No.
Milli-grams	Grams	Milli-grams	Milli-grams	Milli-grams	Milli-grams	Milli-grams	Inter-national units	Retinol equiva-lents	Milli-grams	Milli-grams	Milli-grams	Milli-grams	
4	37	55	101	0.6	628	636	40	12	0.18	0.08	2.3	14	843
4	35	55	97	0.5	607	620	360	42	0.18	0.08	2.3	13	844
29	32	103	118	0.5	489	697	380	44	0.23	0.11	1.4	20	845
170	28	48	130	1.6	635	1,323	520	83	0.19	0.15	2.2	25	846
27	31	88	137	0.9	497	835	360	51	0.05	0.14	2.5	8	847
29	26	140	154	1.4	926	821	330	47	0.17	0.23	2.6	26	848
0	10	5	31	0.2	260	94	0	0	0.03	Tr	0.8	8	849
0	12	37	74	1.4	564	2	2,650	265	0.08	0.19	1.0	12	850
0	20	64	86	3.4	505	12	54,040	5,404	0.06	0.13	0.9	10	851
0	1	4	3	0.1	42	4	Tr	Tr	Tr	0.01	0.1	4	852
0	10	71	47	3.5	401	1,560	40	4	0.05	0.05	0.3	35	853
0	3	48	12	0.8	25	66	30	3	0.01	0.04	0.1	(¹)	854
0	7	34	33	8.1	386	297	160	16	0.67	1.04	3.6	3	855
0	2	54	27	1.5	307	43	3,690	369	0.04	0.10	0.4	15	856
0	7	245	101	6.4	839	126	14,740	1,474	0.17	0.42	0.9	18	857
0	10	277	91	2.9	566	163	14,790	1,479	0.11	0.32	0.8	23	858
0	7	272	94	4.9	740	[72]683	18,780	1,878	0.03	0.30	0.8	31	859
184	3	230	231	1.3	201	763	3,460	675	0.09	0.30	0.5	3	860
0	8	49	70	0.6	346	2	520	52	0.08	0.07	0.9	10	861
0	18	29	41	0.7	896	2	7,290	729	0.17	0.05	1.4	20	862
0	28	32	63	0.5	397	11	24,880	2,488	0.08	0.14	0.7	28	863
0	37	32	41	0.8	278	20	25,750	2,575	0.08	0.21	1.0	26	864
8	29	27	27	1.2	198	74	4,400	440	0.02	0.04	0.4	7	865
0	59	77	133	3.4	536	191	38,570	3,857	0.07	0.23	2.4	13	866

(Tr indicates nutrient present in trace amount.)

Item No.	Foods, Approximate Measures, Units, and Weight (Weight of Edible Portion Only)		Water	Food Energy	Protein	Fat	Fatty Acids		
							Saturated	Mono-unsaturated	Poly-unsaturated
			Grams	Percent	Calories	Grams	Grams	Grams	Grams
	Vegetables and Vegetable Products								
	(continued)								
	Sweet potatoes:								
	Canned:								
867	Vacuum pack, piece 2-3/4 by 1 in- - - - - - - - - - - - -	1 piece - - - - -	40	76	35	1	Tr	Tr	Tr	Tr
	Tomatoes:								
868	Raw, 2-3/5-in diam. (3 per 12 oz pkg.) - - - - - - - - - - -	1 tomato - - - -	123	94	25	1	Tr	Tr	Tr	0.1
869	Canned, solids and liquid - - - -	1 cup - - - - - -	240	94	50	2	1	0.1	0.1	0.2
870	Tomato juice, canned - - - - - - - -	1 cup - - - - - -	244	94	40	2	Tr	Tr	Tr	0.1
	Tomato products, canned:								
871	Paste - - - - - - - - - - - - - - - -	1 cup - - - - - -	262	74	220	10	2	0.3	0.4	0.9
872	Purée - - - - - - - - - - - - - - - -	1 cup - - - - - -	250	87	105	4	Tr	Tr	Tr	0.1
873	Sauce - - - - - - - - - - - - - - - -	1 cup - - - - - -	245	89	75	3	Tr	0.1	0.1	0.2
874	Turnips, cooked, diced - - - - - - -	1 cup - - - - - -	156	94	30	1	Tr	Tr	Tr	0.1
	Turnip greens, cooked, drained:								
875	From raw (leaves and stems) - -	1 cup - - - - - -	144	93	30	2	Tr	0.1	Tr	0.1
876	From frozen (chopped) - - - - - -	1 cup - - - - - -	164	90	50	5	1	0.2	Tr	0.3
877	Vegetable juice cocktail, canned	1 cup - - - - - -	242	94	45	2	Tr	Tr	Tr	0.1
	Vegetables, mixed:								
878	Canned, drained solids - - - - -	1 cup - - - - - -	163	87	75	4	Tr	0.1	Tr	0.2
879	Frozen, cooked, drained - - - - -	1 cup - - - - - -	182	83	105	5	Tr	0.1	Tr	0.1
880	Waterchestnuts, canned - - - - - -	1 cup - - - - - -	140	86	70	1	Tr	Tr	Tr	Tr
	Miscellaneous Items								
	Baking powders for home use:								
	Sodium aluminum sulfate:								
881	With monocalcium phosphate monohydrate - -	1 tsp - - - - - -	3	2	5	Tr	0	0.0	0.0	0.0
882	With monocalcium phosphate monohydrate, calcium sulfate - - - - - - - - - - -	1 tsp - - - - - -	2.9	1	5	Tr	0	0.0	0.0	0.0
883	Straight phosphate - - - - - - - -	1 tsp - - - - - -	3.8	2	5	Tr	0	0.0	0.0	0.0
884	Low sodium - - - - - - - - - - - -	1 tsp - - - - - -	4.3	1	5	Tr	0	0.0	0.0	0.0
885	Catsup - - - - - - - - - - - - - - - -	1 cup - - - - - -	273	69	290	5	1	0.2	0.2	0.4
886		1 tbsp - - - - -	15	69	15	Tr	Tr	Tr	Tr	Tr
887	Celery seed - - - - - - - - - - - -	1 tsp - - - - - -	2	6	10	Tr	1	Tr	0.3	0.1
888	Chili powder - - - - - - - - - - - -	1 tsp - - - - - -	2.6	8	10	Tr	Tr	0.1	0.1	0.2
	Chocolate:								
889	Bitter or baking - - - - - - - - - -	1 oz - - - - - -	28	2	145	3	15	9.0	4.9	0.5
	Semisweet, see Candy, (item 715).								
890	Cinnamon - - - - - - - - - - - - - -	1 tsp - - - - - -	2.3	10	5	Tr	Tr	Tr	Tr	Tr
891	Curry powder - - - - - - - - - - - -	1 tsp - - - - - -	2	10	5	Tr	Tr	(¹)	(¹)	(¹)
892	Garlic powder - - - - - - - - - - - -	1 tsp - - - - - -	2.8	6	10	Tr	Tr	Tr	Tr	Tr
893	Gelatin, dry - - - - - - - - - - - - -	1 envelope - -	7	13	25	6	Tr	Tr	Tr	Tr

¹ Value not determined.

⁷³ For regular pack; special dietary pack contains 31 mg sodium.

⁷⁴ With added salt; if none is added, sodium content is 24 mg.

⁷⁵ With no added salt; if salt is added, sodium content is 2,070 mg.

⁷⁶ With no added salt; if salt is added, sodium content is 998 mg.

⁷⁷ With salt added.

Choles-terol	Carbo-hydrate	Calcium	Phos-phorus	Iron	Potas-sium	Sodium	Vitamin A Value (IU)	Vitamin A Value (RE)	Thiamin	Ribo-flavin	Niacin	Ascorbic Acid	Item No.
Milli-grams	Grams	Milli-grams	Milli-grams	Milli-grams	Milli-grams	Milli-grams	Inter-national units	Retinol equiva-lents	Milli-grams	Milli-grams	Milli-grams	Milli-grams	
0	8	9	20	0.4	125	21	3,190	319	0.01	0.02	0.3	11	867
0	5	9	28	0.6	255	10	1,390	139	0.07	0.06	0.7	22	868
0	10	62	46	1.5	530	[73]391	1,450	145	0.11	0.07	1.8	36	869
0	10	22	46	1.4	537	[74]881	1,360	136	0.11	0.08	1.6	45	870
0	49	92	207	7.8	2,442	[75]170	6,470	647	0.41	0.50	8.4	111	871
0	25	38	100	2.3	1,050	[76]50	3,400	340	0.18	0.14	4.3	88	872
0	18	34	78	1.9	909	[77]1,482	2,400	240	0.16	0.14	2.8	32	873
0	8	34	30	0.3	211	78	0	0	0.04	0.04	0.5	18	874
0	6	197	42	1.2	292	42	7,920	792	0.06	0.10	0.6	39	875
0	8	249	56	3.2	367	25	13,080	1,308	0.09	0.12	0.8	36	876
0	11	27	41	1.0	467	883	2,830	283	0.10	0.07	1.8	67	877
0	15	44	68	1.7	474	243	18,990	1,899	0.08	0.08	0.9	8	878
0	24	46	93	1.5	308	64	7,780	778	0.13	0.22	1.5	6	879
0	17	6	27	1.2	165	11	10	1	0.02	0.03	0.5	2	880
0	1	58	87	0.0	5	329	0	0	0.00	0.00	0.0	0	881
0	1	183	45	0.0	4	290	0	0	0.00	0.00	0.0	0	882
0	1	239	359	0.0	6	312	0	0	0.00	0.00	0.0	0	883
0	1	207	314	0.0	891	Tr	0	0	0.00	0.00	0.0	0	884
0	69	60	137	2.2	991	2,845	3,820	382	0.25	0.19	4.4	41	885
0	4	3	8	0.1	54	156	210	21	0.01	0.01	0.2	2	886
0	1	35	11	0.9	28	3	Tr	Tr	0.01	0.01	0.1	Tr	887
0	1	7	8	0.4	50	26	910	91	0.01	0.02	0.2	2	888
0	8	22	109	1.9	235	1	10	1	0.01	0.07	0.4	0	889
0	2	28	1	0.9	12	1	10	1	Tr	Tr	Tr	1	890
0	1	10	7	0.6	31	1	20	2	0.01	0.01	0.1	Tr	891
0	2	2	12	0.1	31	1	0	0	0.01	Tr	Tr	Tr	892
0	0	1	0	0.0	2	6	0	0	0.00	0.00	0.0	0	893

(Tr indicates nutrient present in trace amount.)

Item No.	Foods, Approximate Measures, Units, and Weight (Weight of Edible Portion Only)		Water	Food Energy	Protein	Fat	Fatty Acids		
							Saturated	Mono-unsaturated	Poly-unsaturated
			Grams	Percent	Calories	Grams	Grams	Grams	Grams	Grams
	Miscellaneous Items								
894	Mustard, prepared, yellow - - - - -	1 tsp or individual packet- - - -	5	80	5	Tr	Tr	Tr	0.2	Tr
	Olives, canned:									
895	Green - - - - - - - - - - - - - - -	4 medium or 3 extra large	13	78	15	Tr	2	0.2	1.2	0.1
896	Ripe, Mission, pitted - - - - - - -	3 small or 2 large - - - - -	9	73	15	Tr	2	0.3	1.3	0.2
897	Onion powder - - - - - - - - - - -	1 tsp - - - - - -	2.1	5	5	Tr	Tr	Tr	Tr	Tr
898	Oregano - - - - - - - - - - - - - -	1 tsp - - - - - -	1.5	7	5	Tr	Tr	Tr	Tr	0.1
899	Paprika - - - - - - - - - - - - - -	1 tsp - - - - - -	2.1	10	5	Tr	Tr	Tr	Tr	0.2
900	Pepper, black - - - - - - - - - - -	1 tsp - - - - - -	2.1	11	5	Tr	Tr	Tr	Tr	Tr
	Pickles, cucumber:									
901	Dill, medium, whole, 3-3/4 in long, 1-1/4-in diam. - - - - -	1 pickle - - - -	65	93	5	Tr	Tr	Tr	Tr	0.1
902	Fresh-pack, slices 1-1/2-in diam., 1/4 in thick - - - - - - -	2 slices- - - - -	15	79	10	Tr	Tr	Tr	Tr	Tr
903	Sweet, gherkin, small, whole, about 2-1/2 in long, 3/4-in diam. - - - - - - - - - -	1 pickle - - - -	15	61	20	Tr	Tr	Tr	Tr	Tr
	Popcorn. See Grain Products, (items 497–499).									
904	Relish, finely chopped, sweet - - -	1 tbsp - - - - -	15	63	20	Tr	Tr	Tr	Tr	Tr
905	Salt - - - - - - - - - - - - - - - - -	1 tsp - - - - - -	5.5	0	0	0	0	0.0	0.0	0.0
906	Vinegar, cider- - - - - - - - - - - -	1 tbsp - - - - -	15	94	Tr	Tr	0	0.0	0.0	0.0
	Yeast:									
907	Baker's, dry, active - - - - - - - -	1 pkg - - - - - -	7	5	20	3	Tr	Tr	0.1	Tr
908	Brewer's, dry - - - - - - - - - - -	1 tbsp - - - - -	8	5	25	3	Tr	Tr	Tr	0.0

Cholesterol	Carbohydrate	Calcium	Phosphorus	Iron	Potassium	Sodium	Vitamin A Value (IU)	Vitamin A Value (RE)	Thiamin	Riboflavin	Niacin	Ascorbic Acid	Item No.
Milligrams	Grams	Milligrams	Milligrams	Milligrams	Milligrams	Milligrams	International units	Retinol equivalents	Milligrams	Milligrams	Milligrams	Milligrams	
0	Tr	4	4	0.1	7	63	0	0	Tr	0.01	Tr	Tr	894
0	Tr	8	2	0.2	7	312	40	4	Tr	Tr	Tr	0	895
0	Tr	10	2	0.2	2	68	10	1	Tr	Tr	Tr	0	896
0	2	8	7	0.1	20	1	Tr	Tr	0.01	Tr	Tr	Tr	897
0	1	24	3	0.7	25	Tr	100	10	0.01	Tr	0.1	1	898
0	1	4	7	0.5	49	1	1,270	127	0.01	0.04	0.3	1	899
0	1	9	4	0.6	26	1	Tr	Tr	Tr	0.01	Tr	0	900
0	1	17	14	0.7	130	928	70	7	Tr	0.01	Tr	4	901
0	3	5	4	0.3	30	101	20	2	Tr	Tr	Tr	1	902
0	5	2	2	0.2	30	107	10	1	Tr	Tr	Tr	1	903
0	5	3	2	0.1	30	107	20	2	Tr	Tr	0.0	1	904
0	0	14	3	Tr	Tr	2,132	0	0	0.00	0.00	0.0	0	905
0	1	1	1	0.1	15	Tr	0	0	0.00	0.00	0.0	0	906
0	3	3	90	1.1	140	4	Tr	Tr	0.16	0.38	2.6	Tr	907
0	3	[78]17	140	1.4	152	10	Tr	Tr	1.25	0.34	3.0	Tr	908

[78] Value may vary from 6 to 60 mg.

Appendix B

1997–2000 Dietary Reference Intakes

1989 Recommended Dietary Allowances

1997–2000 Dietary Reference Intakes (DRI) Recommended Dietary Allowances (RDA)

Age (year)	Vitamin E (mg α-tocopherol)	Vitamin C (mg)	Thiamin (mg)	Riboflavin (mg)	Niacin (mg NE)	Vitamin B$_6$ (mg)	Folate (microgram DFE)	Vitamin B$_{12}$ (microgram)	Phosphorus (mg)	Magnesium (mg)	Selenium (microgram)
Infants[a]											
0–0.5	4	40	0.2	0.3	2	0.1	65	0.4	100	30	15
0.5–1.0	5	50	0.3	0.4	4	0.3	80	0.5	275	75	20
Children											
1–3	5	15	0.5	0.5	6	0.5	150	0.9	460	80	20
4–8	7	25	0.6	0.6	8	0.6	200	1.2	500	130	30
Males											
9–13	11	45	0.9	0.9	12	1.0	300	1.8	1250	240	40
14–18	15	75	1.2	1.3	16	1.3	400	2.4	1250	410	55
19–30	15	90	1.2	1.3	16	1.3	400	2.4	700	400	55
31–50	15	90	1.2	1.3	16	1.3	400	2.4	700	420	55
51–70	15	90	1.2	1.3	16	1.7	400	2.4	700	420	55
Over 70	15	90	1.2	1.3	16	1.7	400	2.4	700	420	55
Females											
9–13	11	45	0.9	0.9	12	1.0	300	1.8	1250	240	40
14–18	15	65	1.0	1.0	14	1.2	400	2.4	1250	360	55
19–30	15	75	1.1	1.1	14	1.3	400	2.4	700	310	55
31–50	15	75	1.1	1.1	14	1.3	400	2.4	700	320	55
51–70	15	75	1.1	1.1	14	1.5	400	2.4	700	320	55
Over 70	15	75	1.1	1.1	14	1.5	400	2.4	700	320	55
Pregnancy	*	+10	1.4	1.4	18	1.9	600	2.6	*	+40	+5
Lactation	+4	+45	1.5	1.6	17	2.0	500	2.8	*	*	+15

* The values for pregnancy or lactation do not change from the normal value for women of the same age.

[a] Values for infants are all Adequate Intakes (AI). The AI for niacin for 0.0–0.5 years is stated as milligrams of preformed niacin, not niacin equivalents.

1997–2000 Dietary Reference Intakes (DRI) Adequate Intakes (AI)

Age (Years)	Vitamin D (micrograms)	Pantothenic Acid (mg)	Biotin (micrograms)	Choline (mg)	Calcium (mg)	Fluoride (mg)
Infants[a]						
0.0–0.5	5	1.7	5	125	210	0.01
0.5–1.0	5	1.8	6	150	270	0.5
Children						
1–3	5	2.0	8	200	500	0.7
4–8	5	3.0	12	250	800	1.1
Males						
9–13	5	4.0	20	375	1300	2.0
14–18	5	5.0	25	550	1300	3.2
19–30	5	5.0	30	550	1000	3.8
31–50	5	5.0	30	550	1000	3.8
51–70	10	5.0	30	550	1200	3.8
Over 70	15	5.0	30	550	1200	3.8
Females						
9–13	5	4.0	20	375	1300	2.0
14–18	5	5.0	25	400	1300	2.9
19–30	5	5.0	30	425	1000	3.1
31–50	5	5.0	30	425	1000	3.1
51–70	10	5.0	30	425	1200	3.1
Over 70	15	5.0	30	425	1200	3.1
Pregnancy	*	6.0	30	450	*	*
Lactation	*	7.0	35	550	*	*

* The values for pregnancy or lactation do not change from the normal value for women of the same age.

[a] Values for infants are all Adequate Intakes (AI). The AI for niacin for 0.0–0.5 years is stated as milligrams of preformed niacin, not niacin equivalents.

1997–2000 Dietary Reference Intakes (DRI)
Tolerable Upper Intake Levels (UL)

Age (years)	Vitamin D (micrograms)	Vitamin E (mg α-tocopherol)	Vitamin C (mg)	Niacin (mg)	Vitamin B₆ (mg)	Folate (micrograms)	Choline (mg)	Calcium (mg)	Phosphorus (mg)	Magnesium (mg)	Fluoride (mg)	Selenium (microgram)
Infants												
0.0–0.5	25	*	*	*	*	*	*	*	*	*	0.7	45
0.5–1.0	25	*	*	*	*	*	*	*	*	*	0.9	60
Children												
1–3	50	200	400	10	30	300	1000	2500	3000	65	1.3	90
4–8	50	300	650	15	40	400	1000	2500	3000	110	2.2	150
9–13	50	600	1200	20	60	600	2000	2500	4000	350	10.0	280
14–18	50	800	1800	30	80	800	3000	2500	4000	350	10.0	400
Adults												
19–70	50	1000	2000	35	100	1000	3500	2500	4000	350	10.0	400
Over 70	50	1000	2000	35	100	1000	3500	2500	3000	350	10.0	400
Pregnancy	50	1000	2000	35	100	1000	3500	2500	3500	350	10.0	400
Lactation	50	1000	2000	35	100	1000	3500	2500	4000	350	10.0	400

* Upper levels were not established due to a lack of data for infants.

Used with permission from *Recommended Dietary Allowances*, ©1989 by the National Academy of Sciences, National Academy Press, Washington D.C., and the first three of the *Dietary Reference Intakes* series, © 1997, and 1998, and 2000 National Academy of Sciences. Courtesy of the National Academy Press, Washington D.C.

1989 Recommended Dietary Allowances (RDA)[a]

| Category | Age (years) or Condition | Energy (calories) | Protein (g) | Fat-Soluble Vitamins | | Minerals | | |
				Vitamin A (μg RE)[b]	Vitamin K (μg)	Iron (mg)	Zinc (mg)	Iodine (μg)
Infants	0.0–0.5	650	13	375	5	6	5	40
	0.5–1.0	850	14	375	10	10	5	50
Children	1–3	1,300	16	400	15	10	10	70
	4–6	1,800	24	500	20	10	10	90
	7–10	2,000	28	700	30	10	10	120
Males	11–14	2,500	45	1,000	45	12	15	150
	15–18	3,000	59	1,000	65	12	15	150
	19–24	2,900	58	1,000	70	10	15	150
	25–50	2,900	63	1,000	80	10	15	150
	51+	2,300	63	1,000	80	10	15	150
Females	11–14	2,200	46	800	45	15	12	150
	15–18	2,200	44	800	55	15	12	150
	19–24	2,200	46	800	60	15	12	150
	25–50	2,200	50	800	65	15	12	150
	51+	1,900	50	800	65	10	12	150
Pregnant		+300	60	800	65	30	15	175
Lactating	1st 6 months	+500	65	1,300	65	15	19	200
	2nd 6 months	+500	62	1,200	65	15	16	200

[a] The allowances, expressed as average daily intakes over time, are intended to provide for individual variations among most normal persons as they live in the United States under usual environmental stresses. Diets should be based on a variety of common foods in order to provide other nutrients for which human requirements have been less well defined.

[b] Retinol equivalent. 1 retinol equivalent = 1 μg retinol or 6 μg β = carotene.

1989 Estimated Safe and Adequate Daily Dietary Intakes of Selected Vitamins and Minerals[a]

		Trace Elements[b]			
Category	Age (years)	Copper (mg)	Manganese (mg)	Chromium (μg)	Molybdenum (μg)
Infants	0–0.5	0.4–0.6	0.3–0.6	10–40	15–30
	0.5–1	0.6–0.7	0.6–1.0	20–60	20–40
Children and	1–3	0.7–1.0	1.0–1.5	20–80	25–50
adolescents	4–6	1.0–1.5	1.5–2.0	30–120	30–75
	7–10	1.0–2.0	2.0–3.0	50–200	50–150
	11+	1.5–2.5	2.0–5.0	50–200	75–250
Adults		1.5–3.0	2.0–5.0	50–200	75–250

[a] Because there is less information on which to base allowances, these figures are not given in the main table of RDA and are provided here in the form of ranges of recommended intakes.

[b] Since the toxic levels for many trace elements may be only several times usual intakes, the upper levels for the trace elements given in this table should not be habitually exceeded.

1989 Estimated Sodium, Chloride, and Potassium Minimum Requirements of Healthy Persons[a]

Age	Sodium (mg)[a]	Chloride (mg)[a]	Potassium (mg)[b]
Months			
0–5	120	180	500
6–11	200	300	700
Years			
1	225	350	1,000
2–5	300	500	1,400
6–9	400	600	1,600
10–18	500	750	2,000
>18	500	750	2,000

[a] No allowance has been included for large, prolonged losses from the skin through sweat.

[b] Desirable intakes of potassium may considerably exceed these values (~3,500 mg for adults).

Expanded List of Serving Sizes for Food Guide Pyramid

Food Group	What Counts as a Serving (includes additional items)
BREAD, CEREAL, RICE, AND PASTA	**GENERALLY:** 1 slice of bread 1/2 hamburger or hot dog bun 1/2 English muffin or bagel 1 small roll, biscuit, or muffin (about 1 ounce each) 1/2 cup cooked cereal 1 ounce ready-to-eat cereal 1/2 cup cooked pasta or rice 5 to 6 small crackers (saltine size) 2 to 3 large crackers (graham cracker square size) **SPECIFICALLY:** 4-inch pita bread 3 medium hard bread sticks, about 4-3/4 inches long 9 animal crackers 1/4 cup uncooked rolled oats 2 tablespoons uncooked grits or cream of wheat cereal 1 oz uncooked pasta (1/4 cup macaroni or 3/4 cup noodles) 3 tablespoons uncooked rice 1 7-inch flour or corn tortilla 2 taco shells, corn 1 4-inch pancake 9 3-ring pretzels or 2 pretzel rods 1/16 of 2-layer cake 1/5 of 10-inch angel food cake 1/10 of 8-inch, 2-crust pie 4 small cookies 1/2 medium doughnut 1/2 large croissant

Food Group	What Counts as a Serving (includes additional items)
BREAD, CEREAL, RICE, AND PASTA *(continued)*	**SPECIFICALLY:** 3 rice or popcorn cakes 2 cups popcorn 12 tortilla chips
FRUITS	**GENERALLY:** a whole fruit (medium apple, banana, peach, or orange, or a small pear) grapefruit half melon wedge (1/4 of a medium cantaloupe or 1/8 of a medium honeydew) 3/4 cup juice (100% juice) 1/2 cup berries, cherries, or grapes 1/2 cup cut-up fresh fruit 1/2 cup cooked or canned fruit 1/2 cup frozen fruit 1/4 cup dried fruit **SPECIFICALLY:** 5 large strawberries 7 medium strawberries 50 blueberries 30 raspberries 11 cherries 12 grapes 1-1/2 medium plums 2 medium apricots 1 medium avocado 7 melon balls 1/2 cup fruit salad, such as Waldorf 1/2 medium mango 1/4 medium papaya 1 large kiwifruit 4 canned apricot halves with liquid 14 canned cherries with liquid 1-1/2 canned peach halves with liquid 2 canned pear halves with liquid 2-1/2 canned pineapple slices with liquid 3 canned plums with liquid 9 dried apricot halves 5 prunes
VEGETABLES	**GENERALLY:** 1/2 cup cooked vegetables 1/2 cup chopped raw vegetables 1 cup leafy raw vegetables, such as lettuce or spinach

Food Group	What Counts as a Serving (includes additional items)
VEGETABLES *(continued)*	**GENERALLY:** 1/2 cup tomato or spaghetti sauce 1/4 cup tomato paste 1/2 cup cooked dry beans (if not counted as a meat alternate) **SPECIFICALLY:** 3/4 cup vegetable juice 1 cup bean soup 1 cup vegetable soup Raw vegetables: 1 medium tomato or 5 cherry tomatoes 7 to 8 carrot or celery sticks 3 broccoli florets 1/3 medium cucumber 10 medium whole young green onions 8 green or red pepper rings 13 medium radishes 9 snow or sugar peas 6 slices summer squash (yellow or zucchini) 1 cup mixed green salad 1/2 cup cole slaw or potato salad Cooked vegetables: 2 spears broccoli 1-1/2 whole carrots 1 medium whole green or red pepper 1/3 summer squash (yellow and zucchini) 1 globe artichoke 6 asparagus spears 2 whole beets, about 2 inches in diameter 4 medium brussels sprouts 2 medium stalks of celery 1 medium ear of corn 7 medium mushrooms 8 okra pods 1 medium whole onion or 6 pearl onions 1 medium whole turnip 10 French fries 1 baked potato, medium 3/4 cup sweet potato
MEAT, POULTRY, FISH, EGGS, DRY BEANS, AND NUTS	**GENERALLY:** 2-3 ounces cooked lean meat without bone 2-3 ounces cooked poultry without skin or bone 2-3 ounces cooked fish without bone 2-3 ounces drained canned fish

Food Group	What Counts as a Serving (includes additional items)
MEAT, POULTRY, FISH, EGGS, DRY BEANS, AND NUTS *(continued)*	**GENERALLY:** Meat alternates (count as *1 ounce,* about 1/3 serving): 1 egg (yolk and white) 1/2 cup cooked dry beans (if not counted as a vegetable) 2 tablespoons peanut butter 1/4 cup seeds 1/3 cup nuts, such as walnuts, pecans, or peanuts 1/2 cup baked beans 1/2 cup tofu Meat/fish products (count as *1 ounce,* about 1/3 serving): 1 ounce lean ham or canadian bacon 1-1/2 frankfurters (10 per pound) 1 frankfurter (8 per pound) 2 ounces bologna (2 slices) 3 slices dry or hard salami 2 ounces liverwurst (2 large slices) 3 pork sausage links 5 canned vienna sausages 1/2 can meat spread (5.5 ounce can) 1/4 cup drained canned salmon or tuna 1/3 cup drained canned clams or crab meat 13 frozen fried breaded clams 4 Pacific oysters or 11 Atlantic oysters 4 medium fried breaded shrimp 1/4 cup drained canned lobster or shrimp
MILK, CHEESE, AND YOGURT	**GENERALLY:** 1 cup milk (skim, low-fat, and whole) 1 cup yogurt (all kinds) 1-1/2 ounces natural cheese 2 ounces process cheese **SPECIFICALLY:** 2 cups cottage cheese 1/2 cup ricotta cheese 1 cup frozen yogurt 1-1/2 cups ice cream

Source: "Using the Food Guide Pyramid: A Resource for Nutrition Educators" by A. Shaw, L. Fulton, C. Davis, and M. Hogbin. U.S. Department of Agriculture: Food, Nutrition, and Consumer Services; Center for Nutrition Policy and Promotion.

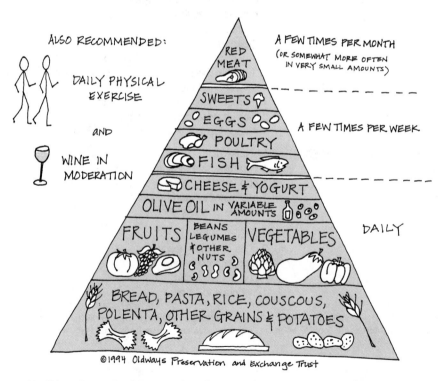

THE TRADITIONAL HEALTHY MEDITERRANEAN DIET PYRAMID

ALSO RECOMMENDED:

DAILY PHYSICAL EXERCISE

AND

WINE IN MODERATION

RED MEAT — A FEW TIMES PER MONTH (OR SOMEWHAT MORE OFTEN IN VERY SMALL AMOUNTS)

SWEETS
EGGS — A FEW TIMES PER WEEK
POULTRY
FISH
CHEESE & YOGURT
OLIVE OIL IN VARIABLE AMOUNTS

FRUITS | BEANS LEGUMES & OTHER NUTS | VEGETABLES — DAILY

BREAD, PASTA, RICE, COUSCOUS, POLENTA, OTHER GRAINS & POTATOES

©1994 Oldways Preservation and Exchange Trust

THE TRADITIONAL HEALTHY ASIAN DIET PYRAMID

REGULAR PHYSICAL ACTIVITY

DAILY:
SAKE, WINE, BEER & OTHER ALCOHOLIC BEVERAGES
(In Moderation and Primarily with Meals)
TEA

MONTHLY

MEAT

SWEETS

WEEKLY

EGGS & POULTRY

FISH & SHELLFISH OR DAIRY

OPTIONAL DAILY

VEGETABLE OILS

FRUITS

LEGUMES NUTS & SEEDS

VEGETABLES

DAILY

RICE, NOODLES, BREADS, MILLET, CORN & OTHER GRAINS

©1995 Oldways Preservation and Exchange Trust

The Healthy Traditional Latin American Diet Pyramid

Six glasses of water a day

Alcohol in moderation, with meals

Occasionally or in small quantities

Meat
Sweets
Eggs

Plant Oils

Fish

Milk Products

Shellfish

Poultry

Daily or less

Beans
Grains
Tubers
Nuts

Fruits

Vegetables

At every meal

Daily Physical Activity

© 1996 Oldways Preservation and Exchange Trust

Chapter 1

1. **Carbohydrate:** Provide energy
   Lipid: Provide energy
   Protein: Provide energy, work as main structure of cells
   Vitamins: Work as helpers
   Minerals: Work as helpers (also some minerals work as the main structure of bone cells)
   Water: Supplies the medium in which chemical changes of the body occur
2. **RDA:** Value that meets requirements of 97–98 percent of individuals
   AI: Value used when there is not enough scientific data to support an RDA
   UL: Maximum safe intake level
   EAR: Value that meets requirements of 50 percent of individuals in a group
3. **Absorption:** Process of nutrients entering the tissues from the gastrointestinal tract
   Enzyme: Substance that speeds up chemical reactions
   Anabolism: Process of building substances
   Peristalsis: Involuntary muscular contraction
   Catabolism: Process of breaking down substances
4. c
5. b
6. a
7. b
8. a
9. a
10. b

Chapter 2

1. **Energy (calories):** 1,300–3,000
   Added sugar: Don't exceed caloric needs
   Fiber: Increase intake
   Total fat: 30 percent or less of calories
   Saturated fat: Less than 10 percent of calories
   Cholesterol: 300 mg or less
   Protein/vitamins/minerals: 100 percent of RDA/DRIs
   Sodium: 2,400 mg or less
2. **Apple:** 1 medium
   Fruit juice: 3/4 cup
   Bread: 1 slice
   Cold cereal: 1 ounce
   Cooked vegetables: 1/2 cup
   Raw vegetables: 1/2 cup (unless leafy)
   Milk or yogurt: 1 cup
   Chicken: 2–3 ounces
   Cooked kidney beans: 1/2 cup
   Eggs: 2
   Peanut butter: 4 tablespoons
   Raw leafy vegetables: 1 cup
   Cooked rice or pasta: 1/2 cup
3. d
4. a
5. b
6. b
7. b
8. b
9. a
10. b

Chapter 3

1. **White bread:** starch
   Whole-wheat bread: starch, fiber
   Apple juice: natural sugars
   Baked beans: fiber, starch
   Milk: natural sugars
   Bran flakes: fiber

Sugar-frosted oats: added sugars
Cola drink: added sugars
Broccoli: fiber
2. b
3. a
4. b
5. a
6. a
7. b
8. b
9. a
10. b

Chapter 4

I. Food	Fat	Cholesterol
1. Butter	X	X
2. Margarine	X	
3. Split peas		
4. Peanut butter	X	
5. Porterhouse steak	X	X
6. Flounder		
7. Skim milk		
8. Cheddar cheese	X	X
9. Chocolate-chip cookie made with vegetable shortening	X	
10. Green beans		

II.

1. b
2. a
3. b
4. a
5. b

III.

1. f
2. g
3. d

4. a
5. c
6. e
7. b

Chapter 5

1. a
2. a
3. a
4. b
5. a
6. b
7. a
8. a
9. a
10. a

Chapter 6

1. b
2. b
3. a
4. a
5. b
6. b
7. a
8. b
9. a
10. a
A. Vitamin B_{12}
B. Thiamin
C. Vitamin C, folate
D. Vitamin B_{12}
E. Vitamin D
F. Niacin
G. Vitamin K
H. Vitamin B_6
I. Vitamin K

J. Vitamins E, C, thiamin, riboflavin, niacin, B_6, B_{12}, folate, panto-
   thenic acid
K. Vitamin C
L. Vitamin A
M. Vitamin D
N. Vitamins A, C, E
O. Vitamins A, C, D
P. Vitamin A
Q. Vitamin D

Chapter 7

1. a
2. b
3. a
4. b
5. a
6. b
7. a
8. a
9. b
10.
A. calcium, phosphorus, magnesium, zinc
B. calcium
C. sodium, potassium, chloride
D. potassium
E. chloride
F. potassium
G. fluoride
H. chloride
I. iron
J. iron
K. selenium
L. cobalt

Chapter 8

1. a
2. b
3. b

4. b
5. a
6. a
7. b
8. b
9. b
10. a

Chapter 9

1. d
2. c
3. a
4. b
5. d

Chapter 10

1. a
2. c
3. d
4. d
5. c

Chapter 11

1. a
2. a
3. b
4. b
5. a
6. a
7. b
8. b
9. a
10. a

Chapter 12

1. b
2. a
3. b
4. b
5. b
6. b
7. a
8. a
9. a
10. a

Chapter 13

1. b
2. b
3. a
4. b
5. b
6. a
7. a
8. b
9. a
10. a
11. b
12. a
13. b
14. a
15. b
16. a
17. a
18. a
19. a
20. b

Glossary

Absorption The passage of digested nutrients through the walls of the intestines into the intestinal cells. Nutrients are then transported through the body via the blood or lymph systems. A few nutrients are absorbed elsewhere in the gastrointestinal tract, such as the stomach.

Acceptable Daily Intake (ADI) An intake level for a food additive that, if maintained each day throughout a person's life, would be considered safe by a wide margin.

Acid-Base Balance The process by which the body buffers the acids and bases normally produced in the body so the blood is neither too acidic nor basic.

Acidosis A dangerous condition in which the blood is too acidic.

Added or Refined Sugars Sugars added to a food for sweetening or other purposes.

Adequacy A diet that provides enough calories, essential nutrients, and fiber to keep a person healthy.

Adequate Intake (AI) The dietary intake that is used when an RDA cannot be based on an estimated average requirement.

Alcohol-Free Wine Wines whose alcohol is removed after fermentation; 99.5 percent alcohol free.

Alkalosis A dangerous condition in which the blood is too basic.

Amino Acids The building blocks of protein. There are 20 different amino acids, each consisting of a backbone to which a side chain is attached.

Amino Acid Pool The overall amount of amino acids distributed in the blood, organs, and body cells.

Amniotic Sac The protective bag, or sac, that cushions and protects the fetus during pregnancy.

Anabolism The metabolic process by which body tissues and substances are built.

Angina Symptoms of pressing, intense pain in the heart area when the heart muscle gets insufficient blood. Sometimes stress or exertion can cause angina.

Anorexia Lack of appetite.

Anorexia Nervosa An eating disorder most prevalent in adolescent females who starve themselves.

Antibodies Proteins in the blood that bind with foreign bodies or invaders.

Antigens Foreign invaders in the body.

Antioxidant A compound that combines with oxygen to prevent oxygen from oxidizing, or destroying, important substances. Antioxidants prevent the oxidation of unsaturated fatty acids in the cell membrane, DNA (the genetic code), and other cell parts that substances called free radicals try to destroy.

Anus The opening of the digestive tract through which feces travel out of the body.

Arterial Blood Pressure The pressure of blood within arteries as it is pumped through the body by the heart.

Atherosclerosis The most common form of artery disease, characterized by plaque buildup along artery walls.

Attention Deficit Hyperactivity Disorder A developmental disorder of children characterized by impulsiveness, distractibility, and hyperactivity.

Baby Bottle Tooth Decay Serious tooth decay in babies caused by letting a baby go to bed with a bottle of juice, formula, cow's milk, or breast milk.

Balanced Diet A diet that avoids excessive amounts of calories or any particular food or nutrient.

Basal Metabolic Needs The energy needs of your body when at rest and awake.

Basal Metabolism The minimum energy needed by the body for vital functions when at rest and awake.

Beta-Carotene A precursor of vitamin A that functions as an antioxidant in the body; the most abundant carotenoid.

Bile A yellow-green liver secretion that is stored in the gallbladder and released when fat enters the small intestine because it emulsifies fats.

Bile Acids A component of bile that aids in the digestion and absorption of fats in the duodenum of the small intestine.

Binge Eating Disorder An eating disorder characterized by episodes of uncontrolled eating or binging.

Biotechnology A collection of scientific techniques, including genetic engineering, that are used to create, improve, or modify plants, animals, and microorganisms.

Blood Cholesterol Cholesterol circulating in the bloodstream. The blood carries it for use by all parts of the body. A high level of blood cholesterol leads to atherosclerosis and an increased risk of heart disease.

Blood Glucose Level or Blood Sugar Level Concentration of glucose in the blood.

Body Mass Index A method of measuring obesity that is a more sensitive indicator than height-weight tables.

Bolus A ball of chewed food that travels from the mouth through the esophagus to the stomach.

Bottled Water Water that is bottled in a sanitary container and sold.

Bran In cereal grains, the part that covers the grain and contains much fiber and other nutrients.

Bulimia Nervosa An eating disorder characterized by a destructive pattern of excessive overeating followed by vomiting or other "purging" behaviors to control weight.

Calorie A measure of the energy in food, specifically the energy-yielding nutrients. The correct name for calorie is kilocalorie.

Cancer A group of diseases characterized by unrestrained cell division and growth that can disrupt the normal functioning of an organ and also spread beyond the tissue in which it started.

Carbohydrate or Glycogen Loading A regimen involving both decreased exercise and increased consumption of carbohydrates before an event to increase the amount of glycogen stores.

Carbohydrates A nutrient group containing carbon, hydrogen, and oxygen, which includes sugars, starch, and fibers.

Carcinogens Substances that initiate cancer.

Cardiac Sphincter A muscle that relaxes and contracts to move food from the esophagus into the stomach.

Cardiovascular Disease (CVD) A general term for diseases of the heart and blood vessels.

Carotenoids A class of pigments that contribute red, orange, or yellow color to fruits and vegetables. Beta-carotene is the most abundant.

Catabolism The metabolic processes by which large, complex molecules are converted to a smaller set of simpler ones.

Cerebral Hemorrhage A stroke due to a ruptured brain artery.

Chlorophyll A pigment in plants that converts energy from sunlight into energy stored in carbohydrate.

Cholesterol The most abundant sterol (a category of lipids), a soft, waxy substance present only in foods of animal origin; it is present in all parts of the body and has important functions.

Chutney A condiment of Indian origin made from fruits, vegetables, and herbs.

Chylomicron The lipoprotein responsible for carrying mostly triglycerides, and some cholesterol, from the intestines through the lymph system to the bloodstream.

Chyme A liquid mixture in the stomach that contains partially digested food and stomach secretions.

Club Soda Filtered, carbonated water to which mineral salts are added to give it a unique taste.

Collagen The most abundant protein in the human body; a fibrous protein that is a component of skin, bone, teeth, ligaments, tendons, and other connective structures.

Colon The large intestine.

Colostrum A yellowish fluid that is the first secretion to come from the breast a day or so after delivery of a baby. It is rich in proteins, antibodies, and other factors that protect against infectious disease.

Complementary Proteins Two protein foods with the ability to make up for the lack of certain amino acids in each other.

Complete Proteins Food proteins that provide all of the essential amino acids in the proportions needed.

Complex Carbohydrates Long chains of many sugars that include starches and fibers.

Coronary Heart Disease (CHD) Damage to or malfunction of the heart caused by narrowing or blockage of the coronary arteries.

Coulis A sauce consisting primarily of a thick purée, usually of vegetables, but also of fruit. Flavorings such as fresh herbs, and a liquid, such as broth, give it taste and body.

Cruciferous Vegetables Members of the cabbage family that contain phytochemicals that might help prevent cancer.

Culture The behaviors and beliefs of a certain social, ethnic, or age group.

Daily Value Nutrient standards used on food labels to allow nutrient comparisons; based on a 2,000-calorie diet.

Denaturation Alteration of a protein's three-dimensional form that renders it useless; can be caused by high temperatures, ultraviolet radiation, acids, and bases, agitation or whipping, and high salt concentration.

Dental Caries Tooth decay. To prevent dental caries, you should brush your teeth often, floss your teeth once a day, and try to limit sweets to mealtime.

Deoxyribonucleic Acid (DNA) The protein carrier of the genetic code in the cells.

Designer Foods Foods that are supplemented with ingredients thought to help prevent disease or improve health.

Diabetes Mellitus A disorder of carbohydrate metabolism characterized by high blood sugar levels and inadequate or ineffective insulin.

Diastolic Pressure The pressure in the arteries when the heart is resting between beats — the bottom number in blood pressure.

Diet The food and beverages you normally eat and drink.

Dietary Fiber Material from plant cells that mostly resists digestion by our digestive enzymes.

Dietary Guidelines for Americans A periodically revised government booklet that gives dietary advice for Americans.

Dietary Recommendations Guidelines that discuss specific foods and food groups to eat for optimal health.

Dietary Reference Intakes (DRI) Nutrient standards that include four lists of values for dietary nutrient intakes of healthy Americans.

Digestion The breakdown of food into its components in the mouth, stomach, and small intestine with the help of digestive enzymes. Complex proteins are digested, or broken down, into their amino acid building blocks; complicated sugars are reduced to simple sugars such as glucose; and fat molecules are broken down into fatty acids and glycerol.

Disaccharides Double sugars such as sucrose (table sugar) and lactose (milk sugar).

Diuretics A category of drugs frequently used to treat hypertension. Diuretics stimulate urination and some may deplete potassium.

Diverticulosis A disease of the large intestine in which the intestinal walls become weakened, bulge out into pockets, and at times become inflamed.

Duodenum The first segment of the small intestine, about one foot long.

Edema Swelling due to an abnormal accumulation of fluid in the intercellular spaces.

Electrolytes Chemical elements or compounds that ionize in solution and can carry an electric current; include sodium, potassium, and chloride.

Embryo The fertilized egg from conception to the eighth week. After the eighth week, it is called a fetus.

Empty Calories Foods that provide few nutrients for the number of calories they contain.

Endosperm In cereal grains, a large center area high in starch.

Energy-Yielding Nutrients Nutrients that can be burned as fuel to provide energy for the body, including carbohydrates, fats, and protein.

Enriched Foods Refined foods, such as white bread, to which nutrients have been added to make up for some of the nutrients lost in processing.

Enzymes Catalysts in the body.

Epiglottis The flap that covers the larynx and the opening to the trachea so that food does not enter during swallowing.

Essential or Indispensable Amino Acids Amino acids that either cannot be made in the body or cannot be made in the quantities needed by the body; they must therefore be obtained in foods.

Essential Fatty Acids Fatty acids that the body cannot produce, making them necessary in the diet: linoleic acid and linolenic acid.

Essential Hypertension High blood pressure.

Esophagus The approximately 10-inch muscular tube that is the part of the gastrointestinal tract that connects the pharynx to the stomach. At the bottom of it is the lower esohageal sphincter.

Estimated Average Requirement (EAR) The dietary intake value that is estimated to meet the requirement of half the healthy individuals in a group.

Exchange Lists for Meal Planning A menu-planning guide that groups foods by their calorie, carbohydrate, fat, and protein content. Each listed food has approximately the same amount of calories, carbohydrates, fat, and protein as another listed food, so that any food can be exchanged for any other food on the same list. The *Exchange Lists for Meal Planning,* developed by the American Diabetes and American Dietetic Association, are used primarily by people with diabetes, who need to regulate what and how much they eat.

Exchange System A tool to plan diets that groups foods by their nutrient and caloric content. Foods within each group have about the same amount of calories, carbohydrate, protein, and fat, so that any food can be substituted for any other food in the same group.

Fasting Hypoglycemia A type of hypoglycemia in which symptoms occur after not eating for eight or more hours, so it usually occurs during the night or before breakfast.

Fat A nutrient that provides 9 calories per gram; a lipid that is solid at room temperature.

Fat-Soluble Vitamins A group of vitamins that include vitamins A, D, E, and K. They generally occur in foods containing fats and are stored in the body, either in the liver or in adipose (fatty) tissue, until they are needed.

Fat Substitutes Ingredients that mimic the functions of fat in foods, and either contain fewer calories than fat or no calories.

Fatty Acids Major component of most lipids.

Fetal Alcohol Syndrome (FAS) A set of symptoms occuring in newborn babies due to alcohol use of the mother during pregnancy. FAS children may show signs of mental retardation, growth retardation, brain damage, and facial deformities.

Fetus The infant in the mother's uterus from eight weeks after conception until birth.

First Trimester The first 13 weeks of pregnancy.

Flavor An attribute of a food that includes its appearance, smell, taste, feel in the mouth, texture, temperature, and even the sounds made when it is chewed.

Flavorings Substances that add a new flavor to a dish or modify the original flavor.

Fluoride The form of fluorine that appears in drinking water and the body.

Fluorosis A condition in which the teeth become mottled and discolored due to high fluoride ingestion.

Food Allergens Those parts of food causing allergic reactions.

Food Allergy An abnormal response of the immune system to an otherwise harmless food.

Food Guide Guidelines that tell us the kinds and amounts of foods we need to eat for a nutritionally adequate diet.

Food Guide Pyramid A food guide developed by the U.S. Department of Agriculture to help healthy Americans follow the Dietary Guidelines for Americans.

Food Intolerance Symptoms of gas, bloating, constipation, dizziness, or difficulty sleeping after eating certain foods.

Food Jag A habit of young children in which they have favorite foods they want to eat frequently.

Free Radical An unstable compound that reacts quickly with other molecules in the body.

Fortified Food A food to which nutrients are added that were not present originally or were present in insignificant amounts.

Fresh Foods Raw foods that have not been frozen, processed (such as canned or frozen), or heated, and that contain no preservatives.

Fructose A monosaccharide found in fruits and honey. Fructose is the sweetest natural sugar.

Fruit Smoothies Frozen blends of fruit, milk, and/or fresh or frozen yogurt.

Gag Reflex A normal reflex stimulated by touching the back of the throat. The gag reflex pulls the tongue back and contracts the throat muscles. If something gets stuck in the back of the throat, the gag reflex often results in vomiting of the food so that the food does not obstruct the windpipe.

Gastrointestinal Tract A hollow tube running down the middle of the body in which digestion of food and absorption of nutrients takes place. At the top of the tube is the mouth, which connects in turn to the pharynx, esophagus, stomach, small intestine, large intestine, rectum, and anus, where solid wastes leave the body.

Galactose A monosaccharide found linked to glucose to form lactose, or milk sugar.

Gelatinization A process in which starches, when heated in liquid, absorb some of the water and swell in size.

Genetic Engineering A process that allows plant breeders to modify the genetic makeup of a plant species precisely and predictably, creating improved varieties faster and easier than can be done using more traditional plant-breeding techniques.

Germ In cereal grains, the area of the kernel rich in vitamins and minerals that sprouts when allowed to germinate.

Glucose The most significant monosaccharide; the body's primary source of energy; also called dextrose.

Glycerol A derivative of carbohydrate that is part of triglycerides.

Glycogen The storage form of glucose in the body; found in the liver and muscles.

Gram A unit of weight. There are 28 grams in 1 ounce.

Growth Spurts Periods of rapid growth.

Health Claims Claims on food labels that state that certain foods or food substances—as part of an overall healthy diet—may reduce the risk of certain diseases.

Heartburn A painful burning sensation in the esophagus caused by acidic stomach contents flowing back into the lower esophagus.

Height/Weight Tables Tables that show an appropriate weight for a given height.

Heme Iron The predominant form of iron in animal foods, heme iron is absorbed and used twice as readily as iron in plant foods.

Hemoglobin A protein in red blood cells that carries oxygen through the body.

Hemorrhoids Enlarged veins in the lower rectum.

Herbs Leaves of certain plants grown in temperate climates.

High-Density Lipoprotein (HDL) Lipoproteins that contain a small amount of cholesterol and carry cholesterol away from body cells and tissues to the liver for excretion from the body. A low HDL level increases the risk of heart disease, so the higher the HDL level, the better. HDL is sometimes called "good" cholesterol.

High-Fructose Corn Syrup Corn syrup that has been treated with an enzyme that converts part of the glucose it contains to fructose to make it sweeter.

Homeostasis A constant internal environment in the body.

Homogenized Characterizes milk or cream in which the fat has been emulsified so the milk and cream are equally distributed throughout the product.

Hormone A chemical messenger in the body.

Hydrochloric Acid A strong acid made by the stomach that aids in protein digestion, destroys harmful bacteria, and increases the ability of calcium and iron to be absorbed.

Hydrogenation A process in which liquid vegetable oils are converted into solid fats (such as margarine) by the use of heat, hydrogen, and certain metal catalysts.

Hyperglycemia High blood sugar levels.

Hypertension High blood pressure.

Hypoglycemia Low blood sugar levels.

Ileum The final segment of the small intestine.

Immune Response The body's response to a foreign substance, such as a virus, in the body.

Incomplete Proteins Food proteins that contain at least one limiting amino acid.

Insulin A hormone that pushes sugar from the blood into the body's cells, resulting in lower, more normal blood sugar levels.

Intrinsic Factor A protein secreted by stomach cells that is necessary for the absorption of vitamin B_{12}.

Iron-Deficiency Anemia A condition in which the size and number of red blood cells are reduced; may result from inadequate iron intake or from blood loss; symptoms include fatigue, pallor, and irritability.

Iron Overload A common genetic disease in which individuals absorb about twice as much iron from their food and supplements as other people. Also called hemochromatosis.

Jejunum The second portion of the small intestine, between the duodenum and the ileum. The jejunum is about 8 feet long.

Ketone Bodies A group of organic compounds that cause the blood to become too acidic as a result of fat being burned for energy without any carbohydrates present.

Ketosis Excessive level of ketone bodies in the blood and urine.

Kwashiorkor A type of protein-energy malnutrition associated with children with insufficient protein intake and who have a preexisting disease.

Lactase An enzyme needed to split lactose into its components in the intestines.

Lactation The process of producing and secreting milk in the mammary glands.

Lacto-Ovovegetarians Vegetarians who do not eat meat, poultry, or fish but do consume animal products in the form of eggs (ovo) and milk and milk products.

Lactose A disaccharide found in milk and milk products that is made of glucose and galactose.

Lactose Intolerance An intolerance to milk and most milk products due to a deficiency of the enzyme lactase. Symptoms often include flatulence and diarrhea.

Lactovegetarians Vegetarians who do not eat meat, poultry, or fish but do consume animal products in the form of milk and milk products.

Lanugo Downy hair on the skin.

Large Intestine The part of the gastrointestinal tract between the small intestine and the rectum. The large intestine is about 2-1/2 inches wide and 5 feet long.

Lecithin A phospholipid and a vital component of cell membranes that acts as an emulsifier (a substitute that keeps fats in solution). Lecithin is not an essential nutrient.

Light Beers Beers with one-third to one-half less alcohol and calories than regular beers.

Limiting Amino Acid An essential amino acid in lowest concentration in a protein.

Linoleic Acid An essential fatty acid, omega-6 fatty acid, found in vegetable oils such as corn, safflower, soybean, cottonseed, sunflower, and peanut oils; increases blood clotting and inflammatory responses in the body.

Linolenic Acid An essential fatty acid, omega-3 fatty acid, found in several oils, notably canola, soybean, walnut, and wheat germ oil (or margarines made with canola or soybean oil) and fatty fish; important for normal growth and development, and seems to protect against heart disease.

Lipid The chemical name for a group of fatty substances, including fats, oils, cholesterol, and lecithin, that are present in blood and body tissues.

Lipoprotein Lipase An enzyme that breaks down triglycerides in the blood into fatty acids and glycerol so they can be absorbed into the body's cells.

Lipoproteins Protein-coated packages that carry fat and cholesterol through the bloodstream. They are classified according to their density.

Low-Alcohol Refreshers A category of drinks made with fruit juices, carbonated water, and sugar, with an alcohol base of wine, distilled spirits, or malt; includes coolers.

Low Birth Weight Baby A newborn who weighs less than 5-1/2 pounds. These babies are at higher risk for disease and experience more difficulties surviving the first year.

Low-Density Lipoprotein (LDL) Lipoprotein that contains most of the cholesterol in the blood. LDL carries cholesterol to body tissues, including the arteries. For this reason, a high LDL level increases the risk of heart disease.

Macaroni Pasta made from flour and water.

Major Minerals. *See* Minerals.

Maltose A disaccharide made of two glucose units bonded together.

Marasmus A type of protein-energy malnutrition characterized by gross underweight and severe food shortage.

Marinades Seasoned liquids used before cooking to flavor and moisten foods; usually based on an acidic ingredient.

Marketing Finding out what your customers need and want, and then developing, promoting, and selling the products and services they desire.

Megadose A supplement intake of 10 times the RDA or AI of a vitamin or mineral.

Megaloblastic Anemia A form of anemia caused by a deficiency of vitamin B_{12} and folate and characterized by large, immature red blood cells.

Metabolism All the chemical processes by which nutrients are used to support life.

Metastasis The spread of a cancer beyond the tissue in which it started.

Mild Fluorosis A condition in which small, white, virtually invisible opaque areas appear on the teeth due to intake of too much fluoride.

Milk Letdown The process by which milk comes out of the mother's breast to feed the baby. Sucking causes the release of a hormone (oxytocin) that allows milk letdown.

Minerals Naturally occurring, inorganic chemical elements that form a class of nutrients. Some minerals are needed in relatively large amounts in the diet—over 100 milligrams daily. These minerals are called major minerals and include calcium, chloride, magnesium, phosphorus, potassium, sodium, and sulfur. Other minerals, called trace minerals or trace elements, are needed in smaller amounts—less than 100 milligrams daily. The trace minerals include chromium, cobalt, copper, fluoride, iodine, iron, manganese, molybdenum, selenium, and zinc.

Mineral Water Water that comes from a protected underground source and contains at least 250 parts per million of total dissolved solids.

Mocktails Drinks made to resemble mixed alcoholic drinks but containing no alcohol.

Moderation Description of a diet that avoids excessive amounts of calories or any particular food or nutrient.

Monoglycerides Triglycerides with only one fatty acid.

Monosaccharides Single sugars such as glucose or fructose. The monosaccharides are the building blocks of other carbohydrates.

Monounsaturated Fat A monounsaturated triglyceride containing mostly monounsaturated fatty acids.

Monounsaturated Fatty Acid A fatty acid that contains only one double bond in the chain.

Morning Sickness Symptoms, especially nausea and vomiting, that occur in some women during the first half of pregnancy. Morning sickness is really a misnomer, since the symptoms can occur at any time of the day.

Myocardial Infarction A heart attack.

Myocardial Ischemia A temporary injury to heart cells caused by a lack of blood flow and oxygen.

Myoglobin An iron-containing protein found in muscles that stores and carries oxygen.

Negative Nitrogen Balance A condition in which the body excretes more protein than is taken in. Negative nitrogen balance can occur during starvation and certain illnesses.

Neural Tube The embryonic tissue that develops into the brain and spinal cord.

Neural Tube Defects A group of defects of the brain and/or spinal cord that are caused by failure of the neural tube to close during early pregnancy. Such defects include spina bifida and meningocele.

Neurotransmitters Chemical messengers released at the end of the nerve cell to stimulate or inhibit a second cell, which may be another nerve cell, a muscle cell, or a gland cell.

Night Blindness A condition caused by insufficient vitamin A in which it takes longer to adjust to dim lights after seeing a bright flash of light (such as oncoming car headlights) at night. This is an early sign of vitamin A deficiency.

Nitrogen Balance The amount of nitrogen taken in by mouth compared with the amount excreted in a given period of time.

Nonalcoholic Malt Beverages Beerlike products with only 0.5 percent or less of alcohol.

Nonessential Amino Acids Amino acids that can be made in the body.

Nonheme Iron A form of iron found in all plant sources of iron and also as part of the iron in animal food sources.

Nonnutritive Sweeteners Sweeteners such as aspartame that contain either no or very few calories.

Nutrients The nourishing substances in food. They provide energy and promote body growth and maintenance. In addition, nutrients regulate the many body processes, such as heart rate and digestion, and support the body's optimum health and growth. The six groups of nutrients are **carbohydrates, lipids, proteins, vitamins, minerals,** and **water.**

Nutrient-Content Claims Claims about the nutrient composition of a food that are written on food labels; regulated by the Food and Drug Administration.

Nutrient-Dense Foods Foods that contain many nutrients for the calories they provide.

Nutrient Density The nutrient content of a food expressed in relation to its calories.

Nutrition A science that studies nutrients and other substances in foods and the body and how they relate to health and disease. Nutrition looks at how you select foods and the type of diet you eat.

Obesity Body weight of 30 percent or more over desirable weight.

Oil A form of lipid that is usually liquid at room temperature.

Oral Cavity The mouth.

Organic Foods Foods that have been grown without nearly all synthetic insecticides, fungicides, herbicides, and fertilizers.

Osteomalacia A disease of vitamin D deficiency in adults in which the leg and spinal bones soften and may bend. Osteomalacia may be seen in elderly individuals with poor milk intake and little exposure to the sun.

Osteoporosis The most common bone disease, characterized by loss of bone density and strength; associated with debilitating fractures, especially in people 45 and older, due to a tremendous loss of bone tissue in midlife.

Overweight Having a Body Mass Index of 25 or higher.

Oxalic Acid A substance found in some plant foods, such as spinach, beet greens, Swiss chard, sorrel, and parsley, that prevents calcium absorption. Oxalic acid is also in tea and cocoa.

Palmar Grasp The ability of a baby from about six months of age to grab objects with the palm of the hand.

Pasteurized Characterizes food exposed to a high temperature for a specific period of time to destroy harmful or undesirable microorganisms.

Pellagra A disease caused by a dietary deficiency of niacin.

Pepsin The principal digestive enzyme of the stomach.

Peptide Bonds The bonds that form between adjoining amino acids.

Peristalsis Involuntary muscular contraction that forces food through the entire digestive system.

Pernicious Anemia A type of anemia caused by a deficiency of vitamin B_{12} and characterized by macrocytic anemia and deterioration in the functioning of the nervous system that, if untreated, could cause significant and sometimes irreversible damage.

Pescovegetarians Vegetarians who will eat fish and shellfish.

Pesticides Chemicals used on crops to protect them from insects, disease, weeds, and fungi.

Photosynthesis A process during which plants convert energy from sunlight into energy stored in carbohydrate.

Phytochemicals Minute substances in plants that may reduce risk of cancer and heart disease when eaten often.

Pincer Grasp The ability of a baby at about eight months to use the thumb and forefinger together to pick things up.

Placenta The organ that develops during the first month of pregnancy that provides for exchange of nutrients and wastes between fetus and mother and secretes the hormones necessary to maintain pregnancy.

Plaque (1) Deposits on arterial walls that contain cholesterol, fat, fibrous scar tissue, calcium, and other biological debris. (2) Deposits of bacteria, protein, and polysaccharides found on teeth.

Point of Unsaturation The location of the double bond in unsaturated fatty acids.

Polypeptides Protein fragments with 10 or more amino acids.

Polysaccharide Another name for complex carbohydrate or long chains of many sugars, including starch and fiber.

Polyunsaturated Fat A polyunsaturated triglyercide, made of mostly polyunsaturated fatty acids.

Polyunsaturated Fatty Acid A fatty acid that contains two or more double bonds in the chain.

Positive Nitrogen Balance A condition in which the body excretes less protein than is taken in. Positive nitrogen balance occurs during growth and pregnancy.

Postprandial Hypoglycemia A type of hypoglycemia that occurs generally two to four hours after meals. This condition has symptoms such as quick heartbeats, shakiness, weakness, anxiety, sweating, and dizziness that mimic anxiety or stress symptoms.

Precompetition Meal The meal closest to the time of a competition or athletic event.

Precursors Forms of vitamins that the body changes chemically to active vitamin forms.

Preformed Vitamin A The form of vitamin A called retinol.

Press Release A written statement sent out to various news media for publicity purposes. The first paragraph often tells the who, what, when, where, and why of your story.

Primary Hypertension A form of hypertension whose cause is unknown.

Primary Structure The number and sequence of the amino acids in the protein chain.

Processed Foods Foods prepared using a certain procedure: cooking (such as frozen pancakes), freezing (frozen dinners), canning (canned vegetables), dehydrating (dried fruits), milling (white flour), culturing with bacteria (yogurt), or adding vitamins and minerals (enriched foods).

Promoters Substances such as fat that advance the development of mutated cells into a tumor.

Proportionality A concept of eating relatively more foods from the larger food groups at the base of the Food Guide Pyramid and fewer foods from the smaller food groups near the top of the Pyramid.

Proteins Major structural parts of animal tissue made up of nitrogen-containing amino acids. Proteins are nutrients that yield 4 calories per gram.

Protein-Energy Malnutrition A broad spectrum of malnutrition, from mild to much more serious cases.

Provitamin A Precursors of vitamin A, such as beta-carotene.

Puberty The time period in which children start to develop secondary sex characteristics and the beginning of the fertile period when gametes (sperm or ovum) are produced.

Publicity Free space or time in various media giving public notice of a program, book, and so on.

Rancidity The deterioration of fat, resulting in undesirable flavors and odors.

Recommended Dietary Allowance (RDA) The dietary intake value that is sufficient to meet the nutrient requirements of 97 to 98 percent of all healthy individuals in a group.

Rectum The last part of the large intestine in which feces, the waste products of digestion, are stored until elimination.

Refined or Milled Processes by which the bran and germ of grain are separated (or mostly separated) from the endosperm.

Retinol One of the active forms of vitamin A. *See also* Vitamin A.

Retinol Equivalent The unit of measure for vitamin A activity in both its forms (retinol and beta-carotene). 1 RE = 1 microgram retinol or 6 micrograms beta-carotene.

Rickets A childhood disease in which bones do not grow normally, resulting in bowed legs and knock knees. Rickets is generally caused by a vitamin D deficiency.

Risk Factors Habits, traits, or conditions of an individual that are associated with an increased chance of developing a disease.

Rub A dry marinade made of herbs and spices (and other seasonings), sometimes moistened with a little oil, and rubbed or patted on the surface of meat poultry, or fish (which is then refrigerated and cooked at a later time).

Saliva A fluid secreted into the mouth from the salivary glands, which contains important digestive enzymes and lubricates the food so that it may readily pass down the esophagus.

Salsa A sauce made from tomatoes, onions, hot peppers, garlic, and herbs.

Satiety A feeling of being full after eating.

Saturated Fat A type of fat containing mostly saturated fatty acids.

Saturated Fatty Acid A type of fatty acid that is filled to capacity with hydrogens.

Scurvy Vitamin C deficiency disease. Scurvy is marked by bleeding gums, weakness, loose teeth, and broken capillaries (small blood vessels) under the skin.

Seasonings Substances used to bring out flavor that is already present in a dish.

Secondary Hypertension Persistently elevated blood pressure caused by a medical problem, such as hormonal abnormality or an inherited narrowing of the aorta (the largest artery leading from the heart).

Secondary Structure The bending and coiling of the protein chain.

Seltzer Filtered, artificially carbonated water that generally has no added mineral salts.

Simple Carbohydrates Sugars including monosaccharides and disaccharides.

Small Intestine The digestive tract organ that extends from the stomach to the opening of the large intestine.

Sparkling Water Any water with carbonation.

Spices Dried bits of bark, leaves, and berries from tropical plants.

Spina Bifida A birth defect in which parts of the spinal cord are not fused together properly, so gaps are present where the spinal cord has little or no protection. Spina bifida is a type of neural tube defect.

Spring Water Water collected as it flows to the surface.

Starch A complex carbohydrate made up of a long chain of hundreds to thousands of glucoses linked together; found in grains, legumes, vegetables, and some fruits.

Stomach A digestive tract organ that prepares food chemically and mechanically so that it can be further digested and absorbed in the small intestine.

Stroke Damage to brain cells resulting from an interruption of blood flow to the brain.

Sucrose A disaccharide commonly called cane sugar, table sugar, granulated sugar, or simply sugar. Sucrose is made of glucose and fructose.

Sugar A form of carbohydrate.

Sugar Alcohols Forms of sugar that contain an alcohol group. They include sorbitol, xylitol, and mannitol.

Systolic Pressure The pressure of blood within arteries when the heart is pumping—the top blood pressure number.

Taste Sensations perceived by the taste buds on the tongue.

Taste Buds Clusters of cells found on the tongue, cheeks, throat, and roof of the mouth. Each taste bud houses 60 to 100 receptor cells. The body regenerates taste buds about every three days. These cells bind food molecules dissolved in saliva and alert the brain to interpret them.

Team Nutrition A project of the U.S. Department of Agriculture to implement the School Meals Initiative for Healthy Children.

Tertiary Structure The folding of the protein chain.

Thermic Effect of Food The energy needed to digest and absorb food. The thermic effect of food is the smallest contributor to your energy needs—from 5 to 10 percent. In other words, for every 100 calories consumed, about 5 to 10 calories are used for digestion, absorption, and metabolism of nutrients.

Thyroid Gland A gland found on either side of the trachea that produces and secretes two important hormones that regulate the level of metabolism.

Tolerable Upper Intake Level (UL) The maximum intake level above which risk of toxicity would increase.

Tonic Water A carbonated water containing lemon, lime, sweeteners, and quinine.

Trace Minerals. *See* Minerals.

Trans Fats Unsaturated fatty acids that lose a natural bend or kink so they become straight (like saturated fatty acids) after being hydrogenated; they act like saturated fats in the body.

Transitional Milk The type of breast milk produced from about the third to the tenth day after childbirth, when mature milk appears.

Triglycerides The major form of lipid in food and in the body; made of three fatty acids attached to a glycerol backbone.

Type 1 Diabetes Mellitus A form of diabetes seen mostly in children and adolescents. These patients make no insulin at all and therefore require frequent injections of insulin to maintain a normal level of blood glucose. Also called *insulin-dependent diabetes.*

Type 2 Diabetes Mellitus A form of diabetes seen most often in older, overweight adults. The diabetic makes insulin but his or her tissues aren't sensitive enough to the hormone and so use it inefficiently.

Unsaturated Fatty Acid A fatty acid with at least one double bond.

Varied Diet A diet that includes many different foods.

Vegans Individuals eating a type of vegetarian diet that excludes eggs and dairy products. The diet relies exclusively on plant foods to meet protein and other nutrient needs. Vegans are a small group, and it is estimated that only 4 percent of vegetarians are vegans. Vegans are also called strict vegetarians.

Very Low-Density Lipoproteins (VLDL) The liver's version of chylomicrons. Triglycerides, and

some cholesterol, are carried through the body by VLDLs, which release triglycerides throughout the body, with the help of lipoprotein lipase. Once the majority of triglycerides are removed, VLDLs are converted in the blood into another type of lipoprotein, called low-density lipoprotein (LDL).

Villi Tiny, fingerlike projections found in the wall of the small intestine that help absorb nutrients.

Vitamins Noncaloric, organic nutrients found in foods that are essential in small quantities for growth and good health.

Water Balance The process of maintaining the proper amount of water in each of three bodily "compartments": inside the cells, outside the cells, and in the blood vessels.

Water-Insoluble Fiber A classification of fiber that includes cellulose, lignin, and the remaining hemicelluloses. They generally form the structural parts of plants.

Water-Soluble Fiber A classification of fiber that includes gums, mucilages, pectin, and some hemicelluloses. They are generally found around and inside plant cells. This type of fiber swells in water.

Water-Soluble Vitamins The vitamins that are soluble in water and are not stored appreciably in the body: vitamin C, thiamin, riboflavin, niacin, vitamin B_6, folate, vitamin B_{12}, pantothenic acid, and biotin.

Whole Foods Foods as we get them from nature.

Whole Grain A grain that contains the endosperm, germ, and bran.

Wine Coolers Mixtures of wine, fruit flavor, and carbonated water or plain water.

Wine Spritzers Drinks made with wine and club soda.

Xerosis A condition in which the cornea of the eye becomes dry and cloudy. A lack of vitamin A often causes this condition. Xerosis also refers to dryness of other body parts, such as the skin.

Xerophthalmia Hardening and thickening of the cornea that can lead to blindness. A deficiency of vitamin A usually causes this medical problem.

Index

Body Mass Index

WEIGHT

HEIGHT	100	105	110	115	120	125	130	135	140	145	150	155	160	165	170	175	180	185	190	195	200	205
5'0"	20	21	21	22	23	24	25	26	27	28	29	30	31	32	33	34	35	36	37	38	39	40
5'1"	19	20	21	22	23	24	25	26	26	27	28	29	30	31	32	33	34	35	36	37	38	39
5'2"	18	19	20	21	22	23	24	25	26	27	27	28	29	30	31	32	33	34	35	36	37	37
5'3"	18	19	19	20	21	22	23	24	25	26	27	27	28	29	30	31	32	33	34	35	35	36
5'4"	17	18	19	20	21	21	22	23	24	25	26	27	27	28	29	30	31	32	33	33	34	35
5'5"	17	17	18	19	20	21	22	22	23	24	25	26	27	27	28	29	30	31	32	32	33	34
5'6"	16	17	18	19	19	20	21	22	23	23	24	25	26	27	27	28	29	30	31	31	32	33
5'7"	16	16	17	18	19	20	20	21	22	23	23	24	25	26	27	27	28	29	30	31	31	32
5'8"	15	16	17	17	18	19	20	21	21	22	23	24	24	25	26	27	27	28	29	30	30	31
5'9"	15	16	16	17	18	18	19	20	21	21	22	23	24	24	25	26	27	27	28	29	30	30
5'10"	14	15	16	17	17	18	19	19	20	21	22	22	23	24	24	25	26	27	27	28	29	29
5'11"	14	15	15	16	17	17	18	19	20	20	21	22	22	23	24	24	25	26	26	27	28	29
6'0"	14	14	15	16	16	17	18	18	19	20	20	21	22	22	23	24	24	25	26	26	27	28
6'1"	13	14	15	15	16	16	17	18	18	19	20	20	21	22	22	23	24	24	25	26	26	27
6'2"	13	13	14	15	15	16	17	17	18	19	19	20	21	21	22	22	23	24	24	25	26	26
6'3"	12	13	14	14	15	16	16	17	17	18	19	19	20	21	21	22	22	23	24	24	25	26
6'4"	12	13	13	14	15	15	16	16	17	18	18	19	19	20	21	21	22	23	23	24	24	25

1997–2000 Dietary Reference Intakes (DRI)
Recommended Dietary Allowances (RDA)

AGE (years)	Vitamin E (mg α-tocopherol)	Vitamin C (mg)	Thiamin (mg)	Riboflavin (mg)	Niacin (mg NE)	Vitamin B$_6$ (mg)	Folate (microgram DFE)	Vitamin B$_{12}$ (microgram)	Phosphorus (mg)	Magnesium (mg)	Selenium (microgram)
Infants[a]											
0–0.5	4	40	0.2	0.3	2	0.1	65	0.4	100	30	15
0.5–1.0	5	50	0.3	0.4	4	0.3	80	0.5	275	75	20
Children											
1–3	5	15	0.5	0.5	6	0.5	150	0.9	460	80	20
4–8	7	25	0.6	0.6	8	0.6	200	1.2	500	130	30
Males											
9–13	11	45	0.9	0.9	12	1.0	300	1.8	1250	240	40
14–18	15	75	1.2	1.3	16	1.3	400	2.4	1250	410	55
19–30	15	90	1.2	1.3	16	1.3	400	2.4	700	400	55
31–50	15	90	1.2	1.3	16	1.3	400	2.4	700	420	55
51–70	15	90	1.2	1.3	16	1.7	400	2.4	700	420	55
Over 70	15	90	1.2	1.3	16	1.7	400	2.4	700	420	55
Females											
9–13	11	45	0.9	0.9	12	1.0	300	1.8	1250	240	40
14–18	15	65	1.0	1.0	14	1.2	400	2.4	1250	360	55
19–30	15	75	1.1	1.1	14	1.3	400	2.4	700	310	55
31–50	15	75	1.1	1.1	14	1.3	400	2.4	700	320	55
51–70	15	75	1.1	1.1	14	1.5	400	2.4	700	320	55
Over 70	15	75	1.1	1.1	14	1.5	400	2.4	700	320	55
Pregnancy	*	+10	1.4	1.4	18	1.9	600	2.6	*	+40	+5
Lactation	+4	+45	1.5	1.6	17	2.0	500	2.8	*	*	+15

* The values for pregnancy or lactation do not change from the normal value for women of the same age.

[a] Values for infants are all Adequate Intakes (AI). The AI for niacin for 0.0–0.5 years is stated as milligrams of preformed niacin, not niacin equivalents.